Mechanical Design

Mechanical Design

Antonino Risitano

CRC Press
Taylor & Francis Group
Boca Raton   London   New York

CRC Press is an imprint of the
Taylor & Francis Group, an **informa** business

CRC Press
Taylor & Francis Group
6000 Broken Sound Parkway NW, Suite 300
Boca Raton, FL 33487-2742

First issued in paperback 2017

© 2011 by Taylor & Francis Group, LLC
CRC Press is an imprint of Taylor & Francis Group, an Informa business

Version Date: 20110831

ISBN 13: 978-1-138-07220-6 (pbk)
ISBN 13: 978-1-4398-1169-6 (hbk)

Library of Congress Cataloging-in-Publication Data

Risitano, Antonino.
   Mechanical design / by Antonino Risitano.
      p. cm.
   Includes bibliographical references and index.
   ISBN 978-1-4398-1169-6 (hardback)
   1. Machine design. 2. Machine parts. I. Title.

TJ230.R57 2011
621.8'15--dc22   2010033566

Visit the Taylor & Francis Web site at
http://www.taylorandfrancis.com

and the CRC Press Web site at
http://www.crcpress.com

Contents

Preface...vii
Symbols...ix

SECTION I *Setting Project Methodology*

Chapter 1 Approaches to Designing and Developing the Product 3

    1.1 Introduction 3
    1.2 Approaches to Designing and Developing the Product.........................3
    1.3 The Main Activity in Designing a Product 5
    1.4 Methodological Evolution in Designing the Product 7
    1.5 Concurrent Engineering 7
    1.6 Life Cycle Design 9
    1.7 Design for X 10
        1.7.1 Property-Objective of Design for X 11
    1.8 Correlating and Choosing Design Tools..11
    1.9 Other Tools and Design Approaches..12
        1.9.1 Knowledge-Based Engineering...12
        1.9.2 Total Quality Management...12
        1.9.3 Reverse Engineering 12

Chapter 2 Mechanical Design and Environmental Requirements....................... 15

    2.1 Introduction 15
    2.2 Design and Environmental Requirements.. 15
    2.3 Design for Environment 16
    2.4 Optimum Environmental Performance 17
    2.5 Implementing Design for Environment and General Guidelines................... 17
    2.6 Product Life Cycle and Environmental Impact................................... 18
    2.7 Environmental Impact of the Product 19
    2.8 Modeling the Life Cycle...20
    2.9 Life Cycle Assessment..23
        2.9.1 Methodological Premises, Property, and Structure 23
        2.9.2 Eco-Indicators to Quantify Environmental Impact 25
    2.10 Introducing Environmental Factors in the Design Process............................26
    2.11 Integrating Life Cycle and Design 27
        2.11.1 The Role of Life Cycle Design...29
    2.12 Integrating Environmental Aspects with Product Development.....................29
    2.13 Environmental Strategies for Product Planning.................................30
        2.13.1 Useful Life Extension Strategies...32
        2.13.2 Reclamation Strategies at End of Useful Life 32
    2.14 Methodological Setup and Design Tools..35
    2.15 The Role of Design for X in Designing for Environmental Requisites...........36
        2.15.1 Design for X for Environmental Strategies.........................39

      2.15.2 Design for X Setup in Product Development 39
  2.16 Tools for Integrated Design: Overview 40

SECTION II   *The Properties of Design Materials*

Chapter 3 Materials for Mechanical Design...45

  3.1 Introduction 45
  3.2 Metallic Materials and Alloys...45
  3.3 Solid State and Structure of Metals..46
      3.3.1 Solidification 47
      3.3.2 Lattice Defects...47
  3.4 Iron–Carbon Alloys..48
      3.4.1 Cast Irons 48
          3.4.1.1 Influence of Common and Special Elements on Cast Irons49
          3.4.1.2 Gray Cast Iron Classification...50
          3.4.1.3 Special Cast Irons Classification 51
          3.4.1.4 Spheroidal Cast Irons Classification......................................51
          3.4.1.5 Cast Iron Alloys Classification 51
          3.4.1.6 Malleable Cast Irons Classification 51
      3.4.2 Steels 52
          3.4.2.1 Manganese Steels...53
          3.4.2.2 Chromium Steels 53
          3.4.2.3 Nickel Steels 54
          3.4.2.4 Tungsten Steels 54
          3.4.2.5 Molybdenum Steels 54
          3.4.2.6 Silicon Steels...54
          3.4.2.7 Less Common Carbon Steel Alloys.......................................55
          3.4.2.8 Special Stainless Corrosion and Heat-Resistant Steels 55
  3.5 Non-Ferrous Metals and Their Alloys 56
      3.5.1   Copper 56
      3.5.2   Copper–Tin Binary Alloys (Cu-Sn Bronzes)....................................56
      3.5.3   Copper–Tin–Zinc Alloys...56
      3.5.4   Copper–Tin–Zinc–Nickel Alloys 56
      3.5.5   Low Lead Copper Alloys 57
      3.5.6   Copper–Lead–Tin Alloys..57
      3.5.7   High Lead Bronzes (Cu–Sn–Zn–Pb) 57
      3.5.8   Aluminum Bronzes 58
      3.5.9   Industrial Bronze...59
      3.5.10 High Resistance Brass...59
      3.5.11 Foundry Brass 59
      3.5.12 Pressure Die-Casting Zinc Alloys...60
      3.5.13 High Aluminum Zinc Alloys 60
      3.5.14 Lead Alloys...60
      3.5.15 Alloys for Cables..60
          3.5.15.1 Aluminum..61
          3.5.15.2 Magnesium 61
          3.5.15.3 Titanium 61
          3.5.15.4 Nickel 62

            3.5.15.5 Chromium 62
            3.5.15.6 Vanadium 62
            3.5.15.7 Cobalt 62
            3.5.15.8 Tungsten (Wolfram) 62
            3.5.15.9 Molybdenum 63
    3.6   An Outline of Polymeric Materials 63
          3.6.1   Polymer Classification 63
                  3.6.1.1   Origin-Based Classification 63
                  3.6.1.2   Classification Related to Temperature Effects 64
                  3.6.1.3   Classification According to Physical Characteristics 64
          3.6.2   Synthetic Resins 64
                  3.6.2.1   Macromolecule Structure 64
                  3.6.2.2   Production 64
          3.6.3   Main Properties of Polymeric Materials 65
                  3.6.3.1   Mechanical Properties 65
                  3.6.3.2   Thermal Properties 65
          3.6.4   Commonly Used Polymers 65
                  3.6.4.1   Cellulose Resins 65
                  3.6.4.2   Styrene Resins 66
                  3.6.4.3   Acrylic Resins 66
                  3.6.4.4   Polyamide Resins 67
                  3.6.4.5   Fluorinated Resins 67
                  3.6.4.6   Phenolic Resins 68
                  3.6.4.7   Amino Resins 68
                  3.6.4.8   Allyl Resins 68
                  3.6.4.9   Polyester Resins 68
                  3.6.4.10 Epoxy Resins 68
                  3.6.4.11 Polyurethanes 69
                  3.6.4.12 Silicones 69
    3.7   Outline of Composites 69
          3.7.1   Matrices 70
          3.7.2   Polyester Matrices 71
          3.7.3   Epoxy Matrices 71
          3.7.4   Phenolic Matrices 71
          3.7.5   Silicone Matrices 71
          3.7.6   Fibrous Stiffening 71
          3.7.7   Glass Fibers 71
          3.7.8   Carbon Fibers 72
          3.7.9   Boron Fibers 73
          3.7.10 Aramid Organic Fibers 73
          3.7.11 Admissible Load and Elasticity Modulus 73
          3.7.12 Multilayer Laminates 73
    3.8   Outline of Ceramic Materials 74

Chapter 4   Characterization of Metals 77
    4.1   Introduction 77
    4.2   Tensile Stress Tests 78
    4.3   Static Elastoplastic Characterization 79
    4.4   Plastic Constituent Link 83
    4.5   Notes about Mechanical Characterization Tests 86

4.5.1   Compression Tests...86
4.5.2   Curvature Tests 86
4.5.3   Shearing Tests 88
4.5.4   Torsion Tests..88
4.5.5   Hardness Tests...88
4.5.6   Brinell Hardness/Number 88
4.5.7   Diamond Pyramid Hardness Number (Vickers)...........90
4.5.8   Rockwell Hardness...91
4.5.9   Knoop Hardness 92
4.5.10   Resilience Tests 92
4.5.11   Standard Bars 93
4.5.12   Other Test Bars 93
4.6   Notes about Technological Characterization Tests 94
4.6.1   Drawing Tests...95
4.6.2   Erichsen Drawing Test 95
4.6.3   Erichsen Drawing Test Modified (UNI 4693)...............96
4.6.4   Socket Drawing Test (UNI 6124) 96
4.6.5   Pomp Drawing Test 96
4.6.6   Bending Test...97
4.6.7   Template Bending..97
4.6.8   Mandrel Bearing Bending 98
4.6.9   Percentage Elongation and the Tetmajer Coefficient 98
4.6.10   Forging Tests 100
4.6.11   Stretch Tests 100
4.6.12   Heading Test..100
4.7   Metallurgic Tests 101
4.8   Tube Tests 102
4.8.1   Enlargement Tests 102
4.8.2   Beading Test (Flanging Test) 103
4.8.3   Compression Test 104
4.8.4   Tensile Stress Test 104
4.9   Tests on Steel Wires 104
4.10   Final Indications 105

Chapter 5   Stress Conditions...107

5.1   Introduction 107
5.2   Mechanical Behavior of Materials 107
5.3   Conditions of Mechanical Stress..108
5.3.1   Simple Tensile Stress...108
5.3.2   Simple Compression..112
5.3.3   Simple Bending of Beams..114
5.3.4   Simple Torsion..116
5.3.5   Simple Shearing 118
5.3.6   Buckling of Column Bars..121

Chapter 6   Fatigue of Materials 127

6.1   General Concepts 127
6.2   Load Characteristics..129
6.3   Fatigue Diagrams 131

6.4 Thermography to Define Fatigue Behavior 133
6.5 Determining the Fatigue Curve 136
6.6 Fatigue Limit by Thermal Surface Analysis of Specimen in Mono Axial
Traction Test 138
6.7 Residual Resistance of a Mechanical Component 140
6.8 The Factors Influencing Fatigue 143
6.9 Safety Factor 145
6.10 Stress Concentration Factor 147
6.11 Fatigue Resistance of Welded Machine Components 153
6.12 Internal Material Damping 155
6.13 Test Machines 157
6.14 Conclusions about Designing a Mechanical Component 158

Chapter 7 Optimal Materials Selection in Mechanical Design 161

7.1 Introduction 161
7.2 Design Requirements and Correlation with Materials 161
7.3 Primary Constraint Selection 164
7.4 Approaches to the Best Choice of Materials 164
7.5 Tools for Selection of Properties 165
    7.5.1 Standard and Weighted Properties Method 165
    7.5.2 Method of Limits and Objectives on Properties 168
7.6 Integrated Selection Tools 170
    7.6.1 Analytical Formulation 170
    7.6.2 Graphic Selection 173
    7.6.3 Application 175
    7.6.4 Research Development 176

SECTION III   Design of Mechanical Components and Systems

Chapter 8 Failure Theories 181

8.1 Material Strength and Design 181
8.2 Stresses in a Three-Dimensional Object 181
8.3 Main Stresses 184
8.4 Mohr's Circles 185
    8.4.1 Simple Tensile Compression Stress Compression (Figure 8.6) 186
    8.4.2 Bending and Shearing (Figure 8.7) 186
    8.4.3 Pure Torsion (Figure 8.8) 186
8.5 Failure Hypothesis 186
    8.5.1 Maximum Linear Strain Hypothesis 188
    8.5.2 Maximum Tangential Stress Hypothesis 188
    8.5.3 Deformation Energy (Beltrami's) Hypothesis 191
    8.5.4 Distortion Energy (Von Mises's) Hypothesis 192
8.6 Failure Hypotheses: A Comparison 195

Chapter 9 Hertz Theory 197

9.1 Introduction 197
    9.1.1 Sphere versus Sphere Contact 208

          9.1.2   Cylinder versus Cylinder Contact 208
     9.2   Application to Rolling Contact Bearings 209
     9.3   Approach 210
     9.4   Equivalent Ideal Stress in Hertzian Contact.. 212
     9.5   Loads on the Ball Bearings 213

Chapter 10 Lubrication 217

     10.1   Introduction 217
     10.2   Prismatic Pair of Infinite and Finite Length 220
     10.3   Calculation of Michell Thrust Bearings... 223
     10.4   Journal-Bearing Pair: Setting Reynolds' Equation and Clearance Height ... 226
     10.5   Infinite Length Journal-Bearing Pair 227
     10.6   Journal-Bearing Pair of Finite Length: Schiebel's Formulae................ 232
     10.7   Journal-Bearing with Finite Length: Bosch's Approximate Computation.... 236
     10.8   Journal-Narrow Bearing Pair: Ocvirk's Hypothesis 237
     10.9   Lubricated Journal with Finite/Infinite Length: Raimondi–Boyd
            Calculation..238
     10.10 Notes on Lubricants.. 246

Chapter 11 Shafts and Bearings... 249

     11.1   Shafts... 249
     11.2   Shaft Measuring 249
            11.2.1   Two-Bearing Shafts (Isostatic).. 250
            11.2.2   Shafts with More than Two Bearings (Hyperstatic)................ 252
     11.3   Deformation Limits and Determination of Elastic Deflections 253
            11.3.1   Assessment of Shaft Elasticity 253
     11.4   Rolling-Contact Bearings (Anti-Friction Bearings)............................. 257
            11.4.1   Classification of Bearings... 257
            11.4.2   Classification by Rolling Body Shape 257
            11.4.3   Classification by Assembly Method 257
            11.4.4   Classification by Thrust Direction 257
     11.5   General Assembly Rules 259
     11.6   Bearing Adjustment... 259
     11.7   Assembly Models 260
     11.8   Friction Bearings 266

Chapter 12 Splined Couplings, Splines and Keys... 271

     12.1   Splined Couplings 271
     12.2   Splines and Tongues 273
     12.3   Transverse Splines 280
     12.4   Spur Gears... 284

Chapter 13 Springs... 287

     13.1   Introduction 287
     13.2   Spring Computation... 288
     13.3   Bending Springs 291
            13.3.1   Leaf Springs.. 296

13.3.2 Leaf Springs with Spring Shackles 297
13.3.3 Characteristics of Leaf Spring Systems 303
13.3.4 Coil Springs..304
13.3.5 Cylindrical Helical Springs 309
13.3.6 Conical Disc Springs... 311
13.3.7 Variable Thickness Disc Springs 313
13.3.8 Diaphragm Spring 316
13.4 Torsion Springs..319
13.4.1 Torsion Bars...321
13.4.2 Cylindrical Helix Torsion Springs ,,,,,,,,,,,,,,,, ...322
13.4.3 Conical Helix Torsion Springs 325
13.5 Compression Springs...329
13.6 Resonant Spring Calculations 333

Chapter 14 Flexible Machine Elements..339
14.1 Introduction 339
14.2 Belts...339
14.3 Sizing Belts...340
14.4 Timing Belts 342
14.5 Precautions for Testing Durability 343

SECTION IV   Design of Components and Mechanical Systems

Chapter 15 Spur Gears..347
15.1 Introduction 347
15.2 Introductory Ideas and Preliminary Definitions 347
15.3 Definitions and Nomenclature..347
15.4 Gear Sizing 352
15.5 Cut and Operating Conditions...353
15.5.1 Ordinary Sizing..354
15.6 Involute Equation...354
15.7 Determining Tooth Thickness...355
15.8 Tooth Base Fillets 359
15.9 Euler–Savary Theorem..361
15.10 Interference and Schiebel's Fillet 362
15.11 Meshing 365
15.12 Expressing Contact Segment and Action Arc 368
15.13 Condition of Non-Interference 370
15.14 Reduction of the Contact Line's Usable Segment Due to Interference 373
15.15 Specific Creep..378
15.16 Modified Gears 381
15.17 German Norms on Modified Gearing 391
15.18 Loads When Engaging Gears..393
15.19 Verification Calculations for Gear Resistance...394
15.19.1 Calculating Tooth Bending 394
15.19.2 Teeth Wear Calculation 399
15.20 Notes on the "American" Norms for Calculating Toothing (Agma).............402

          15.20.1 Bending Test..402
          15.20.2 Wear Verification 403
          15.20.3 Corrective Coefficients for Fatigue 403
     15.21 Design Phases for a Gearing 405

Chapter 16 Press and Shrink Fits...407

     16.1 Introduction 407
     16.2 Fixing Flywheel Components 407
     16.3 Forced Hub-Shaft Shrinking 414
     16.4 The Relationship between Interference and Thermal Variation 421

Chapter 17 Pressure Tubes..423

     17.1 Introduction 423
     17.2 Sizing and Verifying Pipe Strength......................................423
     17.3 Strength Verification for Pressurized Vessels 426
     17.4 Some Oil Pipe Considerations..427
     17.4 Instability in Externally Pressed Pipes................................428
     17.5 Calculations for Flange Bolts 431
     17.6 Fixed Flange Calculations 432
     17.7 Free Flange Calculation...434
     17.8 Flange Bolts 435
     17.9 Notes on Hot Metal Creep..436
          17.9.1 Creep Tests 437
          17.9.2 Microstructural Aspects of the Deformation Phases 440
          17.9.3 Physical Interpretation of Creep.............................441
     17.10 Brief Notes on Testing Creep 442

Chapter 18 Welded Joints..443

     18.1 Introduction 443
     18.2 Welding Defects 444
     18.3 Weld Tests...446
     18.4 Weld Strength 447
     18.5 Static Calculation...447
     18.6 Static Calculation According to Norm CNR-UNI 10011 448
     18.7 Static Calculation—Head and Complete Penetration Weld 448
     18.8 Static Calculation—Welds with Corner Beads.....................449
     18.9 Loads on the Bead 451
     18.10 Fatigue Calculation 453
     18.11 Fatigue Calculation According to Norm CNR-UNI 10011 454
          18.11.1 Fatigue Resistance—σN Curves with $\Delta\sigma$ Stress (Figure 18.17)......454
          18.11.2 Fatigue Resistance—σN Curves with $\Delta\tau$ Stress (Figure 18.18).......455
          18.11.3 Fatigue Resistance—Influence of Thickness on Acceptable Δ........455
     18.12 Spot Welding 456
          18.12.1 Static Calculation 457
          18.12.2 Fatigue Calculation...458
     18.13 Riveted Joint...459

Chapter 19 Couplings..467

    19.1   Preliminary Notions and Types of Joints 467
    19.2   Rigid Joints...468
          19.2.1   Box Coupling/Sleeve Joint 468
          19.2.2   Cylindrical Joints 469
          19.2.3   Disc and Flange Couplings 471
    19.3   Fixed Semi Elastic and Elastic Joints................................475
    19.4   Peg Joints..478
    19.5   Spring Joints 480
          19.5.1   Bibby Joint 480
          19.5.2   Voith Maurer Joint...487
          19.5.3   Forst Joint 488
    19.6   Mobile Joints 490
          19.6.1   Cardan (Universal) Joint 491
          19.6.2   Oldham Joint 496

Chapter 20 Clutches...501

    20.1   Introduction 501
    20.2   Friction Clutches...501
    20.3   Flat Plate Clutches 506
    20.4   Cone and Double Cone Clutches 510
    20.5   Radial Block Clutches 512
    20.6   Surplus Clutches 517
          20.6.1   Cylindrical Rollers 517
          20.6.2   Non-Cylindrical Rollers, Blocked 518
    20.7   Safety Clutches 520
    20.8   Centrifugal Clutches..521

Chapter 21 Brakes..529

    21.1   Introductory Concepts 529
    21.2   Disc Brakes...530
    21.3   Performance and Dimension Analyses 534
    21.4   Drum Brakes 545
          21.4.1   Calculations and Sizing...548
    21.5   Ribbon Brakes 557
    21.6   Differential Ribbon Brakes 560

Chapter 22 Case Study: Design of a Differential..563

    22.1   Introduction 563
    22.2   Review for the Gears Project...564
    22.3   Dimensioning of the Bevel Gears......................................566
          22.3.1   Introduction 566
          22.3.2   Calculation of the Teeth Number 567
          22.3.3   Bending Calculations 568
          22.3.4   Wear Calculation 568
          22.3.5   Dynamic Verification 569
          22.3.6   Bending Verification of the Crown 569
          22.3.7   Dynamic Rim Verification 569

22.4   Project of Planetary Reducer.. 570
        22.4.1   Introduction 570
        22.4.2   Teeth Number Calculation 570
        22.4.3   Bending Calculation.. 571
        22.4.4   Wear Calculation 572
        22.4.5   Dynamic Verification 572
22.5   Differential Gear Dimensions 573
        22.5.1   Calculation of the Number of Teeth.................................. 573
        22.5.2   Bending Calculation.. 574
        22.5.3   Wear Calculation 575
        22.5.4   Dynamic Verification 575
22.6   Measurement of Time Gearing 575
        22.6.1   Introduction 575
        22.6.2   Bending Calculation.. 582
        22.6.3   Wear Verification 582
        22.6.4   Bending and Wear Verification of the External Rim 583
22.7   Dimensioning of the Shaft Pinion and Choice of the Bearings..................... 583
        22.7.1   Calculating Splined Profile 583
        22.7.2   Calculation of Reaction Forces 585
        22.7.3   Calculation of the Shaft.. 588
        22.7.4   Deflection Verification 590
        22.7.5   Choice of Bearings.. 591
22.8   Choice of the Differential Bearings Box 593
22.9   Dimensioning of the Pins 596
        22.9.1   Pins of the Epicyclic Speed Reducer................................. 596
        22.9.2   Maximum Deflection 597
        22.9.3   Specific Pressure 597
        22.9.4   Shear and Bending 597
        22.9.5   Pins of the Satellite Differential.................................... 598
        22.9.6   Dimensioning at Deflection... 599
        22.9.7   Specific Pressure Verification 599
        22.9.8   Verification at Shear and Bending.................................... 599
22.10 Dimensioning of the Box Screws 600
22.11 Dimensioning of the Semi-Shaft 601
        Appendix 606
        Project Data Specification 606

Problems 617

References.. 649

Index.. 651

Preface

This collection is the result of a decade of continual elaboration and revision of teaching notes at the earnest request of students of the machine construction course that has been run for several years now at the Faculty of Engineering of the University of Catania for mechanical engineering students.

Beginning with current training needs, the various parts of the book's arguments are dealt with in such a way as to be accessible to all students of courses in industrial engineering. The goal and the hope are that this collection of notes can be used not only by students of the mechanical engineering course but by all those whose training requires them to acquire the fundamentals of the design of mechanical components.

For this course, the textbooks have always been written by Prof. R. Giovannozzi, (the "maestro"), and the review of notes served mainly to help students get the most out of the texts, supporting them above all in those areas where, by experience, they have the most difficulty. Furthermore, the notes have also served to shorten study time, an important contribution given the requirements of the new teaching regulations.

I have always used Prof. Giovannozzi's texts for reference, because of all the sources consulted, although some inspired me (particularly those by American authors), his proved the most complete in dealing with these subjects; they remain excellent texts, and are certainly stimulating for students. The approach and methodology continue to be extremely valid, and in my opinion cannot be replaced by those proposed in other texts.

The analytical approach to the subjects is still based on algorithms from traditional calculus without reference to more current methodologies, which, however, students do come across in other courses. This choice was made so as not to deprive students of the ability to use simple models and calculations that are reliably effective and helpful at times when more complicated algorithms or well-known commercial programs need to be used.

The aim is still to induce students to be logical, starting by analyzing the physical problem with the most appropriate schematic, and ending with a constructional definition of the component in need of planning.

To guarantee due completeness of subject requirements, it was considered essential on occasion to add references to current norms, or more advanced approaches to calculation (e.g., cogwheel resistance tests, lubrication theory, brake measuring) wherever necessary, compared to the text references. In such cases, however, the calculations suggested in Giovannozzi's text are side-by-side with those obtained by the most recent methods so that students can compare. Often, construction details in the maestro's texts are still quite rightly valid references today.

To comply with the requirements of the new teaching regulations, the principal materials tests and simple stress states are outlined prior to the study of fatigue, which refers to fine-tuning methods developed at Catania's Faculty of Engineering. The hope is that other industrial engineering students—not just the mechanical engineers—can thus benefit from the teachings and procedures of the maestro.

Typical machine construction course subjects/modules occupy the greater part of this book (mechanical system component planning), but two preliminary sections enhance its appeal: the methodological set-up of the project (traditional or more recent), and the project criteria that take into account the environment. These two parts are echoed in a work published in the U.S. authored by myself, together with Prof. G. La Rosa and Ing. F. Giudice, who have collaborated in editing that work.

Finally, as mentioned above, this work should be seen as a collection of notes on lessons conducted by the author and inspired by Giovannozzi's book, without which the notes are inadequate.

"The intelligence of the reader" in using the collection, as the maestro often said, might allow all that is contained in the Giovannozzi manual to conform to current—above all professional—needs.

A case study in which theoretical methods and tools are applied to the planning of real mechanical systems is reported.

Since there is always room for improvement, we welcome suggestions from our readership. Please address these to the author at arisitan@diim.unict.it

Symbols

Symbols

A	Area
a	Distance
B	Coefficient, life
b	Distance, Weibull shape parameter
C	Basic load rating, bolted-joint constant, center distance, coefficient of variation, column end condition, constant, correction factor, heating coefficient, specific heat, spring index
c	Distance
d	Diameter
D	Diameter
E	Energy, error quantity, modulus of elasticity
e	Eccentricity
F	Force
f	Coefficient of friction, deflection, frequency
G	Modulus of rigidity
g	Acceleration due to gravity
H	Heat, power
HB	Brinell hardness
H	Distance, film thickness
I	Integral, mass moment of inertia, second moment of area
J	Mechanical equivalent of heat, polar second moment of area
J_p	Polar second moment of area
K	Stress-concentration factor, stress-correction factor, torque coefficient
k	Endurance-limit modifying factor
L	Length, life
l	Length
M	Moment, bending moment
M_t	Torsional moment
m	Mass, slope, strain-strengthening exponent, factor of safety
N	Normal force, number, rotational speed
n	Factor of safety–load factor, rotational speed
P	Force, unit-bearing load
p	Pitch, pressure, probability
Q	First moment of area, imaginary force, volume
q	Distributed load, notch sensitivity
R	Radius, reaction force, reliability, Rockwell hardness, stress ratio
r	Correlation coefficient, radius
S	Sommerfeld number, strength
s	Distance, sample standard deviation
T	Temperature, tolerance, torque
t	Distance, time
U	Strain energy
u	Unit strain energy
V	Linear velocity, shear force
v	Linear velocity
W	Cold-work factor, load, weight
w	Distance, unit load

X	Coordinate
x	Variate of x
c_i	Surface factor
c_2	Size factor
P_{crit}	Euler critical load
i	Interference
W	Section modulus
α_i	Stress concentration factor
β_i	Fatigue stress concentration factor
σ	Normal stress
τ	Shear stress
τ_{am}	Allowable shear
σ_{am}	Allowable stress
σ_r	Ultimate stress
σ_s	Yield stress
σ_0^1	Fatigue limit for $R =$
η	Notch sensitivity
λ	Slenderness ratio
E_1	Energy required for fatigue fracture
σ_0	Fatigue limit for $R = -1$ (limit stress above which some crystal is plasticized)
σ_i	Minimum stress
σ_s	Maximum stress
σ_m	Mean stress
σ_p	Plastic stress
$\sigma_{0.2}$	Yield stress
ε_p	Plastic strain
R	Tress ratio
N	Current number of cycles
N_f	Numbers of cycles to failure
ε	Strain
ε_0	Strain corresponding to σ_0
α	Coefficient of linear thermal expansion
T	Surface temperature
T_a	Ambient temperature
T_0	"Limit temperature" (corresponding to end of thermo-elastic phase)
ΔT	Temperature increment of the hottest area
E	Modulus of elasticity
υ	Modulus of Poisson
ρ	Density
c_ε	Specific heat for constant strain
c_σ	Specific heat for constant stress
k_c	Thermal convention coefficient
K_m	Thermo-elastic constant
ΔQ_p	Plastic energy liberated as heat
dQ_e	Energy liberated as heat
V	Land volume
V_p	Plastic volume
t_0	Thermo-elastic time (corresponding to T_0)
t_r	Time between thermo-elastic time t_0 and complete test time t_f
$t_f = t_0 + t_r$	Complete test time

Recurrent Symbols for Chapter 15

a	Addendum
u	Dedendum
g	Clearance
m	Diameter pitch
m_0	Diametral tool pitch, base pitch
n	Number of teeth
p	Pitch
pf	Pitch on the base circle
r	Current radius
R	Pitch radius
D	Pitch diameter
ρ	Base radius
θ	Angle of action
δ	Pressure segment
d	Distance center to center
s	Tooth thickness
τ	Gear ratio
φ	Current angle
λ	Addendum tool factor
v	Velocity
x, x'	*Correction factors*

AGMA Symbols
(by J. E. Shigley and C. R Mischke, *Mechanical Engineering Design*, Fifth Edition)

C_a	Application factor
C_f	Surface-condition factor
C_H	Hardness-ratio factor
C_L	Life factor
C_m	Load-distribution factor
C_p	Elastic coefficient
C_R	Reliability factor
C_s	Size factor
C_t	Temperature factor
C_V	Velocity factor (for use in Equation 14.14)
C_v	AGMA dynamic factor
E	Modulus of elasticity
F	Face width
H	Power
H_{BG}	Brinell hardness of gear tooth
H_{BP}	Brinell hardness of pinion tooth
I	Geometry factor
J	Geometry factor
K_a	Application factor
K_f	Fatigue stress-concentration factor
K_L	Life factor
K_m	Load-distribution factor
K_R	Reliability factor
K_S	Size factor
K_T	Temperature factor
K_V	Velocity factor (for use in Lewis equation only)
K_v	AGMA dynamic factor

S_C	Endurance stress
S_c	AGMA surface fatigue strength
S	AGMA bending strength
T	Tooth thickness
V	Pitch-line velocity
Wt	Transmitted (tangential) load
Y	Lewis form factor
Z	Length of line of action
υ	Poisson's ratio
σ	Tooth bending stress
σ_{all}	Allowable bending stress
σ_C	Surface compressive stress
σ_c	Contact stress (AGMA formula)
$\sigma_{c,all}$	Allowable contact stress
Φ	Pressure angle

Part I

Setting Project Methodology

In the two chapters forming Part I of the book, the general concepts on design methodology are introduced, outlining the traditional approach in design, and showing how it has evolved in recent decades, in terms of the methodological frames better suited to the new needs of developing product. Particular attention is given to the imperative need for design to address the requirements of environmental performance. It is shown how these targets can be achieved in the practice of design, through the introduction of procedures and tools to support designers in meeting preservation environmental standards in addition to the conventional requirements.

1 Approaches to Designing and Developing the Product

1.1 INTRODUCTION

In the world of mechanical design, a "product" is a real artefact with a clear physical shape and an engineering application, and therefore falls into that category of technical systems that function on the basis of physical principles and are regulated by the laws of physics.

In very general terms, "design" means any activity that can change existing reality into one whose conditions are preferable. In relation to the technological side of human activity, it becomes a process of the organization and management of the human resources, and the information a design has accumulated during its evolution. In the case of an actual physical industrial product with an engineering application that might be any mechanical system, "design" is a process of the transformation of resources (cognitive, human, economic, and material) into a whole of functional requisites to provide a physical solution (product and system).

Although "design" and "development" of a product are often interchanged, they are frequently complementary, giving rise to the well-known "design and development of a product." This leads to the possible distinction between activities that are specifically "design" in the sense just described, and a more extensive activity that includes designing but embraces something rather larger, starting with identifying a need or a market opportunity and ending with the commencement of production. Sometimes, "product development" refers to an even wider sphere encompassing the entire process of transformation of a market opportunity into a saleable commodity, and therefore includes production, distribution, and commercialization. Thus, a process this complete involves all the main business functions (marketing, planning/design, and production), orienting them according to consumer requirements.

In recent times, the design process and production management have required great innovation, principally to reduce time and resources in the planning, production, and distribution of products whose performances have to be ever better and more diversified.

The evolution of methodological approaches follows in the same track in trying to help the designer face the growing complexity of a design project, and the system of factors that influence it in various ways. In optimizing the specifics of functionality, cost, and reliability, the right compromise is sought to provide an ever-wider performance spectrum.

In this context, among the more significant aspects is the emerging need to view the problem as a whole process of design and development of the product. The conventional approach, which was limited to a final step of analyzing product sales, therefore moves to an innovative approach that takes into account the utilization phase of the product, its life expectancy, and even its replacement.

New design challenges require a systematic approach, with planning integrated with the relative production processes in accordance with the new methods known as concurrent engineering (CE), life cycle design (LCD), design for X (DFX)—design approaches that consider all the life cycle phases, from development of the concept to its replacement, analyzing and dovetailing determining factors, such as quality, cost, producibility, demand, maintenance, and the environment.

1.2 APPROACHES TO DESIGNING AND DEVELOPING THE PRODUCT

This is the process of developing a product which includes that sequence of phases or activities required to create, design, build, and market the product. The study of this process aims to define

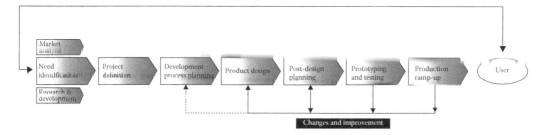

FIGURE 1.1   Product development process.

common schematic paths from the vast variety of possible applications. The objective is to mark out a reference model that translates needs and ideas into technical prescriptions to transform the most suitable resources into useful material products.

Although not even one model exists that includes the huge variety of possible processes for product development, and each process could be considered unique, it is possible to identify activities and elements held in common.

Traditionally, product development basically referred to a sequential process of a few defining moments as reported in Figure 1.1 that combines the suggestions of various authors (Pahl and Beitz, 1996; Dieter, 2000; Ulrich and Eppinger, 2000; Ullman, 2003)*:

*   *Need identification*—Consists of a survey of client needs, identification of user-type and product competitors on the market, evaluation of the most suitable strategies (improving an existing product, developing new technologies). To highlight the close relationship between this phase and knowledge of the market and opportunities from new technologies, Figure 1.1 shows the parallel activities to market analysis, and research and development.
*   *Project definition*—The phase in which the project is approved, marking the real start of product development. It encapsulates company strategies, current market trends, and technological developments in a mission statement describing the market objective of the product, the company objectives, and the main constraints of the project.
*   *Development process planning*—The phase in which the entire process of design and development is planned by deconstructing, planning, and allocating activities, defining and allocating resources (temporal, economic, and human), compiling and assigning information.
*   *Product design*—The phase assembling all the specifically design activities, ranging from defining product requisites to generating the concept, and finally to transforming the concept into a producible system. This phase incorporates a sub-process—design process—described below.
*   *Post-design planning*—The phase that relates to planning the production-consumption cycle. Some authors limit this planning just to the necessities of production. In this case it includes the complete definition and planning to fulfilment of the product, from the component production sequence, to the arrangement of tools and machinery, to planning assembly. Some authors add the necessities of distribution, use, and withdrawal of the product.
*   *Prototyping and testing*—The phase requiring product prototypes to undergo testing to verify how they respond to requirements, their performance levels and reliability. Understandably, this phase is supremely important in terms of any product improvements and consequently of the evolution of the project's solution.
*   *Production ramp-up*—The phase of complete production start-up. It consists of manufacturing the product by the desired production process (which does not happen in the

---

* Pahl G. and Beitz W., *Engineering Design: A Systematic Approach,* 2nd edn. Springer-Verlag, London, 1996; Dieter G.E., *Engineering Design: A Material and Processing Approach,* 3rd edn. McGraw-Hill, Singapore, 2000; Ulrich K.T. and Eppinger S.D., *Product Design and Development,* 2nd edn. McGraw-Hill, New York, 2000; Ullman D.G., *The Mechanical Design Process,* 3rd edn. McGraw-Hill, New York, 2003.

prototype phase because the prototypes are made by other means). The main aim is to verify if the production process is adequate and to fine-tune it should problems arise, as well as to identify any residual defects in the final product. After start-up, there is a further transitional phase to major production and launch of the product on the market.

As highlighted in Figure 1.1, improving the final product is guided by a feedback system that sends information from post-design planning, prototyping and testing, and production ramp-up back to product design (and even back to development process planning). This method of improving the final product, originating from my first experiences in the theoretical planning process, has since been at the heart of the evolution mechanism that leads to the final solution.

Once distributed and commercialized, it is in the hands of the user, who has a close relationship with the initial phase of the whole process; he/she needs to be identified, because he/she interacts with the market and technological innovation, influencing it, and being influenced by it.

1.3 THE MAIN ACTIVITY IN DESIGNING A PRODUCT

It is very unlikely that the product development process models will have anything in common with the models for the sub-process of planning the product because they have different setups and peculiarities. The following, however, are common to the different models:

- The design process is described as a series of activities made up of main stages, each with its own result. There are usually three or four stages (*problem definition*, *conceptual design*, *embodiment design*, and *detail design*).
- The sequence of activities is conceived so that the product proceeds from the abstract to the concrete in order to operate initially in as large a space as possible, and to make the process subsequently converge on a concrete and buildable solution.
- In the design process, the evaluation of results is fundamental, because it makes the process one of developmental transformation based on the interaction of the subsequent cycles of analysis-synthesis-evaluation.

With these premises, designing the product becomes a process of transforming information from a request for a product outline with certain characteristics (requirements, restrictions, user needs, market conditions, and available technology) to the complete description of a technical system able to answer the initial request. This transformation utilizes various resources (cognitive, human, economic, and material), serving to nourish the main phases of the planning process (summarized in Figure 1.2). The planning process consists of a preliminary phase, specifying the problem and developing the project requisites, and planning phases at different levels (concept, systems, and details), as well as evaluation made up of the interactive cycles of analysis—synthesis—evaluation.

The reference diagram, charted in Figure 1.2 to describe the product design process, shows the four most important activities:

- *Problem specification*—In this phase all the information is processed to develop and define in detail the project requisites; product and user requirements, market conditions, and company strategies must all be clarified and used to generate the specifics that will guide subsequent phases.
- *Conceptual design*—Once the project specifics are defined, the ideas to create the product with the necessary requisites must be developed. In this phase the product is purely abstract, a bunch of attributes, which must be incarnated into the concept of product that is first obtained through *concept generation* and then through *concept evaluation*. A number of outline proposals are developed (description of form, functions and principal product characteristics), and these are subsequently evaluated to decide which best satisfy the request.

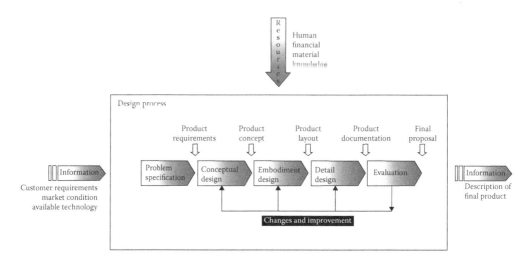

FIGURE 1.2   Product design process.

- *Embodiment design*—Having identified the most appropriate concept, the project idea must be translated into a physical system. The concepts from the previous phase are transformed into an outline layout which defines sub-systems and functional components. Physical elements are arranged to obtain the requested function, and thus the product architecture begins to take form. Furthermore, a preliminary study is carried out on the form of its components and a first choice of materials.

- *Detail Design*—The layout above must now be represented by geometric models and detailed designs. This requires applying methods and support tools to define the design details. The choice of materials, a study of form, a definition of the geometry of the components and assemblies, a plan of the assembly sequence and a definition of connection systems must all be guided by the entire range of product requisites (performance, economic, environmental, etc.). At the end of this phase, some authors foresee the complete planning of the production process, while others suggest including instructions for production, assembly, transport, and use, together with the rest of the documentation.

The diagram in Figure 1.2 shows how important is the extra phase, evaluating how proposed solutions correspond to the project specifics as defined by the problem specification, and guiding any modifications and improvements that can move the design process along to its conclusion.

To this end, the following are carried out: analysis of the critical aspects of the project; a long-range study of durability in relation to environmental factors (socio-economic conditions, consumer taste, competitors, raw material availability), and technological factors (technological progress, decline in performance); verification programs (modeling, first prototypes). Among the latter, modeling product performance is particularly important. It tends to find the simplest methods (analytical, physical, graphical) to compare detailed solutions with engineering targets, generally by number crunching.

Evaluations must, however, take into account any factors of inconvenience resulting, for example, from production, or a change in environmental conditions. The highest-level solutions are the most "robust" in that that their performance is unswayed by factors of inconvenience.

The literature frequently cites a great variety of formal methods as support tools for the various phases described above. They are of the quality function deployment (QFD) type whose purposes are to define project specifics and product requisites; and to develop graphical or physical mock-ups to evaluate and select the concept; to employ techniques of functional decomposition, the use of morphological charts for function-concept mapping, and the various techniques of product generation for development of the detailed project. For a complete panorama of the various formal methods, the literature carries many specific studies.

1.4 METHODOLOGICAL EVOLUTION IN DESIGNING THE PRODUCT

In recent times, the design process and managing production required significant innovation, principally to reduce time and resources in designing, producing, and distributing the product whose performance had to be ever higher and more diversified.

These new necessities highlighted the inadequacy of the sequential nature of the design and development process in two ways:

- Long development times due to sequencing different functions
- Limited capacity to perfect the product because of poor communication between various functions, and the consequent reduced or fragmentary information flow

The product's development models described previously are initially rigidly sequential, but should now be transplanted into a new methodological context that requires simultaneous planning phases of analysis and synthesis in close interaction and in relation to all the other phases of development. The sequential model in Figure 1.1 evolved into the simultaneous/integrated (S/I) model of development as reported in Figure 1.3. In the S/I model, the planning phases of the development process, designing the product, planning the production-consumption cycle, and the evaluation of results, are set up in one sole simultaneous intervention obtaining information from a shared source which takes into account a wide variety of criteria (functionality, producibility, reliability, and costs).

In this new setup, three approaches should be particularly underlined that today constitute a large part of research inherent in design methodologies:

- CE aims to harmonize the increase in product quality, with reduced development times and costs through an ample project team, able to analyze and synthesize simultaneously, as well as to be closely interactive in relation to all the development phases of the product.
- LCD extends design analysis to the entire product life cycle, from production and materials use to mothballing the production line.
- DFX includes a flexible system of integrable methodology and planning instruments, each responsible for a particular product requisite.

1.5 CONCURRENT ENGINEERING

The ever-shortening product life cycle that reflects current market dynamics requires a product development process that continually reduces time and costs. In recent years, however, there has instead been a lengthening in development time due to growing design complexity and the need to involve specialists from diverse disciplines.

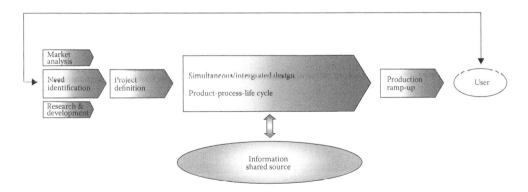

FIGURE 1.3   Simultaneous/integrated product development model.

High quality, high performing products at low cost and in a short time—these are the defining principles required by efficient and competitive industrial production. Putting this paradigm into practice necessitates that product development is structured and managed as a simultaneous and multidisciplinary process carried out by a suitably appointed design team with a wide spectrum of competence. CE arose out of this necessity. It aims to balance an increase in product quality with reduced development times and costs. It could be defined as a systematic approach to the simultaneous and integrated design of products and processes that includes production and post-sales issues. Right from the start, this approach involves the product development team in simultaneously considering all the factors determining product life cycle, from concept development to production end, including the requisites of quality, cost, production programing, and user needs.

The following often-cited essential principles form the basis of CE:

- Emphasis on the role of production process planning and its influence on design process decisions
- Importance of multidisciplinary project teams in product development processes
- Growing attention to client requirements and satisfaction
- Reduction in development times and time-to-market as product success and competitivity factors

These principles are based on common sense and are not radically innovative. CE could be considered as an evolution of the practice of the process of product development based on efficiency criteria. A predictable spinoff of simultaneous design intervention is shorter product development time. Figure 1.4 compares traditional sequential product development to that in CE.

It should be emphasized, however, that the increasing design complexity of the integrated approach can lengthen development time. The main aim of CE is, therefore, to facilitate increasing

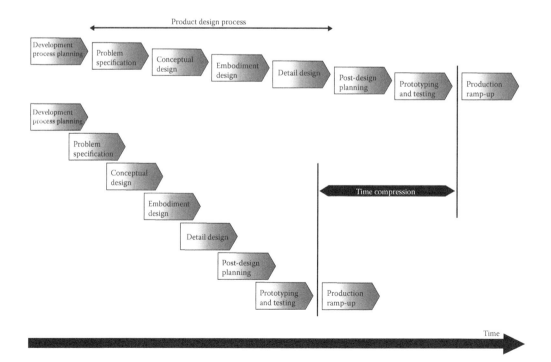

FIGURE 1.4   Product development process typologies: sequential vs concurrent model.

integration between design and manufacturing, leading to greater design complexity, yet reducing times and costs of product development and production.

1.6 LIFE CYCLE DESIGN

LCD means a design intervention that takes into consideration all the product's life-cycle phases (development, production, distribution, utilization, maintenance, production end, and recycling), during the entire design process from defining the concept to developing the detailed project. It incorporates methodologies, models, and instruments that harmonize the evolution of the product from its conception to its production end with a wide range of planning requisites (functional performance, producibility, maintainability, and environmental impact).

Life cycle engineering (LCE) is also very commonly used in the terminology of this sector and includes all the engineering functions (not just the design functions). LCE and LCD are often used synonymously.

As a design approach, LCD has three main aspects:

* Whole life-cycle vision
* Presumes the most effective interventions are those in the first design phase
* Simultaneous analyses and syntheses of various design aspects

This last aspect is shared with design techniques to reduce development times and costs in CE. At least the methodological approach, emphasizing extending the domain of the analysis of design requisites and optimizing performance within product life cycle, distinguishes it from other techniques.

The concept of LCD as proposed by other authors, is summarized in Figure 1.5. The main life-cycle phases of a product in design and development (pre-production, production, distribution, utilization, and production end) must be taken into consideration from the concept phase, because this is the most effective phase and is where the evolution of the design has least economic impact.

The selection of design alternatives must be guided by assessing the product's main success factors, which define the design objectives in relation to all the phases in the whole life cycle:

* Resource use (optimizing materials and energy)
* Planning the production system (optimizing production processes)

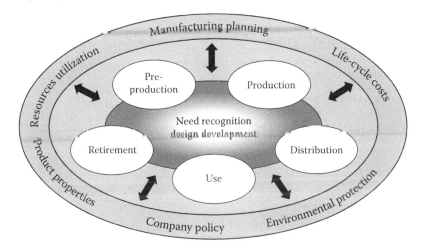

FIGURE 1.5 Life cycle design concept.

- Life cycle costs (optimizing total life cycle cost)
- Product properties (balancing a wide range of product requisites, such as functionality, safety in use, quality, reliability, aesthetics, affordability, ease of production)
- Company policy (respecting company objectives)
- Safeguarding the environment (checks and minimizing environmental impact)

1.7 DESIGN FOR X

As Figure 1.4 shows, structuring the product design process has recently evolved from the sequential model to the concurrent one. Even though some phases might remain sequential, which underlines the huge range of requisites a product needs in relation to the various phases of its life cycle, a structure that tends toward the concurrent model is said to be *design-centered* and is shown in Figure 1.6. Here the principle of simultaneity is applied to the specifically design phases, highlighting the importance of high levels of shared design analysis information. With this setup, development does not need the direct involvement of various company interventions as in the case of CE. By increasing the efficiency of project choices and extending requisites beyond the primary ones, the primary project phases (*conceptual*, *embodiment*, and *design detail*) are actually reinforced by introducing a series of conceptual and analytical tools—diversified according to product requirements—at each design level: performance analysis (PA); design for manufacturing e assembly (DFM, DFA); life-cycle cost analysis (LCCA).

In developing a product, the design-centered model integrates flexible methodology and design tools, each aimed at a particular product requisite and known as design for X.

In a second vision, in which industrial products must pass through every phase of life, and their design is strongly influenced by performance in each phase, there has been growing interest in a new approach known as "design aimed at product properties," also known as design for X (DFX). Each reference to X represents a different product property in one or more phases of the product life cycle. This has led to identifying and classifying design criteria oriented at production, assembly, utilization, and recycling, which aim at developing a complex design tool that takes into account the requirements of all the phases.

DFX can therefore be defined as a design process that pre-establishes specific requisites for the entire life cycle of the product at the design phase.

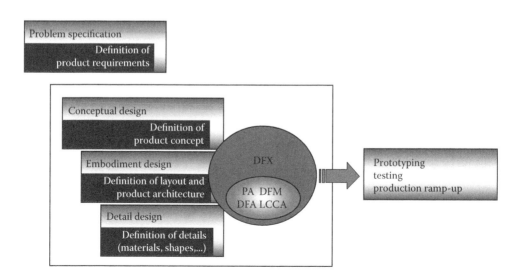

FIGURE 1.6   Design-centered product development model.

1.7.1 Property-Objective of Design for X

The most common DFX property-objectives can be grouped according to the product's life-cycle phases.

For the production phase, these are:

- *Design for producibility/manufacturability*—Design of components and construction systems according to production processes
- *Design for assembly*—Assembly design that facilitates assembly, reducing times and errors
- *Design for variety*—Design oriented at product variety achieved by defining and optimizing the base models so that their architecture is flexible enough to respond to market demand
- *Design for robustness/quality*—Design components to reduce the sensitivity of product performance to externals (production error, unexpected operating conditions) and to guarantee product quality

The following comprise the use phase:

- *Design for reliability*—System and component design to ensure the reliability of the finished product
- *Design for serviceability/maintainability*—Design oriented toward ease of maintenance and repairability, to eliminate possible breakdown
- *Design for safety*—Design oriented toward checking safety standards and preventing misuse during operation

Product retirement includes these phases:

- *Design for product retirement/recovery*—Design oriented to planning strategies for retiring and recycling the product at the end of its life cycle

Other property-objectives are indirect compared to the whole DFX system because they do not relate to any single phase of the cycle. Noteworthy among them are those aimed at safeguarding the environment:

- *Design for sustainability*—Design that aims at balancing the environmental requirements of processes and products with company policy and market constraints
- *Design for environment*—Design oriented to the environmental quality of processes and products

1.8 CORRELATING AND CHOOSING DESIGN TOOLS

The components of the DFX system are in perfect harmony with LCD because they were the reason for its introduction and development, and they can become operating instruments, each one specific to a particular property of the required product.

The choice of DFX instrument "Flexible Machine Elements" in Contents for use in the design phase depends on target requisites, and their use must be planned by taking into account not only the specific design phases in which they might be most efficacious, but also the life-cycle phases they might influence.

To help the designer identify the most suitable tools, it is possible to graphically represent the most appropriate fields of application in relation to the product's life-cycle phases and to the design process.

Figure 1.7 shows how some DFX tools can be correlated in a summary of the product life cycle (horizontal axis) in the phases of component production, assembly, product use and retirement and the design process (vertical axis) during concept development, defining the layout and detailed design.

1.9   OTHER TOOLS AND DESIGN APPROACHES

In addition to the three approaches previously described, other tools have been created to deal with the continually evolving design necessities of engineering. Despite not really being design processes, there are direct spinoffs and so they will be briefly mentioned.

1.9.1   KNOWLEDGE-BASED ENGINEERING

Design systems based on knowledge integrate computer-aided design (CAD) with the baggage of experience and competences of designers via programed systems that define engineering rules and criteria. Models are created which include information about the product in the form of designer rules that are product specific and range from general design rules to guide lines defined by design standards. Given the complexity of knowledge-based engineering (KBE) systems, they are generally based on artificial intelligence (AI) that can manage and process a whole range of specific information.

1.9.2   TOTAL QUALITY MANAGEMENT

Total quality management (TQM) consists of a design process set up to integrate all development activities with the primary aim of satisfying the client. This is done by defining product and process standards, maximizing production efficiency, and creating integrated development teams for efficient product design strongly oriented to client needs.

1.9.3   REVERSE ENGINEERING

Reverse engineering (RE) is a process that starts with an existing product, disassembles it, and defines its design characteristics by analyzing its components. The aim is to acquire all possible

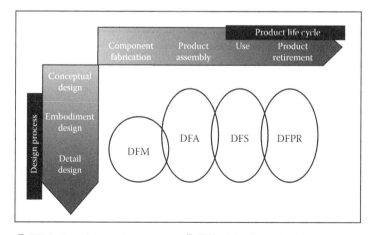

☐ DFM - design for manufacturing   ☐ DFS - design for serviceability
☐ DFA - design for assembly   ☐ DFPR - design for product retirement

FIGURE 1.7   Contextualization of DFX techniques.

information on the product as a basis for re-designing it to improve performance. There are several important phases: product analysis, generating technical data, verifying designs, and developing and applying any modifications. Technical data are generated by developing a complete CAD model of the existing product by using coordinate measuring machines (CMM) that apply electromechanical components, probes, and sensors to accurately map the product's measurements, translating them into CAD data. The model thus obtained can then be re-designed according to the new requisites.

2 Mechanical Design and Environmental Requirements

2.1 INTRODUCTION

Without doubt, industrial activity predominates in the ample panorama of human activity that degrades the environment. The principal countermeasures include monitoring and limiting the consumption of resources; stemming the rising phenomenon of overflowing landfills; conserving as much energy as possible in production processes; definite limiting of all types of gas emissions—premeditated or accidental; and intensifying the reclamation and recycling of resources.

Currently, different factors are motivating producers to adopt policies and tools aimed at safeguarding the environment:

- Legislation aimed at extending the responsibility of producers beyond product commercialization, even to the extent of imposing how to manage life-cycle retirement, with good incentives to reclaim and recycle.
- Introduction of a series of international standards to manage environmental systems and spread the use of certification for environmental-grade products.
- Increasing consumer awareness of the need to safeguard the environment.

Of the various regulations monitoring the environment, the norms that play on these motivating factors are those that are most preventative, as they aim to change company policies regarding protecting the environment by encouraging competition in production. This could mean restrictions at various levels, ranging from requiring manufacturers to bear the costs of waste disposal, in some cases even organizing the collection of their own products after use, to making their products explicitly recyclable. The legislative aim of the latter is to encourage redesign of products in a variety of categories to reduce their environmental impact.

Mechanical designs, being at the root of a large proportion of engineering production, and correlating an extremely wide range of production techniques, should fully embrace the new requirements of legislation.

2.2 DESIGN AND ENVIRONMENTAL REQUIREMENTS

Increased environmental sensitivity has recently generated a push toward optimizing production systems to guarantee higher levels of product eco-compatibility. This has brought about the development of new methodological approaches in design—design for environment (DFE)—that claims that the most effective phases at which to improve environmental impact are design and product development.

As we saw in the preceding chapter, in a similar but more generalized context, life cycle design (LCD) is a design approach that takes into account all the life-cycle phases of the product (development, production, distribution, use, maintenance, retirement, and recycling) from defining the concept to developing the detailed project. It employs methodologies and design tools to harmonize product evolution, from concept to retirement, with a wide range of requirements.

LCD is differentiated from other design approaches in that it favors designs oriented towards environmental requirements. Among its commonly held main objectives, protecting the environment becomes particularly significant where environmental performance is considered over the whole product life cycle.

On a large scale where a multidisciplinary approach is clearly needed and specifically aimed at product design, there seems to be one aspect that needs particular focus: the need to introduce design methods that guarantee efficient product life cycle that optimizes the product's physical properties (architecture, geometry, linkages, components, and materials), as well as an appropriate support service throughout its life cycle, and various ways of recycling the product at retirement. Furthermore, this must involve reducing resource consumption while limiting any type of gas emissions during its life cycle phases.

Such a design intervention requires methodologies and mathematical models, currently non-existent, to handle the complexity of the problem and optimize the product. By improving environmental performance over the whole product life cycle, techniques and tools must be defined that make the product highly eco-compatible as regards the constraints imposed by conventional design criteria (practicality, structural safety, reliability, and quality) and company policy (industrialization, production, and marketing).

2.3 DESIGN FOR ENVIRONMENT

The first attempts at technically reducing the environmental impact of product designs appeared in the mid-1980s. A later phase of greater awareness of the need to safeguard resources followed in the early 1990s, showing widespread agreement on new ideas and experiences, clearly wishing to integrate environmental requirements within traditional design procedures. Hence the terms DFE, green design (GD), environmentally conscious design (ECD), and ecodesign sprang up with the main priority of minimizing product impact on the environment at the design stage. Putting the principles of environmental sustainability into practice directly involves product design and development as vectors for propagating and integrating the new environmental requirements. DFE was the result, and over the past 10 years its definition has become ever more precise. Initially, it was simply a design approach to reduce industrial waste and optimize the use of raw materials, but it has expanded to encompass the management of waste and resources as a much more systematic vision. On a practical level, DFE can be defined as a way to reduce or systematically eliminate any environmental impact that competes within the product life cycle, from extracting raw materials to product retirement. This means evaluating potential impact throughout the entire design process. Apart from the primary aim and life-cycle orientation, there are two other important aspects (Figure 2.1):

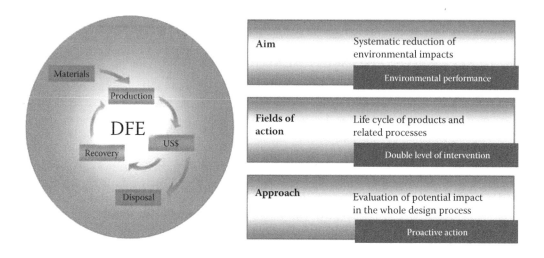

FIGURE 2.1 DFE concept, objectives, and characteristics.

- Intervention on two levels regarding product and process
- Pro-active intervention based on the supposition that the most effective interventions happen at the development stage (creation and design)

It is worth emphasizing that any design interventions to protect the environment cannot disregard the requirements of practicality, reliability, quality, and cost. As long as these requirements are fulfilled, protecting the environment should evolve from simple restrictions to new opportunities and stimuli for innovation.

2.4 OPTIMUM ENVIRONMENTAL PERFORMANCE

The links between the various experiences in DFE can be identified in the common objective of reducing a product's environmental impact during its entire life cycle from design to retirement. The concept of *reducing a product's environmental impact during its entire life cycle* should not be limited to the simple quantification and minimization of direct impact on ecosystems. Rather, it should be understood in the wider sense as optimizing environmental performance, which includes a range of specific aspects:

- Reducing rejects and waste to use resources more efficiently, cut landfill volume and reduce impacts linked to waste management
- Optimum management of materials in terms of their desired performance, their recyclability and minimizing those that are toxic or pollutants
- Optimizing production processes by ensuring energy efficiency and low emissions
- Improving the product with particular regard to its behavior during use so as to reduce resource consumption or the need to use supplementary resources

On these premises it would seem clear how DFE could also bridge two traditionally separate functions: the development of production and environmental management. The aim of DFE is, therefore, to bring these two functions together and emphasize product life-cycle issues that are often ignored.

The aims for improving a product's environmental performance can be approached in two different ways, according to context.

In the case of an already existing production cycle, when a product's life cycle impact needs reducing, is increased technology is generally the answer.

When a product is new, it is possible to set up complete life cycle sustainability from the start, so that the ideal goal is to produce a product whose manufacture, use, and retirement have as little impact on the environment as possible. This could be done by a design with DFE characteristics whose most significant benefits can only be obtained by taking into account the whole life cycle of the product, which, apart from the manufacturing phase, also includes providing of its raw materials, use, and retirement, and the recycling of resources.

2.5 IMPLEMENTING DESIGN FOR ENVIRONMENT AND GENERAL GUIDELINES

Whether the subject for environmental improvement is a product or each single phase in the manufacturing process, incorporating DFE into the design comes about through three sequential phases:

- Scoping—defining the intervention target (product, process, and resource flow) to identify possible alternatives and determine the depth of analysis
- Data gathering—gathering and evaluating the most significant preliminary environmental data
- Data translation—transforming the preliminary data into tools for the design team (from simple guidelines and design procedures to more sophisticated software systems that help obtain environmental data for the design process)

Implementing the second and third phases is carried out by means of two types of tools:

- Support tools for life cycle assessment (LCA) that gather, process, and interpret environmental data
- Support tools for *product and process design*

These tools lead to a more thorough and wider range of phase topics (environmental impact of products and processes, choice of materials ands processes, product or sub-group disassembly, extending and optimizing life cycle, recycling components and materials on retirement). Initially, they might take the form of a series of suggestions and guidelines for the designer that can be summarized as follows:

- Reducing the use of materials, using recycled or recyclable materials, reducing pollutants or toxic materials
- Maximizing the number of replaceable or recyclable components
- Reducing emissions and waste in production processes
- Increasing energy efficiency in production phases and during use
- Increasing reliability and serviceability
- Making the evaluation and recycling of resources easier by facilitating disassembly
- Extending product life
- Anticipating strategies for recycling resources on retirement, facilitating reuse and recycling, and reducing waste
- Checking and limiting costs relative to the improvements in environmental performance
- Adherence to current legislation and awareness of any upcoming changes

Applying these guidelines to the main life cycle phases of the product will enable the environmental opportunities for eco-efficient design and development to be explored.

2.6   PRODUCT LIFE CYCLE AND ENVIRONMENTAL IMPACT

As highlighted in the previous paragraph, the most significant benefits of DFE can only be achieved if they are taken into account at the design stage of the whole product life cycle, which includes phases beyond that of production. Products need to be designed on a "cradle-to-grave" basis, which is the only way to define a complete environmental profile of the product. Only a systemic approach can guarantee design success as well as identify the environmental weak spots of the product, dealing with them efficiently and ensuring they are not transplanted to another life-cycle phase.

As regards the concept "*product life cycle*," some clarification is needed since it is used in different contexts, each with a different meaning. Environmental analysis is oriented towards a meaning linked to physical reality, focusing attention on interactions between the environment and a set of processes that occur from inception right up to product retirement. In this interpretation, "life cycle" means a set of activities or transformational processes each of which compete for resource influx (quantities of material and energy) and for by-product and emission outflow.

To carry out a complete analysis aimed at reducing environmental impact, a number of considerations need to be made outside the production line phase, which includes the pre-production materials phase, product use and post-retirement recycling. All the transformational stages of the resources that compete within the physical life cycle of the product can be grouped according to the following main phases (Figure 2.2):

- Pre-production—When materials or semi-finished products are made to produce components.
- Production—When raw materials are transformed, components are produced and assembled to make the finished product.

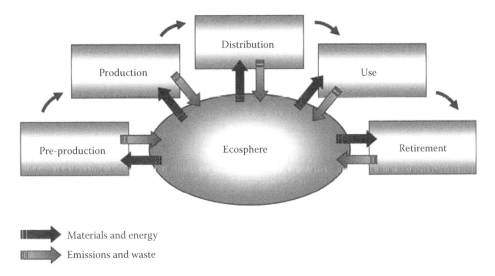

Materials and energy
Emissions and waste

FIGURE 2.2 Life cycle phases and interactions with ecosphere.

- Distribution—Packaging, packing, and transport of the finished product.
- Use—Apart from the use for which the product was designed, this stage includes servicing.
- Retirement—The end of the product's life cycle and open to different options, from reuse to dumping, depending on possible recycling levels.

2.7 ENVIRONMENTAL IMPACT OF THE PRODUCT

Each of the product life-cycle phases interact with the ecosphere, because it is fed by material and energy influx and produces not only by-products and intermediary products that feed the next phase, but it also produces emissions and rejects (Figure 2.2).

The product's impact on the environment is greatest in those life cycle processes that exchange substances or energy with the ecosphere, causing three main effects:

- Depletion—Impoverishment of the resource due to the continual drain on resources throughout the product's life cycle (e.g., impoverishment of mineral reserves and fossil fuels, due to their extraction and their transformation into construction materials and energy).
- Pollution—A set of various emission and pollutant phenomena caused by output into the ecosphere during the life cycle (e.g., dispersion of toxic substances, or phenomena caused by thermal and chemical emissions such as acidification, eutrophication, reduction in the ozone layer, global warming).
- Disturbances—A set of various phenomena that change environmental components due to interaction between the life cycle and the ecosphere (e.g., degradation of the soil, water, and air).

Some of these impact locally whereas others interact regionally, continentally or even globally. This distinction is important because these environmental impacts vary according to geographic location, depending on climate and soil type.

The main environmental impact of product life cycle on the exchange flows of the ecosphere can be summarized thus:

- Material resource consumption and landfill saturation
- Energy resource consumption and the loss of the latent energy of retired products
- Direct and indirect total emissions of the entire product life cycle

Quantifying the impact of the first factor could be done by analyzing the distribution of the volumes of materials involved during the entire life cycle. As for the energy and emissions aspects, a more complete approach is needed, taking into account the energy and emissions contents of the resources and finished products.

2.8 MODELING THE LIFE CYCLE

The complexity of the environmental issue, which has been underlined several times, highlights the fact that a complete evaluation of product performance requires a holistic approach, meaning that *product* refers to all its *life cycle phases*. This is the only approach that makes the effects of design and production-planning choices evident. An appropriate model of the life cycle could therefore be a valid tool for predicting and planning. In addition, modeling a system usually reduces its complexity even though data is lost. Such simplification is necessary in the case of environmental evaluations because of their intrinsic complexity in real life. This is well documented by the ISO 14040 series that deals exclusively with modeling product life cycles to measure their environmental impact and suggests a few basic precautions in constructing models:

- Disassemble the product life cycle into subsystems based on how the product functions.
- Define unit processes at the root of specific functions that compete with input and output resource flows.

With this clear physical–technological approach, the model's behavior can only be described and simulated by mathematical models of limited complexity, as they model linear behavior.

On this basis, modeling the entire system's life cycle is subdivided into elementary functions—activity models—that encapsulate the elementary processes of the main phases of the life cycle. Overall, *activity modeling* tries to define the set of single activities that make up a complex system. These activities could be transformations, handling, production, use, or retirement of material, energy and data resources.

Activity modeling requires a clear prior definition of the main aim of the activity to be modeled, and from which point of view to start developing the model. Both these factors are necessary to define the limits of the system to be modeled and to structure the model, because it must be articulated into subsystems, sequences, operative units, and processes, depending on aims and point of view. Once defined, according to the analysis needs, it is possible to model each activity in the product life cycle. The activity model is the type reported in Figure 2.3 showing resource influx, outflow, and eventual data inflow, should the activity require choice margins in order to be created. The resource influx, whether material or energy, is that produced from previous activity or having come directly from the ecosphere. The outflows are products of activity that are either the main product, secondary sub-products, or various types of emissions into the ecosphere. Once the reference activity model has been defined, transforming the life cycle into an activity model can be done following this procedure:

- Define the system margins.
- Identify elementary processes and functions.
- Identify and quantify the links between elementary activities.
- Evaluate any changes over time in the activities and links between them.

Having developed the activity model, then analyses and simulations of the life cycle can be started and results interpreted. The reference model in Figure 2.3 can be read in different ways according to the type of environmental evaluation required.

If the required life cycle model were only to analyze the material resources involved, a simplified model could be rendered as per Figure 2.4. This model only accounts for materials flow, the influx

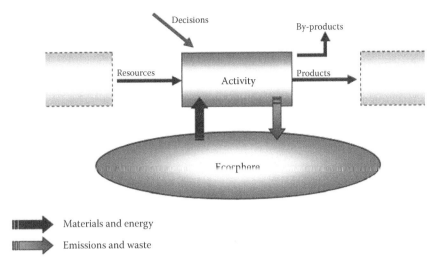

FIGURE 2.3 Reference activity model.

being the resources that feed production, and the outflow being the finished product plus any rejects or waste. For the influx resources, it is necessary to distinguish between:

- Primary or virgin resources originating from the ecosphere
- Secondary or recycled resources

The latter can be further subdivided into:

- Secondary preconsumption resources from rejects or waste of the same activity
- Secondary post-consumption resources from recycled used or retired product

By developing each phase of the life cycle (Figure 2.2) according to primary activity, one obtains an overview of the life cycle of a product and the resource flows that characterize it, as in Figure 2.5 that shows the flow of material resources according to the simplified model in Figure 2.4.

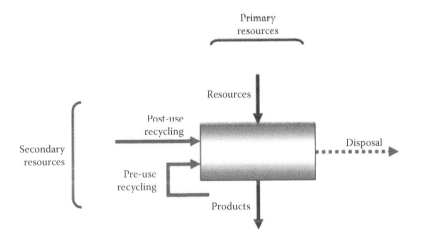

FIGURE 2.4 Flows of material resources.

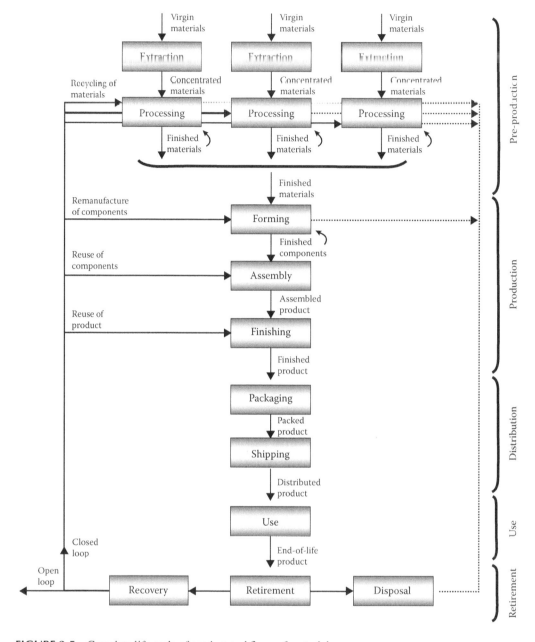

FIGURE 2.5 Complete life cycle of product and flows of material resources.

The first pre-production phase consists of producing materials and semi-manufactured products needed for making the components. Therefore it includes the production phases of all the materials that will make up the final product. Once built, distributed, and used, the product will reach the final phase of retirement.

Breaking down these phases into primary activities, Figure 2.5 provides an overview of the waste flow generated during the production cycle that, for this model, concerns the phases of processing the various materials and forming components, as well as product retirement.

It also provides a complete picture of all the alternatives to dumping the product at life's end and shows how the recycling flows might be distributed within the same life cycle that generated it, by providing the post-consumption secondary resources of the various activities; alternatively, they can be directed externally. As for recycling flows, two types can be distinguished:

- Closed loop—The recycled resources re-enter the life cycle of the same product that generated them, substituting virgin input resources. This might happen by reusing the product at life's end, by reusing certain parts, by remanufacturing certain parts or by recycling materials. Environmentally, these recycling processes increase power consumption and emissions due to the processing and transporting of such large volumes back into circulation, but they also diminish the consumption of resources by lowering the input of virgin materials and reducing the amount of waste materials.
- Open loop—The materials of certain parts of the product at life's end can be used in the production processes of other materials or products that are different to the original ones and not part of its life cycle. This helps recover part of the energy content of the waste materials, saves virgin materials from other production cycles, and generates revenue by selling recycled material.

2.9 LIFE CYCLE ASSESSMENT

One of the main difficulties of reducing the environmental impact of any activity is quantifying it in order to contain it. Although methods for evaluating environmental impact are numerous and they differ in their ability to identify, measure, and interpret the impact, they generally have the following limitations in common:

- They are developed for each single case.
- They cannot predict the reliability and stability of the results through sensitivity analysis.
- There is no basis of a life cycle approach as hitherto considered.

Of the various methods, only the methodology known as life cycle assessment (LCA) gets over the above-mentioned limitations, because it evaluates environmental impact on a set of presuppositions that objectively evaluate resource consumption and waste disposal of a generic company— or rather the whole life cycle of that company's product. This is carried out by determining the amounts of exchange flow between the product life cycle and ecosphere, that is, all the transformational processes involved from the extraction of raw materials to their return to the ecosphere in the form of waste.

2.9.1 METHODOLOGICAL PREMISES, PROPERTY, AND STRUCTURE

The ISO 14040 standards detail the general criteria and methodological structure on the basis of which to carry out the principal phases of a complete LCA. These derive from a set of criteria established by the Society for Environmental Toxicology and Chemistry (SETAC) in the early 1990s. The basic definition of an LCA summarizes SETAC's: an LCA consists of compiling and evaluating inputs, outputs, and the potential environmental impacts of a product through its life cycle. The methodological structure is based on certain premises:

- Environmental interaction analysis of the basic activities of the chain known as the *life cycle* is a cradle-to-grave analysis.
- The life cycle approach is holistic (as discussed earlier).
- Analyzing environmental effects is from a multicriteria viewpoint evaluating the entire panorama of impact categories and environmental damage that could be affected by product-ecosphere interaction.

- Evaluations and activity comparisons are based on a functional unit and thus on the principle of equivalence to product practicality (requiring first defining a practical unit of reference that measures product performance).

The first examples of environmental analysis like the LCA, dating back to the 1970s, basically consisted of quantifying the material and energy resources as well as solid waste. Today, environmental necessities have changed and so too has the level of refinement of life cycle analysis as reflected by the requirements of ISO 14040.

Currently, a complete LCA is structured according to the following four main points:

- *Goal and scope definition* defines the analysis goals and the overall preliminary assumptions on which they are based. The evaluation types are also defined (those aimed at improving the system or comparing alternatives) as are the system limits, the functional unit reference, the assumptions and parameters for inventory and allocation, and the impact categories.
- *Inventory analysis* (LCI) includes the compilation and quantification of the inputs and outputs of the whole life cycle. The data used for this might come from various sources including direct measurements, as well as data from databases and literature.
- *Impact assessment* (LCIA)—Inventory data are expressed as potential environmental impacts, evaluating their size and significance. Conventionally, inventory flows are classified (*classification*) and characterized quantitively in relation to different impact categories (*characterization*). Impact data can subsequently be normalized (*normalization*) and weighted (*weighting*) to obtain single indices of environmental impact.
- *Interpretation* (ISO)/*improvement analysis* (SETAC)—The LCI and LCIA results are evaluated in relation to the prime objectives so that conclusions and improvements can be made.

As Figure 2.6 shows, the structure according to the four main phases described above is common to the LCA method (SETAC) and the ISO norms.

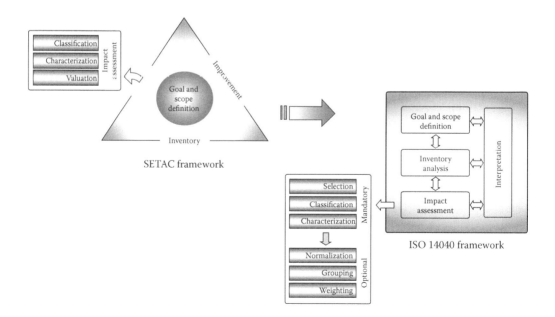

FIGURE 2.6 LCA framework according to SETAC and ISO 14040.

The only significant differences can be found in the impact assessment phase and in the interpretation and improvement phase. In the first phase (LCIA), the SETAC setup differs to the ISO standard because evaluating impact is more rigid, not only in their classification and characterization, but also by concluding with a *valuation* that includes weighting the different types of impact to obtain results that can be compared or aggregated depending on which are used in the final decision.

In the second case, there is a net distinction between *mandatory* and *optional* procedures. The former include selecting which environmental effects to consider *selection, classification,* and *characterization*. Optional procedures include *normalization, grouping,* and *weighting,* which involve processing the results of the characterization phase to obtain synthetic indices that enable a complex evaluation.

As regards the last phase, SETAC differs from ISO:

- In the first case, *improvement analysis* focuses on how to make improvements in environmental performance.
- In the second case, *interpretation* means more extensive intervention that includes sensitivity analyses, uncertainty evaluations of the results, and the formulation of conclusive recommendations.

2.9.2   ECO-INDICATORS TO QUANTIFY ENVIRONMENTAL IMPACT

The environmental impact properties of materials and processes can be evaluated by LCA techniques already seen as a method of identifying and quantifying the resources used, and the waste or emissions generated, through inventory data with a subsequent evaluation of environmental impact.

In the initial setup, LCA does not provide unequivocal conclusions about any given product, material, or process, but relies on appropriate interpretation. This gives rise to the following two problems:

- LCA results are difficult to interpret. Product impact on a single category (global warming, hole in the ozone layer) can be defined, but the total impact on the environment remains unknown. This is due to the lack of unequivocal weightings for each single effect on overall impact.
- Generally, carrying out a detailed LCA is a complex operation requiring masses of data and lots of time. This means that a complete LCA could not be carried out during the first phases of designing a new product.

The *Eco-Indicator 99* method was developed for product design and helps quantify and compare results provided by LCA. The method has proved useful for designers to translate LCA results into easily applicable numerical values called eco-indicators.

Use of eco-indicators can help resolve the applicability of LCA in two ways:

- Identifying single scores which are capable of expressing overall environmental impact
- Developing an eco-indicator database for most common materials and processes

Eco-indicators are based on defining the damage arising from the relationship between impact categories and the damage to man's health and the ecosystem. In particular, the following three types of damage are considered:

- *Human health* considers the number and duration of human illnesses, and the number of years lost due to premature death caused by the environment—climate change, the reduction in the ozone layer, carcinogenic effects, and respiratory effects.

- *Ecosystem quality*—Effects on different animal species, principally on vascular systems and microorganisms—eco-toxicity, acidification, and soil usage.
- *Resources*—Energy required to extract raw materials and fossil fuels and consequent resource impoverishment.

Eco-Indicator 99 provides values for most types of process, summarized as follows:

- Materials production—Eco-indicators refer to the production of 1 kg of material. To obtain these values, all materials production processes were taken into account including raw material extraction and its transport during all the refinement phases. The production of industrial goods (machinery, buildings, etc.) has not been considered.
- Production processes—Eco-indicators are impact values of the primary and secondary treatments and manufacturing processes of various materials, each expressed in an appropriate unit (a significant parameter of the process). Production process values include quotas for emissions and required energy. Even in this phase industrial goods have not been included.
- Transport—Eco-indicators are expressed in tons per kilometer. Emissions caused by fuel extraction and refinement are accounted for as well as the emissions during transport. Usually, the units of measurement are tons per kilometre, but others can be used. An unloaded return journey is also accounted for. The impact of industrial goods (e.g., those relating to producing the truck, building the roads and railways, the availability of cargo planes in airports, etc.) are also included in calculating eco-indicators, as their contribution is significant.
- Energy production—Eco-indicators for electricity and heat take into account fuel extraction and production, the transformation of energy, and the generation of electricity by whichever fuel (in Europe). Apart from their average values for Europe, other eco-indicators have been formulated for other nations according to the technologies used to produce the electricity.
- Retirement and recycling scenarios—Eco-indicators are expressed in kilograms of retired material differentiated according to type and method of waste treatment. Not all products are retired in the same way, so eco-indicators are careful to account for specifics. Paper, glass, and plastic can be collected separately and recycled. If a product is only partly recyclable, it is better off in the city's recycling program. Scenarios for both cases have been evaluated, and in addition, some alternatives introduced such as incineration and some other types of recycling of materials.

2.10   INTRODUCING ENVIRONMENTAL FACTORS IN THE DESIGN PROCESS

Although the most significant questions about the environmental aspects of industrial production have been asked, when it comes to a decent level of environmental awareness, production companies still manifest difficulty in coming up with environmentally sustainable manufacturing. One of the key aspects of this problem is the lack of integrated design principles and methods that meet the environmental requirements of those products at design and management levels. This means that the success factors in product design are still tied to quality and development time, or to those factors connected to market impact.

One of the most important aspects of DFE consists in its ability to bridge two functions that are usually separate—planning and production development—with environmental management. In order to do this, it is essential that planning incorporate the following requirements:

- Life cycle oriented product
- Harmonization of a wide range of requisites
- Simultaneous structure, integrated with design

Only on the basis of these premises can a process of product development be conceived that embodies life cycle sustainability and the ideal of a product, used and retired with the least impact on the environment.

2.11 INTEGRATING LIFE CYCLE AND DESIGN

The life cycle approach can give rise to a qualitative leap in setting up product development so that the product not only harmonizes with the business environment but also with the natural environment. This is confirmed by several observations regarding those factors that impede environmentally oriented product development:

- Little awareness of a product's environmental impact
- Traditional setup of cost-oriented product development
- Lack of a homogeneously and efficiently distributed approach to the environmental requisites of the product throughout the development process

Regarding the first factor—production companies' limited awareness of the environmental impact of their own products—this is historically linked to producers' need to be mainly concerned with impacts at production sites (resource consumption, generation of emissions and waste); these impacts are not directly ascribable to the product and are limited to the production phase. This need meant that the information behind any strategy to improve the environmental quality of the product was missing and this could only have been achieved by a broader vision of the entire life cycle. This problem can be resolved by introducing LCA (introduced in the previous paragraph), which in its simplest form—streamlined LCA—can overcome the inconvenience of an analysis that is too detailed to be carried out during the preliminary phase of product development.

The traditional cost-oriented development process arose out of a "defensive" approach to the environment, where environmental issues were seen as restrictive and generally burdensome, and their potential as the driving force behind value added innovation was disregarded. This is particularly significant if one thinks of the importance of costing and marketing in the development of a product, and of how many obstacles would arise in the absence of any accurate economic analysis or "environmental value" of a product in eco-compatible design. Even in this case, life cycle oriented techniques can come to the rescue, notably life cycle cost analysis (LCCA), environmental accounting, and other techniques that integrate life cycle economic and environmental analysis.

The lack of a thoroughly widespread environmentally oriented approach to the whole development process is crucial. Although it has often been noted that this shortcoming applies mainly to the preliminary phases of product development where there is an absence of methods and tools oriented to the environment, it should also be emphasized that generally design practice lacks an organic approach to the environment throughout the development process, notwithstanding the fact that such an approach is clearly desirable, theoretically.

The life cycle approach, therefore, which as we saw in the last chapter materializes into LCD in a strictly design sense, could form an effective basis for integrating the environment into product development. As will be better explained later, when LCD is expressly oriented to environmental requisites, it becomes an example of design that is totally environmentally oriented and whose methodological setup is a reference for thoroughly integrating environmental factors into product development. In this regard, specifying design objectives and strategies performs a crucial role. Another crucial role is carried out by design for X (DFX), also introduced in the last chapter, which can provide the tools for product design with specific (including environmental) requisites.

The above is summarized in Figure 2.7, which highlights the tools that can help the life-cycle approach overcome any obstacles to environmentally oriented product development.

The same figure also highlights another potential obstacle which is a cross-functional characteristic of both design practice and environmental aspects. It involves the multidisciplinary

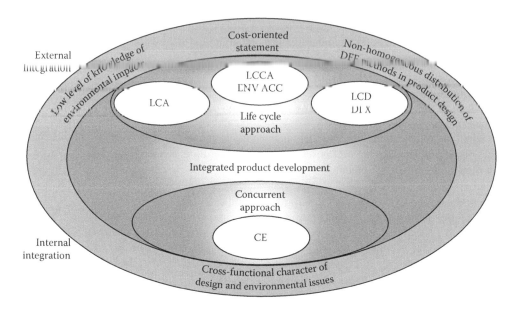

FIGURE 2.7   Environmentally-oriented integrated product development.

nature of the required competences and the transversality of correlated activities with respect to the main company activities (design, production, and marketing). This aspect is well known within the world of the organization and planning of design and product development; it is known as concurrent engineering (CE) and came about to meet the needs of design practice. It consists of a product development approach that aims at fully harmonizing higher quality, shorter times and lower development costs by using an ample design team that can react to analyses rapidly and interconnectedly at all stages of the development process. The effort to integrate environmental aspects with product development can therefore benefit from the organizational structure of CE.

To summarize, and referring to Figure 2.7, full integration of environmental aspects into product development must take place at the following two different but complementary levels:

- External integration—the relationship between product development and factors external to the design team that must be taken into consideration: consumer and market demand, production restrictions, and environmental requirements. As highlighted in the figure, they integrate through the life cycle approach using the tools at their disposal.
- Internal integration—the relationship between functions and competencies within the design team. Necessary for the good management of the cross-functional characteristics of design practice and environmental aspects, these integrate by means of an approach that is simultaneous and concurrent with product development.

Once integration is obtained at this double level, then integrated product development (IPD) in its fullest sense includes integration with the environment. It is interesting to note how IPD can be assimilated with the general concept of improving design solutions for consumer demand and market opportunities. The life cycle approach extends this by putting consumer demand alongside the people involved in the various phases of the product life cycle (manufacturers, distributors, and collectors). A further extension of the base concept of IPD, which includes a response to environmental demands, consists of the basic premise that integrated product design includes environmental requisites.

2.11.1 THE ROLE OF LIFE CYCLE DESIGN

Although in general the environment is just one of a number of aspects included in LCD, it is evident that such a design approach, which simultaneously accounts for a vast range of design parameters and product development costs, could qualify as the best way to implement the DFE if correctly applied. In this light, LCD becomes a systematic approach "from cradle to grave," capable of anticipating an environmental profile complete with goods and services at the design phase.

In the LCD figure in the previous chapter (Figure 1.5), safeguarding the environment was one of the main objectives of design intervention. However, LCD and DFE are often taken to be one concept, as though environmental requisites had become the only objective of LCD. Because this interpretation is restrictive, it would be better to consider DFE as one aspect of LCD and an opportunity to emphasize the environmental side. We will therefore consider life cycle design for environment (LCDFE), a design approach based on a systematic vision of life cycle, which—by integrating environmental requisites into every phase of the design process—requires a reduction in the overall impact of the product, to make it as ecologically and economically sustainable as possible. To this end, the LCD diagram in the previous chapter (Figure 1.5), highlighting environmental safeguards as one of the main objectives of the design, could be interpreted as emphasizing the environmental aspects of each product success factor, even if not specifically oriented to safeguarding the environment:

- Resource use must be planned to account for the environmental efficiency of the flow distribution of all the resources in the entire life cycle.
- Production planning must account for the environmental efficiency of the production cycle by minimizing rejects and waste and optimizing energy efficiency.
- Life cycle costs should include the environmental costs associated with the phases of production, use, and retirement.
- Product properties should give prominence to the environmental aspects of the life cycle, without neglecting the indispensable aspects of performance and economy.
- Company policy should include environmental criteria in accordance with growing market and political sensitivities toward safeguarding the environment and sustainable development.

Notwithstanding the rising prominence of the need to safeguard the environment, designing the life cycle should always maintain its main purpose of integrated intervention that takes into account a wide range of design requisites that are not exclusively environmental. This is an indispensable premise for those who must create a really efficacious final result, or a product whose life cycle might be deemed sustainable in the widest sense, where the use and distribution of resources is optimized, and emissions and waste are minimal, while adequate performance and quality standards are maintained and the economic sustainability of the product is guaranteed.

2.12 INTEGRATING ENVIRONMENTAL ASPECTS WITH PRODUCT DEVELOPMENT

Referring back to the vision of the design process alongside product development proposed in the last chapter, the intermediate design phases are grouped in the product design process to achieve total and homogeneous integration of environmental considerations. This can be done by a series of different interventions, depending on the phase of the development process.

In the preliminary phases of product development—*project definition*, *development process planning*, *problem specification* (see Figures 1.1 and 1.2)—integrating the environment happens through extending the factors that condition the preliminary design setup and defining the specific requisites of the product, including those with an environmental bias (beyond consumer demand and market opportunities), making sure that they carry the right weight with respect to company policy and strategy. This means inputting a data set that is not exclusively environmental but relative to the product life cycle.

Setting up the design phases (the product design process, Figure 1.2) must be guided by the most opportune strategies for encompassing environmental aspects and product life cycle. This will be dealt with in detail in the following paragraphs.

During the main phases of the design process and beginning with *conceptual design* and, in particular, in the phases of *embodiment design* and *detail design,* the design must aim at and harmonize an ever wider range of design requisites as provided for in the LCD. Here, DFX tools can meet these requisites, each oriented to a specific product requirement and each able to emphasize environmental requisites (to follow).

The *post-design planning* phase that follows the *product design* phase (Figure 1.1) must integrate with it, as provided for in concurrent design. This integration must occur by extending post-design planning to the planning to fully encompass the entire life cycle including production, distribution, use, and product retirement in such a way that the life cycle approach is fully encompassed. Planning the entire life cycle—production, consumption, retirement—is exactly how the DFX tools will be introduced.

2.13   ENVIRONMENTAL STRATEGIES FOR PRODUCT PLANNING

Broadly speaking, environmentally efficient product life cycles are those that impact least on the environment. In previous paragraphs, we have seen how they depend on exchanges between the ecosphere and the physical/chemical flows of the technological processes of the life cycle. For a complete environmental analysis, not only must the materials flow in the life cycle be identified, but also energy and emissions flows.

Environmental strategies aimed at reducing the wide range of environmental impacts arising from phases of the life cycle are numerous and are differentiated by type of intervention.

As the environmental efficiency of a product depends directly on its design, it is fundamentally important that the strategy be closely tied to the main parameters of the design. Not all environmental strategies can be assimilated into design strategies. Some materialize through interventions not directly connected to design choices.

Referring again to the main impact of a product on the environment (consumption of material resources and saturation of landfills, consumption of energy resources and the loss of energy contained in retired products, total direct and indirect emissions relative to the whole product life cycle) we can conclude that design interventions to attenuate the behavior of product life cycle in terms of the environment should try to optimize the distribution of the flow of material and energy resources, waste, and emissions, creating favorable conditions to achieve the following:

- Reduction in the volume of materials employed and extension of their life
- Closure of the resource flow cycle by recycling
- Minimization of emissions and energy consumption during production, use, and retirement

To fully achieve these conditions, changes are needed in product design and process design. The latter, due to its primary importance, is increasingly considered intimately connected to the process of product development (according to the above CE), and is not part of the remit of the objectives cited here. Instead we will focus on product design as a material entity, and therefore as a system of material components designed to make up a functional system corresponding to certain prerequisites. This side of product entity directly results from the choices made during the design phases of development (conceptual, embodiment, and detail design) whose parameters are directly related to the physical dimensions of the product: materials, form and size of components, system architecture, interconnections and joints. These product entity dimensions within its life cycle are expressed by the flow of material resources. Everything, therefore, is referable to the first of the three main aspects of a product's impact on the environment—the consumption of material resources.

This partial vision of the environment issue might seem limited. In reality, its reach is very wide and does not exclude from environmental evaluation the impact of energy consumption, or that of emissions, throughout the product life cycle. As regards the impact of production energy and its emissions, these are generally connected to the volumes processed or to specific process parameters that depend on the physical properties of materials and component geometry; therefore they too can be managed through the choices made during product design by defining the materials and main geometrical parameters that also condition the choice of processes and the way they are carried out.

Focusing on two materials flows, and therefore on a physical dimension of the product, improvement of the environmental performance of its life cycle can be achieved by applying two main strategies, as summarized in Figure 2.8:

- *Useful life extension strategies* that evaluate the use of materials and all the other resources needed to make the product (maintenance, repair, up-dating, adaptation)
- *End of life strategies* that close the materials cycle and help reclaim some of the resources used to make the product (reuse of systems and components, recycling materials in the primary production cycle or in external cycles)

Both types of strategy have great potential for environmentally optimizing resource flow in the product life cycle. Extending useful life exploits product resources by avoiding the consumption of other resources to make substitute products. Reclaiming the product or parts thereof at the end of life means that components or materials can be used for new products, reducing the consumption of virgin resources and reducing energy resources and waste.

These strategies increase the intensity of resource use in making the product and so raise their value. This is a substantial difference compared to another important environmental strategy, highlighted in Figure 2.8 and still linked to the materials side of the product: *resource reduction strategies*. These include all the interventions and choices that reduce materials and energy use, and so are ascribable to a wide range of expedients that not only concern product design but also its planning.

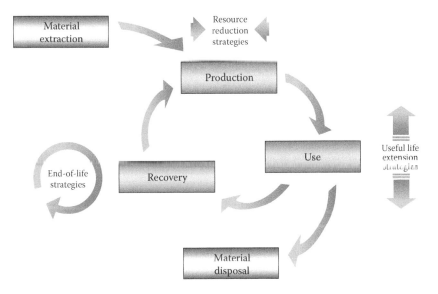

FIGURE 2.8 Environmental strategies in product life cycle.

Attention will later focus on the two main types of environmental strategy. The next paragraph highlights how these strategies can be integrated into a methodological structure for designing the product, and introduces the tools that can help the designer.

2.13.1   Useful Life Extension Strategies

During their useful life, products are subjected to use, maintenance, and the repair of worn out or damaged components. The opportunities offered by these activities can only be evaluated after an accurate assessment of environmental implications: if maintenance and/or repair have significant environmental impact it might be appropriate to retire the product, unless its substitution by making another product has a still higher environmental cost.

In design, the requisites of durability (ability to retain performance levels over time) and maintainability (predisposition to maintenance to restore initial performance levels) are particularly significant. From an environmental efficiency perspective, these requisites should not be indiscriminately maximized. Raising durability should only be to a certain level, above which environmental impact is greater than substituting the product (e.g., because it consumes more resources during use than a new more efficient replacement).

Maintenance, which is all those activities of periodic intervention and substitution of parts, is very important for limiting the environmental and economic costs of repair as well as the impact of dumping and the production of a substitute. To ease maintenance, cleaning should be carried out regularly, parts need to be accessible, appropriate tools must be available, and there should be some system for monitoring the state of parts and components. Other elements that could extend useful life include updating (technological evolution and cultural mutation) and adaptability (products with fast obsolescence and composed of a high number of reconfigurable components).

Here follows a summary of the strategies to extend useful life:

* Maintenance—component cleaning, monitoring and diagnosis, substitution of parts subject to wear and tear
* Repair—regeneration or substitution of parts subject to damage or wear
* Adaptation and updating—reconfiguring components to adapt them to customer needs or to a change in environment, and substitution of parts due to obsolescence

Below we report the design expedients that facilitate the strategies:

* Maintenance—Facilitate access to the parts that need cleaning; predispose and facilitate access and substitution of the parts that wear out more quickly; group components by physical/mechanical properties, reliability and common functions; predispose diagnostics for the parts requiring maintenance, and provide diagnostics for all critical components
* Repair—Predispose and facilitate the removal and reassembly of critical components subject to deterioration or damage; design standardized parts and components
* Adaptation and updating—Design modular and reconfigurable architecture for adaptability to different environments; design multifunctional products to adapt to the technological and cultural evolution of the end-user

2.13.2   Reclamation Strategies at End of Useful Life

There are different opportunities at the end of a product's useful life to exploit the resources employed in its production; functionality may be recouped (entire product or part/s) to do the same or different tasks, subject to collection and transport, or subject to reprocessing to restore the product to its original condition; material or energy content might be exploited via recycling, composting, or incineration. Some of these expedients, such as reusing the product or some of its parts and remanufacturing, could arguably be extending the useful life of the product, although the distinction between the strategies during use (introduced in the previous chapter) and those after use should be remembered. Even reuse could be considered a post-use strategy, some authors describing it as a phase in itself.

From this viewpoint, it is important to conceive of and design easily disassembled products, making it easy and economical to separate parts or materials and therefore reuse components or remanufacture products, recycle parts or materials for recycling, for isolation (toxic or harmful), for composting, or for incineration.

Reused products, whether entirely or in part, can be collected and aimed at the same use or another use requiring less restrictive requisites. Modifications vary from cleaning to disassembly and reassembly with some new parts. As for remanufacturing, or renovating worn out products to almost new, it is very important that their removal be made easy; likewise their substitution and parts interchangeability within the same line of products.

If the objective of disassembly is recycling, good economics is sensitive to the variation in raw materials prices and the costs of collection and recycling. Minimizing the times and costs of processing should be economical both in safeguarding and exploiting the reclaimed materials (the purer the materials, the more they conserve their original performance and the higher their market value). An alternative to disassembly is crushing, sifting and cleaning product components (so as not to compromise their characteristics).

One way to extend materials' useful life as an alternative to recycling is by reclaiming energy content through incineration or other appropriate processes. In both cases, the environmental advantage is twofold: first, landfill impact is avoided, and second, non-virgin resources are made available, avoiding those impacts that arise out of the need to produce a corresponding amount of material and energy originating from the raw material.

Recycling cost trends are defined by a series of variables: the costs of collection, transport and warehousing, the cost of disassembly or crushing, the cost of raw materials or landfilling, and the value gained in terms of recycled product/material (linked to its purity). Apart from the economics, not all materials are recyclable in the same way: it may happen that the performance characteristics of the recycled material are considerably different to those of the raw material, or that a technologically easy to recycle material might require abundant energy, compromising environmental convenience. The criteria for choosing materials that facilitate recycling orient the designer towards more easily reclaimed materials in terms of performance, avoiding composites and additives, or to materials that reclaim energy content efficiently and have little environmental impact.

All in all, the strategies for reclaiming resources at the end of useful life correspond to different potentials for environmental benefit, which in turn depend on the different levels of incidence of life cycle recovery flows (Figure 2.5). A complete panorama of the entire range of reclamation strategies can be made by clearly distinguishing between reclaiming the whole product or its parts, and this can be done in various ways (Figure 2.9).

- *Reuse* product if it is in optimum condition and can guarantee performance.
- *Remanufacturing of product*—If the product is in good condition it could be reprocessed, regenerated and reused.
- *Recycling of product*—If the product is made up of recyclable or compatible materials it could be entirely recycled.

If the product contains parts that can be reutilized, reprocessed, or recycled, it must be disassembled. This in turn leads to four different reclamations:

- *Reuse of components*: The reutilizable parts should be separated from the product, re-sent to production, tested and reassembled into new products
- *Remanufacturing of components*: Other parts could be reprocessed before being reassembled into new products.

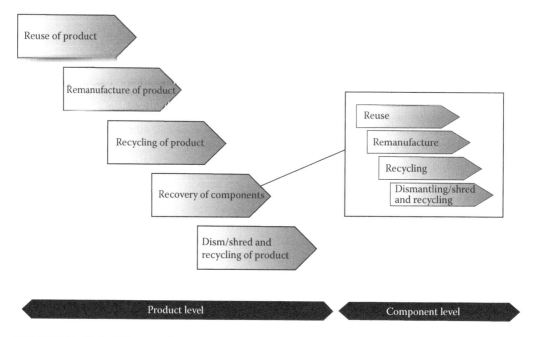

FIGURE 2.9   End-of-life strategies.

- *Recycling of components*: Some parts can be directly recycled.
- *Dismantling, shredding, and recycling of components:* Parts containing volumes of recyclable materials can be shredded so any valuable materials are separated from those sent to landfills.

- *Recycling of materials:* If the product contains small volumes of recyclable material it can be shredded to extract these materials.
- *Disposal*: If there are no valuable parts, it is sent to landfill.

In this clear hierarchical distinction of the possible options at the end of useful life of a product, some clarification may be needed of "remanufacturing." This refers both to the product and its components and means reutilization by means of intermediate processes of regeneration and reconditioning without recourse to substantial substitution or repair.

This whole range of possible recovery strategies was developed in hierarchical order according to highest environmental benefit. This distinction is intended to define only the environmental side of reclamation. The best solution is again that which harmonizes environmental requisites with time and costs. The overall reclamation plan at the end of useful life must recover maximum resources through the minimum use of resources and minimum emissions. This confirms the validity of the systemic approach recommended since the introduction, which stated that to correctly analyze the environmental problems associated with product life cycle, it is indispensible to evaluate all the resource and emission flows competing at each phase of the cycle.

Accordingly, reclamation can be formulated as follows: for any given product (or design proposal), the reclamation plan must efficiently balance any reclamation outlay against the relative profits. Outlay and profit can be seen strictly in terms of economics, or in terms of resources used and reclaimed.

Plotted as a graph (Figure 2.10), the most favorable conditions are evident.

The curve trends are typical of the functions they express. Reclamation outlay (*disassembly, testing, remanufacturing,* and *repair*) becomes prohibitive as the extent of reclamation deepens.

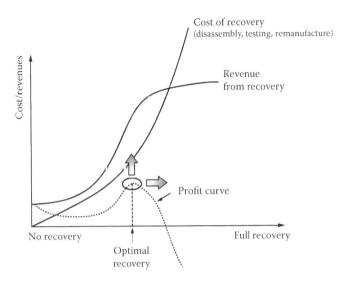

FIGURE 2.10 Recovery planning optimization.

However, revenues tend to stabilize after a certain level. Consequently the profit curve trend (given the differences between the two preceding curves) is shown in Figure 2.10. Optimizing reclamation means designing the product and planning its reclamation such that:

- The curve peak is shifted toward greater reclamation.
- Profits increase at the maximum point.

The above considerations outline a series of expedients for a design intervention that favors reclamation strategies at the end of useful life, facilitating the disassembly of the construction system and the amount of resource reclamation. On the basis of the most well-known construction standards (GE Plastics, 1992; ICER, 1993; VDI, 2002) and some significant contributions from the literature, the following summarize the main design factors that define the architecture of the product: choice of materials, system of joints/seams, architecture layout.

- Choice of materials—Minimize the number of different materials; make underassembled components and connected parts irreversibly of one material or of compatible materials; use both recyclable as well as recycled material; avoid toxic materials.
- System of joints/seams—Minimize the number and type of joints; use rapidly reversible joints above all for connecting valuable components or incompatible components; foresee access to joints; use joint materials compatible with the connected parts; foresee separation by breakage to speed up disassembly; simplify and standardize matching components.
- Architecture layout—Minimize the number of parts; as far as possible make the architecture modular, separating component functions; allocate non-recyclable parts so they are easily separable; allocate valuable parts for accessibility and easy reclamation; avoid metal inserts or stiffeners in plastic components; foresee and highlight access points and preferred lines of separation and breakage.

2.14 METHODOLOGICAL SETUP AND DESIGN TOOLS

Design strategies are crucial in the life cycle approach. They are able to express environmental product requisites into its design. It should be stressed, however, that the most efficient and appropriate environmental strategies for a specific design can only be chosen after the design objectives have been accurately expressed as product requisites.

The environmental strategies to improve product life cycle introduced in the previous paragraph are of two types (extending useful life, reclamation at end of life), and can in practice be retraced to design strategies or expedients that guide the designer's choices in developing the design. Table 2.1 summarizes the most significant design expedients, classified according to the main design parameters that require intervention. Furthermore, the design parameters are distinct according to whether they relate to system design (architectural characteristics, layout, and inter-component rapport) or component design (materials, form, and geometric parameters). The same table also directly correlates each design strategy and the environmental strategies that are mutually viable.

Intervention level: System, components.

Design parameters: Layout, component relationship, materials, form, dimensions.

Design expedients: Minimize component number, optimize modularity, ensure system designs use multifunctional and updateable components, foresee component accessibility, reduce connections, reduce connection variety, ease disassembly, reduce rare, polluting or hazardous materials, increase biodegradable and low impact materials, reduce materials variety, increase materials recyclability and compatibility, specify and mark materials, optimise performance, durability and reliability, reduce mass.

Environmental strategies: This provides an outline of the methodological setup through which environment can be integrated into design practice.

Defining of environmental requisites

Choice of most appropriate environmental strategies to fulfill requisites

Identification of design strategies that best marry with the predefined environmental strategies

Identification of the design parameters that require intervention and at which levels

The result of this methodological setup, as highlighted in Figure 2.11, is the ability to see all the design parameters together, thereby enabling identification of the environmental strategies appropriate to the requisites and helping manage any conflicts between these strategies. Designs aimed at different environmental objectives often clash. Another advantage of this overview is that it is the first stage of clarifying the links between environmental choices and those inspired by conventional criteria. In any case, those environmental choices must harmonize with the whole range of requisites, managing any conflict that might arise from design orientations with different objectives.

2.15   THE ROLE OF DESIGN FOR X IN DESIGNING FOR ENVIRONMENTAL REQUISITES

Although DFE is sometimes thought to be part of the DFX system, in reality it is better viewed as an approach to design, not so much an active design tool as a design philosophy with a profound influence on the way in which industry faces up to the environment. As a design approach, it requires tools that cement premises and objectives. Several DFX tools can cover this role effectively.

A particularly interesting aspect of DFX tools is their specificity, which allows the design to be broken down into manageable sections; this is a bonus, since environmental requisites are an added complication to an already weighty subject. Each DFX tool is characterized by methods, procedures, and models that process very specific data. A well-chosen set of DFX tools can deal with specific sections of a design, each one of which can be handled by the most expert of the design team.

TABLE 2.1
Design Parameters, Design Strategies, and Environmental Strategies

Design Level	Design Parameters	Design Strategies	Environmental Strategies					
			Useful Life Extension			End of Life Recovery		
			(ES1)	(ES2)	(ES3)	(ES4)	(ES5)	(ES6)
System	Layout	Minimize number of components	✓	✓			✓	✓
		Optimize modularity	✓	✓			✓	✓
		Design multifunctional and upgradable components			✓	✓	✓	✓
		Plan accessibility to components	✓	✓			✓	✓
	Relations between components	Reduce number of connections	✓	✓	✓		✓	✓
		Reduce variety of connecting elements	✓	✓	✓		✓	✓
		Increase ease of disassembly	✓	✓	✓		✓	✓
Component	Materials	Reduce unsustainable and hazardous materials	✓	✓				✓
		Increase biodegradable and low-impact materials						✓
		Reduce material variety						✓
		Increase material compatibility and recyclability						✓
		Specify and label materials						✓
	Form	Optimize performance, resistance, and reliability	✓		✓		✓	✓
		Design for easy removal	✓	✓	✓	✓	✓	✓
		Reduce mass	✓		✓		✓	✓
	Dimensions	Optimize performance, resistance, and reliability	✓		✓		✓	
		Design for easy removal	✓	✓	✓	✓	✓	✓

Note: ES1 = maintenance; ES2 = repair; ES3 = updating/adaptation; ES4 = direct reuse; ES5 = reuse of parts; ES6 = recycling.

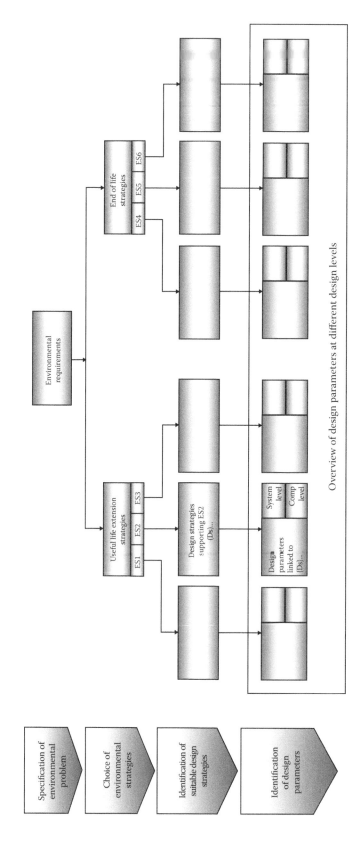

FIGURE 2.11 Implementing environmental strategies into the design process.

This breaking down of the problem and the design itself is at the heart of modern methods of product design and might be a useful resource in helping integrate environmental and traditional needs. At the same time, excessively specific but separate designs can delay or even halt a design solution that is balanced and effective, doable and economically viable.

2.15.1 Design for X for Environmental Strategies

Among various DFX tools, some are certainly applicable to the two intervention strategies for improving the life cycle environmental performance identified earlier:

- Those aimed at making it easier to keep the product working well so as to extend its useful life: *design for maintainability* and *design for serviceability* (DFS). Because the latter takes into consideration all the servicing operations (diagnosis, maintenance, and repair), it therefore includes the former.
- Those aimed at design governing end of life processes, as they aim at reducing the impact of retirement and at reclaiming resources: *design for product retirement* (DFPR)/*recovery*. More specifically, *design for remanufacturing* and *design for recycling* deal with two end of life outcomes.

In both cases, the tools try to intervene directly with the most significant design parameters linked to product architecture and component characteristics.

A third type of DFX is *design for disassembly* (DFD), but because it is often aimed at end of life recovery, it is sometimes thought to be integral to *design for recovery and recycling*. Nevertheless, DFD is transverse to both environmental strategies, extending useful life and end of life recovery. In both cases it helps ease disassembly.

2.15.2 Design for X Setup in Product Development

Using DFXs to resolve environmental requisites implies adopting *design-centered* product design, introduced in the previous chapter, which proves the most appropriate of the tools, as the design phases of importance occur almost simultaneously. This model, and the integration of environment, are shown in Figure 2.12, which closely follows Figure 1.6.

Problem specification is a preliminary design phase when the basic requisites of the product are defined. So too are the objectives of the development process, and any correlated problems and project requirements or restrictions. This is the phase where environmental objectives are also established, identifying the associated variables and analyzing the problems arising from achieving those objectives. The importance of this preliminary step has already been underlined; the most appropriate and effective environmental strategies can only be selected after the design objectives have been expressed as product requisites. It is obvious that in formulating environmental requisites, the life cycle approach is indispensable. It is in fact an added dimension that significantly expands the project's objectives.

Following the problem specification phase are the main design phases, not rigidly in order but mostly simultaneously:

- *Conceptual design*—Once the design objectives are defined and expressed as product requisites, design ideas need to be developed to achieve them. In this phase, a certain number of outline proposals are defined (by means of an initial description of the functions and main characteristics of the product), and are then classified by merit. As interventions oriented at the environment are much more effective the sooner they are made in the development process, this is the moment to introduce the environmental strategies most appropriate to the pre-established objectives. This can be done according to the setup described in the previous paragraph.

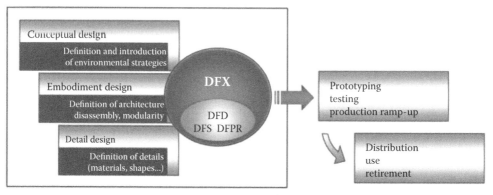

FIGURE 2.12   Use of DFX tools for integrating the environmental aspects into product development process.

- *Embodiment design*—Once the preliminary concept has been defined and the most effective environmental strategies expressed as design expedients, the next phase consists in producing a complete description of the product idea. The concepts formulated in the previous phases are developed, their applicability is verified, and finally expressed as an outline layout of the product that defines subsystems and functional components. In this phase the DFX tools to support environmental strategies intervene by applying methods and instruments that help correctly define product architecture, with particular emphasis on the properties of disassembly and modularity that aid servicing and reclamation.
- *Detail design*—The architecture of the product developed in the previous phase is expressed in complete and detailed solutions. In this phase the DFX tools to support the environmental strategies intervene by applying methods and instruments that help correctly define design details. The choice of materials, the study of form, defining the geometry of the components, and joint systems must not only be guided by performance and economics but by environmental requisites.

After the design phases when the most appropriate DFX tools intervene (DFD, DFS, DFPR), come the last phases of product development: *prototyping*, *testing*, and *production ramp-up*. However, the final verification of the product's environmental behavior that can be approximated by commonly available instruments (such as LCA) and purpose-built meters, requires that the entire life cycle be concluded with the phases of distribution, use, and retirement. Only then will it be possible to evaluate the effective environmental performance of the product and identify any redesign necessary.

2.16   TOOLS FOR INTEGRATED DESIGN: OVERVIEW

Thanks to their variety, the most noteworthy DFXs offer a flexible system of tools for integrated design oriented at a huge variety of product requisites in specific phases of the life cycle or transverse to the entire cycle. Among them, *design for manufacturing* and *design for assembly* (in terms of the necessities of the production phase), and *design for reliability*, *design for maintainability*, and *design for quality* (for the correct functioning and quality of the product during use), have already been shown to have reliable methodological structures and models.

To integrate environmental aspects into product design, the DFXs that support environmental strategies must obviously take on a critical role in this amply flexible system of tools for integrated design. How they integrate between themselves is also critical for a final solution that succeeds in interpreting the various requisites.

DFX tools are most effective in the design phases of embodiment design (in defining functional units and the layout) and detail design (defining component form, dimensions, and materials). In the previous phase of conceptual design they are more limited to applying fundamental criteria that could help the designer choose his first ideas. This particular phase is by is nature creative, encouraging designers to use more appropriate tools such as *concept generation methods*. From an environmental viewpoint there are tools other than DFXs that can help, such as techniques based on analyzing different environmental categories in context, compared to the life cycle; they can help formulate the best opportunities for implementing those aspects in product planning. Independently of its specificity, the role of *industrial design* in this phase of design development is crucial, particularly for environmentally oriented products.

Before the design phases, the problem specification phase is very important for setting up correct design strategies. One of the most commonly used aids is quality function deployment (QFD), used to express consumers' needs and desires regarding product requisites, taking into account market competition and identifying the relationship between these requisites. A derivative, Green QFD, is a method of preliminary setup of product development that integrates consumer demand with the environmental and economic demands of the life cycle.

Subsequently applied to an environmental issue, environmental FMEA (E-FMEA) is clearly derived from the well-known and widely used mode effect analysis (FMEA), a technique developed

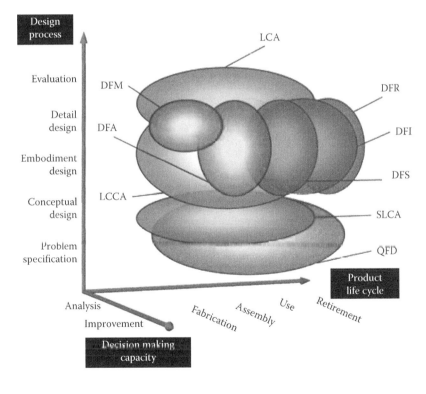

DFD - Design for disassembly
DFS - Design for serviceability
DFR - Design for recovery
DFA - Design for assembly
DFM - Design for manufacturing

QFM - Quality function deploiment
LCA - Life cycle assessment
SLCA - Streamline LCA
LCCA - Life cycle cost analysis

FIGURE 2.13 Integrated design: tools and techniques.

for identifying the potential failures of a design that could come about as a result of the functions required of the product being developed. Based on the same methodological setup, but adding the life cycle approach, E FMEA differs from FMEA, because instead of being oriented to identification of failure in technical functions, it focuses on failure in obtaining the objectives and environmental performance of the product.

All these tools should stand alongside those of traditional engineering design, oriented at the product's prime performance factors, among which the most evolved and widely used are those based on finite element analysis (FEA). In tandem with computer-aided design (CAD) software, solid parametric models can be made of components and systems, and together are able to create a virtual prototype of the product being developed, which can reduce times and costs.

For a complete and efficient environmentally oriented design all these tools must be IPD managed in the terms described earlier and shown in Figure 2.7. Here, DFX tools are not the only support for the main environmental strategies; there are also tools for the environmental and economic analysis (LCA, LCCA) of the life cycle.

Figure 2.13 shows most of the different tools described above. It classifies them by the design phase in which they are most applicable and by the life cycle phase which they will most influence. Added to these two dimensions, a third characterizes the tools by the information they provide to the designer, and therefore by their potential to help in decision-making (solely analysis—improvement and optimization).

Part II

The Properties of Design Materials

Part II provides general information on the materials used in mechanical design that, although overlapping other course areas, is not necessarily available to all industrial engineering students.

The need to know about these properties, even to a minimum of certain characteristics, has merited a whole chapter. For the same reason, summaries of the characterization tests for materials and mechanical component stress statuses are included in Chapters 4 and 5.

Chapter 6, which stands on its own, is dedicated to characterizing materials subject to dynamic stress and fatigue design, particularly regarding high cycle rates.

Chapter 7 concludes Part II and describes how designers can make rational choices about efficient materials according to product performance requirements, given that the range of materials is ever wider.

3 Materials for Mechanical Design

3.1 INTRODUCTION

The creative co-operativeness of builders together with the continuous intellectual activity of their employees results in the design and construction of what is more or less rationally planned. Everything is summarized into graphics of the geometrical shapes and dimensions of the components to be built.

The science of proportioning the parts and structures designed is not generally sufficient for engineers to evaluate the behavior of the materials used. For this reason, it is necessary to deal with topics related to them so as to facilitate the choosing of the right materials.

It should be easy, therefore, to understand that good designers need to know about the real behavior of materials, necessitating a deep understanding of them, as their application requires a working knowledge of their intrinsic properties. Thus, the development of topics related to the properties of materials is crucial to choosing them according to the requirements of their application in the planned system.

This study is devoted to the main classes of materials for use in engineering and, considering the vastness of the subject, it simply aims to give students some information and a brief outline of those materials most widely used in industry. When choosing materials, one should refer to the UNI tables to which we briefly refer as examples. As already stated, the choice of materials at the planning stage has to take into account environmental parameters, for which we recommend specialized reference books. The general rule is that when materials have comparable characteristics, the right choice is the one with the most technical data. For instance, in the field of mechanical design, issues connected with fatigue are very common, whereas data related to the fatigue resistance of materials is not so easily obtainable.

3.2 METALLIC MATERIALS AND ALLOYS

Metals are substances that have:

- Metallic alloys
- Monatomic molecules
- Good electrical and thermal conductivity
- Metallic shine/brightness
- Malleability and ductility

When combined with oxygen they form oxides.

Technically pure, they are rarely used in mechanical constructions, as their properties do not comply with the functional requirements of projects. Iron, for instance, which is a fundamental element for steel alloys (white iron/cast iron and steel), in its pure state—at room temperature—has poor mechanical properties:

- Fracture load under tensile stress $Rm = 180$–250 N/mm^2
- Yield load under tensile stress $Rs = 100$–140 N/mm^2

- Brinell hardness HB = 500–700 N/mm^2
- Stretch $A\% = 40$–50%
- Neck-in $Z\% = 70$–80%

The importance of this material, though, derives from the structural changes it undergoes as temperature varies, and the fact that certain alloy elements modify these changes, thereby satisfying practically all the main requirements of industry.

For these reasons, metallic alloys, made of various metals or metals and metalloids, have been created.

3.3   SOLID STATE AND STRUCTURE OF METALS

It is known that all metals are solids at room temperature, except for mercury, and their generally monoatomic molecules form a well-defined geometric grating typical of solid substances (crystalline structure).

Any substance with a crystalline structure is considered solid if its atoms—whether one or more chemical species, i.e., pure or composite—are distributed according to defined rules of geometric regularity.

An ideal crystal, therefore, can be defined as an aggregate of atoms placed in space according to a periodic law. The periodic repetition of the atoms might be different in ways that are studied by geometric crystallographers.

Any 3-D lattice can be imagined as made up of a series of parallel and equidistant plains containing all nodes, called reticular plains, the smallest portion of which on the plain is the transverse

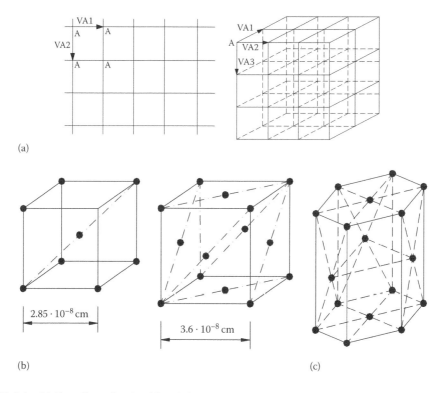

FIGURE 3.1   (a) Two-dimensional grid and three-dimensional lattice. (b) Atomic arrangement in crystal lattices of iron α (left) and iron γ (right). (c) Compact hexagonal cell.

couple and in space is the transverse tern, also called unitary or elementary cells that represent the simplest geometric form in a 3-D lattice. A 3-D lattice is the indefinite repetition of the elementary cell in space.

Metals are solids with metallic bonds that, overall, are positively charged, being made up of metallic atoms that have lost one or more peripheral electrons. The lost electrons form an "electron cloud," which, interacting with the positively charged metal atoms, gives metals cohesion. The strong electrical and thermal conductivity of metals is due to the remarkable mobility of these electrons.

The most common metal lattices are cubic or hexagonal, with nodes positively charged and immersed in an electron cloud of valence electrons (Figure 3.1c)

3.3.1 Solidification

When a pure metal or an alloy is cooled to a solid, a certain number of atoms in the preferential cooling areas place themselves within the mass along the lattice lines (crystallization centers or nuclei). Later, other atoms approaching these crystallization centers are caught and kept, thus increasing the number of atoms (growing phase). As the crystals grow, they collide, until they form into the metal or alloy lattice, but with different orientations. These constitute the crystalline grains of the metal or alloy. Their mechanical and technological properties mostly depend on the dimensions of these grains.

The size of the crystalline grains is influenced by cooling velocity; the slower, the fewer the crystals and consequently the bigger the grains; vice versa, rapid cooling causes the formation of many crystals that form small crystalline grains.

Clearly, as the initial nuclei grow bigger, they eventually touch. In the contact areas (borders) of the crystalline grains, all the impurities gather, and it is in these areas that corrosion and fusion phenomena start.

During solidification, significant thermal energy is released that hinders further solidification and crystallization in those areas; thus, it begins and continues in other areas to form dendritic or branched structures.

3.3.2 Lattice Defects

It has to be admitted that the real structure of solids differs from the idealized one above, in that solids exhibit numerous defects that are generally local deviations from the regular crystalline structure, nearly always generated by a non-uniform energy distribution in the crystals' atoms.

A perfectly structured crystal is only possible at absolute zero. At any other temperature, because of non-uniform energy distribution among the atoms, a certain number of them have enough energy to originate a defect.

Because of the thermal agitation of each atom, there are three normal vibration motions. Clearly, the overall energy contributed by each atom by its thermal agitation must be quantified, and that energy can vary finitely.

Energy variations related to an atom's thermal agitation occur by the absorption or emission of quanta of elastic waves, or phonons, which spread throughout the lattice of the atom. The absorption of a phonon by an atom causes it to vibrate, and so not have a fixed position in space.

Phonons propagate in crystals in all directions, colliding with each other and the lattice atoms, conserving the energy and quantity of motion.

In special conditions, phonons can interact with some electrons of the crystal's atoms by pulling them away from their normal position and leaving holes that create dot-like imperfections in the electron structure as a whole.

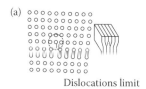

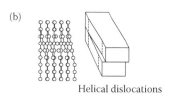

FIGURE 3.2 Dislocation: *linear* (a) and *helical* (b).

The most common defects in metals can be:

- Dot-like, if the area affected by the defect is limited to a few atoms. There are frequent lattice vacancies and interstices
- Linear, if the irregularity appears for the whole length of the crystal, along a well defined line
- Planar, in sections of the crystal
- 3-D, when the area involved in the imperfection extends in three directions, including volume of the lattice itself

An aggregation of defects represents a discontinuity in the regular distribution of the atoms in the lattice. These clusters are referred to as "dislocated," and they can be *linear* (Figure 3.2a) or *helical* (Figure 3.2b).

3.4   IRON–CARBON ALLOYS

As mentioned above, pure iron at room temperature has poor mechanical properties. In order to improve its characteristics, it is used in alloys with various elements. Among the iron alloys, those containing carbon are the most significant and most widely used in industry. They can be grouped according to carbon content:

- Cast iron—carbon content ranging from 2.06% to 6.67%, generally non-pliable, used in casting
- Steels—carbon content ranging from 0.008% to 2.6%, generally more pliable, the great majority of iron alloys industrially employed are steels

The above-mentioned alloys, apart from other alloy elements, contain carbon to varying degrees; exact knowledge about binary iron–carbon is the basis of the study of all iron alloys.

It should be noted that the carbon in iron alloys might be elementary carbon, or iron carbide Fe_3C (cementite).

3.4.1   CAST IRONS

As mentioned above, cast irons are alloys of iron and carbon (2.06%–6.67%). To a lesser extent, Si, Mn, P, and S are also present in different proportions.

Generally the structural elements are:

- Ferrite, in very thin layers
- Graphite (free carbon)
- Cementite, combined carbon (iron carbide Fe_3C)

For further information about binary Fe-C, refer to a materials technology course.

Cast irons are widely used in industry, for their widely differing properties. In fact, castings can be very hard or pliable, or they may be made resistant to special groups of stresses (mechanical, chemical, and thermal).

The field of cast irons is very wide, and the most important and well known cast irons in current use are:

- White cast iron
- Common gray cast iron with graphite layers
- Special cast iron or alloyed
- Malleable white heart cast iron
- Malleable black heart cast iron
- Cast iron with spheroidal graphite

This study will include the types of castings most commonly required by industry. Considering the endless uses to which these castings are put in the construction of mechanical components, each needs a varying percentage of alloy element to be able to make the required material:

- Good resistance to tensile stress
- Excellent compression resistance
- Good bending resistance
- Good shock resistance
- Good wear resistance
- Easily worked by machine tools
- Good pressure resistance
- Good corrosion resistance

3.4.1.1 Influence of Common and Special Elements on Cast Irons

To enhance various chemical, physical, and mechanical characteristics of cast irons it is necessary to add certain elements. The following list includes some of them with their influence on the alloy.

- *Silicon* (Si) metalloid, avid for oxygen, is a reducing agent; its main effect is to favor iron carbide decomposition (cementite and so to graphitize carbon). Si of 10% in cast iron increases heat resistance, whereas from 3% to 6% increases electrical resistance.
- *Manganese* (Mn) combines with sulfur to form manganese sulfide, which being insoluble in the metal, goes straight into clinker/slag, while what has not combined with sulfur acts as a stabilizer for cementite. When manganese is present in cast iron, it tends to heighten its mechanical characteristics.
- *Phosphorus* (P) is present in all cast irons in proportions that vary up to about 1.30%. It forms iron phosphide Fe_3P, which is remarkably hard. Phosphorus increases fluidity and smoothness in cast iron and acts as a graphitizer. It is required when thin castings have to be produced.
- *Sulfur* (S) prevents the graphitization of carbon, which causes blowholes and shrinkage in the castings, reducing their smoothness. Adding manganese can reduce these undesired effects.

- *Nickel* (Ni) has a graphitizing effect, increasing the hardness of the material without affecting the workability of castings by machine tools.
- *Chromium* (Cr) reduces graphitization, favoring the combination of carbon and iron. Quite often this element is added with nickel; their combined action gives the alloy high resistance and sufficient hardness without reducing its workability by machine tools.
- *Copper* (Cu) is as graphitizing as nickel, and when added to silver cast iron, it prevents carbide formation.
- *Molybdenum* (Mo) is a slightly anti-graphitizing element, it therefore limits carbide production. It increases wear resistance, and when associated with chromium, has extremely fine grains and its mechanical properties are significantly increased.
- *Titanium* (Ti) is more graphitizing than silicon; increases resistance to wear, heat, and tensile stress, and at the same time favors workability by machine tools.
- *Vanadium* (V) stabilizes carbides, tends to reduce grain size, increases heat resistance and hardness in tempered zones.
- *Tin* (Sn) improves tensile stress resistance and increases hardness.
- *Aluminum* (Al) helps cast irons endure high temperatures in sulfurous environments, and favors nitriding.
- *Calcium* (Ca) is an active deoxidizer, it is added as briquettes to furnaces as calcium silicide (CaSi).
- *Tellurium* (Te) is an anti-graphitizer, it increases cooling and reduces internal tension in the casting.
- *Tungsten* (W) prevents graphitization, favoring the formation of carbides; high quality cast irons with high resistance to tensile stress, compression, shock, and bending are obtained.
- *Magnesium* (Mg) is mainly used in the production of spheroidal cast irons (spheroidized graphite). In normal cast irons it is used as a desulfurizer.

The more elements in an alloy, the more complex are their interactions: for in-depth analysis refer to a metallurgy course.

3.4.1.2 Gray Cast Iron Classification

The use of these materials is quite common in mechanical constructions, mainly in molten castings for their high fusibility, fluidity, and creep. Like other cast irons they endure compression very well, but tensile stress less well.

Gray cast iron castings contain lamellar graphite and are classified by tensile stress/breaking load, measured in newtons per square millimeter (Table 3.1).

TABLE 3.1
Gray Cast Irons

G 100 UNI 5007-69	Minimum breaking load unit	Rm 100 N/mm^2
G 145 UNI 5007-69	Minimum breaking load unit	Rm 145 N/mm^2
G 195 UNI 5007-69	Minimum breaking load unit	Rm 195 N/mm^2
G 245 UNI 5007-69	Minimum breaking load unit	Rm 245 N/mm^2
G 295 UNI 5007-69	Minimum breaking load unit	Rm 295 N/mm^2
G 345 UNI 5007-69	Minimum breaking load unit	Rm 345 N/mm^2

TABLE 3.2
Special Cast Irons

Gh 130 UNI 5330-69	HB 130–180	Use: automobile construction
Gh 170 UNI 5330-69	HB 170–220	Use: hubs, boxes
Gh 190 UNI 5330-69	HB 190–240	Use: metal rings, valves
Gh 210 UNI 5330-69	HB 210–250	Use: fittings, machine tools
Gh 230 UNI 5330-69	HB 230–280	Use: hubs, boxes

3.4.1.3   Special Cast Irons Classification

Special cast irons are obtained by using an expedient in the smelting process to suit their specific use. Cast irons used in the automobile industry belong to this category. They are classified according to Brinell hardness (Table 3.2).

3.4.1.4   Spheroidal Cast Irons Classification

Spheroidal cast irons are the best known cast iron alloys because of their good mechanical characteristics due to spheroidal graphitization. They are classified according to their minimum tensile stress/breaking load, Rm, in newton per square millimeter their minimum yield load under tensile stress unit, Rs, in newton per square millimeter, their percentage elongation $A\%$, and their hardness HB (Table 3.3). They also have fairly good fatigue resistance because they are less sensitive to intaglios.

3.4.1.5   Cast Iron Alloys Classification

Highly mechanically resistant cast iron alloys ACI 30, ACI 40, ACI 50, ACI 60, ACI 70, and ACI 80, contain Ni, Cr, and Mo.

Wear-resistant cast iron alloys NI HARD 1, NI HARD 2, and NI HARD 3, contain Si, Mn, Ni, and Cr.

Corrosion-resistant cast iron alloys NI RESIST type 1a, type 2, types 2a-2b, type 3, type 4, and type 5 Minvar, contain Si, Mn, Ni, Cu, and Cr.

Heat-resistant cast iron alloys: SILAL with 6% Si; ENDURON with 16%–35% chromium; NICROSILAL with 18% Ni, 2% Cr, 5% Si, and 1.8% C.

3.4.1.6   Malleable Cast Irons Classification

Malleable cast irons are obtained by annealing pig iron to malleabilization. They are classified according to their minimum tensile stress breaking load, Rm, in newton per square millimeter, their minimum yield point unit, Rs, in newton per square millimeter, their percentage elongation $A\%$, and their hardness HB (Tables 3.4 and 3.5).

TABLE 3.3
Spheroidal Cast Irons

GS 370-17 UNI 4544-7	$Rm = 370 - Rs = 230 - A\% = 17 - HB = 128–176$
GS 400-12 UNI 4544-7	$Rm = 400 - Rs = 250 - A\% = 12 - HB = 140–197$
GS 500-7 UNI 4544-7	$Rm = 500 - Rs = 320 - A\% = 7 - HB = 168–236$
GS 600-2 UNI 4544-7	$Rm = 600 - Rs = 370 - A\% = 2 - HB = 188–264$
GS 700-2 UNI 4544-7	$Rm = 700 - Rs = 440 - A\% = 2 - HB = 225–296$
GS 800-2 UNI 4544-7	$Rm = 800 - Rs = 480 - A\% = 2 - HB = 243–345$

TABLE 3.4
Malleable Cast Irons (White Heart or European)

GMB 345 UNI 3779-69	$Rm = 345 - Rs = 180 - A\% = 7 - HB = 125–200$
GMB 390 UNI 3779-69	$Rm = 390 - Rs = 216 - A\% = 10 - HB = 125–200$
GMB 440 UNI 3779-69	$Rm = 440 \quad Rs = 250 \quad A\% = 7 - HB = 150–210$
GMB 490 UNI 3779-69	$Rm = 490 - Rs = 294 - A\% = 5 - HB = 170–230$
GMB 540 UNI 3779-69	$Rm = 540 - Rs = 323 - A\% = 4 - HB = 190–240$
GMB 636 UNI 3779-69	$Rm = 636 - Rs = 382 - A\% = 3 - HB = 210–250$
GMB 685 UNI 3779-69	$Rm = 685 - Rs = 490 - A\% = 2 - HB = 240–285$

3.4.2 STEELS

Steels are ferrous alloys suited to plastic working having over 2% carbon. Due to their mechanical characteristics, chemical composition and type of use please refer to unification. The Italian norm referring to steels is UNI EN 10027.

Due to their properties, carbon steels are unparalleled in certain areas, but two characteristics limit their use in modern technology:

- High critical cooling velocity, which makes it impossible to temper large pieces in depth.
- Brittleness in comparison with steels of equal hardness alloyed with special elements.

The above-mentioned drawbacks have necessitated the use of steels containing higher or lower percentages of alloy elements (alloyed steels) for many applications.

A general classification grouped steels (by composition and usage) into non-alloyed steels and steel alloys according to the presence of alloy elements. Referring to the UNI Tables, the following is an outline showing the variety of steels.

Steel	*non-alloy*	*quality*	*not for thermal treatments*
			for thermal treatments
		special	*not for thermal treatments*
			for thermal treatments
	alloy	*quality*	*not for thermal treatments*
			for thermal treatments
		special	*not for thermal treatments*
			for thermal treatments

Generally, every country has its own classification (unification) of steels with different designation criteria; Italy has UNI, the United States has AISI and ASTM, and Germany has DIN, which means that any reference to steel must specify its classification.

TABLE 3.5
Malleable Cast Irons (Black Heart or American)

GMN 345 UNI 3779-69	$Rm = 345 - Rs = 180 - A\% = 12 - HB = 125–200$
GMN 365 UNI 3779-69	$Rm = 365 - Rs = 216 - A\% = 14 - HB = 120–160$
GMN 440 UNI 3779-69	$Rm = 440 - Rs = 255 - A\% = 7 - HB = 150–210$
GMN 490 UNI 3779-69	$Rm = 490 - Rs = 294 - A\% = 5 - HB - 170–230$
GMN 540 UNI 3779-69	$Rm = 540 - Rs = 323 - A\% = 4 - HB = 190–240$
GMN 636 UNI 3779-69	$Rm = 636 - Rs = 382 - A\% = 3 - HB = 210–250$
GMN 685 UNI 3779-69	$Rm = 685 - Rs = 490 - A\% = 2 - HB = 240–285$

In the spirit of the aim of these notes please refer to the tables in Volume I of Giovannozzi's book whose classifications are outdated but still useful, especially for their practical summaries. It is not difficult to link the type of the steel to the different unifications

Without dwelling on the steel descriptions, the information on steel alloys is useful and it summarizes the effects each ligand has on the mechanical, metallurgical, and technological characteristics of the steel, facilitating choice of the most suitable steel at the design stage.

3.4.2.1 Manganese Steels

In annealing steel, manganese spreads out among the single structural constituents, melting in the ferrites and carbides.

Manganese causes the following property variations in steel:

- Decreases critical cooling velocity and consequently increases temperability
- Increases durability and hardness during the annealing and heat treatment phases
- Improves durability/toughness at all stages of treatment
- Increases sensitivity to superheating
- Improves resistance to wear and tear following impact or compression

Mn is also used for manufacturing steels that have high levels of durability and yield limits as well as good weldability.

3.4.2.2 Chromium Steels

Compared to manganese, chromium has a greater affinity to carbon. In steel it causes the following property changes:

- Decreases critical cooling velocity and consequently increases temperability
- Increases durability and hardness during heat treatment
- Improves durability/toughness at all stages of treatment
- Decreases sensitivity to superheating
- Increases tempering stability
- Increases hardness and wear resistance
- Increases corrosion and hot oxidation resistance
- Increase in surface hardness by nitriding

Chromium steels are largely used for their high resistivity and for permanent magnets. Moreover, chromium is the most important alloy element in the manufacture of heat-resistant steels, stainless steels, and steels for tools.

3.4.2.3 Nickel Steels

Nickel steel is qualitatively similar to manganese steel, producing the following property changes:

- Decreases critical cooling velocity and consequently increases temperability
- Increases durability and hardness during heat treatment and annealing
- Improves durability/toughness at all stages of treatment
- Decreases sensitivity to superheating
- Increase in corrosion and hot oxidation resistance

In addition, nickel improves the electrical and magnetic characteristics of steel, and affects expansion coefficient and the modulus of elasticity.

Nickel steels are widely used in the field of special steels, especially for stainless, construction, and tool steels.

3.4.2.4 Tungsten Steels

Tungsten in steel produces a series of special carbides with various compositions and the following property changes:

- Decreases critical cooling velocity and consequently increases temperability
- Decreases sensitivity to superheating
- Increases tempering stability
- Increases heat resistance
- Increases hardness and wear resistance
- Decreases tendency to tempering fragility

Because of these characteristics, tungsten steel is widely used for tools (fast steels) of any kind and for permanent magnets.

3.4.2.5 Molybdenum Steels

The distribution of molybdenum among the structural constituents of steel is identical to that of chromium and tungsten. It produces the following effects:

- Decreases critical cooling velocity and consequently increases temperability
- Decreases sensitivity to superheating
- Increases tempering stability
- Increases heat resistance
- Increases hardness and wear resistance
- Eliminates the tendency to tempering fragility
- Is corrosion resistant to certain chemical agents

Molybdenum steels are used for tools, the construction of components operating at high temperatures, stainless steel for special uses, steels for permanent magnets, and nitriding steels whose core must not become brittle at persisting 500°C.

3.4.2.6 Silicon Steels

In annealing steel, silicon is probably found dissolved in ferrite. Typically, silicon in iron–carbon alloys causes the tendency to graphitization; it has a remarkable hardening effect and clearly decreases cold deformability, which is almost zero for silicon content slightly higher than 3%. Hot deformability is also greatly decreased, the alloys whose silicon content is over 6% being practically unforgeable.

Silicon in steel produces the following properties:

- Decreases critical cooling velocity and consequently increases temperability
- Increases durability, and especially elastic limit, in every treatment, with some loss of toughness
- Slightly increases insensitivity to superheating
- Increases tempering stability
- Increases wear resistance
- Resists hot oxidation
- Increases electric resistance and decreases hysteresis losses

Silicon steels are used in the manufacture of springs, electrical resistors, dynamos and transformers, and components that have to resist hot oxidation.

3.4.2.7 Less Common Carbon Steel Alloys

There are many types of steels, as the UNI tables show, that have small amounts of other ligands that make them insensitive to carbidization or render the structure more stable, giving it improved mechanical properties.

Therefore, apart from the above-mentioned ligands, tiny amounts of the following elements may be present:

Vanadium gives steel good mechanical characteristics on heat treatment and increases its elastic limit. It is used in tool and construction steel for high temperatures. In tiny quantities, it is used in many types of steel to limit sensitivity to superheating.

Aluminum increases steel hardness through nitriding, electrical resistance, and hot oxidation resistance. In the field of special steels, it is widely used in nitriding steel and high temperature electrical resistors.

Cobalt improves tempering and heat hardness, as well as significantly increasing coercive force in tempered steel. It is widely used in fast steels and for some permanent magnets.

Copper improves steel's corrosion resistance, but causes a strong tendency to heat brittleness. It is only used in some stainless steels.

Sulfur in tiny amounts is always present in steel, originating from raw material ores. 0.1% or higher quantities can help workability by machine tools, which is why it is used in some automated production steels.

Titanium is added to some stainless steels to prevent intergranular corrosion. It can make steel resistant to harmful sensitization phenomena, such as the formation or precipitation of chromium carbides with consequent decrease in mechanical corrosion resistance.

Niobium is present in small amounts in corrosion-resistant steels and, like titanium, confers "protection" characteristics, especially in high temperatures. NB, carbides prevent the precipitation of chromium carbides.

3.4.2.8 Special Stainless Corrosion and Heat-Resistant Steels

Stainless steels are the most widely used alloys in mechanical construction. They contain more than 18% chromium and often more than 8% nickel. Stainless steels are widely used in the chemical and oil industries and generally in all components exposed to particularly aggressive corrosion environments.

When titanium (Ti) or molybdenum (Mo) are added to chromium and nickel steels, they have better mechanical properties and, above all, are resistant to high temperatures (over 500°C) when creep might limit or prevent certain resistance characteristics. In Giovannozzi's Volume I there are tabled some characteristic steels with the use indications. By the chemical and mechanical characteristics reported it is not difficult to find the corresponding steel of the unification (UNI, ASME, DIN, etc.)

3.5   NON-FERROUS METALS AND THEIR ALLOYS

3.5.1   Copper

Copper is one of the most widely used industrial materials after iron, and used for semi-finished products either destined to become electric conductors, or which require good deformability, good weldability, and the absence of brittleness in reducing environments. It has the following properties:

- High thermal and electrical conductivity
- Good hot and cold deformability
- Good mechanical properties
- Good corrosion resistance
- Ease of welding and brazing
- 50 HB (Brinell hardness)
- 216 N/mm^2 tensile stress/breaking load (annealed)
- 260–440 N/mm^2 tensile stress/breaking load (work-hardened)
- $A\% = 30\%$–45% elongation

3.5.2   Copper–Tin Binary Alloys (Cu-Sn Bronzes)

There is a wide range of copper–tin binary alloys, but only the most widely used are unified (1) alloys. As an example, G-CuSn10 and G-CuSn12 (UNI 7013, ASTM B30, BS 1400, DIN 17656, ISO 1358) alloys are recommended for all uses where wear and corrosion resistance are required. They are used to manufacture bearings, shaft bushes and helical steel cogwheels coupled to worm gears. They are durable in corrosive and marine environments, and are therefore used in shipbuilding and the petrochemical industry (turbo impellers, valves).

Alloys with high levels of Sn (from 14% to 20%) are used in machine components subject to great friction and wear and tear, such as thrust bearings, pump piston rings, pistons and slip blocks subject to friction and heavily loaded bushes, and bearings.

3.5.3   Copper–Tin–Zinc Alloys

Known worldwide as gunmetal, Cu–Sn–Zn alloys are made of two binary bronzes by substituting two or four Sn units with as many zinc units, whose percentage does not significantly modify either the binary bronze mechanical characteristics or corrosion resistance. The most widely used alloys of this type are used for manufacturing heavily loaded bearing shells, marine and rolling mill drive-shaft jackets, pump bodies, and gate valves for hot fluids (up to 225°C).

3.5.4   Copper–Tin–Zinc–Nickel Alloys

Adding nickel to Cu–Sn–Zn alloy improves its mechanical characteristics without changing corrosion resistance; tables of the unifications (UNI 7013, ASTM B30, BS 1400, DIN 17656, ISO 1358) report these types of alloys. They are used in large machine gears, large rolling mill bolts, steam cylinder piston rings, and bearing cases. In shipbuilding these alloys are prominent, being used for pump bodies and impellers, flanges, and fittings, as well as flanges and valves for the oil industry.

3.5.5 Low Lead Copper Alloys

Tiny amounts of lead are added to Cu–Sn–Zn alloys to increase their workability by tool machines and to improve smelting quality. To improve the mechanical characteristics of these bronzes, gas must be avoided during castings. There is a wide range of these alloys.

- *BZn2* is used in the production of flanges or castings that need brazing, is unified in law UNI 1698.
- *BZn6* is used in electromechanical castings, brasses, and bearings; its wear resistance is mediocre.

In the tables of the unifications for these materials as UNI 7013, ASTM B30, BS 1400, DIN 17656, ISO 1358, designation, chemical, physical, and mechanical characteristics are reported

- *G Alloy* is used to produce valves and valve parts for high pressure steam, carburettor components, injectors, brasses and bearings that are highly stressed, and structural castings for electronic equipment.
- *90/10 Industrial* is used to produce valves, valve parts for high pressure steam, pump bodies and impellers for petrol and crude oil, carburettor components, injectors, brasses and bearings that are highly stressed, and structural castings for electronic equipment.
- *C Alloy* is largely employed in car engine components, brasses and bearings, in paper machine parts, and in steam valves and fittings.
- *Sn12Ni1, Industrial 5* is nickel industrial bronze with 12% tin, which is used for high speed gears and bearings, oblique cogwheels, and highly loaded bearings.
- *Industrial 86/14* is used to produce high-speed gears and bearings subject to high stress and also for oblique cogwheels.

3.5.6 Copper–Lead–Tin Alloys

Cu–Pb–S alloys are preferred to binary bronzes for their anti-friction properties, their tool workability and corrosion resistance. The tables of the unifications (UNI 7013, ASTM B30, BS 1400, DIN 17656, ISO 1358) report the most widely used alloys. The uses of these alloys can be summarized as follows: anti-friction bearings without white metal linings, with medium loads, in contact with untoughened surfaces at medium speed and with little lubrication.

Only G-CuSn10Pb10 can be used with mediocre wear results, at high loads and speeds. It is used to produce bearings, brasses, slip blocks for hot and cold rolling mills, machine tools and farm machinery, diesel engines, electrical machinery, and rolling stock. Its excellent anti-corrosive properties widen its applicability to machinery for the chemical, mining, oil, and paper industries.

3.5.7 High Lead Bronzes (Cu–Sn–Zn–Pb)

High lead bronzes are particularly suitable for sealed casting, thanks to their high zinc and even higher lead content. They have tool workability comparable to dry brass. The uses of some of the tables of the unifications are as follows:

- *G-CuSn5Zn5Pb5* (85.5.5.5) is the most common brass alloy worldwide with excellent foundry characteristics, sea water and low acid water resistance, good mechanical and anti-corrosive resistance, firmness, and workability. It is used to produce hydraulic and pneumatic valves, pump bodies and impellers for low-acid water and wine, flanges and fittings for ships, bearings, carburettors, injectors, gauge components, and machine components for the paper industry.

- *SAE 660* is an ideal alloy for valves and hydraulic and pneumatic valve components, for water pump bodies and impellers, meter and gauge components, and also for water and steam fittings; it can be used to produce bearings subject to moderate stress and untoughened shafts only superficially hardened.
- *G-CuSn7Zn4Pb6* is widely used for valves, gauge and meter components, carburettors, water pump bodies and impellers, and in particular for various water and steam fittings.
- *G-CuSn3Zn10Pb7* is widely used for flanges and expansion joints, as well as for hydraulic valves, taps, and sanitary fixtures.
- *BZn7* is used for less extreme but similar uses to those of *G-CuSn5Zn5Pb5* (85.5.5.5) alloy.
- *R Bronze* is widely used for ornaments, building fixtures, fittings, taps and water valves, railway line clamps, bearings and brasses with and without white metal linings.
- *Red Alloy* is used for sand casting, and recently used in *cire perdue* casting.
- *Valve Alloy* is used in the production of air–gas–water fittings, for pump components, valves, and low-pressure fittings.

3.5.8 ALUMINUM BRONZES

Growing design requirements have created new interest in foundry aluminum bronzes. Noted industrially for their resistance to impact and wear, erosion and cavitation, aluminum bronzes are highly corrosion proof, and are unique in maintaining elevated mechanical characteristics at high temperatures where copper alloys are useless.

- *ACF 60* is widely used in a molten state, and forged, in shipbuilding, aeronautics, and the automobile and mechanical industry in general, because it maintains a high tensile stress load to cutting, good resistance to heat oxidation and high fatigue resistance. It has elevated anti-sparking properties, low magnetic permeability, and high corrosion resistance to aggressive agents, and therefore this alloy is used to manufacture products for the chemical, petrochemical and liquid gas industries, as well as in the ground extraction of oil.
- *G-CuAl9Fe3* is widely used in shipbuilding for stressed components in steering gear, winch stabilizers, pump bodies and components, industrial engineering plant for cryogenics, pump components for acids and alkali, components for water control systems, electrical engineering plant for forged machine/auto accessories and terminal board production, as well as for bearings and brasses/shells, cylinder and rolling mill nuts, and for machinery and equipment subject to high wear and tear.
- *G-CuAl11Fe4* is used for large sliding surfaces and molds for stainless steel and plastic materials, pasta dies, and pump parts for use in acids, alkalis, and seawater.
- *G-CuAl11Fe4Ni4* contains nickel, which increases corrosion resistance. Besides the previous alloy uses, it is also used to produce gears, camshifts, valves, turbine impellers, and for some types of propellers. International norms classify this alloy as UNI G-CuAl11Fe4Ni4; DIN 17656 GB-CuAl10Ni BS 1400 AB2; ASTM B 148 C95500 ISO 1338 Cu-Al 10 Fe5Ni5.
- *QQC 390* is used in shipbuilding for propellers, stressed components in steering gear, stabilizers, and hoists in corrosive environments. The alloy is unified in DIN 17656, denominated GB-CuAl10Ni and with the ASTM B148 C 95800 norm.

Other important foundry aluminum bronzes are *Inoxida 3*, *Inoxida 53*, *Inoxida 72*, and *Inoxida 90*.

3.5.9   INDUSTRIAL BRONZE

Used in the production of bearings, brasses/shells, cylindrical and conical gears, and molds for stainless steel and plastic materials. It is workable by machine tools and has good forgeability.

3.5.10   HIGH RESISTANCE BRASS

Brass containing 10%–20% zinc makes very plastic and easily deformable alloys that when cold, are golden, and known as pinchbeck, or tomback, and used for costume jewellery. They have low mechanical characteristics.

Of Cu–Zn 10, Cu–Zn 15, and Cu–Zn 20, the latter is most widely used in industry. Brass Cu–Zn 37 is the most widely used in Europe for pipes, bars, and cables/wires. Brass Cu–Zn 40, commercially called "Muntz brass," is used for finishing parts while hot. It is frequently used to make radiator plates in condenser systems. It has better mechanical characteristics than the other brasses. The alloys in the table comply with the qualitative requirements of national and foreign unifications.

- *OTS 1* is used to produce valve stems, brasses/shells, bearings, weapon and propeller pars, pump bodies, and special fittings for sea use.
- *OTS 2* is used to produce axles, propeller liners, shells, bearings, and valve stems for mechanical application.
- *OTS 2 for propellers* is used to produce propellers.
- *OTS 3* has similar applications to OTS 2.
- *Grade A SAE 430* is used to produce brasses/shells, bearings, gears and weapons parts.
- *SAE 430 Grade B* is used to produce parts of hydraulic cylinders, locking nuts, large valve stems, and slow but heavily loaded bearings.
- *SAE 43* is widely used in all brass marine applications other than in the construction of mountings and processed machine components, valve stems and weapons components.

3.5.11   FOUNDRY BRASS

The alloy range in tables of the unifications cover every type of cast obtained by mold pouring or pressure die casting, and the following in particular (referring to UNI):

- *DELTA (green 1st, green 2nd, A, B, C)* is used to produce handles and knobs, ironmongery, and ornamental and decorative casts.
- *TONVAL* is used for its excellent pourability/runnability to produce handles and knobs, sanitary fixtures, radiators, and gas valves.
- *SAMO* adds low pressure mold pouring to Tonval's excellent qualities. It is used for taps and valves, and for high-pressure castings.
- *Decorated DELTA* is obtained by pressure die-casting and suitable for decorative items, naval decoration, chandeliers, and building fasteners.
- *Silica brass* is largely used to produce small gears, burner components, carburettor bodies, handles and knobs, and brush holders in electrical equipment.
- *Special silicon brass* is used for *cire perdue* castings and for ornamental castings.
- *DELTA B* is used for taps, valves, nozzles, gas systems, nuts and bolts, ironmongery, knobs and handles, and also small gears and for high-pressure castings.
- *Smeltered press* is used in pressure die-casting and ideal for products with a polished finish.

3.5.12 Pressure Die-Casting Zinc Alloys

Three alloys are produced:

- *ZAMA 12* offers the best properties of hardness and tensile stress resistance.
- *ZAMA 13* has good collision resistance, and corrosion resistance when exposed to severe weather conditions.
- *ZAMA 15* offers good tensile stress resistance and collision resistance, and dimensional stability over time. It possesses good corrosion resistance even in critical weather conditions.

3.5.13 High Aluminum Zinc Alloys

These are used instead of copper and aluminum alloys, compared to which they offer advantages in terms of costs, melting characteristics and subsequent finish, while maintaining fundamental mechanical characteristics.

- *ZA 8* can be sand or mold cast, even if its main use is in hot chamber pressure casting. It is more durable than Zama alloys, has excellent slip resistance at variable temperatures from 100°C to 150°C, and possesses a high level of hardness.
- *ZA 12 denominated ILZRO12* offers the following advantages: higher durability than gravity cast non-iron alloys except for aluminum and manganese brasses; lower specific gravity than copper alloys and other zinc alloys; excellent workability by machine tools; gravity casting of complex pieces impossible to produce with other alloys; as an anti-friction alloy, it can substitute for brass in the production of small brasses/shells.
- *ZA 27 denominated ILZRO27* has high breaking resistance by tensile stress, yield, and wear. Its workability by machine tools together with its suitability for surface treatments (electroplating, chromatization painting, and anodization) and weldability make it a highly versatile and usable alloy, considering its inexpensiveness. In some cases it can compete with silver iron and with some copper and aluminum alloys.

3.5.14 Lead Alloys

These include the following alloys:

- *Antimonial lead*—Antimony may range from 0.5% to 20%, and in complex alloys there are sometimes tin, arsenic, copper, selenium, etc. Antimonial lead alloys are used for accumulators, lead shot, for chemical systems but in limited quantities (valves, pump impellers), and plastic resin molds. Lead–stibium alloys are used to produce antimonial lead wires destined for making small-caliber bullets.
- *Special alloys and mother alloys*—This range of alloys and mother alloys is varied and includes the following: Pb/Na, Pb/Ca, Pb/Li, Pb/Bi, Pb/Se, Pb/Te, etc., used for refining; lead–silver alloys for anodes; lead–calcium for batteries; Pb/Sb mother alloys with over 20% Sb; Pb/As mother alloys with As over 5%; Pb/Sn/Cd, Pb/Sn/Sb, etc., mother alloys. These materials are used in various shapes and dimensions for surfaces, different cross-section billettes, ingots, plates, sheets, etc.

3.5.15 Alloys for Cables

There is a wide range of alloys for wires with different alloys: tin, antimony, tellurium. These alloys are obtained with highly refined lead by special treatments. The following indications summarize the effects of the various metals when used as alloys, together with their properties.

3.5.15.1 Aluminum

Aluminum has a silver white color, a melting point of 659°C, specific gravity of 2.69 kg/dm^3, HB 15–40 Brinell hardness. Aluminum forms alloys with copper, silicon, magnesium, manganese, and zinc. These alloys, tempered for a few days (aging), acquire better mechanical properties. Aluminum alloys are the lightest alloys; the best known for plastic workability are the following:

- *Avional* is used in aeronautics for highly stressed but light parts (struts, connecting rods, hooks, etc.), it has tensile stress resistance equal to 470–500 N/mm^2 and a percentage elongation $A = 6\%–10\%$.
- *Chitonal 24* (trade name of Avional 44 plated with aluminum) is used only in aeronautical building.
- *Ergal 55 and Ergal 65* are only used extruded, have high mechanical tensile stress resistance $Rm = 560–580$ N/mm^2, and are used in aeronautics for highly stressed components.
- *Duralumin* or *Duralite* has good tensile stress resistance 400–450 N/mm^2, resists acid corrosion, does not oxidize in air, and can be laminated or extruded. Used in aeronautical and automobile construction (engine connecting rods, brake blocks, bicycle rims, etc.).

Among the main foundry aluminum light alloys we include:

- *Inafond S 13* or *Silafond* contains 5%–13% Si, has reasonable corrosion resistance, and low tensile stress resistance $Rm = 200$ N/mm^2.
- *Silumin* contains 13% Si, $Rm = 160–340$ N/mm^2, HB = 120 Brinell hardness, and mediocre corrosion resistance.
- *Anticorodal* contains 2% Si, 0.65 Mg, and 0.70 Mn, easily weldable. Corrosion resistant and good mechanical resistance.
- *Corrofond* contains up to 7% Mg, has good corrosion resistance and medium mechanical resistance.
- *Termofond C12T* is used for mold castings, has poor corrosion resistance but reasonable mechanical resistance.

3.5.15.2 Magnesium

Magnesium is silver white, its melting point is 650°C, its tensile stress breaking load is 80–120 N/mm^2, and it is the lightest of all the metals used in industry. Its alloys are defined as ultralight and have good mechanical characteristics. These alloys use Al, Zn, Si, and Ni; their specific gravity ranges from 1.7 to 2 kg/dm^3, they are moderately workable by machine tools, have good pourability and corrosion resistance, on average their Brinell hardness is 50 HB, and they have a tensile stress breaking load of about 200 N/mm^2. They are available on the market in sheets, sections, wires, pipes, and forge ingots. On account of their lightness, they are used in aeronautics, for motorbikes, and electrical appliances. Among these ultralight alloys are the following:

- *Elektron*, also called *Atesina*—Besides magnesium, it contains 3%–10% Al and up to 3% Zn. This alloy improves its tensile stress breaking load that reaches 380 N/mm^2 when undergoing plastic working.
- *Magnalium*—This alloy has properties similar to brass when it contains 10% Mg, and similar to bronze when it contains 20% Mg.

3.5.15.3 Titanium

Titanium has a white silver color, its melting point is 1725°C, specific gravity 4.5 kg/dm^3, tensile stress breaking load 460 N/mm^2, and HB (Brinell hardness) 80–100. It has excellent mechanical characteristics and is corrosion resistant. In alloy with 4% V and 16% Al, it achieves Rm = 1000 N/mm^2 when subjected to thermal treatment.

Titanium alloys have a tensile stress breaking load almost double the base metal. The alloy, Rem-Cru 130 A, with 8% Mn, has an Rm of 1000 N/mm^2. This metal also alloys with Al, St, Zr, Cr, V, and Mo and is used in aeronautics, missiles and engines (gas turbine components).

3.5.15.4   Nickel

Nickel is gray with shades of green, its melting point is 1445°C, specific gravity 8.5 kg/dm^3, tensile stress breaking load Rm 600–1000 N/mm^2, and 80–200 (HB Brinell hardness). It is strong, pliable, malleable, forgeable and weldable, as well as air unalterable, fresh- and sea-water resistant, hardly vulnerable to acids. It is an element of many stainless alloys (nickel–chromium steel). The maraging steels, used in the aerospace industry (Rm = 2000 N/mm^2), contain 18% Ni.

- *Monel* contains 65%–70% Ni and 25%–30% Cu, and is used for pump parts, filters, and valves in corrosive environments.
- *Costantan, nickeline, manganin* contain different percentages of copper. They are used for electrical resistors; in particular, costantan is used to produce thermocouples.
- *Nichromium* (80% Ni and 20% Cr) is used to produce thermocouples.
- *40-Inconel* (85% Ni, 13% Cr, 6.5% Fe) is oxidation and corrosion resistant up to 1100°C, it is used to make gas turbine components, foundry vessels, and for thermal treatments.
- *Hastelloy* (82.5% Ni, 24% Mo, 3% Fe) is used in the nuclear industry.
- *Nimonic* (37%–74% Ni, 41%–1% Fe, 18%–20% Cr, 2%–20% Co, and other elements such as Ti, Al, etc.) is used for gas turbine components such as flame tubes, turbine blades, valves for internal combustion engines, diesel engine parts, and casting molds.

3.5.15.5   Chromium

Chromium is a shiny white metal, its melting point is 1875°C, specific gravity 7.2 kg/dm^3, 70 HB (Brinell hardness). It is fragile when cold, deformable when hot, is corrosion resistant, and neutral to all gases at room temperature. It alloys with Al, Fe, Ag, Ni, Co, and Pb. In alloys with Ni, whose percentage may range from 0.5% to 25%; it is used as a protective lining against corrosion, and in stainless steel.

3.5.15.6   Vanadium

Vanadium is a silvery white metal, its melting point is 1726°C, specific gravity 5.70 kg/dm^3, HV 70 (Vickers hardness). It is a rare element, with good strength, pliability, and corrosion resistance. It is alloyed with iron to produce steel that can resist superheating. If reduced to powder it can be compressed and vacuum sintered at 1400°C–1500°C.

3.5.15.7   Cobalt

Cobalt is a light gray metal, its melting point is 1493°C, specific gravity 8.9 kg/dm^3, 124 (HB Brinell) hardness. It is a relatively tough element, pliable and strong, highly magnetic, and with good mechanical properties. It alloys with Cr and W to make stellites, materials that are wear and corrosion resistant. It is used for high-speed steel tools.

3.5.15.8   Tungsten (Wolfram)

Tungsten is a silvery white metal with the highest melting point of 3140°C, 19.3 kg/dm^3 specific gravity and 270 HV (Vickers hardness). It is used, at various percentages ranging from 12% to 23%, in alloys with Fe, C, and other elements such as Cr, V, Mn, and Mo for high-speed steel (tools). As a carbide it is used, together with cobalt carbide, to produce very hard small plates through sintering (Widia). These used in machine tools, permit high cutting velocity.

3.5.15.9 Molybdenum

Molybdenum is a silvery white metal, its melting point is 2662°C, it is quite hard, malleable during hot and cold processing, corrosion resistant, and has excellent electrical conductivity. It is used to alloy some special Ni–Cr–Mo steels. When reduced to powder and sintered, it is used as a high temperature electrical resistor in cathode tubes, incandescent lamps and quartz lamps.

3.6 AN OUTLINE OF POLYMERIC MATERIALS

The widespread use of these materials in the mechanical industries requires brief notes on only the most common, describing some of their properties and possible uses.

- Plastic materials are organic substances, also called synthetic resins, with a macromolecular structure obtained by chemically binding simple molecules called *monomers*.
- When monomers bind with two or three other molecules they produce large molecules called *polymers* (when made of identical monomers) and *copolymers* (when they are the result of two or more different monomers).
- Polymers may be formed by *polymerization*, *polyaddition*, and *polycondensation*. In the first two cases the simple molecules associate without eliminating other products; in the third, products are eliminated (usually water).

The degree of polymerization influences mechanical, technological, physical and chemical characteristics and also affects the plastic thus produced.

3.6.1 Polymer Classification

The huge variety of plastics requires classifying somehow, even if there has been no unification.

3.6.1.1 Origin-Based Classification

A classification based on origin groups these materials as follows:

- Natural plastics
- Artificial plastics
- Synthetic plastics

Natural plastics have different origins:

- Animal—bovine and ovine horns, ivory, tortoise, shellac (insect secretion)
- Vegetable—gum arabic, copal (from mineral deposits), dammar (from trees), turpentine distilled from the California pines, sealing wax (colophony/rosin based), carnauba wax (or leaf-wax)
- Mineral—bitumen (from oil distillation), asphalt (crude bitumen), mineral wax (wax from mineral deposits), ceresin (from refining ozokerite), montan wax (from extracting bituminous lignite), paraffin (mixture of aliphatic hydrocarbons), amber (from tree resins)

Artificial plastic materials can derive from:

- Cellulose—celluloid, cellulose butyrate acetate, acetate, cellulose propionate, ethyl-cellulose, cellophane
- Casein, galalith
- Polyisoprene—ebonite and the semi-ebonites

Among the synthetic plastics there are polystyrene, its ABS copolymers and resins, polymethil-methacrylate, polyvinylchloride and its copolymers, polyamides, polyethylene, ionomers, polypropylene, fluorinated resins, polyacetal, polycarbonate, modified polyphenylene oxide, polyimides, phenolic resins, urea resins, melamine resins, allyl resins, polyester resins, epoxy resins, polyurethane silicons, and reinforced plastic.

3.6.1.2   Classification Related to Temperature Effects

Plastic materials can also be categorized as thermoplastic and thermosetting substances:

* Thermoplastic materials become plastic by heating and are hot-working.
* Thermosetting materials, after heating and remaining at that temperature for a certain length of time, become permanently hard, a state that cannot be altered.

3.6.1.3   Classification According to Physical Characteristics

Plastics can also be classified according to their physical characteristics:

* Fibrous polymers
* Elastic polymers
* Plastic polymers

Natural fibers such as wool, silk, linen, jute, etc., which are macromolecular fibers, must be included with fibrous polymers. Among the synthetics, the cellulose fibers of rayon must be cited: nylon, delfion, lylion, perlon (obtained by polycondensation of polyamids), terital (polyester), vinyon (copolymer), euraclil, leacril, orlon and acrilan.

Elastic polymers are usually called rubbers or sometimes elastomers, and can either be natural or artificial. The materials used are vulcanized, that is, rubbers heat-treated in autoclave or in molds (vulcanization).

In the mechanical industries, plastic polymers are highly rated, especially cellulose derivatives. Celluloid possesses great tensile strength, high resilience and is easy to work, yet its high flammability is a hindrance to its use. Another widely used material is cellulose acetate, used for film; it is not flammable, but has low resilience.

3.6.2   Synthetic Resins

3.6.2.1   Macromolecule Structure

Resins are used because their mechanical and thermal properties result from their macromolecular structure.

* Linear macromolecules, formed by a linear chain of monomers that range from 1,000 to 10,000 units and are thermoplastic.
* Reticulated macromolecules made of polymeric chains. When there are strong bonds between main chains, the material's characteristics depend on the number of these bonds, reducing movement of the various macromolecule segments and of one chain compared to another; this reciprocal movement occurs even at high temperatures. Poorly reticulated materials are elastic over wide temperature ranges. Strongly reticulated materials are very rigid even at high temperature and are not plasticizable.

3.6.2.2   Production

Three chemical processes are used to obtain these macromolecular compounds:

* Polymerization
* Polyaddition
* Polycondensation

The resins obtained through these processes must have the following complementary additives added:

- Plasticizers (phthalates, glycerine, camphor) add flexibility
- Lubricants (stearates or stearic acid) aid flow and easier molding
- Stabilizers (metallic salts) avoid macromolecule degradation at high temperatures
- Pigments, which are organic or inorganic insoluble substances, give color and make the mass opaque or translucent
- Fillers (made of wood flour, fiberglass, asbestos, mica, calcium carbonate, etc.) give the mass the desired properties and characteristics
- Anti-deoxidants reduce the degradation of polymeric chains, caused by oxygen in the air, and lessen decoloration and yellowing

3.6.3 MAIN PROPERTIES OF POLYMERIC MATERIALS

3.6.3.1 Mechanical Properties

These are revealed through tests carried out according to international norms that refer to:

- Tensile strength in N/cm^2
- Elongation percentage
- Tensile strength elasticity modulus in N/cm^2 (norms UNI 5819, ASTM D.638-72, DIN 53455)
- Impact strength (resilience) in J/cm^2
- Rockwell hardness

3.6.3.2 Thermal Properties

- Vicat Degree (°C) defines the temperature at which a needle with a set cross-section, subjected to a certain load, penetrates the material to a set depth as its temperature gradually increases. This temperature refer to softening under load.
- Martens Degree (°C) defines the temperature at which a test bar bends under a set load as temperature gradually rises (norms UNI 4281—DIN 53458).
- Linear shrinkage percentage is the linear shrinkage a test bar undergoes when removed from the mold after molding and left to cool down at room temperature for a set minimum time (norms UNI 4285—ASTM D696-70 53464).
- Linear expansion factor in cm/°C (norms UNI 6061—ASTM D 696-70).
- Mass thermal capacity (specific heat) in kilojoule per kilogram degree Celsius (norm UNI-CIM 0025).
- Combustibility is the flashpoint of a material at high temperatures (norms UNI 4286, 4287–ASTM 635-74).
- Melt index (kg/10 min), is the mass of material at set temperature and load that is expelled through a nozzle of fixed dimension in a fixed time. This measures material fluidity (norms UNI 5640—ASTM 1238-73).

3.6.4 COMMONLY USED POLYMERS

This section deals with the most common polymers with reference to their use and distinctive properties.

3.6.4.1 Cellulose Resins

- *Celluloid*: high tensile strength, excellent resilience and workability, limited use due to high flammability
- *Cellulose acetate*: non-flammable with excellent resilience, used to make film for motion pictures and photographs, and also for toys

- *Ethylcellulose*: excellent dielectrical rigidity
- *Cellulose propionate*: good mechanical characteristics, used in auto accessory molding, and for television and radio parts
- *Cellulose acetate butyrate* (CAB): obtained by cellulose esterification with acetic acid and anhydride, and with butyl acid and anhydride; when plastified with phthalic esters it becomes a transparent thermoplastic, strong and weatherproof
- *Cellulose nitrate* (CN): also known as nitrocellulose, produced by cellulose esterification with nitric acid. The esters with nitrogen content are used for explosives and paints
- *Formaldehyde casein* (CS): an artificial plastic, also known as galatite, produced by hardening casein with formaldehyde, insoluble and infusible; heat molded at softening temperatures of 170°C–180°C, sensitive to humidity and heat, good mechanical properties and low voltage dielectrical stiffness; opaque, cream colored, dyeable; used to produce buttons, combs, buckles, umbrella handles, door handles, fountain pen casings, etc
- *Vulcanized rubber*: an artificial material of rubber hardened by sulfur vulcanization, also known as ebonite; a thermoelastic, softens when heated, resembling elastic rubber; because of excessive hardness can only be machined with Widia; good chemical strength and weatherproof; used for antacid tanks, battery case linings, wind instrument mouthpieces, medical equipment, pipe mouthpieces, etc

3.6.4.2   Styrene Resins

Most are obtained by polymerization of vinyl benzene (also known as styrene), ethylene and propylene.

- *Polystyrene* (PS), also known as styrofoam, has low specific gravity and good compression resistance; used for packing and thermal insulation

There are many polystyrenes with mechanical, thermal, and other specific properties: odorless, lightproof, rapidly pressable, fluid, different characteristics for various uses. Moreover, polystyrene can be mixed with its copolymers to produce materials with greater impact strength.

- *Styrene acrylonitrile copolymer* (SAN): styrene and polymerized acrylonitrile; rigid thermoplastic resin, impact resistant, transparent and solvent-proof
- *Styrene butadiene copolymers* (SB): polymerization of styrene and butadiene in emulsion. Proportion dependent, they can behave as thermoplastic resins and are used as additives; those consisting of 70%/30% butadiene styrene have elastomer characteristics and are used for synthetic styrene rubbers
- *Styrene-elastomer copolymers*: polymerization of styrene with an elastomer in solution; they possess high impact strength
- *ABS resins*: acrylonitrile, butadiene, and styrene; obtained from emulsion polymerization of styrene and acrylonitrile with polybutadiene, or by heat-activated mixture of copolymers butadiene acrylonitrile and styrene acrylonitrile. They are translucent thermoplastics, soft and highly impact and heat resistant

3.6.4.3   Acrylic Resins

Acrylic acid derivatives producing various copolymers for paints, enamels, emulsions, and starches. They are grouped into polyacrylates, polymethacrylates and polyacrylonitrile.

- *Polymethacrylate* (PMMA) is produced in suspension with the monomer dispersed in water and rapidly polymerized at about 100°C at atmospheric pressure with radical activators. Via this process, molding and extrusion resins can be obtained

- *Polyacrylonitrile* has no plastics applications, but used to make fibers and synthetic thread
- *Polyvinylchloride* (PVC) is produced by suspension polymerization or water emulsion; thermoplastic, rigid, transparent, solvent-proof, non-flammable resin

3.6.4.4 Polyamide Resins

There are numerous polymers and they are divided into two groups: linear polymers obtained from an aromatic diamine reacting with an aromatic dianhydride, and reticular polymers.

The thermal characteristics of the first group (linear polymers) permit them to conserve their mechanical properties up to 500°C: good stiffness, excellent electrical and chemical properties, and wear resistance; molding only via special processes.

Polymers in the second group (reticular polymers) are well suited to molding.

A host of polyamide compounds (PA) can be made, each by a different process. The most common uses of polyamide resins are thread extrusion, injection molding, or as components of paints and inks.

- *Polyethylene* (PE) is obtained from ethylene monomer polymerization; thermoplastic, translucent, soft, semi-crystalline resin, impact resistant but low solvent resistance.
- *Ionomers* is obtained from high-pressure ethylene copolymerization; in the resulting copolymer, carboxylic ionization is induced to produce steady reticulation at room temperature, as occurs in thermosetting. These materials are remarkably hard and stiff at ordinary temperatures, becoming fusible once again with heat and remoldable, like thermoplastic materials. They have good impact strength, chemical resistance and fairly good dielectric characteristics. They are used for packaging, flasks, and bottles.
- *Polypropylene* (PP), also known as Moplen is made from propylene polymerization melted in an organic vehicle (naphtha) with a catalyst (titanium tri-chloride and alkyl-aluminum); thermoplastic, translucent resin with a regular crystalline structure, heat proof, tough, and with low specific gravity.

3.6.4.5 Fluorinated Resins

These contain a high quantity of fluoride bound to the carbon chain.

- *Polytetrafluoroethylene* (PTFE), also known as Teflon is obtained by polymerizing tetra-fluoroethylene in a water emulsion with catalysts; thermoplastic with crystalline structure, translucent, soft, non-flammable, difficult to work, not strong, used for electrical insulation, for chemically inert protective linings
- *Polychlorotrifluoroethylene* (PCTFE) is produced by polymerizing trifluorochloroethylene in a water emulsion with oxidizing activators; thermoplastic, transparent, soft, non-flammable, not strong, heat proof
- *Polyfluoroethylenepropylene* (PFEP) is a thermoplastic with a weak crystalline structure, translucent, soft, non-flammable, not strong, easy to work
- *Polyvinylfluoride* is obtained by high pressure (1000 bar) polymerization in water dispersion; thermoplastic, highly weather proof, used for protective coating of external building panels
- *Polyacetates* is produced by polymerizing pure and anhydrous formaldehyde in heptane solution with acid initiators; thermoplastic, rigid, very tough, strong with non-transparent crystalline structure
- *Polycarbonates* (PC) is only used industrially, produced from bisphenol by definile carbonate reverse esterification; thermoplastic resins, very impact resistant, tough, heat proof, inert, transparent, non-flammable
- *Polyarylether* is a thermoplastic with high resistance to temperature of distortion, excellent resilience and chemical resistance; for molding, extrusion, and vacuum-molding

3.6.4.6 Phenolic Resins

Known as Bakelite (PF), this refers to all polymers obtained from mixing phenol with other substances (aldehydes, ketones, furfural, and natural resins). Many of these products are used to make adhesives, paints, varnishes, and putties; of no use as plastics. The most common phenolic resin is obtained from condensing phenol and formaldehyde.

These resins are also called phenolic plastics. They can be thermoset and non-infusible; their fusibility, however, depends on condensation reaction.

Phenolic molding resins are obtained from mixtures of novolac and hexamethylenediamine which, added to other reinforcing materials, produce a great number of impact-proof resins and for electrical applications.

A great variety of laminated plastics for decoration, electrical insulation, and protection are produced from resols by adding reinforcing materials dissolved in alcohol.

3.6.4.7 Amino Resins

Amino resins are produced by condensing formaldehyde with amino-like substances (urea, melamine, aniline, etc.). Also known as amino plastics, these resins are used as protective coating adhesives, paper additives, and starches, but mostly these resins are used in molding.

- *Urea resins* (UR) is produced by condensing urea in alkaline aqueous solution with excess formaldehyde, in addition to pigment fillers (cellulose), plasticizers, and lubricants; produces a light-colored resin that is thermosetting, chemically inert, and has good electrical properties
- *Melamine resins* (MR) is produced by condensing melamine in aqueous solution with excess formaldehyde in addition to fillers (cellulose, asbestos, silica, etc.); produces a light-colored resin that is hard, chemically inert, thermosetting, and has good electrical properties

3.6.4.8 Allyl Resins

Allyl resins, also known as DAP, are obtained from allyl alcohol, they are thermosetting resins with good electrical properties even at high temperatures. Reinforced with glass, fabric, or paper fibers, they are chemically and mechanically resistant and they show a good resistance to the heat. They are used in the electrical, electronic and space industries.

3.6.4.9 Polyester Resins

There is a wide range of polyester resins used as thermosetting materials and thermoplastic resins for fibers, film, paint components, lacquers, and coatings. The two main groups of polyester resins are:

- Saturated polyesters are obtained by reverse esterification followed by condensation to polymerization level 80; this produces linear structured fabric fibers and membranes
- Unsaturated polyesters are made from a mixture of monomer styrenes and unsaturated polyester resins, they are used to impregnate fabrics or glass fiber felt

3.6.4.10 Epoxy Resins

Epoxy resins (EP) with a complex reticular structure, are produced indirectly from epichlorohydrin and bisphenol. Some high molecular compounds are used to produce plasticizers, stabilizers, adhesives, putties, and protective coatings.

Because of their reticular structure, epoxy resins are infusible and also used to encapsulate electrical equipment. Complete cross-linkage can be achieved with additives that react with the hydroxyl or epoxy groups, or by adding catalysts that induce a condensation reaction between the epoxy groups. Amines such as diethylenetriamine and meta-phenylenediamine are used as reactants, and triethylamine tertiary amine is used as a catalyst.

3.6.4.11 Polyurethanes

Polyurethanes (PUR) are used for foams, synthetic rubbers, fibers, filaments, paint resins, adhesives, coatings, and thermoplastic resins for molding and extrusion. The foams are obtained by the rapid reaction of polyethers and toluene di-isocyanate with amino catalysts. Rubbers are obtained from polyesters and naphthyl-isocyanate, fibers from glycols and hexamethylene di-isocyanate. Thermoplastic resins are ozone-, solvent- and oil-proof. They are tough, flexible, and abrasion proof.

3.6.4.12 Silicones

Silicones can be grouped into three different categories: oils, rubbers, and resins. They include polyorganosiloxanes, silicon resins (polysiloxanes) that are temperature stable and weatherproof, have good dielectrical characteristics and are hydrophobic. They are used as varnishes to impregnate electrical equipment for temperatures up to 600°C.

3.7 OUTLINE OF COMPOSITES

The history of composites started with the first commercially available type of fiberglass. By the end of the 1970s, advanced composites were produced that were reinforced with carbon, boron and Kevlar (manufactured by Du Pont) fibers instead of glass fibers. Since then, their use in mechanical construction and aeronautics has increased steadily. In today's automobile manufacturing, they are a significant proportion of the materials.

Composites are artificial materials made from two or more materials. Because of their special characteristics, they are often preferable to metals, as demonstrated by their widespread use in the aeronautics, nautical, and automobile industries and in telecommunications and sports.

These materials typically have two components:

- A homogenous part called the matrix
- A fibrous or particle part called the reinforcement

The fibrous part can have different shapes that classify them into two broad categories:

- Particle composites, which are reinforced by small, similar sized particles
- Fibrous composites, reinforced by round sections whose length is much greater than their width

Generally, the matrix is made of ceramic, metal, or plastic. Nowadays, polymer matrices are the most successful composites.

The most widely used polymer composites are shown in the following figures.

- Matrices with random discontinuous fibers whose mechanical characteristics are the same in different directions (Figure 3.3a)
- Matrices with parallel discontinuous fibers (Figure 3.3b)
- Matrices with parallel continuous fibers (Figure 3.3c)

Regarding polymer matrix composites, a distinction may also be made between thermoplastic matrices and thermosetting matrices.

At present, thermosetting matrices are largely used in the structural field, although it is commonly thought that in the future, thermoplastic matrix composites predominate.

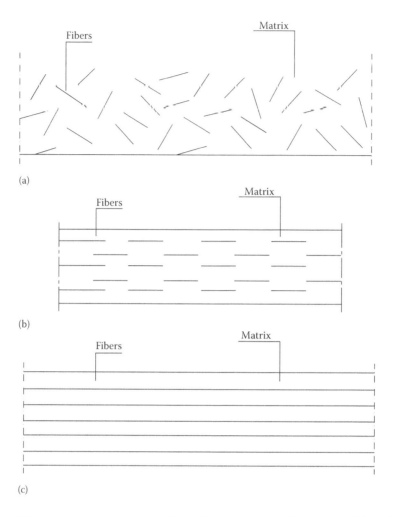

FIGURE 3.3   (a) Random discontinuous fibers; (b) parallel discontinuous fibers; (c) parallel continuous fibers.

In manufacturing composite fibers, the principle is to produce faultless parts to improve their durability. Thanks to their small size, these fibers show out-of-the-ordinary perfect structure that, together with the intrinsic properties of the constituent materials, ensures high mechanical durability, high elastic modulus, very low specific gravity and linear elastic behavior to breaking point.

3.7.1 Matrices

Matrices provide a stable shape to the designed part; otherwise, despite their high values of elastic modulus and durability, they would have no practical use. Therefore, apart from including the fibers, the matrix gives shape to the part and also protects the fibers from the external environment. As already mentioned, there are metallic, ceramic, and organic matrix composites.

Those with metallic matrices possess high resistance to temperatures (300°C), greater hardness, good isotropy, and better compression. They are flammable, though they are good electrical conductors. Metallic matrices are generally alloys of magnesium, aluminum and titanium.

Until now, ceramic matrices have been used only in special cases.

Polymeric matrices are those that possess the best mechanical durability and an excellent elongation behavior; among these matrices, the polyesters are most widely used.

Thermoplastic matrices have high deformation to the breaking point, high impact strength, fatigue resistance, durability, storage stability, short molding cycles, can be further transformed, and are easy to repair. These materials are poorly resistant to solvents, cannot be fiber coated because of the high viscosity of the matrix, and they are expensive compared to thermosetting matrices.

Thermosetting matrices have the following advantages: easily wettable, high thermal stability, and resistance to chemical, environmental, and solvent agents, low polymer creep value after cross-linkage, and lower costs compared to thermoplastics. These composites have shorter storage stability, longer molding cycles, low elongation to breaking values of the reticulated polymer, low impact strength, and are toxic.

Thermosetting matrices will be further outlined below. They are in a more or less viscous liquid state before use. In this state they do not undergo cross-linking, for which catalysts and hardening elements are necessary in polyester matrices and in other types of matrices, respectively.

3.7.2   POLYESTER MATRICES

Polyesters represent about 90% of thermosetting plastics employed in reinforced plastics. Because they cross-link easily at room temperature, they are widely used in the nautical industry and in the production of containers for liquids.

3.7.3   EPOXY MATRICES

Epoxy matrices have better physical, chemical, and mechanical properties compared to polyesters. Their excellent elongation to breaking point enhances their mechanical properties. They are more expensive than polyester matrices and harder to use. They are used in advanced aerospace aeronautics and in the sports industry.

3.7.4   PHENOLIC MATRICES

Because of their high heat resistance (250°C), phenolic matrices are widely used in the aerospace, electrotechnical, electronic, and automobile industries.

3.7.5   SILICONE MATRICES

Silicone polymers are inorganic polymers structurally similar to the carbon based ones, where silica takes the place of carbon. They are remarkably heat resistant up to 450°C–500°C. Composites made with these matrices are used in the components of supersonic planes and electrical appliances.

3.7.6   FIBROUS STIFFENING

The main function of fibrous stiffening is to improve durability. It produces anisotropy, though, because its resistance to tensile stress in the direction of the fibers is quite different to the perpendicular one. However, being artificial fibrous materials, single parts are made by aligning the fibers in the direction of the main forces.

Below is an outline of the properties typical of glass, carbon, boron, and Kevlar fibers.

3.7.7   GLASS FIBERS

A draw-plate with hundreds of holes of 1 mm diameter, placed at the bottom of a furnace, is used to draw the fibers. Before coiling them on reels, the fibers are coated so that each single fiber avoids reciprocal contact and there is correct adhesion between fiber and matrix.

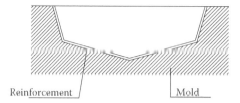

Reinforcement   Mold

Formation of composite by brushstrokes of resin on the reinforcement fiber

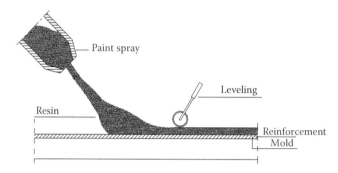

Formation of a sheet of composite material
by spraying the resin on the reinforcement and rolling of the upper surface

FIGURE 3.4   Formation of composite by brushing and by spraying.

The roving and yarn fibers can make felts or threads for both thermoplastic and thermosetting matrices.

The most widely used stiffening among low cost composites is made of E glass fibers that are cheaper than those of S glass. E glass fibers have 75,000 N/mm^2 Young's modulus, 3450 N/mm^2 tensile strength, 3.9 deformation percentile to breaking, 2550 kg/m^3 volume mass, and a cylindrical shape of 10 μm diameter. M glass has 110,000 N/mm^2 Young's modulus, while S glass is 30% more elastic and resistant compared to E glass. Finally, R glass shows equal heat resistance at 400°C to that of E glass at room temperature.

ABS, nylon, polyamide, polyolefin, and polyacetyl resins are among the thermoplastic resins to couple with glass fibers. Epoxy, polyether, and phenolic resins are used for thermosetting.

There are various fiberglass production technologies, but one of the simplest is to impregnate a felt pad or a mat laid on a mold with a brush, or spray and roller (Figure 3.4).

Among the most advanced techniques are extrusion and molding, but other techniques include coiling and pultrusion (a method similar to that of extrusion for resins, but the material in this case is not pushed but pulled).

3.7.8 CARBON FIBERS

Carbon fibers are obtained by graphitizing rayon or polyacrylonitrile (PAN) organic fabric fibers at over 2000°C in an inert atmosphere. During the process, the fibers are stretched to increase their Young's modulus, but this reduces their strength.

Types C1 and C3 are on the market: the first has high modulus and low strength carbon fibers (defined HM, high modulus); the latter has low modulus and high strength (defined HT, high tensile strength).

Compared to glass fibers, these have high elastic modulus but low volume mass and thermal dilatation coefficient. They are lighter, with special mechanical characteristics.

3.7.9 BORON FIBERS

Boron fibers are obtained by inserting a 0.01 mm diameter tungsten filament at 1100°C into a mixture of boron trichloride BLC 3 and hydrogen. The boron produced in the consequent reaction ($2BCl_3 + 3H_2 = 2B + 6HCl$) settles on the tungsten filament. The boron fiber diameter is 0.1 mm.

These fibers have modulus of elasticity and tensile strength greater than steel. However, as these fibers' diameter is ten times larger than the others', their use is limited, as they cannot be bent much because they risk breaking.

Boron fiber is marketed with a coating of silica chloride and is called Borsic, while boron aluminum composite is sold in sheets or tapes coated with a resin for brazing and used for the selective reinforcement of metallic structures.

3.7.10 ARAMID ORGANIC FIBERS

The most widely used fiber is Kevlar by Du Pont, which has good mechanical characteristics and low volume mass, reasons for its use in the most advanced composites.

3.7.11 ADMISSIBLE LOAD AND ELASTICITY MODULUS

Taking the simplest composite of continuous fibers placed parallel to each other and incorporated into a one-dimensional matrix, knowing the percentage and kind of components, the admissible load K_c and elasticity modulus E_c of the material in the direction of the fibers can be found using the following formulae:

$$K_c = \frac{K_m V_m + K_f V_f}{V_m + V_f},$$

$$E_c = \frac{E_m V_m + E_f V_f}{V_m + V_f},$$

where K_m and K_f are admissible load of matrix and fiber, E_m and E_f are the corresponding elasticity modulus values, and V_m and V_f are the matrix and fiber volumes.

The above parameter determination is based on the variation and shape of the composite components (matrix and fiber) undergoing load in the direction of the fibers.

3.7.12 MULTILAYER LAMINATES

The development of multilayer laminates arose from requirements of the aeronautics industry where these materials are now in widespread use.

Producing a laminate requires an impregnated material cut into sheets of the desired shape, layered according to a prescribed fiber direction and number of layers. They are then pressed, generally in an aluminum or steel mold with smooth surfaces to obtain a satisfactory end product finish. The laminate is then placed in a vacuum bag (650–700 mm Hg) of nylon sheeting and put in an autoclave for polymerization. This process could take place in the mold, although the autoclave is operationally easier. Polymerization is carried out at temperatures and pressures that vary according to the case. For instance, polyamine resin composites require a temperature of 590K and a pressure of 1.5 MPa (about 15 atm).

The following table gives a general idea of the mechanical characteristics of different materials according to weight:

Material	Specific Gravity	Young Modulus (GPa)	Failure Tensile Stress (MPa)	Yield Stress (MPa)	Modulus/ Specific Gravity 109 m	Failure Tensile Stress/Specific Gravity 106 m
SAE 1010 steel	7.87	207	365	303	2.68	4.72
AISI 4340 steel	7.87	207	1722	1515	2.68	22.3
AL 6061-T6 aluminum alloy	2.70	68.9	310	275	2.6	11.7
AL 7178-T6 aluminum alloy	2.70	68.9	606	537	2.6	22.9
Ti–6 Al–4 V titanium alloy	4.43	110	1171	1068	2.53	26.9
High-strength carbon fiber–resin	1.55	137.8	1550	–	9.06	101.9
High modulus carbon fiber–resin	1.63	215	1240	–	13.44	77.5
Fiberglass E–epoxy	1.85	39.3	965	–	2.16	53.2
Kevlar fibers 49–epoxy	1.38	75.8	1378	–	5.6	101.8
Boron fibers–6061 Al	2.35	220	1109	–	9.54	48.1

3.8   OUTLINE OF CERAMIC MATERIALS

Mechanical construction also uses ceramic materials. The following paragraph is a brief outline to give engineering students an idea of what they are and how they are used. For greater depth, it is advisable to refer to specialist courses or specific literature. The composition of ceramic materials includes metallic and non-metallic components with ionic or covalent bonds (oxides, carbides, nitrites, etc.).

Traditionally, ceramic materials have always been used, especially where heat protection and resistance were necessary. More advanced ceramic materials have been used in various fields ranging from semi-conductors to dental prostheses. The difference between the traditional and the more advanced ceramic materials is due to their use and not their characteristics, because even the traditional ones have evolved to the point of being considered advanced.

The main characteristics of ceramic materials are high heat resistance, hardness, and high elastic modulus, but low toughness. Mechanical resistance varies according to type ranging from 80 N/mm^2 (magnesium oxide, glass silica) to a maximum of 1000 N/mm^2 (silicon nitride). The elasticity modulus is variable as well as is breaking point; all parameters are closely related to the percentage of porosity volume.

Current ceramics production technology is making zirconium alloys with characteristics that vary according to the alloying component as with steel alloys.

Advanced ceramic components are obtained through a sintering process including mold filling, firing, and expulsion, as they cannot be melted, or because of the high temperatures required, or because of decomposition.

Because of their characteristics, ceramic materials are used to make gaskets, heat exchangers, ballistic protection, thermal shields and cutting tools, unlubricated ball bearings, mountings for the electronics industry, piezoelectric materials, etc.

Mechanical characterization is related to the compression resistance, bending at two or three points, Young's modulus, hardness, breaking, and fracture point.

Ceramic materials are therefore ideal for static applications at extremely high temperatures (turbine blade protection). Their toughness is relatively low, and this makes them vulnerable in terms of fatigue resistance.

The following table shows some data on the mechanical characteristics (compression breaking load, tensile strength, Young's modulus, density) of some ceramic materials in widespread use in industry.

	σ_{rc}	σ_{rt}	E	ρ
	[MPa]	[MPa]	[GPa]	[g/cm^3]
MgO – PSZ	1200	300	200	5.9
HP Si3N4	2000–3500	400–580	280–310	3
RB SiC	2000–3400	310	410	3.10
HP SiC	1000–1700	400	350–440	3.15
Al2O3 999	2200–2600	380	330–400	3.96

Tailoring physico-mechanical characteristics in the way in which new steel alloys are created (adding alternative ligands) is fascinating for materials scientists, who will undoubtedly come up with new ceramic materials for more challenging applications.

4 Characterization of Metals

4.1 INTRODUCTION

Metals are the most notoriously exploited materials in mechanical construction—iron, copper, aluminum, and magnesium alloys; more specifically, steels, cast irons, bronzes, brass, and light as well as ultra light alloys. Moreover, non metallic materials also have to be taken into account, such as wood, plastics, rubbers, ceramic materials and compounds (fiberglass PRVF, carbo-resin C1 PRFC1, carbo-resin C3 PRFC3, Boron resin, Aramidiche resin).

Each of these big classes of materials has different physical–mechanical properties, crucial in defining their best possible utilization in the sphere of construction. There are several tests, all conducted in laboratories, that enable us to spot their main features; they will only be dealt with briefly, as they are already topics of other courses.

All the tests that mechanical construction materials have to pass are divided into mechanical and technological. There are also tests aimed at assessing the behavior of machine components in terms of strain or deformation tolerated without the components actually breaking (non-destructive tests).

Mechanical tests are carried out by subjecting materials to increasing force, until they break down; in this way, we are able to determine their resistance and deformability. The numerical outcome enables us to express judgments about the condition of materials and their possible utilization. Obviously, such tests take into account their effects and seldom their causes, and so they may be further complemented with other proofs, analyses, and surveys.

Technological tests, however, enable us to determine the aptitudes materials have in terms of different processes. These tests, though required by law, are nevertheless limited, and the numerical results are dimensionless, unlike those produced by mechanical tests. The results obtained by technological tests can only express subjective judgments.

The mechanical characterization of any material requires knowing its resistance parameters (yield point, breaking load), as well as those attesting to its deformability. If we consider two elements of equal geometrical dimension stressed by an equal tensile stress load, we understand how deformations differ if the material is steel or rubber. When there is a deformation along the axis of the load, the parameter that links stress with deformation is the elasticity modulus E of the material, which is totally different for the two materials.

To characterize the deformation of the material, however, the E value might not be sufficient. As we shall see, it could result from a tensile stress test that connects deformations according to the axis of the load, but it is also necessary to define the normal movements of the axes, that is to say, link the deformations related to the normal directions of the load axis to the corresponding stresses.

The parameter that allows such connection in the elastic field is the Poisson modulus ν; for isotropic materials this is linked to the tangential elasticity modulus G through the ratio:

$$G = \frac{E}{2(\nu + 1)},$$

which for steels ($E = 21{,}000$ N/mm^2, $\nu = 0.3$) is about 8,000 N/mm^2.

So, if x is the direction along which the load works:

$$\varepsilon_y = \varepsilon_z = \nu\varepsilon_x = \nu\frac{\sigma}{E}.$$

These parameters, as will be noticed later, are indispensable for coping with any calculation, either through traditional or more updated methods as FEM (Finite Element Method).

Conventional mechanical tests of tensile stress, compression, bending, shearing, and torsion, are carried out according to the UNI norms, both in terms of the procedures and the dimensions of the test samples. UNI 555/1961 and UNI 1986 guidelines indicate how big the sample bars of the materials being tested have to be.

4.2 TENSILE STRESS TESTS

Tensile stress tests are carried out on circular section bars, with the following parameters (Figure 4.1): the cylindrical segment length L_c, the workable segment L_0, and the heads fixed to the test machine. The following ratio is at work:

$$L_c = L_0 + 2 d_0,$$

d_0 being the diameter of the test bar.

   UNI 551/61 Table UNI 551/61 classified test bars as:

* Short proportional, those with $L_0 = 5.65 \sqrt{S_0}$ (that is, $L_0 = 5 d_0$)
* Long proportional, those with $L_0 = 11.3 \sqrt{S_0}$ (that is, $L_0 = 10 d_0$)

Subsequently, UNI classified as "normal" only the test bars with $L_0 = 5.65$.

At the end of the test, these are the resistance coefficient values of greatest concern:

* Unitary yield point $R_s = \sigma_s = F_s/S_0$ N/mm^2
* Maximum unitary load $R_m = \sigma_m = F_m/S_0$ N/mm^2

where F_s and F_m stand for, respectively, the yield point and breaking load of the material.

   It should be pointed out that for many materials the yield limit is made evident by the discontinuity of the elongation load curve. In other materials such discontinuity might not come to the fore, and then the yield limit would be determined conventionally in the following way: test bars undergo gradually increasing stresses, and deformations are gauged each time the test load is taken off. After several tests have been performed as described, it occurs that when the load is taken off, a permanent

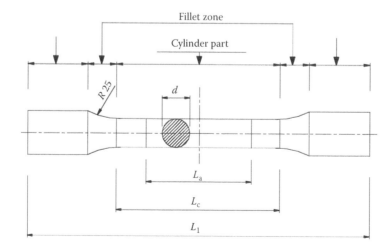

FIGURE 4.1   Circular section bars for tensile stress tests.

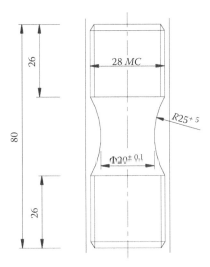

FIGURE 4.2 Test bars for grey cast iron.

0.2% deformation remains in the test bar; this is known as the conventional yield point, F_s (0.2%), and the corresponding unitary yield point is obtained through the ratio:

$$R_s = (0.2\%) = \frac{F_s(0.2\%)}{S_0}.$$

For gray cast iron castings (UNI 5007/62), test bars are obtained from cylindrical bars with a 30 mm diameter, cast separately: dimensions are shown in Figure 4.2, whereas for spheroidal cast irons, bars measure as in Figure 4.3.

For tensile stress tests, the types of test bars are shown in Figure 4.4, with the corresponding symbols and descriptions.

4.3 STATIC ELASTOPLASTIC CHARACTERIZATION

Prior to a brief description of the other tests, let us remind ourselves of how important the classic tensile stress test is, as it can provide useful information for predicting the behavior of materials, even under abnormal conditions.

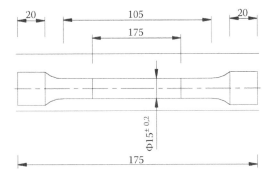

FIGURE 4.3 Test bars for spheroidal cast irons.

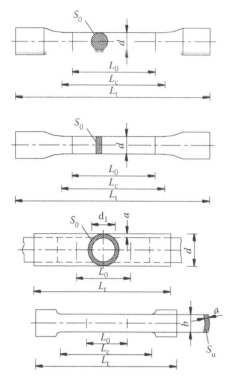

Symbol	Description
d	Diameter of the cylindrical specimen or external diameter of the tube
a	Thickness of the specimen or of the tube
b	Width of the plate specimen
d_1	Internal diameter of the tube
L_0	Initial length between the marks
L_c	Length of the part with equal section
L_t	Total length of the specimen
L_u	Length between the marks after failure
S_0	Original area of the specimen (before the beginning of the test)
S_u	Area of the specimen following the failure

FIGURE 4.4   Different types of tensile test bars.

The static characterization of materials is essential in terms of elastic and also plastic design. Unlike what used to happen until a few decades ago, nowadays design of plastic is spreading, subject to deep knowledge of the material's behavior at each phase. This design philosophy is increasingly important in all those high technology applications where it becomes necessary to reduce weight, bulk, and costs, even as regards significant improvements in safety standards.

In the end, because of the ability to predict with reasonable certainty the maximum load that might act on a structure, and knowing enough about how the material behaves when it is acted on, from yield to breaking point, it is quite reasonable to arrange that under an extreme load the structure approaches its breaking limit, rather than being far from it.

As a matter of fact, it is often costlier to oversize a structure, rather than adjust it to the loads it sustains and replace it fully or in part as soon as the "damage gauge" nears the breaking point.

Having pinpointed the need to characterize materials from a static point of view even beyond the yield limit, it is necessary to assess if and how much the traditional stress–strain curve assessment is appropriate. This curve (called the "engineering curve") is obtained by putting, in ordinate and abscissa respectively, the stress and deformation values obtained as follows from the monotonic tensile stress test on standard geometry bars as shown in Figure 4.5.

The tension is given by the ratio between the instantaneous load applied and the nominal resisting section, which is the one with the just-made test bar. The deformation is given by the ratio between the elongation ΔL_0 of the gauge length and its original length L_0.

These measurements generate the classic curve as in Figure 4.5, where the characteristic tensions are the yielding tension and the "ultimate" that, as we shall see, is definitely not the real maximum tension on the test bar.

These tension and deformation values correspond to the idealization that, moment by moment, at each point within the "workable" volume of the bar, tension is distributed totally independent of

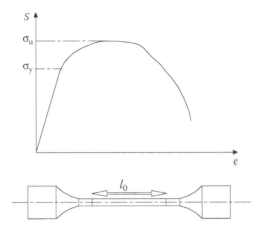

FIGURE 4.5 Tension engineering curve.

its position. This hypothesis is realistic only if yield is obtained at all the points in a cross section; in fact, after such an occurrence, the tension–deformation status begins to "thicken" in this cross section.

This phenomenon in a bar segment is called necking (the limited segment is indeed the neck of the distorted bar), and its trigger occurs close to the maximum load/stress condition, which on the engineering curve is identified when the ultimate tension is reached.

Under these conditions, it is clear that the deformation calculated from the elongation of the workable segment represents only an average value that locally can differ a great deal. Taking this into account, it is obvious that the characterization from the engineering curve is only significant in the elastic field.

In any case, it is worth noting that the tension for any characterization, once a yield or breaking criterion is established, is the equivalent (the one almost universally used is the Von Mises). This does not invalidate what we have been saying, as up to yield, the only non-null tension of the strain tensor corresponds to the axial tension defined as load/area.

Things change when yield is exceeded, because the localized necking begins. A more precise characterization in the plastic domain than the previous one, is feasible hypothesizing that stress and deformation are focused more in one cross section (the neck) than in any other, but that within this cross section distributions are constant and still characterized by a unique non-null value in the strain tensor and in plastic deformation.

According to these hypotheses, tension derives from the ratio between load and instantaneous area (diameter needs measuring throughout the test), whereas deformation must be reasoned differently. In changing from an average deformation reading over the total length to one localized at the neck, the expression that defines this deformation is as follows

$$\varepsilon_z = \frac{\Delta l_0}{l_0}\bigg|_{l_0 \text{ inf } \textit{initesimo}} = \int_{l_0}^{l} \frac{dl}{l} = Ln\left(\frac{l}{l_0}\right).$$

If it was possible to measure length l that the infinitesimal segment, l_0, adopts during necking, then it would be possible to assess total deformation according to this formula. However, plastic deformation in the neck area is so much greater than the elastic deformation, that it might be considered insignificant. This is all the more so because the characterization we are about to describe concerns only plastic behavior, so we can substitute total deformation by approximating it to the plastic one and applying a property of the former, that is, the incompressibility that follows any plastic deformation.

To preserve the length volume l_0 across the neck, a and a_0 being the outer radiuses of the distorted and undistorted segments respectively, we obtain

$$l\pi a^2 = l_0 \cdot \pi a_0^2 \Rightarrow \frac{l}{l_0} = \left(\frac{a_0}{a}\right)^2 \Rightarrow \varepsilon_z = \ln\left(\frac{l}{l_0}\right) \cong 2\ln\left(\frac{a_0}{a}\right).$$

It is clear that, by measuring the diameter of the necked section, from the hypotheses made, both the tension and deformation appropriate to a plastic characterization of the material through the so-called "true curve" is more realistic by far than the previous one, although not devoid of some inaccuracy.

In particular, the hypotheses that make the true curve characterization unrealistic are related to having only axial tension and deformation, as well as to the constancy of these distributions in the necked section. A more accurate characterization would quantify all the distributions within the stressed neck, which in reality is not monoaxial but triaxial. Once the complete tension status is known, it is easy to deduce the trend of equivalent tension and to use this datum to build a significant curve. A thorough assessment of the complete distributions of tension and deformation within the neck to date remains an unresolved issue, although some analytical models can reconstruct tension trends that are very close to the real ones. Without going into too much detail, because the triaxiality of the stress is due to geometrical modification, it is logical to expect it would come into play in the expressions of the much sought distributions; in particular, one of the worthiest models in this respect, by the American Bridgman, states the tension distribution laws as a function of the radial abscissa r of the neck:

$$\sigma_{eq}(r) = \frac{\sigma_{zAVG}}{\left(1 + \frac{2R}{a}\right) \cdot \log\left(1 + \frac{a}{2R}\right)}\left(1 + \log\frac{a^2 + 2aR - r^2}{2aR}\right)$$

$$\sigma_r(r) = \sigma_\vartheta(r) = \frac{\sigma_{zAVG}}{\left(1 + \frac{2R}{a}\right) \cdot \log\left(1 + \frac{a}{2R}\right)}\left(\log\frac{a^2 + 2aR - r^2}{2aR}\right),$$

where

- σ_{zAVG} is the average axial tension, that is, the instantaneous ratio between load and area
- R is the instantaneous curvature radius of the profile corresponding to the neck
- A is the maximum value r can have, or the instantaneous semi-diameter of the neck

The typical layout of an engineering curve as well as the corresponding true curve and equivalent tension curve is shown in Figure 4.6.

When apparent maximum tension is reached (ultimate tension) in an engineering curve, reduction of the resisting cross section means that the actual tension does not stabilize or decrease, but continues to grow, as does the plastic deformation. The average axial tension (upper curve) becomes almost linear just after necking starts, and grows until the breaking value σ_f, which is significantly greater than the ultimate tension (around twice or three times as much for some steels). This estimate, however, is slightly excessive; indeed, assessing the overall tensional status by using mathematical models such as the previous one, and then calculating the neck's

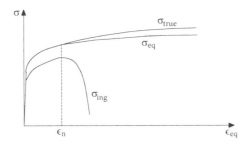

FIGURE 1.6 Tension engineering curve, true curve and equivalent curve.

ideal maximum tension, we obtain trends similar to that of the equivalent lower tension curve in Figure 4.6. The breaking values of the estimated equivalent tensions, as stated before, with reference to medium quality steels, turn out to be around 15%–30% lower than those for the average axial tension. Also, as far as plastic deformation is concerned, the values assessed locally on the cross section by the logarithm of the area percentage reduction become, under breaking conditions, almost an order of greatness higher than the average values calculated with the elongations over the bar's working length. For instance, for AISI304 steel test bars, with an identical nominal diameter of 9 mm and gauge length of 55 mm, extensibility is 15%–20%, whereas the plastic extensibility reaches around 130%.

4.4 PLASTIC CONSTITUENT LINK

In the plastic field it is no longer possible to define univocally a connection between stress and deformation tensors as in the elastic field; rather, there is, between the two quantities, an incremental relationship that can be reproduced analytically. The outcome of mathematically modeling such behavior is a set of differential equations that can only be dealt with by a code to finite elements.

The following relations completely define a model:

- The yield function, closely linked to the yield criterion, and then to the selection of the equivalent tension, which defines the domain of the possible tensional statuses:

$$f = f(\sigma, R, X) = (\varepsilon - X)_{eq} - R - \sigma_Y = 0$$

- The so-called "normality rule" or "flow rule," which states the ortogonality between the strain increment and the yield surface:

$$d\varepsilon^p = \frac{\partial f}{\partial \sigma} d\lambda$$

- Strain hardening laws or work hardening laws, which define yield change as plastic deformation increases. These exist in various forms in the literature, and are mostly necessary for fitting experimental data.
- The "consistency condition" that imposes a condition on yield so that plastic deformation can occur. The condition is that yield is independent of tension, or that a tension increase does not modify the shape or position of the yield surface (hardening depends either on deformation or plastic work).

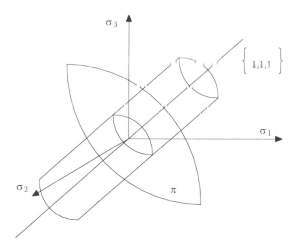

FIGURE 4.7 Cylindrical main stresses domain.

In its simplest form, yield is the difference between the equivalent tension (Von Mises criterion is almost always the only one used in the elastic field for construction materials) and yield stress. In this way, a material with no hardening is one whose "true curve" is only horizontal with height equalling yield stress. When a material hardens (as do all real construction metals), then its true curve is the one shown in Figure 4.6 and at every moment the equivalent tension is given by combining the yielding and hardening tensions, the former being obtained for several types of hardening identified by the corresponding laws. If we represent the domain of points that make yield null in the space of the main stresses, the result is a cylinder with its axis on the trisectrix of the positive octant (Figure 4.7).

Since, according to Von Mises, the equivalent tension is not affected by the hydrostatic stress status (a deflecting stress tensor and another obtained by its combination with a hydrostatic one, feature the same equivalent tension), the cylindrical domain can be easily replaced by its intersection with the plane with the above-mentioned trisectrix as its normal. This cross section is perfectly circular, and provides a simple visual interpretation of plastic behavior (Figure 4.8).

Whichever the stress increase tensor, in order for it to determine any plastic deformations, its deflecting component must have main tensions that locate a point lying on the aforementioned circumference, which in the meantime might have shifted or enlarged in the tension space.

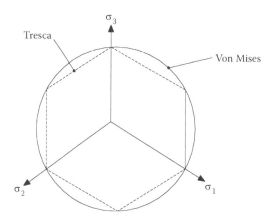

FIGURE 4.8 Tresca and Von Mises domains.

If the point representing the stress increase falls within the circumference, then there is an elastic discharge. At the start (with "fresh" material), this circumference is almost centered and has a radius equal to $\sqrt{2/3}\sigma_y$. As the load gradually increases from zero, as the point with the main deflecting stresses approaches the circumference, up to touching it, then further stress increments generate plastic deformations, which in their turn generate changes in the yield surface through hardening, so that circumference radius and translations increase.

Isotropic hardening, R (a scalar quantity), corresponds to radius increases, while kinematic X (a tensor, deflecting quantity), corresponds to translations of the center Figure 4.9.

Under plastic stress, the flow rule requires the deformation vector be perpendicular to the yield surface at the spot where the stress is applied, whereas the increase modulus of the plastic deformation vector (in the main reference system), is given by the term "$d\lambda$" also called "plasticity multiplier."

It is possible to demonstrate that this multiplier is also equal to a parallel one that distorts the equivalent tension (root of adding the squares of the main deformations), also called equivalent plastic deformation. By obtaining the stress tensor through the known elastic link and the elastic deformation (difference between total and plastic deformation), through some simple though realistic hypotheses, the plasticity multiplier $d\lambda$, or equivalent plastic deformation, can be expressed as follows:

$$\varepsilon_{eq} = \lambda = \sqrt{\frac{2}{3}\left(\varepsilon_1^2 + \varepsilon_2^2 + \varepsilon_3^2\right)}.$$

In expressing yield, the Von Mises operator who associates a scalar "equivalent" to a tensor, acts on the difference between stress tensors and kinematic hardening.

Deriving yield to a major stress, and considering that now the plasticity multiplier is an equivalent plastic deformation, the following expression of the normality rule can be written

$$d\varepsilon_i^p = \frac{3}{2}\frac{s_i - X_i}{(\sigma - X)_{eq}}d\varepsilon_{eq},$$

in which the plastic deformation increase toward i-ma is proportional to the ratio between the main deflecting stress in the same direction and the equivalent stress. The two hardening function variables are then left to be defined, and their evolution with the deformation "history."

The term hardening a is the couple of functions that contain the isotropic hardening scalar R, and the kinematic hardening vector X; then from a dimensional standpoint, it is a stress quadrivector:

$$\alpha = \begin{Bmatrix} R(r) \\ X(\mathbf{a}) \end{Bmatrix}.$$

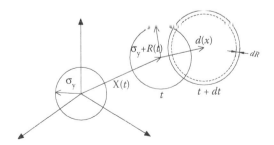

FIGURE 4.9   Isotropic hardening, R and kinematic tensor X.

The equivalent plastic deformation ε_{cq} is defined as the dual deformation to the equivalent stress in the direction the deformation is moving, so:

$$\int_0^p \sigma_{eq} \cdot \varepsilon_{eq} \cdot d\varepsilon_{eq} = \int_0^{\varepsilon_{ij}^p} \sigma_{ij} \cdot \varepsilon_{ij}^p d\varepsilon_{ij}^p.$$

An important outcome is that "natural plastic deformation" coincides with the plasticity multiplier on which the evolution of all plastic behaviors is based.

Ultimately, this means that the two hardening stresses, R and X, depend on the scalar variable ε_{cq} and the tensorial variable ε, respectively, and so the characterization of the plastic behavior of any material is obtained by the experimental detection of hardening laws related to plastic deformations, as well as to the first yield stress.

It is obvious that to "make a system" out of these relationships, some of which are of incremental type among vector quantities, it is necessary to exploit numerical methods, the most effective currently being the finite elements' discretization.

As to the form of hardening functions, Lemaitre proposes some standardized practical laws, assuming that for metallic materials both hardenings are characterized by an early phase, immediately on yielding, in which the nominal tensions R and X (relative to the nominal gross cross section) grow to saturation, and a second phase, which coincides with the maximum load, where the two functions, while tending asymptotically to a value (that of "saturation"), can be considered nearly constant and equal to the aforementioned value.

The following expressions describe the hardening function according to Lemaitre and are subject to experimental characterization in the elastoplastic field:

$$X_{ij} = X_\infty \left[1 - e^{-\gamma a_{ij}} \right],$$

$$R = R_\infty \left[1 - e^{-bp} \right].$$

To experimentally define the parameters of these laws (which reasonably reproduce the experimental trends but are not generated by reproducible physico-mathematical analyses), it is sufficient, within certain deformation limits, to carry out mono-axial static tests, since the only non-null deformation coincides with natural plastic deformation, measurable through tests; thus it is quite simple to obtain X∞, γ, R_∞, b. In practice, as inferred from Figure 4.10, it is necessary to statically load the test bar with tensile stress, alternating phases of compression, limited to the elastic phase only (in case of plasticization these compressions must be halted), with an immediate new phase of tensile stress.

4.5   NOTES ABOUT MECHANICAL CHARACTERIZATION TESTS

4.5.1   COMPRESSION TESTS

In the compression tests, bars must be cylindrical with a diameter d_0 of at least 20 mm and a height of $3\ d_0$.

4.5.2   CURVATURE TESTS

In curvature tests, bars can have different cross sections (squared, circular, rectangular).

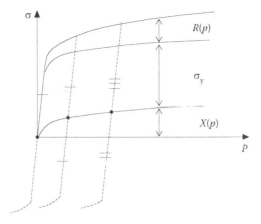

FIGURE 4.10 Statically load with tensile stress and compression stress phases.

Test bars are mounted with their ends on freely rotating rollers, loading the bars in the center, in the case of tough materials, or with two symmetrical loads in the case of fragile materials (Figure 4.11). The detectable data is:

- The unitary yield point under curvature given by the ratio between the yield moment M_{fs} and the resistance modulus under curvature W_f of the bar:

$$R_{fs} = \sigma_{fs} = \frac{M_{fs}}{W_f} \ \text{N/mm}^2.$$

- The unitary break point under curvature, given by the ratio between the maximum moment, M_{fm}, and the resistance modulus under curvature, W_f, of the bar:

$$R_{fm} = \sigma_{fm} = \frac{M_{fm}}{W_f} \ \text{N/mm}^2.$$

The information about the deformability of materials stressed by bending is given by

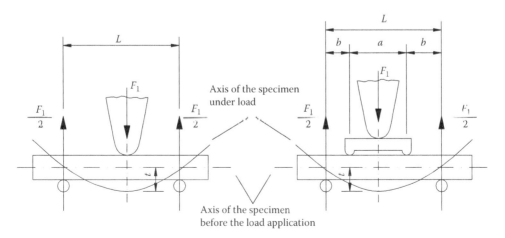

FIGURE 4.11 Bars for curvature tests.

- The total arrow f in millimeters, measured in between the props at the break point.
- The percentage arrow f_L given by the ratio between the arrow f and the distance L between the bar props;

$$f_L = 100 \frac{f}{L}.$$

4.5.3   SHEARING TESTS

Shearing test bars are cylindrical and rarely with square cross sections; their size depends on the device being tested.

4.5.4   TORSION TESTS

There are no standard guidelines that define the size and shape of the bars.

4.5.5   HARDNESS TESTS

Hardness tests are based on the principle of penetration by a harder body. The main hardness-gauging methods, such as Brinell, Vickers, Rockwell, and Knoop are based on this principle.

4.5.6   BRINELL HARDNESS/NUMBER

The Swedish engineer J.A. Brinell's method consists in penetrating the test material with a thermic-treated steel sphere (penetrator) to which a force F is applied, and measuring the diameter of the crater (spherical segment) left by the penetrator after the force is removed (Figure 4.12).

The Brinell number symbol is HB and its value is given by the following expression:

$$HB = 0.102 \frac{F}{S},$$

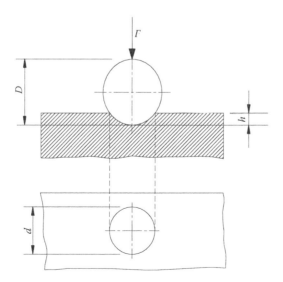

FIGURE 4.12   Penetrator for Brinell hardness.

where F is the load expressed in N, S is the crater surface in square millimeters, and 0.102 the conversion factor $1N = 0.102$ kgf.

In conclusion:

$$HB = \frac{2 \cdot 0.102 \cdot F}{\pi D \left(D - \sqrt{D^2 - d^2} \right)},$$

where D is the diameter of the penetrator and d is the diameter of the crater.

This last expression is of value only if the load applied is a function of sphere diameter (penetrator) according to

$$F = nD^2,$$

where F is the load applied in kgf, or in newton, D is the sphere diameter in millimeters (it may be either 10 or 5 or 2.5), whereas n is the quality coefficient of the materials. As a general guide, the values of n are

- For iron/steel alloys $n = 20–30$
- For copper alloys $n = 5–10$
- For light alloys $n = 2.5–5$

These values require a constant d/D ratio $= 0.375$ that enables the test results to be compared. In theory, this particular ratio represents the perfect one. If the ratio constancy is met, the penetration angle β (Figure 4.13) corresponds to 136°.

The spheres can either be of hardened steel or sintered material, their Vickers hardness being $KV \geq 850$ kg/mm². The norms recommend avoiding carrying out the Brinell test on materials with a hardness greater than 450 HB.

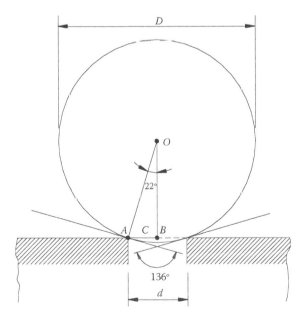

FIGURE 4.13   Details of the Brinell hardness penetrator.

Testers have found a relationship that converts Brinell number into tensile strength, a relationship of proportionality:

$$Rm = n\text{HB},$$

where the ratio n has values:

$$n = 0.36 - 0.34.$$

These conversions are acknowledged (EURONORM) for steels, apart from the austenitic ones:

- For perlite cast irons $Rm = 0.12\,\text{HB}$

- For copper and its alloys $Rm = 0.55\,\text{HB}$

The hardness indicator of an ordinary test only reports the value, for instance:

$$\text{HB} = 250 \text{ (steel)}.$$

For a normal test, the sphere diameter is 10 mm, and as the material being tested is steel, the value of $n = 30$, consequently the load to be applied, given by $F = nD^2$, turns out to be $F = 30 \times 10^2 = 3000$ kg.

For a non-normal test, using a 5 mm diameter sphere, the load $F = 30 \times 25 = 750$ kg, so to the right of the symbol HB is written, in order, sphere diameter, load, and load duration. Example:

$$\text{HB } 5/750/15 = 200.$$

4.5.7 DIAMOND PYRAMID HARDNESS NUMBER (VICKERS)

The Vickers method consists of introducing a pyramid-shaped, square-based, diamond penetrator, with a 136° dihedron angle between two opposite faces (Figure 4.14).

The Vickers number is obtained from

$$\text{HV} = 0.102\,\frac{F}{S},$$

whereas the final formula is

$$\text{HV} = 0.1891\,\frac{F}{d^2},$$

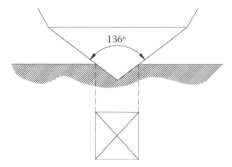

FIGURE 4.14 Vickers hardness diamond penetrator.

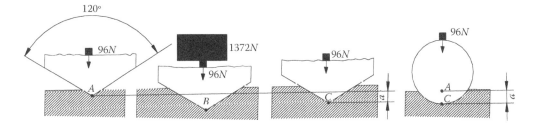

FIGURE 4.15 Rockwell hardness penetrator.

where d is the diagonal of the crater left by the penetrator (Figure 4.14).

An ordinary Vickers test is carried out with load $F = 294$, $N = 30$ kgf, whereas load duration must be 15 sec.

4.5.8 ROCKWELL HARDNESS

This takes its name from the American inventor Rockwell, whose idea was to insert a standard penetrator into the surface of the test piece in two subsequent moments, and measure the residual growth e of the penetration depth.

The Rockwell hardness is a function of residual growth e, and equal to 0.002 mm.

The standard penetrators he used were:

- A conical diamond penetrator with a 120° vertex angle (Figure 4.15)—the hardness number obtained with this penetrator is defined by the symbol HRC.
- A spherical tempered-steel penetrator whose hardness is greater than 850 HV, 1/16" diameter equal to 1.5875 mm (Figure 4.16)—the hardness number obtained with this penetrator is defined by the symbol HRB.

As for the test procedure, the penetrator is placed on the surface of the test piece and a 10 kgf load is applied, the depth gauger is set to zero, then a further 140 kgf load is applied in 3–6 sec so as to obtain a total load of 150 kgf for the HRC or 100 kgf for the HRB.

The HRC hardness is expressed by

$$HR = 100 - e.$$

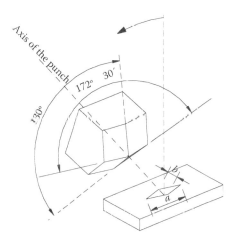

FIGURE 4.16 Knoop hardness penetrator.

Whereas for the HRB, this number is given by

$$HRB = 130 - e.$$

4.5.9 KNOOP HARDNESS

The Knoop method exploits a pyramid-shaped, rhomboid-based diamond penetrator with vertex angles of 172° 30′ and 130°, respectively (Figure 4.16).

The Knoop equation is

$$HK = \frac{F}{S} = 14,228\frac{F}{d},$$

where F is the applied force, S is the rhomboid crater surface projected on the plane perpendicular to the load, d is the rhomboid longer diagonal, expressed in micrometers, F is the applied load in kgf 10^{-3} (Figure 4.17).

4.5.10 RESILIENCE TESTS

Resilience tests are performed on all those materials that undergo dynamic stress in use. Indeed, resilience is defined as the resistance ability of a strained material. These are mostly collision deflection tests. They are carried out with a pendulum (Charpy) made up of a pedestal, two uprights across which a horizontal crosspiece articulates a bar, at the end of which a knife-ending ram is fixed.

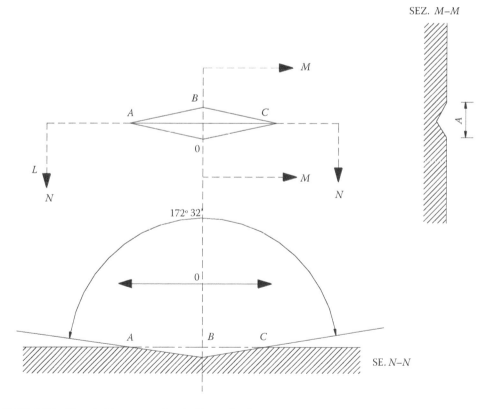

FIGURE 4.17 Rhomboid crater surface in Knoop hardness test.

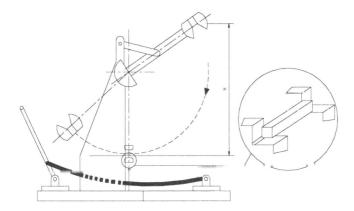

FIGURE 4.18   Pendulum Charpy for Resilience tests.

The ram is situated at a predetermined height H with respect to the bar's resting plane (Figure 4.18). The ram, at height H, has potential energy: $E_p = P \cdot H$ (P ram weight). After release, it collides with the test bar at a velocity of about 5–7 m/s, right opposite the notch made in the bar, Figures 4.19 and 4.20 turning kinetic energy, gained while falling, into strain energy and, finally, into break energy.

Resilience is indicated by the symbol K, and is defined by the ratio between the energy absorbed by the bar to break it and the fracture cross section:

$$K = \frac{L_a}{S} \text{ measured in } \frac{J}{cm^2}.$$

Ordinary pendulums have a ram (including the shaft) and a fall height H providing a potential energy of 30 kgf $m = 300$ J, enough to break the bars of any material. Should the bar not break, it is necessary to specify: resilience $> 300/0.8$ J/cm^2 or $> 300/0.5$ J/cm^2, depending on the bar type.

4.5.11   STANDARD BARS

MESNAGER (UNI 3212) for non-ferrous materials (Figure 4.19)
CHARPY (UNI 4431) for steel (Figure 4.20)

4.5.12   OTHER TEST BARS

Izod test, standardized in England. The test bar has a square 10×10 mm section, which is constrained as in Figure 4.21.

Schnadt test uses the Charpy pendulum against the test bar as in Figure 4.22. The ram hits the bar on the side where a hole has been made, inside which there is a 5 mm steel pin, so reducing the resistant cross section. With this test, fiber compression breaks are avoided.

FIGURE 4.19   Standard bars for non-ferrous materials.

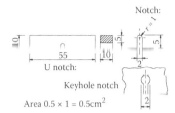

FIGURE 4.20   Standard bars for steel.

In the collision tensile stress test, the test bar is set up on the back of the ram (Figure 4.23). In this way, it breaks under collision tensile stress, when the plates screwed to the bar collide with the two uprights of the device.

4.6   NOTES ABOUT TECHNOLOGICAL CHARACTERIZATION TESTS

It has already been stated that the technological properties of materials relate to their aptitude to react to external forces. The main technological tests for metallic materials will be examined. Some of these tests are standardized, while others, though not officially recognized, are quite widespread anyway.

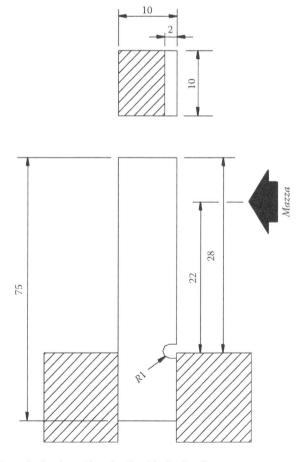

FIGURE 4.21   Specimen for Izod test (Standardized in England).

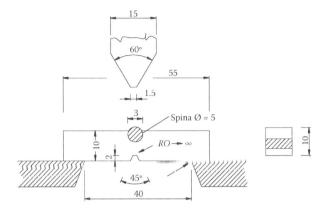

FIGURE 4.22   Specimen for Schnad test.

4.6.1 DRAWING TESTS

These tests aim to assess the aptitude of sheet metal to be turned into a hollow body without breaking. Obviously, this aptitude varies according to sheet type.
Italian norms provide for three different types of drawing tests:

- Erichsen drawing test, symbol I_E (UNI 3037):
- Erichsen drawing test modified, featuring the symbol I_E (UNI 4693)
- Socket drawing, featuring the symbol I_B (UNI 6124)

4.6.2 ERICHSEN DRAWING TEST

The test performing device is shown in Figure 4.24. It is made up of a spherical 20 mm diameter punch, a template on which the sheet is laid, and a blank holder A that presses against the bar to prevent it from creasing during the test.

The metal sheets are 0.2–2 mm thick and big enough to accommodate a minimum 70 mm circle.

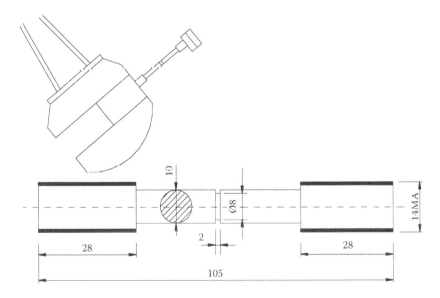

FIGURE 4.23   Pendulum Charpy in Schnad test.

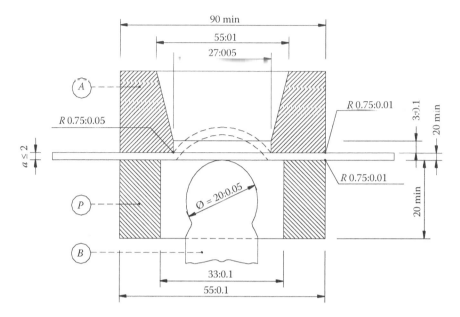

FIGURE 4.24 Erichsen drawing test.

The sheet is placed between the template and blank holder; first the punch is brought into slight contact with the sheet, then, either by a hydraulic or manual system, the punch is pushed forward at a 0.2–0.4 mm/min speed, until it generates a fracture in the sheet that has turned into a spherical surface. At the end of the test, the momentum of the punch expressed in millimeters is taken as the drawing index and is represented by I_E.

4.6.3 ERICHSEN DRAWING TEST MODIFIED (UNI 4693)

This test is performed much like the previous one with two exceptions: the punch is pushed forward at a velocity of 5–20 mm/min and sheet thicknesses are 0.5–2 mm.

4.6.4 SOCKET DRAWING TEST (UNI 6124)

The test device is identical to the one for the ordinary Erichsen test, only the punch is different. No longer spherical but cylindrical (Figure 4.25), tests are carried out on sheet disks of increasing diameter: 55, 56, 58, 60, 62, 64, 66, 68, and 70 mm. The procedure is the same, only the punch speed varies from 60 to 120 mm/min, until the test bar breaks.

4.6.5 POMP DRAWING TEST

Here, the test bars are sheet metal disks with a central hole. The punch is cylindrical with a central pin that enters the sheet hole for centering (Figure 4.26).

The testing stops when the test bar hole, now enlarged, shows the earliest edge cracks in a radial direction. The material quality index is given by

$$Ac = \frac{d_u - d_0}{d_0} 100,$$

d_u being the test bar hole diameter at the end of the test, d_0 the test bar hole at the beginning.

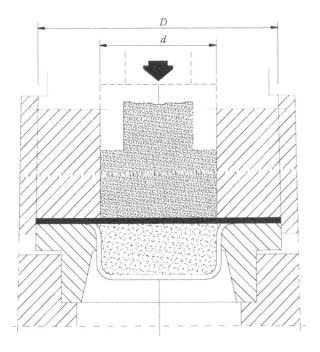

FIGURE 4.25   Socket drawing test.

4.6.6   BENDING TEST

This test is compulsory for the steel bars in reinforced concrete, for the sheet metal in boilers, and so on. The two following bending tests are possible.

4.6.7   TEMPLATE BENDING

The test is shown in Figure 4.27. The test bar lies on a template and is then bent to a V by the appropriate punch. The template V has an angle of 60°–70° and an opening of at least 125 mm.

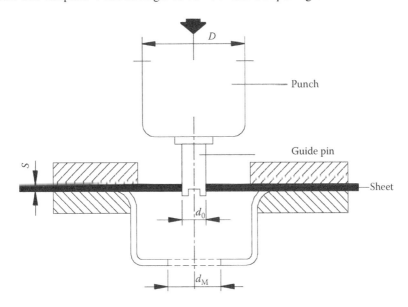

FIGURE 4.26   Pomp drawing test.

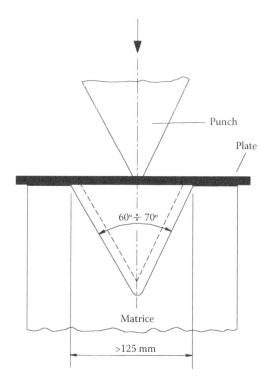

FIGURE 4.27 Template bending test.

4.6.8 MANDREL BEARING BENDING

The test consists of bending a full-section prismatic or cylindrical bar lying on two roller bearings by using a mandrel (Figure 4.28).

The mandrel diameter as well as that of the two roller bearings must be indicated in the test specification. The distance between the two bearings must be: $D + 3a$, D being the mandrel diameter and a is the bar thickness.

Pressure is applied by the mandrel until one of the two bar edges forms an angle a with the extension of the other (Figure 4.29).

After bending, if the edges are parallel $\alpha = 180°$, two cases are possible:

- Bending with a certain aperture D_0 (Figure 4.30);
- Block bending, the edges brought into contact (Figure 4.31).

Test results are given after examining the status of the outside of the bent part. They are positive if no cracking occurs.

4.6.9 PERCENTAGE ELONGATION AND THE TETMAJER COEFFICIENT

We define the bending test elongation as the percentage given by

$$\frac{L_u - L_0}{L_0} 100,$$

L_u being the width of the arc AOB after the test, whereas L_0 is the same section before the test (Figure 4.32).

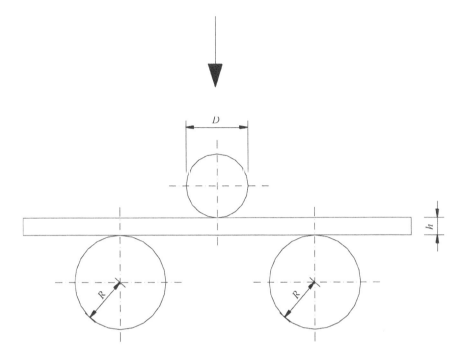

FIGURE 4.28 Mandrel bearing test (at beginning).

The Tetmajer number is shown with the symbol Cp; it can be demonstrated that

$$Cp = \frac{L_u - L_0}{L_0} 100 = \frac{100}{n+1},$$

where $n = D/a$, D being the mandrel diameter by which the bar is bent and a the test bar thickness. Notice that the expression above is missing the angle α, which does not affect the test at all, whereas

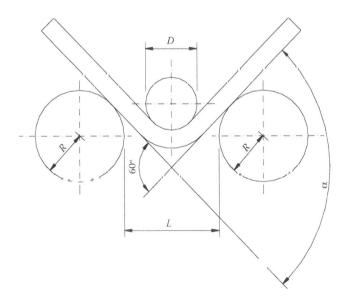

FIGURE 4.29 Mandrel bearing test (at end).

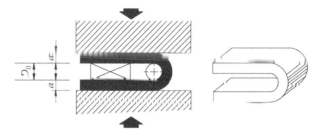

FIGURE 4.30   Mandrel bearing at D_0 aperture.

the bar curvature radius R is essential, since bar cracking depends on it. The expression $p = 100/(n + 1)$ can indeed be written as

$$Cp = \frac{50a}{R}.$$

4.6.10   FORGING TESTS

Forging is when metals undergo deformation through hot work procedures and when several different tests are carried out. The most important are described below.

4.6.11   STRETCH TESTS

This test measures the deformability of a metallic material. It is carried out on a prismatic bar that for steel has the following dimensions (Figure 4.33):

$$L_0 = (1.5 - 2)b_0 \quad b_0 = 3a_0.$$

4.6.12   HEADING TEST

This is carried out on metallic materials that become rivets. It is performed on a cylindrical test bar with diameter d_0 and height h_0. The results are given by the heading number:

$$A = \frac{h_0 - h_u}{h}100,$$

where h_0 = initial bar height, h_u = final bar height (Figure 4.34).

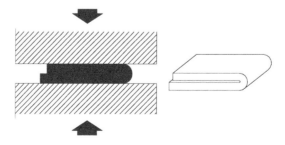

FIGURE 4.31   Mandrel bearing at contact.

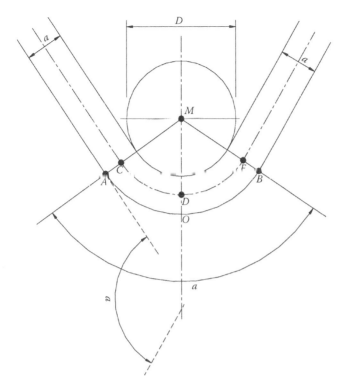

FIGURE 4.32   Scheme to define the percentage elongation.

4.7   METALLURGIC TESTS

Non-standardized tests are carried out in foundries. However, they provide reassuring information about metal quality.

The colability test is the most important. Colability is the aptitude of a liquid alloy to fill the die in which it is poured. The measure of this aptitude is given by the alloy's capacity to fill a spiral-shaped tube and fill notches cut into the spiral. Its colability is assessed (Figure 4.35) by the notch number reached by the head of the alloy.

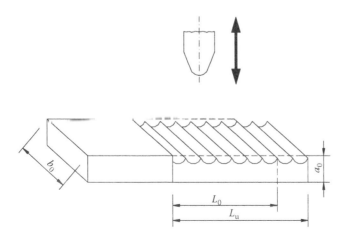

FIGURE 4.33   Bar of steel for stretch tests.

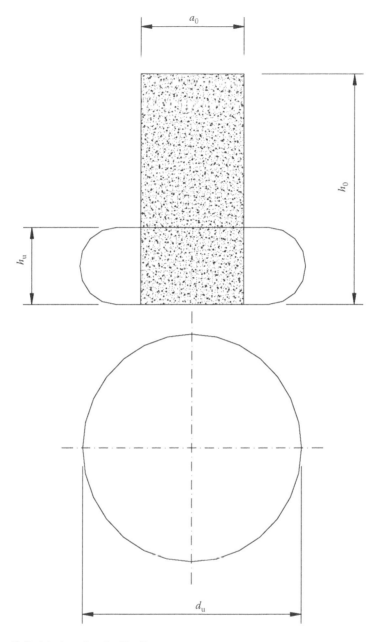

FIGURE 4.34 Cylindrical test bar for Heading test.

4.8   TUBE TESTS

4.8.1   ENLARGEMENT TESTS

Enlargement tests are carried out on quality tubes with external diameters $De \leq 140$ mm and thicknesses $s \leq 8$ mm. The test is performed by forcing a punch into the tube hole to a depth of 30 mm (Figure 4.36). The test is considered positive when the section of distorted tube shows no cracking.

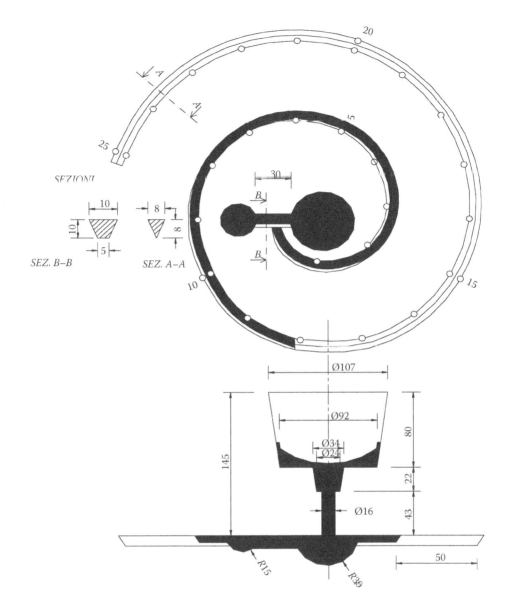

FIGURE 4.35 Colability test.

4.8.2 BEADING TEST (FLANGING TEST)

This is carried out on tubes with the following geometrical characteristics (De and thicknesses s):

- $De = 108 \div 140$ mm $\quad s < 0.06\, De$
- $De = 51 \div 108$ mm $\quad s \leq 0.08\, De$
- $De = 51$ mm $\qquad\quad s \leq 0.10\, De$

This test is carried out by placing the 100 or 200 mm long test bar, inside a steel ring (Figure 4.37) and exerting a force with a press onto an appropriate punch inside the tube. The flange angle obtained can either be 90° or 60°.

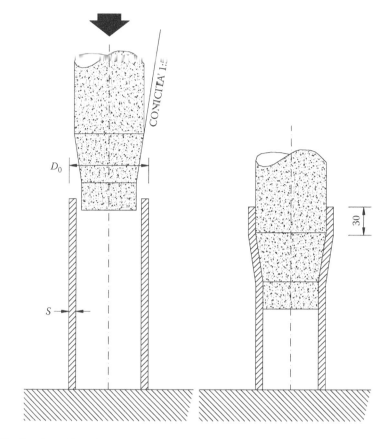

FIGURE 4.36   Specimen for enlargement tests.

4.8.3   COMPRESSION TEST

This is performed on a 50 mm long tube section, placed between the two walls of a press and compressed (Figure 4.38). The test is valid if the bent part of the tube has no cracking.

4.8.4   TENSILE STRESS TEST

The tensile stress test is carried on a section of tube by inserting a steel mandrel into each end (Figure 4.39). The section length is

$$L_0 = 5.65\,S_0 \text{ with } S_0 = \frac{(De^2 - Di^2)}{4} \text{ as tube section.}$$

This test is used to assess the unitary yield point Rs, the maximum unitary load Rm, and the percentage elongation $A\%$ after breaking.

4.9   TESTS ON STEEL WIRES

Steel wires undergo the following tests:

- Tensile stress
- Alternate bending

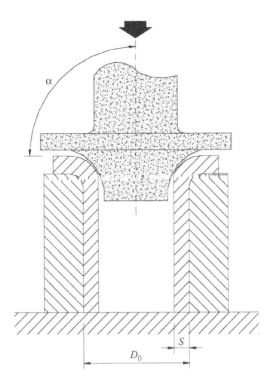

FIGURE 4.37 Specimen for flanging tests.

- Winding
- Torsion

4.10   FINAL INDICATIONS

In order to characterize any material mechanically, in accordance with resistance parameters, static tests can provide early indications of the resistance parameters of mechanical components under loads that are constant in time.

In practice, nearly all machine components are liable to loads that vary in time, so a characterization based only on static tests, such as those so far examined, is not thorough enough to provide all the information needed for design choices. Therefore, it is necessary to resort to tests that in some way define material resistance to various dynamic stresses, that is, stresses that change over time. These issues and suitable tests are the topics in Chapter 6 on the fatigue of materials.

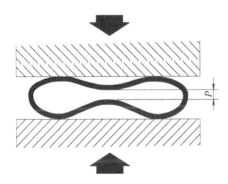

FIGURE 4.38 Specimen for compression tests.

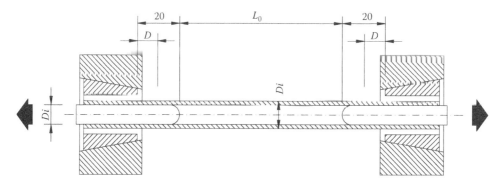

FIGURE 4.39   Specimen for traction tests.

Other tests described in other courses are non-destructive. Generally, they are carried out directly on the machine component and aim to investigate its real status without breaking it or causing its destruction. These tests are quite useful when defining or diagnosing the deformation/tension statuses of mechanical components.

The following is a list of the most common:

- Penetrating liquids
- Fragile varnishes
- Ultrasounds
- Acoustic emission
- X-rays
- Thermography
- Extensometry

5 Stress Conditions

5.1 INTRODUCTION

In order to better utilize and, above all, draw the students' attention to, the differences between the actual mechanical component and the simple, theoretical model calculation, here is a summary of the notions students in construction sciences already have

The simple formulae quoted are those essential to the design of the most common machine components. At the same time, experimentation is essential in ascertaining the effectiveness of any design or component through simple models that can only approximate for all the factors impinging on it.

Though often necessary, oversimplification can often lead to mistakes and show the following:

- Component shape does not always correspond to easily referable geometries.
- The actual spatial distribution of tensions is approximate.
- Errors are made when assessing dynamic and vibration phenomena.
- Additional thermal phenomena are often neglected.
- Possible internal stresses are often overlooked.
- Materials are handled as isotropic or at least with known characteristics in limited directions.
- Possible odd stresses are not taken into account.
- More generally, a whole series of approximations is made, on which we are forced to rely for each case.

At the same time, we would like to underline that in carrying out the modeling, we bypass any inconvenience by introducing adequate safety coefficients, usually on the basis of the designer's experience as well as his/her own sensitivity.

5.2 MECHANICAL BEHAVIOR OF MATERIALS

As previously stated, any study aimed at producing mechanical components and complex structures takes place gradually. First, it is necessary to determine the nature and intensity of the interacting forces, then a model is made as physically and geometrically "close" to the design element as possible, and finally the calculation method is chosen to assess the tensions and distortions of the component or structure.

Tension and distortion are calculated by Building Science Mechanical Design methods that are based essentially on Isotropic Body Resilience Elasticity Theory. However, the isotropy hypothesis of bodies is only valid on a macroscopic scale, as they are made up of small crystal grains, each with their own strongly anisotropic structure.

Assessing the intensity of the internal forces that stress materials (response of materials to external forces applied to them) leads to a rational apportioning of those forces to enable exploiting resistance characteristics.

As shown in other courses, there are materials with both resilient (elastic) and non-resilient (anelastic) behavior. The former are significantly resilient (tensions and distortions linked to a linear law), whereas the latter are negligibly resilient (no linearity between stress and distortion).

The materials considered resilient are those that, under stress from external forces, within certain limits, warp, and when those forces cease, distortion ends, restoring initial geometry. Anelastic materials such as lead, wax, clay, etc., behave the opposite way. In fact, perfectly elastic and anelastic materials do not exist.

5.3 CONDITIONS OF MECHANICAL STRESS

From the above, it can be deduced that, to preserve machines and their components for a long time, it is necessary to find the right balance between external forces so that once these forces are applied, they generate tolerable distortion.

The forces at work within a machine component can be the following:

- A permanent or static moment—Those forces whose maximum value is reached depending on their increase in time and which are then constant.
- Impulsive—Those forces that reach maximum instantaneously (crashes).
- Alternative—Those forces ranging between maximum and minimum values by a periodic law and which then repeat.
- Static at constant temperature—Those static forces at a constant temperature subject to viscous slide (the load is applied at different temperatures according to the nature of the materials under stress).

The forces applied to a body's axis generate stresses that vary according to direction and type of action in relation to the following:

- Tensile stress—The applied force coincides with the axis of the body and tends to elongate it (axial stress).
- Compression—As above, but the strain tends to shorten the body.
- Buckling—The strain lies in the plane that passes through the body's axis as well as in the plane perpendicular to it, leading to a buckling of the body.
- Torsion—The strain lies in the plane perpendicular to the axis, not passing through it, tending to twist the body.
- Shear—The strain lies in the plane of the cross section, forcing it to shear from the adjacent section.

Furthermore, stresses can be:

- Simple—The stress works in one direction only (monoaxial).
- Compound—Stresses operate simultaneously (multi-axial).
- Derived—An indirect reaction is provoked by the primary stress, e.g., in durability, shearing and buckling bring about the removal of material.

5.3.1 SIMPLE TENSILE STRESS

Any material exposed to external axial tensile stress behaves according to the degree of applied force. Internal response depends on the interactions between each single crystal grain in the material.

In contrast with the Theory of Elasticity for isotropic bodies, which predicts uniform distribution of tensions, they are actually distributed unevenly owing to the anisotropic crystalline structure of metals (Figure 5.1).

Thus, there are areas where actual stress exceeds that of real resistance, and in these areas plastic distortion can occur. All this might at first sight look dangerous, but these distortions are extraordinarily important since they get the metal grains to align themselves, and consequently guarantee a redistribution of specific stress, enhancing global resistance and, in particular, the conventional elastic limit (assessed with a tensile stress test).

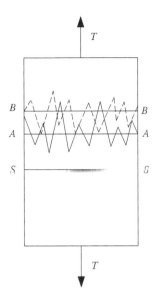

FIGURE 5.1 Real qualitative stress distribution.

In the simplest of hypotheses, imagine a small bar sustaining tensile stress force F (Figure 5.2). When early distortions affect the material, it starts reacting within. On every square millimeter s of the cross section S, perpendicular to the geometrical axis of the bar, the material reacts with σ perpendicular unit tensions that tend to counterbalance the F force.

This equilibrium can be written as follows

$$\sigma \cdot \Sigma_s = F,$$

that is $\sigma \cdot S = F$, and

$$\sigma = F / S,$$

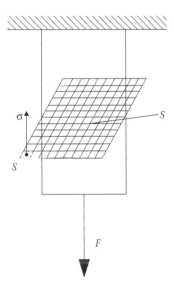

FIGURE 5.2 Ideal stress distribution.

an equation that enables us to determine the perpendicular unit load σ, as a function of the force F and the reacting section S. As the load is expressed either in kilograms or in N and the section in square millimeters, σ is measured either in kilograms per square millimeters or newton per square millimeter.

To assess stress limits (those that could provoke component breakage), tests are performed on the materials in the following way.

The machines used for stress limits plot graphs that interpret the law of the variation of force F and the elongation ΔL of the material. Notice that the first segment OP of the curve in Figure 5.3 is rectilinear; where F_p intersects the curve, and P is the total load at the limit of proportionality. So, in this segment, Hooke's Law or the law of proportionality is

$$\sigma = E \cdot \varepsilon,$$

where σ is the nominal unitary tension, E is the standard elasticity or Young's modulus for the material, and $\varepsilon = \Delta L/L$, the specific distortion of the material.

Because

$$\sigma = \frac{F}{S} \quad \dots \quad \varepsilon = \Delta L / L,$$

then

$$\frac{F}{S} = E \frac{\Delta L}{L},$$

so

$$E = \frac{F \cdot L}{S \cdot \Delta L} \text{ in kilograms per square millimeters or newton per square millimeters.}$$

The modulus of elasticity E can be defined as that force able to double the length of the unitary solid section, provided the material allows such elongation. The section PE shows a slight curvature—the first permanent distortions that add up to the elastic ones.

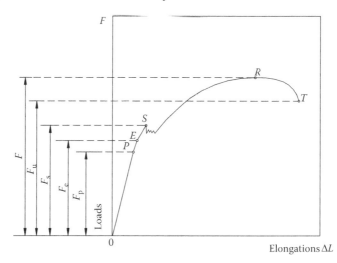

FIGURE 5.3   Qualitative load–elongation curve in traction test.

Point E of the chart represents the limit of elasticity, which is hard to assess because loads cannot be incremented infinitesimally. The ordinate F_e represents the total load to the limit of elasticity. The unitary load to the limit of elasticity is expressed as follows

$$\sigma_e = \frac{F_e}{S}.$$

In the ES segment, permanent distortions increase more rapidly. Right on point S, where the chart shows discontinuities, lies the yield limit of the material, whose yield unitary load is obtained by:

$$\sigma_s = \frac{F_s}{S}.$$

The stress $\sigma_r = F_r / S$ represents the unitary breaking load of the material.

The irregular progress of the load, at S, is caused by the internal flow of crystal grains, located on planes sloping at 45° to the planes perpendicular to the body's axis. Consider the bar in Figure 5.4, being subjected to tensile stress by load F. Let us consider section S_1, sloping by angle α to the perpendicular section S, the strain F applied to the center of gravity of S_1; it will split into two components, a perpendicular N and a tangential T, whose values are

$$N = F\cos\alpha \quad T = F\sin\alpha.$$

In the simplest event of uniform distribution, the tensions in S_1 will be

$$\sigma = \frac{N}{S_1} \quad \ldots \quad \tau = \frac{T}{S_1},$$

and since

$$S_1 = \frac{S}{\cos\alpha},$$

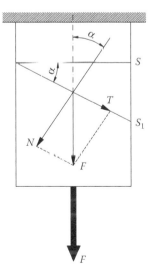

FIGURE 5.4 Normal and tangential components in a bar under load.

we obtain

$$\sigma = \frac{F \cdot \cos^2 \alpha}{S} \qquad \tau = \frac{F \cdot \sin \alpha \cdot \cos \alpha}{S},$$

Maximum σ is when $\alpha = 0$ in the sections perpendicular to the body's axis or force F.

To determine the maximum value of the tangential tension τ, it is necessary to derive the function and equalize it to zero. Then

$$\cos \alpha \cdot \cos \alpha - \sin \alpha \cdot \sin \alpha = 0.$$

Equality only occurs when $\alpha = 45°$. Then, τ_{max} occurs in planes sloping at $45°$ to the body's axis.

The reaching of the τ_{max} (critical tangential tension) in the material's crystal grains that have this orientation, causes the sliding of the grains, initiating the first distortion. However, flow is limited by the surrounding grains with different orientations and so is subject to a lesser tangential tension than the critical one. If load increases, so does internal tension and the number of grains likely to flow. When the number of grains is too high, great plastic distortion occurs (section SR) up to breakage.

To be certain that the external stress on machine components brings about only plastic deformation, it is only necessary to establish a σ_{amm} (admissible unitary load) smaller than the value of σ_c (unitary elastic load). Being difficult to assess, a σ_{amm} as a fraction of the σ_r breaking load is preferred. With safety degree n or safety coefficient $1/n$, depending on the type of stress and nature of the material, then

$$\sigma_{amm} = \frac{\sigma_r}{n} = \frac{F_r}{n \cdot S} \, N/mm^2,$$

represents the safely admissible unitary load.

Safety degree n, for static loads, is likely to show values ranging from 1.5 to 5, but even higher values (ropes) depending on the element or component and calculation hypotheses.

As above, subjected to a pulling force, any material is inevitably bound to distort (Figure 5.5).

Distortion indexes are represented by

$$A\% = \frac{L_u - L_0}{L_0} \cdot 100 \quad \text{elongation percentage,}$$

$$Z\% = \frac{S_0 - S_u}{S_0} \cdot 100 \text{ striction percentage,}$$

from an elastic standpoint:

$$E = \frac{\sigma}{\varepsilon},$$

perpendicular elasticity or Young's modulus and elongation, with the symbols above, turns out to be

$$\Delta L = \frac{F \cdot L}{S_0 \cdot E}.$$

5.3.2 SIMPLE COMPRESSION

Compression exists when it operates along a body's axis, tending to shorten the fibers of the body itself (Figure 5.6).

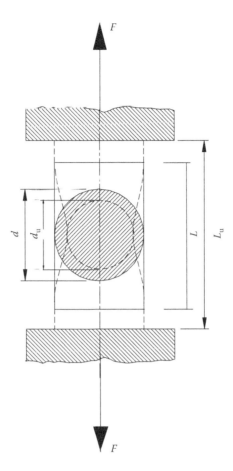

FIGURE 5.5 Qualitative deformation under traction load.

As for tensile stress, the equation of stability under compression is obtained from

$$F = \sigma_c \cdot S.$$

For which

$$\sigma_c = \frac{F}{S} \text{ N/mm}^2.$$

For stability and safety, σ_c has to be less than or, at the most, equal to the admissible load or safety under σ_{amm} compression, then

$$\sigma_c \leq \sigma_{amm}.$$

For steels, σ_{amm} compression can be considered equal to σ_{amm} tensile stress. This does not apply to cast iron components that have high resistance to compression, but much less to tensile stress; in fact, cast iron compression σ_{amm} is around four times greater than for tensile stress. For the shortening of a body under compression by strain F with section S:

$$\Delta L = \frac{F \cdot L}{S \cdot E_c},$$

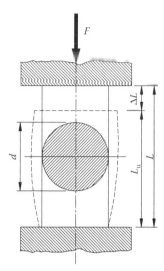

FIGURE 5.6 Qualitative deformation under compression load.

where F is the strain expressed in newton, L is the original body length in millimeters, S is the initial cross section in square millimeters, and E_c is the perpendicular modulus of elasticity under compression that adopts values equal to E, the perpendicular modulus of elasticity under tensile stress.

5.3.3 SIMPLE BENDING OF BEAMS

A rectilinear bar is lying on A and B, loaded with two equal forces F at the end of each overhang (Figure 5.7).

In a cross section a–a between A and B, the resultant force acting to the left or right will become a couple whose moment is given by

$$M_x = P \cdot d.$$

Consider the beam section dl (Figure 5.8), between S and S_1; let y–y be the stress axis (track of the stress plane of the loads acting on the beam). Most of the time the y–y axis coincides with the symmetrical axis of the section perpendicular to the bar axis.

Under the bending moment M_x, a curvature occurs in the longitudinal fibers y distant to the x–x axis due to the relative rotation of the two sections. Owing to this curvature, the upper fibers of axis z elongate while the lower ones shorten. Obviously, the greatest elongations and shortenings occur in the fibers most distant to the x–x axis, that is, at y_{max}.

The x–z plane fibers maintain their initial length. This plane is called the neutral plane.

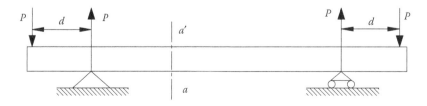

FIGURE 5.7 Pure bending in the central part of a rectilinear bar.

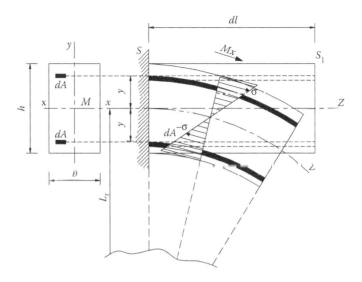

FIGURE 5.8   Qualitative deformation of fibers in pure bending.

If dA is the area of a fiber located at y distance from the neutral axis and σ_f is the corresponding unit tension, the tension developed in the dA fiber will be

$$\sigma_f \cdot dA.$$

Since the algebraic sum of unit tensions must be nil:

$$\int \sigma_f \cdot dA = 0.$$

To satisfy the hypothesis of the preservation of plane sections and proportionality between unit stress and distortion, the σ_f unit tension in an ordinary fiber will be proportional to its y distance from the neutral axis, so multiplying and dividing the previous equation by y:

$$\frac{\sigma_f}{y} \int y \cdot dA = 0.$$

Since σ_f/y = constant, the equation becomes

$$\int y \cdot dA = 0.$$

This equation represents the static moment of the section in relation to the neutral axis that, being zero, confirms that this axis is the center of gravity.

As the moments generated by internal unit tensions must be equal to the external agent moment, M_x, equilibrium is

$$\int \sigma_f \cdot y \cdot dA = M_x,$$

so, multiplying and dividing by y,

$$\frac{\sigma_f \cdot \int y^2 \cdot dA}{y} = M_x,$$

given that $\int y^2 \cdot dA$, the J_x inertial moment is

$$\frac{\sigma_f \cdot J_x}{y} = M_x \quad \text{(Navier's formula)}.$$

As maximum internal stress occurs in the fibers most distant from the neutral axis, that is, in y_{max}, the previous equation becomes

$$\frac{\sigma_{f\,max} J_x}{y_{max}} = M_x.$$

As $J_x/y_{max} = W_x$, where W_x is the modulus of deflection resistance, the previous equation becomes

$$\sigma_{max} = \frac{M_x}{W_x},$$

a formula that shows the simple deflection stability equation.

To design and statically verify the resistant cross section, the following condition must be satisfied:

$$\sigma_{f\,max} \leq \sigma_{f\,amm}.$$

5.3.4 SIMPLE TORSION

Consider a cylindrical solid of length L, fixed in a frame but with a cross section at one end free to rotate (Figure 5.9).

M_t is the twisting moment that works on the freely rotating cross section at L from the fixed one. In an internal cylindrical section of the solid with radius r, consider the fiber P–P_1. Applying a turning moment, the free end will rotate by angle θ, with respect to the fixed one. Thus, point P remaining fixed, point P_1 will have rotated by θ. So, the P–P_1 fiber will have undergone a relative shift γ (Figure 5.9).

Due to proportionality:

$$\tau = G \cdot \gamma,$$

where G is the tangential modulus of elasticity and γ is the fiber shift.

From Figure 5.9 it can be inferred that

$$s = L \cdot \gamma,$$

where

$$\gamma = \frac{s}{L}.$$

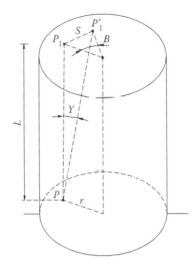

FIGURE 5.9 Qualitative deformation of fibers under pure torque.

Since $s = r\theta$ (Figure 5.9), the tangential stress is

$$\tau = \frac{Gr\theta}{L}.$$

In elements of area dA of the cross section, internal forces with value τdA come into play; owing to the distributive symmetry of tensions, they generate two elementary moment pairs equal to $\tau\, rdA$, which in equilibrium will be

$$M_t - \frac{G\theta \int r^2 dA}{L} = 0.$$

As $r^2 dA$ represents the polar moment, J_p, of the cross section, the previous equation can be written

$$M_t - \frac{G\theta J_p}{L} = 0,$$

where

$$\theta = \frac{M_t L}{G \cdot J_p},$$

is the equation of torsional distortion. Dividing the first and second members of the equation by L, we get

$$\frac{\theta}{L} = \frac{M_t L}{LGJ_p} = \frac{M_t}{GJ_p}.$$

This ratio represents the rotation angle of the mobile section with respect to the fixed one, located at a unit distance. The θ/L ratio is called the torsion unit angle.

Replacing $\tau = Gr\theta/L$ by θ from the previous equation, we get

$$\tau = \frac{GrM_tL}{LGJ_p} = \frac{M_tr}{J_p}.$$

Peripheral τ_{max}, where maximum r is $r_{max} = d/2$ is

$$\tau_{max} = \frac{M_tr_{max}}{J_p},$$

and since $J_p/r_{max} = W_t$ the torsion resistance module, the previous equation can be written

$$\tau_{max} = \frac{M_t}{W_t}.$$

This equation is used both for the design and checking components undergoing simple torsion. The following general condition must always apply:

$$\tau_{max} \leq \tau_{amm}.$$

5.3.5 Simple Shearing

A solid is subject to shearing when in each cross section the resultant external force lies in the plane passing through its center of gravity, tending to shift it with respect to the adjacent cross section.

Consider the rectangular cross section of the bar in Figure 5.10, and suppose shear is uniformly distributed in the cross section with area A, then unitary tangential stress τ is expressed by

$$\tau = \frac{T}{A}.$$

It is possible to demonstrate that in a rectangular cross section, tangential stress along an i–i chord is given by the formula:

$$\tau_{i-i} = \frac{TM_i}{J_nb},$$

where T is the applied force, M_i is the static moment of dashed section A_i with respect to axis n–n, J_n is the inertial moment of section A with respect to the same axis, and b is the length of the chord.

Since

$$J_n = \frac{bh^3}{12} \text{ and } M_i = A_iy_g,$$

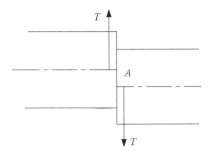

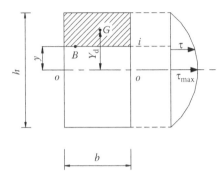

FIGURE 5.10 Qualitative deformation of fibers under pure shear.

where $A_i = b(h/2 - y)$ and $y_g = 1/2(h/2 + y)$, then M_i becomes

$$M_\mathrm{i} = \frac{b}{2}\left(\frac{h^2}{4} - y^2\right).$$

By substituting in the $\tau_{\mathrm{i-i}}$ formula:

$$\tau_{\mathrm{i-i}} = \frac{\dfrac{Tb}{2}\left(\dfrac{h^2}{4} - y^2\right)}{\dfrac{1}{12}bh^3 b}.$$

Simplifying:

$$\tau_{\mathrm{i-i}} = \frac{6T}{Ah^2}\left(\frac{h^2}{4} - y^2\right).$$

This equation shows that tension varies (proportionally to M_i) by a parabolic law, while T, J_n and b remain constant. $\tau_{\max}$ for $y = 0$:

$$\tau_{\max} = \frac{\dfrac{6Th^2}{4}}{Ah^2} = \frac{3T}{2A}.$$

For circular sections (Figure 5.11), consider a chord i–i that forms angle α with radius r. The tangential stress along i–i is

$$\tau_{i-i} = \frac{TM_i}{J_n b_i}.$$

The static moment of the lined section is

$$M_i = \frac{b_i^3}{12},$$

and the moment of inertia is

$$J_n = \frac{\pi r^4}{4}.$$

Furthermore, in Figure 5.11

$$b_i = 2r \cos \alpha.$$

Substituting everything into the tension equation:

$$\tau_{i-i} = \frac{\dfrac{2Tr^3 \cos^3 \alpha}{3}}{\dfrac{\pi r^4 2r \cos \alpha}{4}}.$$

Simplifying:

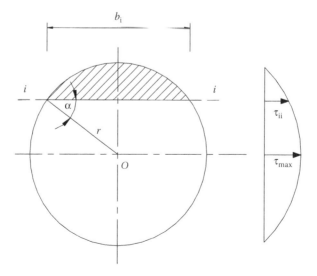

FIGURE 5.11   Qualitative tangenzial stress distribution.

$$\tau_{i-i} = \frac{4Tr^2 \cos^2 \alpha}{3Ar^2} = \frac{4T}{3A} \cos^2 \alpha.$$

τ_{max} is obtained when $\alpha = 0$; for $\cos \alpha = 1$, τ_{max} is

$$\tau_{max} = \frac{4T}{3A}.$$

Two points, A and B, separated by L, and part of a fiber in a prismatic solid subject to shearing, will undergo a relative creep $\gamma = AB_1/AB$.

With reference to proportionality, we can write

$$\tau = G\gamma.$$

For isotropic bodies, the tangential modulus of elasticity is

$$G = \frac{2E}{5}.$$

It is known that creep breaking point in isotropic materials is

$$\tau_r = \frac{4\sigma_r}{5}.$$

If the same ratio applies to safety loads:

$$\tau_{amm} = 4/5\sigma_{amm}.$$

5.3.6 BUCKLING OF COLUMN BARS

For slim bars with compression loads parallel to their axis, verifying resistance cannot be carried out using the previous simple formulae. When loads reach a certain value, deformation is not negligible and is sufficient to modify the stress parameters and, in particular, the bending moment generated by the non-coincidence of the load vector with the bar axis

In these conditions, there is no longer any linearity between the external force and deformation, but minimal force increments bring about deformations that increase the bending moment till the bar breaks.

In fact, the non-coincidence of the load vector with the bar axis can occur for various reasons, such as vibrations, transverse forces, and lack of material homogeneity. Instability manifests itself by almost instantaneous structural breakdown.

The theoretical study of instability leads to the definition of a P_{crit} axial load below which the structure does not break down.

Referring to Figure 5.12, this moment can be expressed as follows

$$M = EJ\frac{d^2y}{dx^2} = -P(y+a),$$

where E is the elastic modulus of the material, $J = S\rho^2$ is the minimum moment of inertia of the bar with: $S = bh$ the area of the line section, and ρ the minimum radius of gyration.

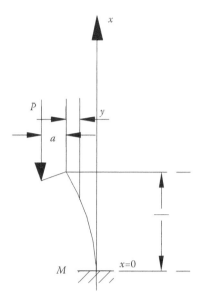

FIGURE 5.12   Scheme of a column bar.

If there is no eccentricity ($a = 0$), given $k^2 = P/EJ$, the integral is

$$y = C_1 e^{kx} + C_2 e^{-kx},$$

or else

$$y = A \sin kx + B \cos kx.$$

The constants A and B are determined by the boundary conditions and, in particular, by the constraints of the solid.

In the case of a bar hinged at both ends, with zero movement:

$$y = 0 \ \text{for} \ x = 0,$$

$$y = 0 \ \text{for} \ x = l,$$

which, replaced in the equation of the elastic line, determine $A \sin kl = 0$ and, excluding the obvious solution:

$$\sin kl = 0, \quad kl = \pi,$$

that is

$$\frac{Pl^2}{EJ} = \pi^2,$$

and so the load can be written as

$$P_{\text{crit}} = \frac{\pi^2 EJ}{l^2}.$$

In the case of a bar embedded at one end and free at the other, the conditions at the limits are

$$\frac{dy}{dx} = 0 \ \text{ for } x = 0 \text{ (no rotation)},$$

$$y = 0 \ \text{ for } x = l \quad \text{(no slide)}.$$

and the solution leads to $Kl = \pi/2$ which corresponds to

$$P_{\text{crit}} = \frac{\pi^2 EJ}{4l^2}.$$

Critical load values can be found in the same way for all other constraint cases.

If l_0 is the free inflexion length at the *Eulerian* critical load, the more general expression is

$$P_{\text{crit}} = \frac{\pi^2 EJ}{l_0^{\,2}},$$

where
$l_0 = l$ for bars hinged at both ends
$l_0 = 2l$ for bars embedded at one end and free at the other
$l_0 = 0.5l$ for bars embedded at both ends
$l_0 = 0.707l$ for bars embedded at one end and hinged at the other.

If the moment of inertia J is written as a function of the radius of gyration with $\sigma_{\text{crit}} = P_{\text{crit}}/S$, the previous expression becomes

$$\sigma_{\text{crit}} = \frac{\pi^2 E \rho^2}{l_0^{\,2}}.$$

The critical stress value must be less than σ_{amm} of the material, that is a fraction less than the yield point (σ_s) (according to the safety coefficient s).

The ratio $l_0/\rho = \lambda$ is the *slenderness* of the bar.

The equation above can be written as

$$\sigma_{\text{crit}}\lambda^2 = \pi^2 E = \text{cost} \quad \text{(for each material)},$$

in the $\sigma_{\text{crit}} - \lambda^2$ plane, it represents an equilateral hyperbola and highlights that the more slender the bar, the smaller the critical load and *vice versa*.

If on the chart the line $\sigma_{\text{crit}} = \sigma_{\text{amm}} = \sigma_s/s$ is drawn, it is possible to identify the field of critical loads (Figure 5.13). Favoring safety, the security norms replace the line with curves that qualitatively coincide with the hyperbola. The norms propose several calculations, also based on experimental data, that can assess critical loads.

Safety coefficient values s depend both on materials and the operating conditions of the bar. Generally for steels, values ranging from four to seven can be used, depending on steel type.

In the case of a bar embedded at one end and free at the other with an eccentric load ($a \neq 0$):

$$M = EJ \frac{d^2 y}{dx^2} = -P(a + y).$$

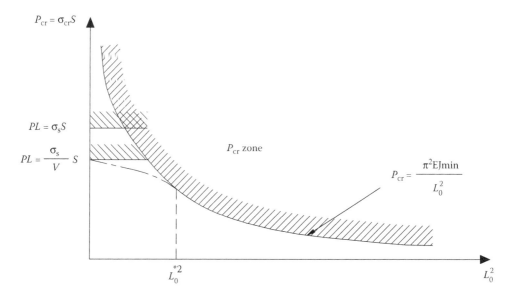

FIGURE 5.13 Euler's curve.

This solution adds to the homogeneous integral above creating a special one:

$$y = A\sin kx + B\cos kx - a.$$

With the boundary conditions:

$$\frac{dy}{dx} = 0 \text{ for } x = 0 \text{ (no rotation)},$$

$$y = 0 \text{ for } x = l \quad \text{(no slip)},$$

and assuming $\beta = kl$, we get

$$A = 0 \quad B = \frac{a}{\cos\beta} \quad y = a\left(\frac{\cos Kx}{\cos\beta} - 1\right).$$

At point B, the absolute value is

$$\left(\frac{dy}{dx}\right)_{x=1} = \left[\frac{ka\sin kx}{\cos\beta}\right]_{x=1} = \frac{ka\sin kl}{\cos\beta} = ktag\beta.$$

With serial development,

$$tg\beta = \beta + \frac{\beta^3}{3} + \cdots$$

and taking into consideration the first two terms:

$$y_0 = 2 \frac{Pa^2}{EJ} \left(\frac{a}{3} + l + \frac{Pl^3}{6EJ} \right).$$

The bending moment, which now is variable, has its maximum value at point M:

$$M_{max} = -EJ \left(\frac{d^2y}{dx^2} \right)_{x=0} = EJ \frac{k^2 a}{\cos\beta} = \frac{Pa}{\cos\beta}.$$

Developing in series:

$$\cos\beta = 1 - \frac{\beta^2}{2} + \cdots \qquad \frac{1}{\cos\beta} = 1 + \frac{\beta^2}{2} + \cdots$$

and the second term:

$$M_{max} = Pa \left(1 + \frac{Pl^2}{2EJ} \right).$$

Maximum stress as an absolute is compressive and bending stress combined:

$$\sigma = \frac{P}{S} + \frac{M}{W},$$

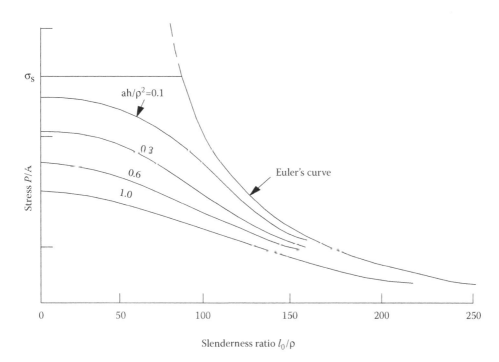

FIGURE 5.14 Euler's curve and ec/k^2 parameterized curves.

with W as the minimum resistance modulus of the cross section.

The same procedure can produce expressions for distorted as well as maximum stress for other constraint cases. Diagrams like those in Figure 5.14 can be produced for various cross sections (h minimum distance of the most external fiber to the neutral axis) as well for several shapes and materials, parameterized by ah/ρ^2 in which it is possible to identify theoretical fields of criticality

6 Fatigue of Materials

6.1 GENERAL CONCEPTS

Machine components do not usually break because of static stress, but because of dynamic stress. In other words, dynamic stress is where load over time varies, which means that the break is not entirely due to the maximum load, but also to the minimum

Everyone knows that to break a soft iron wire, it is no good pulling to stretch to a break, which would be static tensile stress, but alternately bending it one way then the other (cycles) will cause a steep rise in temperature followed by breakage due to plastic deformation.

This example shows how the break is caused by two factors: load variability, and plasticity whose effect is enhanced by the repetitiveness of the load cycle. Generally, high cycle fatigue breaks depend on load inconstancy, so when the number of cycles has reached plastic deformation, a fatigue break follows.

A test sample subjected to tensile stress (Figure 6.1) shows this "characteristic."

This characteristic is a macroscopic behavior of the workplace under load. It provides no information on its regular behavior that might produce different characteristics. When a material experiences an external force, not all its crystals reach their elastic limit at the same time due to a number of irregularities (crystal orientation, dislocation, etc.). Consequently, the elastic deformation of some crystals will lead to the plastic deformation of others.

With reference to a metal test sample under load, its behavior can be subdivided into phases as follows:

  I. The load is so low that all the crystals are elastically stressed.
 II. The load is such that most crystals deform elastically, but in some parts there is plastic and elastic deformation together, so when the sample is unloaded the metal regains its shape. Macroscopically, the sample has behaved perfectly elastically.
III. The load is such that some crystals deform elastically together with plastic deformation so that once unloaded the sample does not regain shape; it is permanently deformed.
 IV. The load has reached levels where the plastic deformation is such that most crystals are plastically deformed. The elastically deformed ones steadily decrease as the load continues to be applied, and when the load ceases, the permanent deformation becomes all the more obvious.

The yield value represents the transition between phases II and III. In some apparently perfect metals, phase I is very limited.

The internal tensions between the zones deformed elastically and plastically give rise to an "after-elastic" effect in polycrystalline metals. This effect is linked to elastic hysteresis.

If the load oscillates and there is high cycle fatigue, the internal structure could change even if the deformation is phase II. It is possible that the break starts where a crystal has slid and then broken. In that case, there are severe tensions at the junctions of certain atomic bonds, and as the oscillations continue, junction cohesion breaks down. The cumulative effect of the breakdown of atomic bonds increases after the first appearance of discontinuity—micro-cracks—and continues to a visible break in the structure.

So, if the defect was localized in a tensile stress test, there would be the same overall result. If the test sample was made so small as to contain only one defective crystal, we would find a

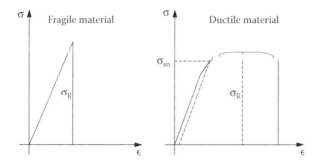

FIGURE 6.1 Qualitative stress–strain curve for perfectly fragile material and for perfectly plastic material.

qualitative trend of the same type but with different elastic limit (in the classical sense) values. In other words, plasticization would have a local elastic limit different to the macroscopic elastic limit intended as the total external average load tension capable of producing 0.2% permanent deformation.

The average nominal tension at a local level is manifest in the first plasticization (micro-plasticization), which under dynamic stress will lead to breakage. Therefore, high cycle fatigue can be defined as that tension (phase II) where there is no micro-plasticization anywhere in the material. In all probability, this definition of fatigue limit refers to oscillation resistance. As experience demonstrates, for non-symmetrical alternating stress at its limit, the test sample would never break or at least would only break beyond an infinite number of cycles. In this case, a higher stress value would be needed to break the material.

So, if it were possible to predict the tension of the first micro-plasticization, it would be possible through a static test to define the fatigue limit for symmetrical stress ($R = -1$), R being the ratio between the lower and upper tension ($R = \sigma_i / \sigma_s$). Therefore, it may be deduced that fatigue is a function of the definition of "infinite life." The corresponding tensions are, therefore, only conventional values. In other words, a test sample under an oscillating stress ($R = 0$) equal to the oscillation resistance would break, perhaps after a great deal more oscillations than those defined for infinite resistance (conventionally $N = 2 \times 10^6$).

It is well known that on plastic deformation in an homogeneous material there is a local increase in temperature due to crystal viscous sliding whose lattice will remain deformed.

The macroscopic characteristics of a workplace subject to tensile stress shows that the deformation energy is made up of an elastic and a plastic part. The elastic energy can always be recouped when the external load is removed (phases I and II), but the plastic energy cannot (phases III and IV). If the test sample is unloaded during macroscopic plastic deformation, it is possible to recoup some of the energy required for elastic deformation, thereby restoring some of the crystals to their steady state, given that the tension needed to deform all the crystals was insufficient. The return curve slope is almost identical to the up-slope. A model that can best simulate the test sample behavior is a spring with a viscous damper. Supposing most of the test sample (the part initially plasticized) was brought to its breaking point (lack of bonds), there would be a smaller active area with an overall rigidity variation but with the same elasticity modulus (the crystal structure has not changed because there was no permanent deformation).

However, it is impossible to recoup the energy permanently that deformed the crystals. Part of this energy was the heat created at the deformation site. If it were possible to follow the thermal evolution of all the crystals, it would be possible to identify the "defective" one because it would be the first to heat up. The fatigue limit for alternating symmetrical stress (oscillation resistant) is lower than in any other relationship between extreme tensions and in particular the value of resistance from the start; that is, the static tension before micro-plasticization corresponds to the material's oscillation.

Evaluating the point before micro-plasticization would reveal the critical tension, after which any further stress would cause breakage.

The energy required to produce breakage can be divided into two parts; that from the elastic part is recoupable and relates to the crystals that remained whole after unloading, but that from the plastic part is not recoupable because it relates to the permanently deformed crystals that cannot return to their original state and to the heat energy produced by the permanently deforming crystals.

An energy equation for the load period shows that the external work (done by the machine) is in part used to increase the internal energy of the material and in part given up externally in the form of heat.

E_i is the variation of internal energy and is made up of three parts: E_1 is elastic energy, E_2 is the one that produces internal temperature variations in each component, and E_3 is the one that produces the plastic deformations necessary with the variation of micro-spaces. E_1 is reversible, but the others are not. E_2 and E_3 come into being only after the material yields microscopically and macroscopically. It should be noted that yield tension is different for each. Only in the ideal case of a material that is entirely homogeneous (or with all its molecules equally defective) do the two values (microscopic and macroscopic) coincide.

Generally, the microscopic value of yield tension (relative to a single crystal) is different. Microscopic plasticization starts at different times. Macroscopically, it is visible when all the flat surface crystals become plasticized, but microscopically, it might have started at completely different times.

A first approximation of reversible plastic deformation energy can be obtained by subtracting from external work L_e (carried out by the machine during the tests), the heat given up externally Q_e, and E_1 and E_2:

$$L_e - Q_e - E_1 - E_2 = E_3.$$

6.2 LOAD CHARACTERISTICS

Load fatigue depends on the minimum and maximum tension values. Generally, for every maximum tension (over σ_s) there is a minimum one (below σ_i) for simplicity referring to a sinusoidal trend of load against time, which leads to the test sample breaking after a certain number of cycles and is different for every different pair for minimum and maximum tensions.

In practical tests, instead of referring to the above tension couple, reference can be made to parameters dependent on them, such as average tension $\sigma_m = (\sigma_s + \sigma_i)/2$ and oscillation amplitude $\Delta\sigma = (\sigma_s - \sigma_i)/2$.

The first person to realize that fatigue depended on upper and lower tension was Wöhler. His first attempts subjected each homogeneous test sample to sinusoidal mono-axial stress as in Figure 6.2 and then in Figure 6.3, which shows gradually decreasing values of each value pair (maximum applied tension—ordinate, x) against the cycle number (abscissa, y) that broke the sample.

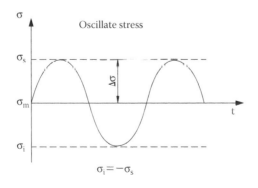

FIGURE 6.2 Sinusoidal mono-axial stress.

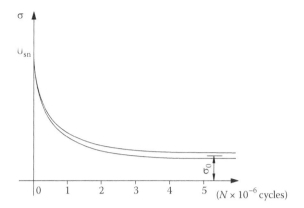

FIGURE 6.3 Qualitative Wöhler's curve.

He noted, therefore, that there was an applied tension value (minimum and maximum) at which the sample never broke. The literature calls it "oscillation resistance" (σ_0) and it corresponds to the ratio between the minimum and maximum loads and is equal to -1 ($R = \sigma_i / \sigma_s = -1$). Using other load pairs with other average tension σ_m and amplitude $\Delta\sigma$ pairs, curves of the same type are obtained. In particular, using minimum tensions equal to zero, and maximum tensions over zero (oscillating load) ($R = \sigma_i / \sigma_s = 0$), the "original resistance" is obtained.

Obviously, if the maximum tension is equal to breaking point load, the sample would break on the first cycle (static resistance). In mechanical design, the acceptable static load limit equates to the yield load, so a stress equal to the yield load would break the sample on the first cycle. Conventionally, for steels, it is assumed that if the sample does not break after 2×10^6 cycles, it will never break. The number of "infinite life" cycles is something else and for other materials. For light alloys, it could correspond to 500 million cycles. In practice, it is difficult to define a fatigue limit for light alloys, which nearly always reach their breaking points and whose spread is high.

Plotting Wöhler test data can be done differently. The most common method is semi-logarithmic for high cycle values (x). The approximated curve has three parts. The first, defined by yield load, runs parallel to x as far as $N = 10^3$; then the curve inclines downward at a straight slope of k (depending on the physical and geometric characteristics of the material) as far as $N = 2 \times 10^6$, where it becomes horizontal and runs to infinity.

The literature reports fatigue data in logarithms (logs), so it is easy to verify that for the intermediate zone ($N = 10^3 - 2 \times 10^6$) the equation $\sigma N^k = $ constant is valid, given the fatigue limit and the straight-line coordinates (Figure 6.4). At $N < 10^3$, the break occurs at high loads (relatively big

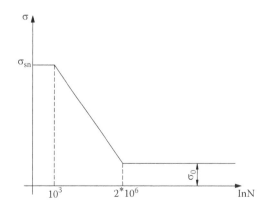

FIGURE 6.4 Three different zones in the Wöhler's curve.

deformations) and low cycles. This is called oligocyclic fatigue where characterizing the material is best done through deformation tests rather than by load tests, which are done for high cycle fatigue.

Using an everyday test machine, a whole Wöhler curve can be created by defining the average tension applied to the test sample (preload) and oscillating the subsequent loading by $\Delta\sigma$.

On the grounds that apparently identical test samples behave differently under fatigue, (the spread of test results is much higher than in static tests), at least five samples are used for the same load. Figure 6.5 shows the type of results obtained. It should be noted that as the load approaches the fatigue limit, spread is ever bigger. For one cycle, spread is minimal.

Spread, which can be due to a variety of factors as will be seen, can significantly influence fatigue resistance, so it is possible to define curves with different breaking point probabilities. The highest curve in the diagram has a 100% breaking point probability, while the lowest has 0%, all related to the same number of cycles. The reference curve is the central one with 50%.

Wöhler curves can represent "infinite life" (refer to fatigue limit values) or time (refer to the tension defined by the Wöhler curve intersecting the cycles vertical line corresponding to the desired duration).

Obviously, to know the exact behavior of the material in all load conditions, a Wöhler curve to infinity would be needed for each of the maximum and minimum load combinations.

In practice, approximated diagrams for fatigue are used for the overall behavior of materials (fatigue diagrams). They are created by simulating the behavior of the material in three conventional situations: load yield, resistance to oscillation, and resistance from origin.

Each of these three diagrams can be made using data from any of the curves of non-breaking probability. All the above count for characterizing the material, so tests must be carried out on samples that are theoretically without surface defect.

6.3 FATIGUE DIAGRAMS

In the literature, there are fatigue diagrams that can help create a fatigue design by applying the parameters that characterize its dynamic stress.

Goodman-Smith's (Figure 6.6) is the classic that plots average tension (x) against upper (+) and lower (–) tension (y). To construct the diagram, the x-axis is for the yield load by tensile stress (+) and by compression (–), points A and B. The intersection of the straight line parallel to the x-axis passing through A and B with a straight line at 45° are points V and T. At average tension zero, the

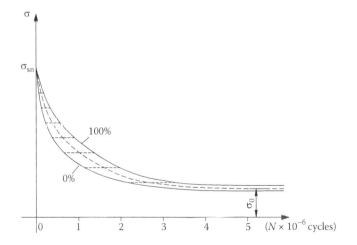

FIGURE 6.5 Qualitative dispersion in the Wöhler's curve.

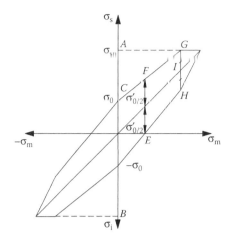

FIGURE 6.6 Goodman-Smith's diagram.

resistance to oscillation (σ_0) is reported, points C and D. Finally, at average tension half of the resistance from origin (σ_0^1) (point E), the resistance from origin is point F on the x-axis. Point G is at the intersection between the line joining C and F and the straight line $\sigma_s = \sigma_{sn}$. H is on the vertical with G, and I is between them. In such a way, the third quadrant should be considered.

The zone defined by the dashed line (Figure 6.6) is where stress does not lead to 100% breakage if the behavior of the material corresponds to a 100% Wöhler curves.

Analogously, Pohl and Bach (Figure 6.7) diagrams and Weyrauch–Kommerell (Figure 6.8) diagrams can be made, although the latter are not common in the literature.

A diagram very commonly used in practice (Haigh curve, Figure 6.9) because it is very direct with its test procedure, plots average tension σ_m (x) against $\Delta\sigma$ (y). It also highlights oscillation resistance, resistance from origin, and static resistance (yield or breakage).

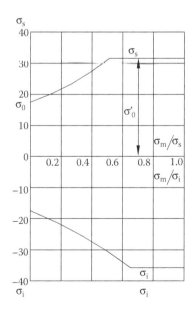

FIGURE 6.7 Pohl and Bach's diagram.

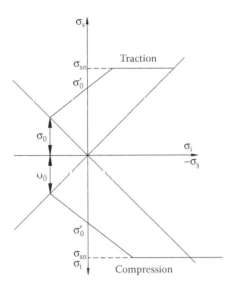

FIGURE 6.8 Weyrauch–Kommerell diagram.

Designers do not always have diagrams available that can outline the general behavior of a material. More often than not, the breaking and yield points are available, and if one is lucky the oscillation fatigue too. In this case, an approximated Haigh diagram can be drawn up (*AB* straight line in Figure 6.9) that favors stability.

All the diagrams above are drawn with many fewer points compared to the infinite number required to represent the complete performance of the material. In reality, such diagrams would have demarcation lines that were not regularly straight or curved because they would show the irregularities of the behaviors of each single test sample. In the unreal hypothesis that all the test samples were equal (same surface roughness, same crystal structure, etc.) one would produce identical Wöhler curves (zero spread) with regular lines.

6.4 THERMOGRAPHY TO DEFINE FATIGUE BEHAVIOR

Researchers at the University of Catania's Machine Construction Department have highlighted how it is possible to obtain quasi-Wöhler curves with a limited sample number (even only one).

Experimentally, samples under fatigue show rising surface temperature (T) with cycle number (N) and applied stress (σ_i). In the first phase, the maximum temperature rises linearly with increasing cycles as much as the load exceeds the fatigue limit. In the second phase, the temperature rises more rapidly, and in the third phase there is total plastic deformation and breakage (Figure 6.10).

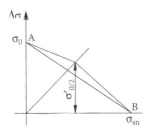

FIGURE 6.9 Haigh's diagram.

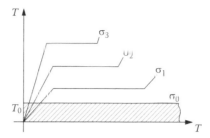

FIGURE 6.10   Qualitative surface temperature T with cycle number N under applied stress.

The phenomenon has been studied for several metals (steels, light alloys, and cast irons) under different loads (resistance from origin, tensile stress/compression, and bending). Preliminary studies have highlighted how temperature variation correlates to the energy of plastic damping of the material that remains below the fatigue limit but grows with cycle number above that limit. Where the initial temperature gradient as a function of applied tension intersects with the x-axis (zero plastic energy), this is the material's fatigue limit (Figure 6.11).

This technique has some advantages over the traditional one on account of a drastic reduction in testing time (days to hours), sample number (in practice 3), low results spread, and its non-destructive nature. So, for instance, during prototyping, speed is important when investigating parameters to improve fatigue. In studying materials, for example, hardening and tempering profoundly modify metallurgic and mechanical characteristics as regards dynamic response, so control of the material's parameters becomes critical, and traditional techniques for evaluating fatigue are unthinkable within a reasonable time. Thermography leads to extrapolating fatigue very quickly, allowing variations in thermal treatment and the consequent production of "new" material in a controlled manner.

Fatigue research into the parameters that can influence it is a significant sector. Further complications arise when tension values, after long experimentation, trigger results with high spread. Moreover, the fatigue limit in this way does not correspond to an infinite number of cycles but to a specific number depending on the material.

Local plasticization, due to internal and external structural defects, causes temperature increases up to a thermal equilibrium between the heat produced and the heat dissipated externally.

According to one of the more popular physical hypotheses, breakage by fatigue happens when internal energy E_i reaches a constant that is characteristic of every material ($\int E_i \, dN = $ constant), depending on mechanical and thermal characteristics.

According to the first law of thermodynamics, and excluding any transition process, the internal energy level per cycle can be expressed as the difference between the energy supplied E_w and the energy converted into heat Q_c:

$$E_i = E_w - Q_c.$$

In a model of material mechanical behavior, internal energy can be thought of as the sum of two energies, E_e (elastic deformation energy) and E_p (plastic deformation energy):

$$E_i = E_e + E_p.$$

E_e represents that energy absorbed by the material without deforming permanently and under repetitive stress, and is linked to the material's internal damping under applied stress lower than the fatigue limit. The material is able to absorb energy E_e for a certain number of cycles. The temperature rise is practically insignificant.

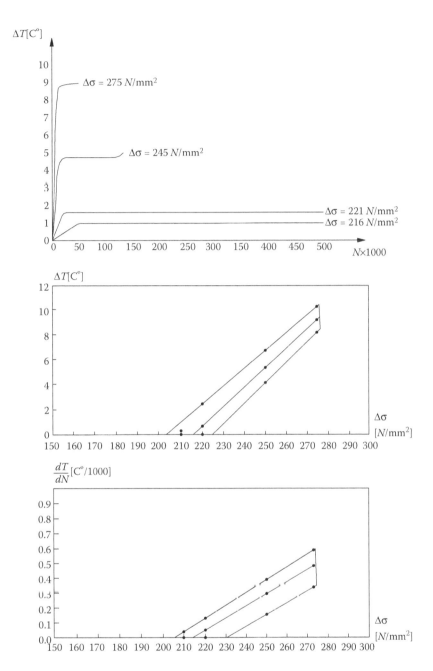

FIGURE 6.11 Fatigue limit by $T–N$ curve and by $(dT/dN)–N$ curve.

E_p represents the energy that gives rise to permanent deformation. It has different values for stress over the fatigue limit and is linked to "damage" (see Section 6.6). Fatigue life in energy terms is

$$\int E_p \, dN = \text{constant.}$$

So, fatigue limit σ_0 is that stress that produces no plastic deformation anywhere in the material ($E_p = 0$), such that for low values and those equal to "damage" it is nil and consequently life is infinite. Above the fatigue limit, the energy absorbed is greater than E_c and the stress is such as to cause permanent deformation, meaning a shorter life.

Supposing that the ratio between plastic deformation and elastic limit deformation is constant (an acceptable hypothesis for useful materials), and that specific mass and specific heat are constant, then

$$a\sigma^2 - \Delta T = b,$$

the constants a and b depending on the mechanical and thermal characteristics of the material.

Taking into account that for stress below the fatigue limit the temperature variation ΔT is negligible compared to the increments over σ_0, then for $\sigma > \sigma_0$, the previous equation becomes

$$a(\sigma^2 - \sigma_0^2) \approx \Delta T \qquad \text{con } \sigma_0 = \sqrt{b/a}.$$

For values of $\sigma > \sigma_0$, the temperature increments are due to local plastic deformation as a result of reaching the corresponding local plastic stress level, even for average applied tensions that are very different to average elastic limit of the material.

6.5   DETERMINING THE FATIGUE CURVE

The energy equilibrium equation highlights how the plastic deformation energy per cycle is practically proportional to the quantity of heat, Q_c, and the temperature of the exposed surface. The temperature integral of the sample surface equals (depending on the coefficient of thermal transmission) the energy limit, E_c, which characterizes the material from the viewpoint of fatigue limit. It has been verified that the energy limit is as high as the fatigue limit, such that their ratio is practically constant for different materials.

Given that $\int E_p \, dN$ is a constant that characterizes a material or a certain mechanical component, it can be evaluated by applying a prefixed tension, σ_0, over the fatigue limit and up to breakage. Figure 6.12 shows that $T = f(N)$ whose integral Φ is proportional to the energy limit E_c. In practice, depending on fatigue limit (σ_0) or lower tension values, the thermal variations are slightly higher than the theoretical one (zero). The T values needed in calculating Φ are the temperature increases above

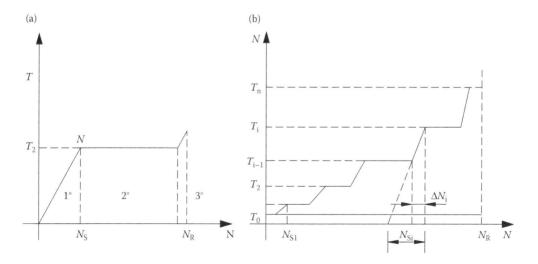

FIGURE 6.12 $T = f(N)$ whose integral is Φ proportional to the energy limit E_c.

T_0, the fatigue limit. So, it is better to define fatigue limit first to avoid using temperature increases from loads below the fatigue limit that are not due to plastic deformation but to background noise.

By determining the fatigue limit for each tension, the second phase temperature T_2 and the stabilization cycle number N_s can be found.

Knowing Φ, being proportional to E_c, the number of cycles to breakage N_r can be calculated for any tension. Figure 6.12a shows that the first phase temperature rise is practically linear, since the second phase temperature is constant and the third phase cycles are low, so

$$\int_0^{N_r} F_p dN = \Phi = T_2 N_s / 2 + T_2 (N_r - N_s),$$ (6.1)

from which N_r can be found for a certain tension without needing to reach breaking point, and interrupting the test a little above N_s. For Φ, it is better to refer to temperature curves in which the three phases are well defined, therefore to tensions higher than the fatigue limit and not too close to the yield tension. If load is close to yield tension then the three phases are not well defined, since breakage occurs with rising temperature to a certain specific cycle number N_r. In this case, Equation 6.1 becomes:

$$\Phi \approx T_2 N_r / 2.$$ (6.2)

For tensions with the three phases well defined, the stabilization temperature is reached quickly compared with the break temperature (N_s is negligible compared to N_r), and N_r is a little higher than the second phase cycles ($T_2 = Constant$). In this case, Equation 6.1 becomes:

$$\Phi \approx T_2 N_r = const.$$ (6.3)

In other words, the equation represents n equilateral hyperbola.

For materials with a hypothetical link between tension and temperature, it is evident that Equations 6.1 and 6.3 can define tension–cycle number.

If the fatigue limit is obtained by varying load as in Figure 6.12b, Φ is equal to the integral of curve $T = f(N)$ for N_r. In this case of Equation 6.1, N_s (N_{si}) is expressed by (Figure 6.12)

$$N_{si} = N_i T_i / (T_i - T_{i-1}).$$ (6.4)

Considering the numerous fatigue data available, these tests on metals convalidated the link between energy limit and fatigue limit and by the "rapid Risitano method" defined the whole fatigue curve to completion of the factors that characterize the dynamic behavior of machine components.

The observations and instructions reported above can rapidly define the whole fatigue curve of a material or prototype using relatively simple tests without recourse to complex test benches that work at high frequencies.

Obtaining a fatigue curve from a few pieces, practically the number for a static test, provides good results, because during prototyping the components' morphology and mechanical characteristics can be "tweaked" to respond best to the actual stress they will experience.

For the relatively simple cases of structural components, for which it is easy to take a tension reading at the point under most stress (highest temperature), the fatigue limit can be found in situ under real loads. The same procedure can also provide a measure of durability by the energy required to break a sample structural component.

There are fast methods in the literature for finding the characteristic fatigue curve of a material. They are often based on experimental data rather than on physical reality. Although not entirely accurate, the fatigue limit values of Prot and Locati can provide a useful shortcut.

The Prot method examines temperature, a parameter not usually observed, by observing the sample or machine component under stress.

6.6 FATIGUE LIMIT BY THERMAL SURFACE ANALYSIS OF SPECIMEN IN MONO AXIAL TRACTION TEST

In a recent work[*] by A. and G. Risitano, a new method to estimate the fatigue limit was proposed based on the observations reported in paragraph 6.1.

In terms of the thermal behavior of the material during a tensile static test, we can distinguish two phases (Figure 6.13):

- The first phase in which all crystals are deformed in an elastic field (zone I in Figure 6.13). In this phase, the relation stress–strain is linear.
- The second phase in which not all crystals are deformed in an elastic field, and only some are deformed in the plastic field (zone II in Figure 6.13).

In the first phase, the behavior of the material follows the thermo-elasticity theory. As is already known, for an elastic body the stress–strain law can be written as

$$\{\sigma\} = [C](\{\varepsilon\} - \{\alpha\}\Delta T),\tag{6.5}$$

where α is the thermal elongation coefficient, ΔT is the temperature variation of the material, and C is the stiffness matrix.

Under the hypothesis of adiabatic phenomena, the first thermodynamic law leads to the following relation:

$$\rho c_\varepsilon \Delta T/T = [(\{\varepsilon\} - \{\alpha\}\Delta T)\delta[C]/\delta T - (\{\alpha\}+) - \Delta T\delta\{\alpha\}/\delta T)[C]]\Delta\{\varepsilon\},\tag{6.6}$$

where

ρ is the material density
c_ε is the specific heat for constant strain
T is the Kelvin temperature

For a static traction test of isotropic material, the stress σ_m is constant at any point of the specimen and Equation 6.6 can be written as:

$$\Delta T = -K_m T\sigma_m.\tag{6.7}$$

In this phase, the surface temperature shows a linear decrement during the application of the load.

If for the applied stress σ_p (lower than $\sigma_{0.2}$) a plasticity condition in some defected point (small volume dV_p) of the material is reached at this point, Equation 6.7 is no longer valid and the effect of the dissipative plastic deformation is associated with a thermal energy dQ_p that changes the linear trend of the temperature during the application of the load. Therefore, there will be a part $(V-dV_p)$ of the volume V that will follow the linear elastic law, and a part, dV_p, which in a hypothesis of an ideal plastic condition, adds heat, i.e., $dQ_p = \sigma_p d\varepsilon_p$.

---

[*] Risitano, A. and Risitano, G., L'importanza del "parametro energetico" temperatura per la caratterizzazione dinamica dei materiali, Workshop "Progettazione a Fatica di Giunzioni Saldate (...e non) — Sviluppi teorici e problemi applicativi", Forni di Sopra (UD), March 9 and 10, 2009, *Frattura ed Integrità Strutturale*, 9, 123–124, 2009.

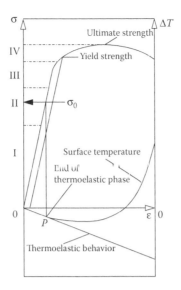

FIGURE 6.13 Qualitative engineering stress–strain curve and temperature–strain curve.

With the increase of the stress, the plastic deformation in the volume dV_p grows and at the same time, in other crystals, the plasticization process begins and consequently also the heat contribution.

In this phase, the quantity of heat is a function of the stress, of the local deformation law, of the increase in time of deformation, and of the distribution in the volume V of the plasticization process during the time.

In the literature, there are many analytical and numerical models that link the damage with energy and thermal variation for fatigue loading and for static tensile test and the hypothesis concerning the plasticization model, internal heat, and external transmission, can help to find qualitative solutions only.

If we had to capture the first plastic heat in relation to the external average stress applied, we could find the stress that produces the first plastic damage in the specimen and therefore the conventional fatigue limit for $R = -1$. In fact, if this stress were applied in a fatigue test of the specimen, the specimen would fail; instead, for fatigue stress under this value, the specimen would not reach the failure point.

This remark suggests that the fatigue limit can be found by the individualization of the first plasticization at which an internal heat is released and, consequently, a surface increment of the temperature follows.

Therefore, by a static traction test, using adequate instrumentation and test conditions, it is possible to find the fatigue limit stress σ_0 (average value of the external applied stress) through the determination of the "temperature limit" T_0, where T_0 is the temperature for which in a stress (strain) versus temperature curve the linear trend finishes; thus, for the first time, a variation of the derivate is noted (Figure 6.13).

Following the above, it is possible to have an experimental definition of temperature T_0 and consequently of the zone where the plasticization appears for the first time. The increase in temperature depends on the material characteristics (for steel at an ambient temperature of 23°C there is an increase of approximately 0.1°C, for each 100 N/mm² of stress), on the sensibility of the thermal sensors, on the image analysis quality, and on the signals synchronization of the three parameters (load, elongation, and surface temperature).

As a general indication, we can say that during the first phase (perfectly elastic) for which at any point of the surface $\Delta T = - K_m T \sigma_m$, the temperature variations are proportional to the thermo-elastic coefficient K_m (approximately 3.3×10^{-12} [Pa^{-1}] for steel and 10×10^{-12} [Pa^{-1}] for aluminum alloy) and to the point value of the stress. The temperature variations are more evident near the high stress concentration point.

The new method previously reported was used in different cases to verify the fatigue limit σ_0 finding values in sound agreement with those determined by means of traditional methods or Rigitono's methods for $R = -1$. An application of the method to the steel Fe 36 notched specimen is reported here.

As an example, in Figure 6.14a, the load versus time for a notched steel Fe 36 specimen and the corresponding curve temperature versus time are reported. The temperature was that of the surface point near the hole boundary. In the same figure, the temperature found by Equation 6.6 for equal value of load velocity and with the test values of $\varepsilon_0 = 0.0014$ and $\varepsilon_p = \Delta l_p / l_0 = 0.04$ is reported.

In this case, the gradient's change is very evident, and it is easy to determine T_0 and consequently, the corresponding load [5.8 kN]. The fatigue limit load, determined by traditional method for $R = -1$, was 5.7 kN. Figure 6.14b shows four of the circa 2400 recorded images from the beginning to the end of the test.

6.7   RESIDUAL RESISTANCE OF A MECHANICAL COMPONENT

An important practical issue is to analyze breakage risk, linked to the lifespan of a mechanical component that has been subjected to stresses for a certain time over the fatigue limit. Below the fatigue limit (zone I) no plasticization is generated; even if the load is repeated and alternated, the sample or component will never break. It has been experimentally demonstrated that cyclic stress below the fatigue limit "trains" the dynamic characteristics of the material to improve.

If the stress is over the fatigue limit at cycle number n_1, and applied in the same way that produced the fatigue curve, the material is damaged and the sample's lifespan is no longer that of the original.

If the load is the same and N_1 is the number of cycles to breakage, residual life is $N_1 - n_1$. The sample's lifespan has been shortened by n_1 / N_1. (Figure 6.15).

$$\Sigma n_i / N_{i+} = 1.$$

Having been stressed by σ_1, it is further stressed by cycle number n_2 and tension σ_2; the life used up is n_2 / N_2. The sum of the two percentages i of life used up equals to 1, its limit according to Miner's equation:

$$\Sigma n_i / N_{i+} = 1.$$

For a given time curve, the previous relationship defines a new curve, provided no other damage has been inflicted (stress over the fatigue limit).

Given the load history of the type reported in Figure 6.16, it is possible to go back in time, highlighting load *bars* by cycle number for which a given tension exceeds a fixed value. In the case of vehicle components, e.g., gearbox cogwheels, load history can be deduced from a reference test that defines the percentage of lifespan used up by the vehicle in its various conditions of use (motorway, urban streets, etc.). Normalized bar charts of maximum tension give an immediate idea of durability margins if the Wöhler curve of the material is plotted on the same chart.

With Miner's equation, one can go back to the residual life for a certain number of cycles, e.g., to N_{eq}, which defines fatigue limit (2×10^6). One might suppose that the damage produced by a series of loads over the fatigue limit is the same as an equivalent load σ_{eq} applied for a number of cycles n_{eq}:

$$\Sigma n_i / N_i = n_{eq} / N_{eq}.$$

The new number of cycles n_{eq}, calculated with N_{eq} relating to fatigue resistance σ_0, can identify the new straight line parallel to the durability straight line (in a logarithmic or semi-logarithmic

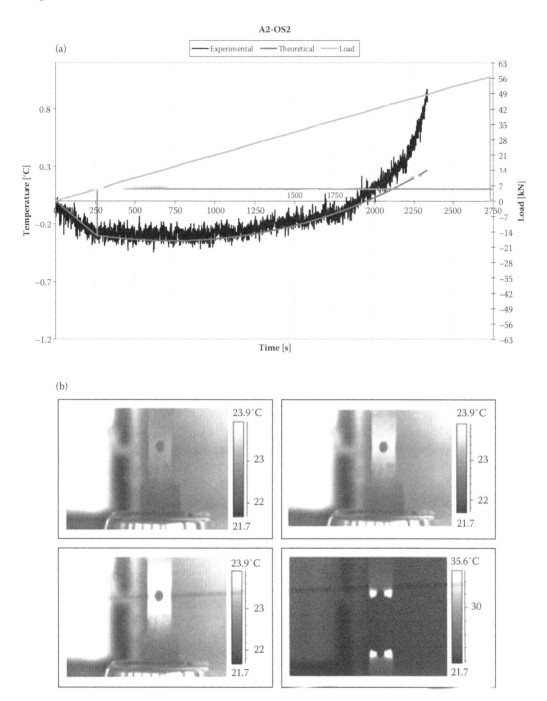

FIGURE 6.14 Load vs time for a notched steel Fe 36 and curve temperature vs time (a) and temperature images from to beginning to the end of a traction test (b).

diagram). This straight line intersects parallel to the y axis at $N = 2 \times 10^6$ in $\sigma_0{}^*$, which is the new value of the fatigue limit (Figure 6.17).

---

* Risitano, A., Clienti, C., and Risitano, G., Determination of fatigue limit by mono-axial tensile specimens using thermal analysis; *Key Engineering Materials*, Vols. 452–453, pp 361–64, 2011.

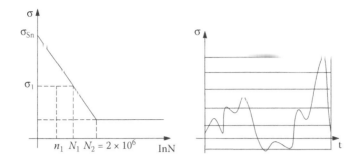

FIGURE 6.15   Stress is over the fatigue limit at cycle number n_j, and qualitative stress time diagram.

The value n_{eq} is the cycle number for a load equivalent to a "life" N_{eq}, which alone would create the same damage of various loads at different cycles n_i.

Using Miner's hypothesis, residual life can be evaluated and a new curve drawn (straight line in the log–log diagram) against time after the material has been subjected to a certain load history.

If the damage of various loads at different cycles n_i is $D = \Sigma n_i / N_i$, the equivalent tension applied for an equivalent cycle number n_{eq} should produce the same damage D (and produce a new curve):

$$n_{eq} = N_{eq}D.$$

Even if the hypothesis of linear damage has practical validity, it is also contradictory (see Figure 6.17 relative to a single load) because the curve would have a different yield value (straight line intersecting the x-axis), which is macroscopically not true. One might conclude that the new limit could depend on "trained" tension values that, now they are over the fatigue limit, would be damaging loads. The first contradiction might be overcome by applying Manson's hypothesis that advises intersecting the new time line with the x value of yield tension. There would be a new lower fatigue limit compared to that of Miner.

In practice, load history is as shown in Figure 6.16 (left) with a casual trend. This is due to discrete loads and consequently cycles over time; these loads are shown in the bar chart in Figure 6.16 (right). From the Wöhler curve of the material (n_i/N_i) of life is used up, and so $n_{eq}/N_{eq} = \Sigma n_i/N_i$ and the new curve that defines durability. From n_{eq}, the new Wöhler curve can be plotted.

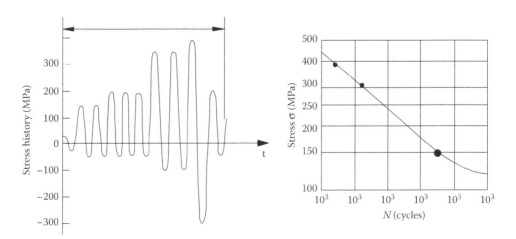

FIGURE 6.16   Qualitative load history and cumulate curve.

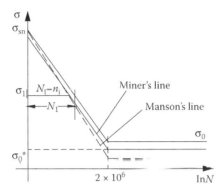

FIGURE 6.17 Qualitative Miner's curve and Manson's curve.

In the semi-logarithmic diagram the life section of the Wöhler curve can be approximated by the equation $\sigma_k N$ = constant, where k is the coefficient of material characteristics, so

$$N_i / N_{eq} = (\sigma_{eq} / \sigma_i)^k,$$

and

$$n_{eq} = \Sigma_i n_i / (\sigma_{eq} / \sigma_i)^k,$$

from which the new tension is σ_{eq}.

Another way of dealing with variable stress is Gassner's method that adapts Wöhler's curve, derived from loads with constant amplitude up to breaking point, by comparing load history to the "equivalent" Wöhler curve—a Gassner curve.

6.8 THE FACTORS INFLUENCING FATIGUE

It has already been stated that fatigue breaks occur due to the formation of internal or external micro-cracks because of load cycles that repeat until they fracture the sample. Any microscopic imperfection, therefore, can trigger a fracture. Perfectly identical materials from a structural point of view might have surface details that are quite different and therefore will have different fatigue limits. The finer the finish, the less likely are imperfections, and thus the higher the fatigue limit. With mirror finishes, given the same working conditions, fatigue limits are higher.

The graph on the left in Figure 6.18 taken from a textbook (Giovannozzi Vol. 1) shows values for coefficient c_1 of reduced fatigue limit (curve parameter of surface finish) against break load. Notice immediately how fatigue limit depends also on break load. Harder steels (generally with higher break loads) are more at risk because they are more sensitive to imperfections.

All of the above highlight that any heat or mechanical treatment capable of reducing imperfections will improve fatigue resistance given the same conditions. Heat treatments that improve the material's structure are certain to improve fatigue resistance. In some cases, treatments can give rise to internal tension that is very positive in terms of fatigue resistance if the material is exposed to opposing external tensions. Heat treatments generally modify material hardness (linked to break load), which does not provide significant improvements in dynamic resistance.

So, one might also deduce that sample or component size can influence fatigue resistance, which would be correct if one thought qualitatively about the origin of fatigue breaks. As regards material structure, the higher the mass, the greater the chance of structural defects.

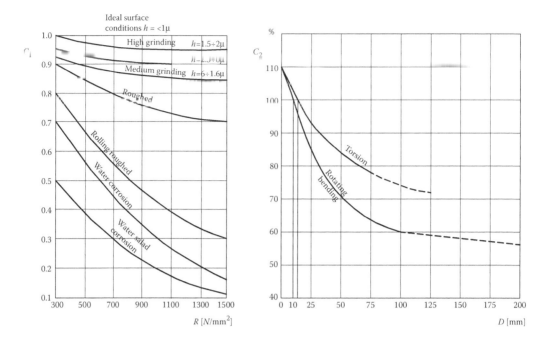

FIGURE 6.18 Coefficients of reduced fatigue limit.

The same applies to the external surface; the chance of surface microdefects is all the higher, the greater is the external surface. Smith's diagram (Figure 29, page 108 of [1]) summarizes this.

Stress type affects materials differently and should be seen in terms of their probability of creating micro-plasticization. In the case of tensile stress, given that the whole sample is subject to it, when a discontinuity reaches plasticization, a fatigue break will start.

Under micro-plasticization, twisting and bending are more difficult as the surface is under most stress. Alternate flat bending is even more difficult because the most stressed points are on opposite sides of a cylindrical sample (Figure 6.19).

In the graph on the left (Figure 6.18), the curve parameters are the type of stress. The reduced fatigue limit is greater the larger the diameter. The same graph that shows the c_2 value of dynamic resistance to use, provides an idea (just an idea) of the reduced percentage of break load according to stress type for the same diameter.

The graph also shows the reduction coefficient to apply to the yield load according to stress type. This type of trend might also predict fatigue limits. The lower curve is for bending. In practice, because of the uncertainties of fatigue results due to a variety of factors, not least the internal tension caused by the bending, it is better to refer to the lower curve.

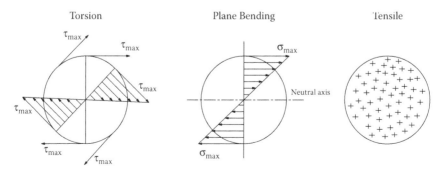

FIGURE 6.19 Qualitative stresses for torsion, bending and tensile.

Mechanical treatments such as shot blasting or rolling can improve fatigue resistance, reducing or even eliminating surface defects that could trigger fatigue fractures. Analyzing the surface of a sample broken due to fatigue generally shows completely different zones. First, there is no plastic deformation. If the stress is due to normal alternating tension, there will be a shiny zone due to the surface having been hammered, and this is the first zone to fracture and trigger further fracturing. There is a second noteworthy zone that is opaque, and as its cross section is smaller it simply snaps off.

As an example, Figure 6.20 shows the case of a gudgeon pin broken by alternating flat bending. Two sides are smooth, an effect produced by hammering prior to the break, and there is a central rough-lined zone typical of snapping off. Surface zones that are visibly different occur in components subjected to alternate twisting. By microscopy, the break zones generally show different aspects that are due stress fatigue.

6.9 SAFETY FACTOR

For static stress, calculating safety levels is immediate. The ratio between ideal tension and the reference load (yield or breakage) provides a foolproof measure of the danger from breakage. Things change if stress is inconstant and variable over time. In this case, one must refer to the three possible alternatives. If cycle number is lower than 10^3 (low cycle fatigue), the safety coefficient can be calculated as if it were static. If the measurement is over time, the safety coefficient must be calculated by referring to the inclined part of the Wöhler curve as mentioned on the subject of the Miner hypothesis (Figure 6.17).

If the stress is infinite, the safety coefficient could be worked out by inserting values for average tension and $\Delta\sigma$ variation into the Goodman-Smith equation and comparing it with the one that intercepts average tension with the dashed line at the edge of the diagram.

The safety coefficient could also be defined as the ratio between the maximum average tension, which defines the x-axis, and the maximum working tension. This would provide a different value for the safety factor. A third safety coefficient might refer to minimum tension. In other words, according to stress type, there would be different values, all giving the idea of danger from breakage—values that would have to be unique. Such values in some cases might be unreal or distorted. It is sufficient to think about the case of a medium tension stress close to a yield tension with very small variations. The ratio of amplitude $\Delta\sigma$ for that value of medium tension in Smith's diagram at $\Delta\sigma$ of work (for the null hypothesis) would produce high values (theoretically infinite) of safety coefficient, whereas in reality the sample is close to breaking.

FIGURE 6.20 A pin broken by alternating flat bending.

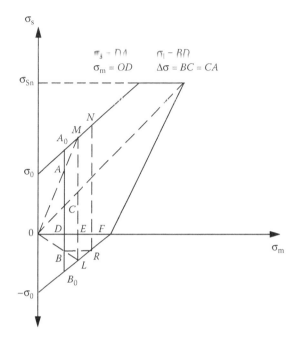

FIGURE 6.21 Smith's diagram for the determination of the safety coefficient.

To avoid any values that are quite different to the safety coefficient, a measure of the danger of breakage, it is calculated that by the time of breakage the sample has experienced proportional increments both of lower as well as higher tension. This means that plotting the tension values into Smith's diagram breakage can be predicted from the intersection of axes.

According to this hypothesis, a single value represents the relationship between real state parameters and those generated by the Smith diagram. The ratio between maximum, minimum, and average tension, and $\Delta\sigma$, is the same. In this way it is better to choose the dynamic stress parameter rather than some other parameter.

Graphs can be plotted in the various diagrams that are defined by a parameter that can directly provide the safety coefficient once the fatigue values are known (Figure 6.21).

Figure 6.22 is an example. The outside dashed line has a safety coefficient of 1, whereas the inside dashed line is greater than 1. To calculate the safety coefficient, the ratio between the material's stress limit (as defined earlier) and the corresponding work tension must be defined, and this is corrected by multiplying the coefficients c_1 and c_2 (< 1) and then dividing everything by β_i, the

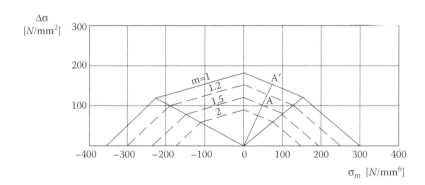

FIGURE 6.22 Parametric m curves in Haigh's diagram.

coefficient of the notch effect, which, as we shall see, is a characteristic of the component, taking into account its shape and material.

Referring to Smith's diagram in Figure 6.21:

$$m = \frac{ML}{AB\beta_i} c_1 c_2,$$

with c_1, c_2, and β_i as above. When loads are variable, resistance from the origin and to oscillation that are used to make the Smith diagram must be calculated before correcting the Wöhler curve, according to Miner.

When designing for time (cycle number $10^3 - 2 \times 10^6$) by calculating n_{eq}, the Wöhler curve corrected by c_1, c_2, and β_i provides the residual life and consequently its durability. Smith diagrams can be made with infinite life resistance or with timed resistance (cycle number) in just the same way as for safety coefficient.

6.10 STRESS CONCENTRATION FACTOR

Until now, nothing has been said about how shape could have an effect on fatigue resistance. Our reasoning is based on the influence of the internal and external (finish) structure of the sample as well as its diameter. This latter hints at the idea of geometrical size and for good reason; the shape of a component can influence its behavior under dynamic stress.

Studying tension by rigorously applying the elasticity theory to a component with a different shape to Saint-Venant's produces different tension values. So, the general equations of elasticity theory (equilibrium and congruence limits) must be applied. For reasonably uncomplicated geometrical shapes, Neuber provides a solution for tension state by calculating tension along the whole perimeter and arriving at a maximum value (Figure 6.23).

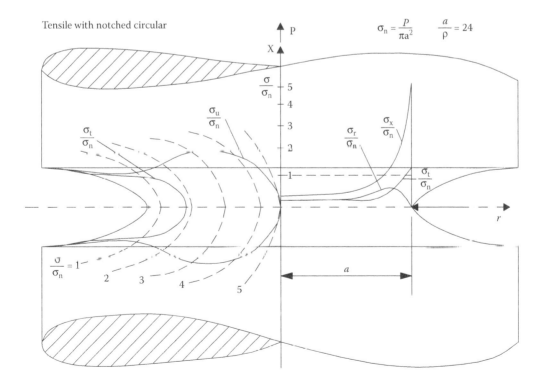

FIGURE 6.23 Stress concentration in a circular notched shaft.

By applying Neuber's method to flat notches, we obtain what the literature calls the stress concentration factor, the ratio of maximum theoretical tension and Saint-Venant's nominal tension, calculated at the same point:

$$\alpha_i = \frac{\sigma_t}{\sigma_n}.$$

The stress concentration α_i, which accounts only for the component's shape, evaluates increments of tension in a point compared to the average tension (nominal) in that point.

For complicated shapes, those you find in practice, it is not always possible to find an analytical solution to tension. There are experimental or numerical methodologies that describe tension trends in every point and so arrive at the maximum tension.

Among the numerical methods, there is the Finite Elements Method (FEM) and the Boundary Elements Method (BEM) that is a part of other courses. These discrete methods apply the hypothesis of perfect elasticity as much as possible to the boundary of the notch. The calculation extrapolates boundary results.

Among the experimental methods, the photo-elastic one provides good results in uncomplicated cases for increments of tension compared to normal values. Even strain gauges can help get more accurate results.

When experimental methods require real α_i values, the samples must have perfect elasticity, or better still, no plasticity whatsoever, such as chalk. Static tests on chalk can provide values for the increments of tension due to notches compared to nominal tension.

Generally, every time the axis or surface varies, the tension flow changes and so do the tension values, which in fragile materials means a change from nominal values to values which are α_i times higher.

One could also say that the higher the variation in tension flow, the higher is α_i. The stress applied to a notched component depends not only on tension, but also in most cases on three main tensions (given a particular breakage theory) and an ideal tension. To release the shape factor from breakage theory, it is defined in the literature as the ratio of nominal tension to maximum tension, even for multi-axial tensions.

Figure 6.24a and 6.24b show examples of Neuber diagrams for some cases of notch from a textbook.

With notch, there are increases in tension that for fragile materials push them to breakage, or for more ductile materials such those used in mechanics, to plasticization. In designing mechanical components, one should avoid concentrations of tension by using linkages or geometries that lighten the load. Figure 6.25 shows some examples of how to improve fatigue.

In fragile materials, notches increase tension by α_i times the nominal values and breakage occurs for $1/\alpha_i$ times the break load of a similar example without notch. So, as soon as the breaking point is reached, it breaks off because it cannot deform itself before this. The fracture propagates immediately to surrounding areas because that section has lost its resistance.

In ductile materials there is a different fracture process in which there are local deformations before breaking point. Under static tension, the material deforms plastically once its elastic limit is reached. In ideal plastic conditions, stress would remain constant in that point, and as load increased, the surrounding area would reach plasticity and things would progress to breaking point with large global deformation.

If the sample has notch, the static load needed for breaking point is higher than for the same sample with no notch. This is due to hardening of the material around the notch with a consequent impediment to deformation.

The situation is different for dynamic stress. Once plasticity is reached even in ductile materials, the capacity to share tension diminishes, making it almost "brittle," if the stress is variable with time. So, the material is unable to exploit fully its ductility.

For material without plasticity $\beta_i = \alpha_i$, and for plastic material $1 \leq \beta_i \leq \alpha_i$.

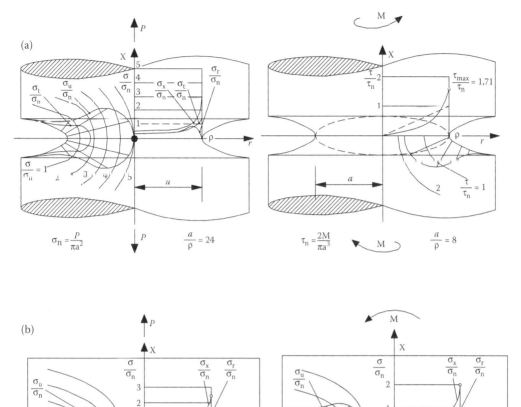

FIGURE 6.24 Examples of Neuber diagrams

So, β_i can be defined as the ratio between effective tension, because of which in dynamic stress conditions the material breaks, and the relative nominal tension:

$$\beta_i = \frac{\sigma_e}{\sigma_n}.$$

By effective tension, we mean the tension at the start of plasticization in the point of most stress, which depends on the plastic deformation of the material.

In the literature, η is defined as sensitivity to notch and is the ratio between the increase in effective tension compared to nominal tension and the increase in theoretical tension compared to nominal tension:

$$\eta = \frac{\sigma_e - \sigma_n}{\sigma_t - \sigma_n}.$$

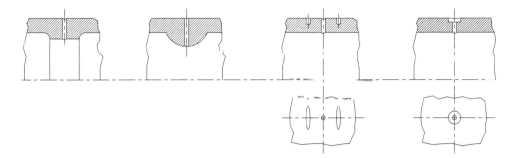

FIGURE 6.25 Examples to improve fatigue resistence.

The sensitivity to notch factor is unfortunately not constant, and depends on the elasto-plastic characteristics of the material. In practice, the value of β_i for a certain material can be obtained as the ratio between the fatigue limit (oscillation resistance) of samples without notches and that for notches with the same cross section under oscillating stress (Figure 6.26).

Because the case for static stress is different to the one for oscillating stress, imagine it is made of a constant average tension that has no effect on fatigue, and an oscillating tension that does have an effect on fatigue resistance. As an example, Table 6.1 contains general indications about the sensitivity to notch of several materials.

The diagram in Figure 6.27 also shows the sensitivity to notch trend for the same notch radius for different metals.

With values from the table and diagram, the coefficient of the notch effect can be calculated:

$$\beta_i = 1 + \eta(\alpha_i - 1),$$

which is what is needed for designing.

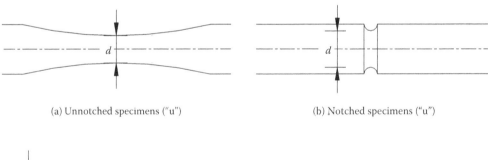

(a) Unnotched specimens ("u")   (b) Notched specimens ("u")

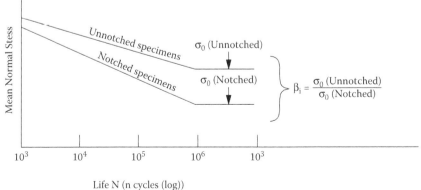

Life N (n cycles (log))

FIGURE 6.26 Wöhler's curve to evaluate β_i.

TABLE 6.1
General Indications about the Sensitivity to Notch of Several Materials

Material	η
No alloyed steels	$0.40 \div 0.80$
Alloyed steels	$0.65 \div 0.80$
Hardening and tempering steel	$0.80 \div 1.00$
Austenitic steel	$0.10 \div 0.35$
Alluminum alloy	$0.40 \div 0.75$

The table and diagram data highlight that materials are all the more sensitive to notch ($\eta = 1$, $\alpha_i = \beta_i$); with notch, the breaking point is higher and correspondingly hardness is greater.

It should be noted that as the notch radius decreases, the material's sensitivity to notch diminishes, because as the notch's dimensions approach those of the imperfection in the lattice, that imperfection predominates. In other words, the material is naturally notched.

Heywood's study uses a formula that summarizes these concepts:

$$\beta_i = \alpha_i / (1 + a / (b \cdot r)^{1/2}),$$

where a depends on the type of material, b the notch shape, and r the length of notch.

Giovanozzi's text contains diagrams that provide values of β_i for practical cases of variously loaded machine components. The same text contains Peterson's diagrams, useful for fatigue design and for calculating β_i.

Peterson writes about the notch sensitivity trends of different types of material with the notch as parameter (Figure 6.28).

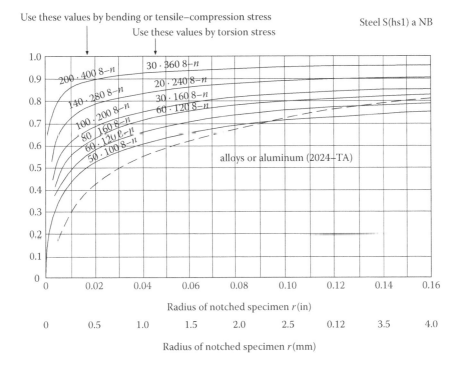

FIGURE 6.27   Notch sensitivity trends of different types of material.

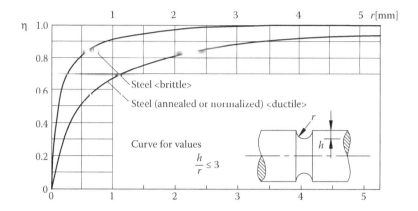

FIGURE 6.28 Notch sensitivity according to Peterson.

So, the value α_i is reported for different types of notch, being the ratio between ideal tension (distortion energy for ductile materials and the maximum tangential tension for fragile materials) and nominal tension as a function of the geometric parameters that characterize the notch with the "geometry" of the parameter of the curve (Figure 6.29).

For three-dimensional cases and deriving α_i from the breaking point hypothesis, the text includes diagrams for fragile and ductile materials (for the plane curve, the two hypotheses have the same value α_i).

So, as regards the danger of fatigue breakage, higher performance steels do not always provide a proportionally higher resistance to fatigue. The advantages are not so evident as to justify the choice of more costly materials. The increase in fatigue resistance can be relatively limited compared to the increase in cost of any given component.

To this point, everything has been about high cycle fatigue resistance. Low cycle fatigue cannot be based on the status of tensions, but on effective deformation. Component breakage must be seen in relation to the slow deformation cycles to which the material is subjected, so the approach to its capacity to resist is different. Generally, β_i values are initially precautionary compared to the equivalent parameters in a low cycle system.

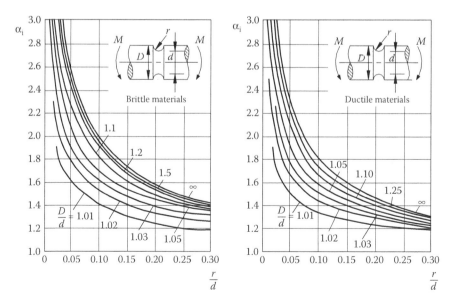

FIGURE 6.29 Example of Peterson's diagram.

6.11 FATIGUE RESISTANCE OF WELDED MACHINE COMPONENTS

Calculating the fatigue resistance of welded machine components follows the fatigue calculation already cited. In evaluating the safety coefficient, Smith's diagrams for the materials (smooth samples) can be referred to, taking into account all those parameters that compete in weakening the weld. The parameters are summarized thus:

- Effect of the material's physical deterioration close to the weld bead as a result of the high temperature needed for welding and its subsequent cooling
- Effect of weld quality in terms of bead imperfections
- Effect of the bead shape (convex, concave, etc.) and the relative connectors between the two welded parts
- Effect of the geometry of the welded parts
- Effect of the geometrical sizes of the weld (length, thickness, etc.)

The first factor is a function of steel type and more precisely steel "stability" during welding, which usually causes enlargement of the crystalline grains with a weakening in the zone close to the bead. Even the precipitation of carbide ligands, important for material durability, is a candidate for the first factor.

The second factor is a function of the welder (qualifications) and the conditions in which the weld is carried out. A perfect weld shows no defects such as lack of penetration, blowholes, etc.

The third and fourth factors together produce the notch effect and influence most the fatigue resistance of the weld.

The fifth factor accounts for the probability of having large volumes of material be defective, a function of the size of the components being welded together.

Bobeck's calculation covers all of the above. He refers to Smith's diagrams for samples welded (V) for an A37 steel (Figure 6.30).

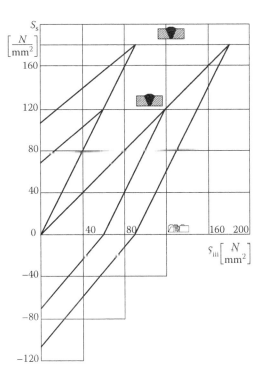

FIGURE 6.30 Smith's diagrams for samples V welded joints.

To calculate the degree of safety, Bobeck does not use proportional increase but notes σ_{max} and σ_{min} during operation, and works out the corresponding average tension σ_m, so that on the Smith diagram the breaking point $\Delta\sigma_L$ corresponds to σ_m. Before comparing $\Delta\sigma_L$ with $\Delta\sigma_w$ during working, he multiplies it by coefficient c that takes into account the various causes of decreased durability, so the safety coefficient becomes:

$$m = c\,\frac{\Delta\sigma_L}{\Delta\sigma_W} \quad \text{with } c = c_0 \cdot c_1 \cdot c_2 \cdot c_3 \cdot c_4.$$

- c_0 takes into account the durability of the base metal, 1 for an A37 steel, >1 for other steels.
- c_1 takes into account the welder's skill, 1 for competent/qualified welders, >1 for the others.
- c_2 takes into account the shape and orientation of the bead (local shape factor) >1 according to the table in Figure 6.31.
- c_3 takes into account notch effects due to the position of the bead in relation to the shape of the join (geometric shape factor)—c_3 >1 according to the disturbance trend of tension flow.
- c_4 takes into account the sizes of the join components, c_4 >1. Figure 6.31

Often, when designing welds fatigue, diagrams like Bobeck's are not available (diagrams for steel and weld type), so the deterioration of the material's characteristics due to the welding process must be taken into account.

This increased fragility of the material reduces durability and fatigue resistance; even residual tension, a physiological result of welding, contributes.

In the absence of Smith's diagrams, not so in Bobeck's case, one could refer to a fatigue homothetic diagram of the base metal, using a reduction coefficient smaller than 30% for those steels less prone to becoming more fragile. So, stability is the prime concern, especially closest to variations in limited loads where, compared to fatigue limit, yield tension is sensitive to increasing fragility.

The fatigue calculation for pipe welding is carried out when the installation is subject to internal pressure variations that could create significant stress cycles, or when its geometry might be subject to vibration because of circulating fluids or because of its proximity to other vibrating systems. In this case, the fatigue calculation should follow the appropriate fatigue design with adequate analysis of the

C_2 Values for corner joints with different forms of welding and other solicitation (Bobeck)						
	Welding corner			Concave welding with a vertex or with opposing points		
Forms welds	Simple plan	Double plan	Double concave	Simple	Double with slit	Double without slit
Tensile-compression ..	0.4	0.6	0.7	0.7	0.7	0.9
Bending..........................	0.2	0.8	0.9	0.6	0.8	0.9
Shear.............................	0.4	0.6	0.7	0.7	0.8	0.9

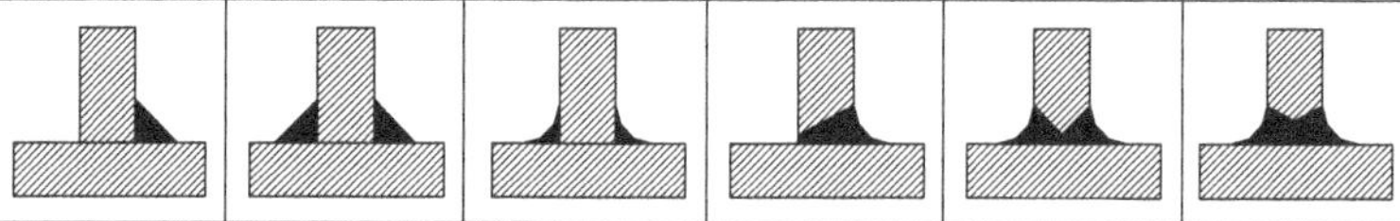

FIGURE 6.31 Values of c_2 according Bobeck's indications.

factors that could influence weld fatigue (geometry and local shape), especially when there are pipe joins, and then be followed by the coefficients that take into account all the factors bearing on fatigue.

6.12   INTERNAL MATERIAL DAMPING

We have already seen that if a material is stressed by a tension below the fatigue limit, plastic deformation is not produced within the material, and even for cyclic stress there are no significant increases in temperature. Nevertheless, due to friction between the material's constituents, even elastically, deformation lags behind stress. There is a certain hysteresis, which explains why work lost to friction in a cycle is not zero. As a consequence, there is a slight rise in temperature that is difficult to detect if instruments are not very sophisticated. This slight rise is easily mixed up with background noise.

The increase in temperature is due to friction, and its amount could provide an idea of the material's behavior to fatigue.

Substantial increases in temperature are only noted when micro-plasticization begins, so the heat produced is not only due to the natural lag between tension and deformation but also to plastic deformation that gradually becomes more evident (compared to elastic deformation) as load is repeated (see Figure 6.32, from a work of Prof. Calderale).

For a steel of fatigue limit 255 N/mm^2, Figure 6.32 shows a parameter proportional to damping per unit area and cycle on the y-axis; on the x-axis is the number of cycles. The curve parameters are the maximum applied stress and show that damping is qualitatively linked to the surface temperature of the sample.

If stress is below the fatigue limit, the only heat produced is due to elastic hysteresis, so the conditions for breakage are never reached. If stress is above the fatigue limit, progressive plasticization produces substantial temperature rises that are easily measurable. As the temperature suddenly rises as soon as the fatigue threshold is passed, so internal yield undergoes a similar behavior.

Consequently, internal yield measurements can provide information about fatigue behavior in the same material.

Supposing the applied tension varied according to the sinusoidal law $\sigma = \sigma_0 \sin \omega t$ and was lower than the elastic limit at every point on the curve, the corresponding deformation would be sinusoidal but with a phase shift $\varepsilon = \varepsilon_0 \sin (\omega t - \varphi)$.

By this hypothesis, elastic hysteresis can be calculated per cycle and per volume:

$$A = \int \sigma d\varepsilon = \int \sigma_0 \sin \omega t \varepsilon_0 \cos (\omega t - \varphi) \, d\omega t.$$

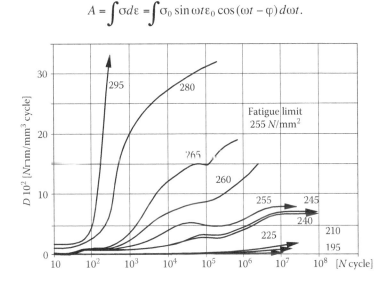

FIGURE 6.32   Damping trend for a steel.

Notice, too, that on the $\sigma - \varepsilon$ plane, the curve is elliptical with a semi-axis along the straight line ξ of angular coefficient σ_0/ε_0, where $a = (1 + \cos \varphi)^{1/2}$ and $b = (1 - \cos \varphi)^{1/2}$ along the normal η. The work calculated above is equal to the area of the elipse:

$$A = \pi \sigma_0 \varepsilon_0 \sin \varphi.$$

The maximum elastic deformation energy per unit volume is

$$E = \frac{1}{2} \sigma_0 \varepsilon_0.$$

Relative yield is the ratio between the energy dissipated per cycle, A, and the potential elastic energy E:

$$\Psi = A/E = 2\pi \sin \varphi.$$

Relative yield can be measured through the logarithmic decrease of deformation. The logarithmic variation of oscillation amplitude over time is defined as a logarithmic decrease:

$$dl = d(\ln \varepsilon_0) = d\varepsilon_0/\varepsilon_0,$$

and by means of the potential elastic energy:

$$dl = d\varepsilon_0/\varepsilon_0 = dE/2E = A/2E = \Psi/2.$$

In the perfectly elastic hypothesis, relative yield Ψ equals 2 dl.

When the elastic limit is exceeded and the material plasticizes, the area of the cycle becomes ever bigger and according to perfectly plastic behavior (Giovannozzi, vol. I, p. 141. Figure 80) being the σ-ε a parallelogram area, it could equal $4\varepsilon_{pl}\sigma_0$, and so relative yield would be

$$\Psi = 8\varepsilon_{pl}/\varepsilon_0.$$

Summarizing:

- In the first case for ideal elasticity $\Psi = 2\pi \sin \varphi$
- In the second case for ideal plasticity $\Psi = 8\varepsilon_{pl}/\varepsilon_0$

Analyzing the two results, the value for relative damping is constant for every cycle, but the second value varies slightly as plastic deformation progresses, 0 being for no plastic deformation and 8 being high deformation.

Instead of looking at a single unit, one looks at the whole volume; relative yield per cycle remains constant because all the points are subject to the same stress (elastic field). For plastic deformation, the overall value per cycle varies because the plasticization reached by every point progresses with every cycle. So, it can be deduced that internal yield provides an idea of the capacity of a material to plasticize prior to breakage. Fragile materials have relative internal damping values close to zero, while plastic materials with large deformations prior to breakage have large values (Figure 6.33).

These materials are capable of great local deformation prior to breakage and so can absorb stress points due to imperfections or notches.

Internal yield can be measured through calculating dl, by oscillating the sample calculating the logarithmic variation of every cycle's amplitude, or by measuring temperature variation after

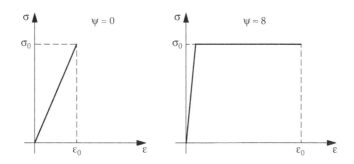

FIGURE 6.33   Values of Ψ for perfectly fragile material and for perfectly plastic material.

subjecting the sample to oscillating loads. Since it is practically impossible to take measurements in precise spots, the entire sample must be considered, so the load must be such that it is mostly constant across the whole sample. So, in Foppl's tests to find internal yield to twisting, he used thin pipes that maintained a constant tangential tension and he applied a constant twisting moment.

The remarks about internal yield confirm the validity of research into fatigue by measuring temperature increase. In addition, the rapid rise in yield (and temperature) over resistance to oscillation further confirms validity (Figure 6.32).

6.13   TEST MACHINES

Fatigue test machines can be classified into those with fixed loads and those with fixed deformation.

In test machines with fixed loads, applied stress always reaches sample breakage, which happens faster than in machines testing for fixed deformation. Once breakage starts and since the load is constant, the resistant cross section gets smaller faster. In this case, sample breakage reduces rigidity with a proportional reduction in the maximum applied load.

If, for example, a machine applies a tensile stress on the test sample, it is always equal to the ratio of load P to cross-section A; once breakage starts, it progresses quickly because the area diminishes and so the tensile stress $\sigma = P/A$ increases. In the case of an imposed deformation regulated by $\Delta l = Pl/EA$, which is constant, P/A must be constant, and therefore the breakage occurs more slowly because it adapts to the load.

Another way of categorizing machines is to subdivide them by how loads are applied: oil pressured, electromagnetic, and mechanical.

The first testing machines were mechanical, as the loads applied were weights, but today the most common are oil pressured, which apply pressure by means of pumps regulated by electric valves that are able to operate in extreme conditions depending on test frequency. Electromagnetic machines are used when test loads are more modest. If thermographics are used to test for fatigue, fairly uncomplicated servo-actuators can work at low frequencies.

A third category is the type of load applied to the sample. There are machines that bend (rotary or flat), apply tensile stress, and twist. Some machines can even apply a combination of the above (bending and twisting).

The main components of a test machine are: chassis, load applicator, and gauge. The chassis is as heavy as the loads applied. The applicator of loads (time variable) is particularly delicate as regards the gauge and timing. Generally, the higher the load, the lower the frequency the machine can operate at, assuming that the machine's activation power is a constant. The electronic torquemeter and dynamometer are particularly important. These machines are able to apply loads according to any law and therefore reproduce "real" load histories.

Figures 6.34 to 6.38 show some fatigue test machines of the categories reported above.

For further particulars and a more detailed description, refer to Giovanozzi, vol. 1, which carries an ample number of test machines (some manufacturers no longer exist, having been taken over by others).

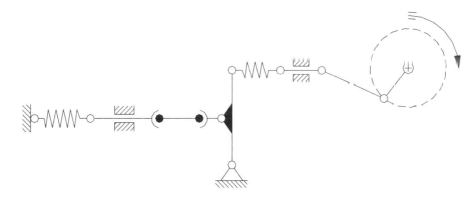

FIGURE 6.34 Scheme of a fixed drive test machine with an elastic mid-section. Structural elements: Small horizontal pulsator PH made by Schenck-Erlinger.

From a mechanical viewpoint, it is interesting that today's test machines are very similar (structure, load application, gauge system, etc.) to those of the past. The change is in the electronics (control systems, acquisition systems, etc.) that have been simplified in their use and programming.

Electronic control systems and electric valves have enabled the use of oil-pressured machines in which loads are applied via hydraulic circuits.

6.14 CONCLUSIONS ABOUT DESIGNING A MECHANICAL COMPONENT

Nearly all mechanical components (apart from the external frame or gearbox) are subject to fatigue. This chapter has provided general information on how to design or inspect components.

In the following chapters, we will look at how to include stress and deformation (springs, cogwheels, shafts, etc.) in the calculations. On implementing the design or inspecting it, the concepts in this chapter must be included.

With reference to the symbols thus far employed, the design or verification of a component part of a mechanical system must undergo the following phases:

1. System analysis and identification of the functions and constraints of components
2. Power, torque, and forces at play
3. Variation with time of power and torque
4. Kinetics chart
5. Determining the torque and forces operating on the component (load history)
6. Choice of materials (knowing static and dynamic behavior)

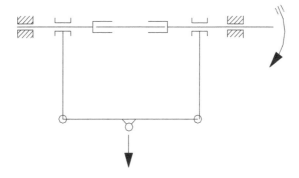

FIGURE 6.35 Scheme of non-resonating rotational bending test machine.

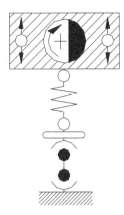

FIGURE 6.36 Scheme of resonating operation with centrifugal stimulation Structural elements: Vertical pulsator PV made by Schenck-Feder.

7. Structural calculation chart (axes with constraints and forces)
8. Component dimensions (verification) or hypothetical dimensions (design)
9. Choice of calculation instrument (Saint-Venant, finite elements, finite differences, etc.)
10. Determining maximum and minimum amounts of stress and deformation
11. Assumption of breakage hypothesis depending on material
12. Choice of safety coefficients m in relation to system employment
13. Static verification of the component by comparing maximum ideal tension σ_i with allowable tension σ_a (6)
14. Fatigue verification after calculating maximum σ_s and minimum σ_i ideal tensions σ_s (7) from the fatigue diagram
15. Assumption of the coefficients c_1, c_2, β_i and calculation of the fatigue safety coefficient

For a design, optimizing results means repeating the calculations from point (7) on.

When using today's processing techniques, experience teaches us that it is always better to make a first approximation (or inspection) by simple processing (beam theory, Saint-Venant), which forms

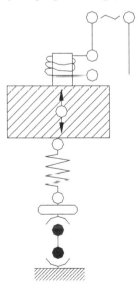

FIGURE 6.37 Resonating operation with electromagnetic stimulation Structural elements: Amsler Vibrofore, Testronic 7001 made by Russernberger-Muller.

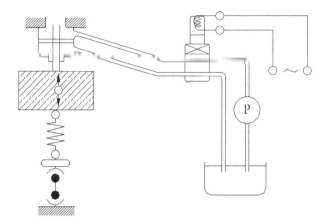

FIGURE 6.38   Scheme of resonating operation with electro-hydraulic stimulation. Structural elements: Pax 750 made by Schenck.

the basis for subsequent processing that will take into account the whole range of tension and defor-mation and deal with the component's behavior if necessary, given the constituent curves of the material beyond the classic elastic field.

It is on the basis of this concept that subsequent chapters will deal with studying the mechanical components of a system covering points (7) to (13).

Design of components is ever more rigorous and must take into account the criteria in Chapter 2, when function becomes a condition of environmental respect (recycling, reuse, etc.). The design route should run through points (6) to (15) with the addition of durability tests and environmental type. This procedure (6) to (15) should be repeated for all materials applicable to making the component.

7 Optimal Materials Selection in Mechanical Design

7.1 INTRODUCTION

The ever-growing variety of materials for engineering applications widens design possibilities but, coupled with the huge number of pre-requisites that make the choice more difficult, these usually imply tough, multi-criteria, problem-dependent optimization. To help designers address this issue, several systematic methods have been proposed over the years, aimed at selecting the best materials and work processes. As will be seen, quantity selection methods change depending on whether applied to the single properties of materials or to a mix of different properties, and according to which functionality is to be optimized, relating these to other project variables.

There are simple tools for multi-objective analysis, with other analytic and graphic instruments supporting these methods to make a rational, effective selection of materials, depending on the various performances required; consequently, they are of vital help to designers who have a wide range of materials to choose from.

When developing a new project, an optimum sequence for selection of materials can be described as follows:

- Defining product functions (single or system component), formulation of required functions into requisites for each component correlated to property requirements.
- Matching required properties to as many materials as possible, to rule out all those that do not respond to essential requirements (primary constraint selection, or *screening*).
- Analyzing materials that have passed screening and trade-off assessment among project objectives, such as guaranteed performance, producibility, and economic sustainability of the constructed solution, in order to find all those that best satisfy planning needs (or *ranking*).
- Further refining selection, taking into account detailed features of the solution, planning the whole production cycle, and ensuring full compliance with building directives.
- The selection process develops right from the first design concept, and evolves with design detail, working first on a wide range of alternatives and finally arriving at the best solution with detailed design development.

7.2 DESIGN REQUIREMENTS AND CORRELATION WITH MATERIALS

Any component in an engineering system has to carry out one or more elementary functions, withstand static or variable stress, absorb shocks, soften vibrations, permit relative movement, transmit heat, and be electrically isolated. The requirements that express such functions are summarized according to the following typologies:

- Project function parameters, load tolerance, pressure resistance, thermal flow transmission.
- Constraints that regulate how functions are performed, ensuring demands are met (constraints on load tolerance, acceptable deformation), in accordance with some restrictions (geometric and shape constraints, service condition constraints, such as temperature range or other environmental conditions).
- Objectives peculiar to the needs of efficiency functions (lightness, economy, safety, their interactiveness) that qualify as the best solution.

By referring to the functional requirements governing the selection of component materials, their definition needs an analysis of the exact functions a system will carry out and how these will be performed. These requirements can be various, depending on the system. For engineering systems, the following are the primary requisites:

- Mechanical–structural requirements—those that ensure the structural stability of components, so they are referable to all mechanical phenomena they undergo while working. The best known in this category are static load resistance, rigidity, and resistance to fatigue, to creep phenomena, to breakage and to wear.
- Requirements linked to the physical properties of a component—how it reacts to thermal, electric, magnetic, and optical phenomena.
- Requirements linked to environmental conditions—how a component reacts to external agents in the working environment that may cause wear and tear. Among the commonest are resistance to corrosion, oxidation, attack from acids, and other agents.
- Manufacturing and budget requirements—these refer to the most effective and economical way to build the components. They closely depend on the type and variety of forms to be built and on other properties required by the components (finishing type and quality, manufacturing tolerances).

In defining the planning phase, all these requirements become functional parameters, constraints, and objectives. Satisfying any of these three categories depends on the properties of the materials used to build the components. They are classified as follows:

- Physical properties—density, viscosity, porosity, permeability, transparency, reflectivity
- Mechanical properties—elasticity and rigidity parameters (moduli of longitudinal, tangential, and hydrostatic elasticity, Poisson's coefficient), stress-deformation curve parameters (elastic limit, yield or breakage point, elongation or ductility), because of variations in load, fatigue limit, fracture tolerance, shock resistance, creep parameters (high temperature deformation), hardness, and coefficient of friction
- Thermal properties—maximum and minimum service temperatures, melting temperatures, specific heat, coefficient of thermal expansion, conductivity, emissivity
- Electrical properties—resistivity, dielectric potential
- Chemical properties—composition, resistance to corrosion and decay processes (weather, sea water, acids, high temperature gases, UV radiation) oxidation, flammability
- Properties linked to processability—possibility of casting, ease of molding by deformation or removal, potential to carry out thermal, hardening, or welding treatments

The correspondence of component requirements to material properties depends on the relation between physical-mechanical phenomena that govern the requirements and properties of the materials involved in each of these phenomena. Table 7.1 shows some of the best-known sources of malfunction and structural yield as well as properties of materials that react to these phenomena.

TABLE 7.1
Failure Phenomena and Material Properties

Material Properties

Failure Phenomena	Maximum Tensile Strength	Yield Tensile Strength	Yield Comprehensive Strength	Fatigue Properties	Ductility	Impact Energy	Elastic Modulus	Creep Velocity	Toughness	Electrochemical Potential	Hardness	Thermal Expansion Coefficient
Yield		✓										
Buckling												
Creep			✓									
Brittle fracture					✓	✓	✓	✓	✓			
Fatigue	✓			✓								
Corrosion	✓									✓		
Loads under corrosion									✓	✓		
Fatigue under corrosion				✓						✓		
Wear											✓	
Thermal fatigue								✓				✓

7.3   PRIMARY CONSTRAINT SELECTION

In choosing materials, primary constraints refer to essential requirements any project must satisfy, represented by three main types of conditions materials must be provided with:

- Quantity limits—one or more properties of a material must be given a lower or upper limit.
- Quality limits—as above, but where the property limit is not quantifiable, and so corresponds to one or more properties matching a minimum or maximum quality level (low, medium, high).
- Boolean conditions—the primary requirement is the response of a certain material to a particular property, expressible only in Boolean terms (the material satisfies/does not satisfy the requirement).

For instance, a high-temperature pressure tank in contact with aggressive chemical agents, made by bending and welding, has the following screening constraints: minimum quantity limit at highest operational temperature; quality limit to tolerate chemical agents (high corrosion resistance); Boolean constraints on producibility (compatibility with bending and welding operations).

Since these definitions are fixed for the properties of materials, screening consists of a strict selection that rules out any materials unable to satisfy all conditions simultaneously.

7.4   APPROACHES TO THE BEST CHOICE OF MATERIALS

Once all unsuitable materials have been ruled out applying primary constraints, the other materials need ranking on the basis of their effectiveness in satisfying project requirements. There are two approaches to the selection of materials; these can be done in various ways and with different tools, depending on the level of complexity:

- Selection based on material properties—done by working only on their properties (material selection is separate to other project variables)
- Selection coupled with other project variables—the design is such that it requires integrating other variables (material selection cannot be separate from such variables)

In the former case, the project requirement allows a choice that must satisfy only a set of requirements referable to one or more properties that, one by one, ensure the functionalities needed, by satisfying all constraints or any specific goals, these too referable to properties peculiar to the material.

This is the case of project interventions that tend to solve problems that include particularly simple functions and constraints without complex goals. A decomposable project requirement derives, that is, its primary variables can be handled separately, allowing choice of materials independently of other project parameters (geometry and forms), and in most instances, after those parameters have been defined.

Separating the choice of materials from other parameters is generally a one-time possibility to make things simpler for the designer. However, when problems are more demanding, an approach like this slackens the corrective process as, in these cases, rigidly separate choices imply the need to repeat these interventions until a proper solution is found that harmonizes all variables.

In these cases an effective project intervention is needed that considers primary variables and uses different formulations for selecting materials, more suitable for finding a solution that responds simultaneously to as many requirements as possible, harmonizing forms, geometries, and materials.

Apart from function parameters and constraints, requirements for selection of materials almost always become specific goals a designer fixes at the start, in order to find solutions that, in addition to ensuring all functionalities required in accordance with constraints, optimize a few key aspects that involve more properties of a material and other variables.

7.5 TOOLS FOR SELECTION OF PROPERTIES

In dealing with the best selection only on the basis of properties of materials, it is necessary to interpret the requirements linked to the functions needed and any other constraints, as indications of what properties a material ought to have.

A ranking of materials can be made based on two different criteria, depending on how materials and constraints are referable to their properties:

- If they just identify properties coming into play, it is possible to make a choice selecting those materials that feature the most suitable values of these properties (higher or lower, depending on the property).
- If requirements and constraints are expressed by minimum or maximum limits to impose on some properties, or by specific values these properties are to have, the choice can be made ranking materials on the basis of their compliance with these limit values.

7.5.1 STANDARD AND WEIGHTED PROPERTIES METHOD

The first case is typical of a preliminary search for materials, when the design process is still taking the first steps, and very little information is known, being limited to primary requirements. Continue to consider a pressure tank (Figure 7.1), imagine making a pre-survey of materials to make a compressed-air tank that can withstand a maximum service pressure of 1.5 MPa (15 bars), with temperatures up to 100°C, perhaps with the presence of water.

As requirements and constraints related to other variables are not considered in this preliminary phase (tank capacity, shape, bulk, and geometric parameters such as length, diameter, wall thickness) the problem involves the following set of requirements, each referable to specific properties of the material:

- Resistance to load due to pressure (yield point)
- Rigidity (Young's modulus)
- Resistance to breakage (tenacity)
- Stability at maximum working temperature (service temperature, thermal expansion)
- Stability and endurance at service environment conditions (water resistance)

Additional requirements, generally considered in accordance with efficiency and on economic grounds, can also include the following:

- Weight reduction (density)
- Cost containment (cost per unit)

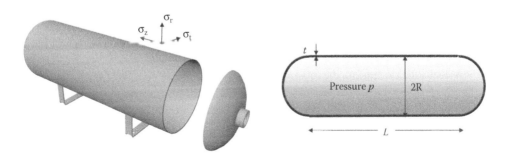

FIGURE 7.1 Example: pressure vessel.

Some of these requirements become primary constraints for the initial step of screening of materials to be selected: the top service temperature of the material must have a minimum value of 100°C, the material must be water resistant.

The assessment of most suitable materials to be used after screening begins consists of a ranking based on the relation among the six remaining properties, differentiating those with high values (yield point, Young's modulus, resistance to fracture) from those with reduced values (thermal expansion, costs).

As in general a material with optimal values in some of these properties might also feature critical values in others, for a better comparison it is advisable to use a couple of useful instruments. Specifically, an expedient can be used to normalize values that properties assume for each material, and then compare these sets of values, also resorting to graphic supports.

Standardization is necessary, as the properties under scrutiny are varied and take on values that are very different from one other. If a set of n_M materials $\{M_i\}_{i=1,...,n_{PM}}$ is to be compared to a set of n_{PM} properties for the $\{PM_j\}_{j=1,...,n_{PM}}$ ranking, standardization can be obtained by the following expressions:

$$N_{ij} = \frac{V_{ij}}{\max[V_{ij}]_{i=1,...,n_{PM}}} \cdot 100, \tag{7.1}$$

$$N_{ij} = \frac{\min[V_{ij}]_{i=1,...,n_{PM}}}{V_{ij}} \cdot 100, \tag{7.2}$$

where V_{ij} is the quantity of the j-th property for the i-th material and N_{ij} is the corresponding standard value. Equation 7.1 is used for properties to maximize, and Equation 7.2 for those to minimize. In this way, in either case the best standard value is the higher, which is closer to the top value, equal to 100.

To compare the materials in Table 7.2, the related equations are applied and standard property values are obtained that can be graphed (Figure 7.2).

This type of graph shows an overall picture of how materials respond to compared properties. In the case illustrated, it is clear how the carbon or inox steel solution turns out from the start to be the most interesting. Comparison results, however, are not always clear. At any rate, for a rational evaluation it is possible to resort to the weighed sum of standard properties by the expression:

$$\gamma_i = \sum_{j=1}^{n_{PM}} \alpha_j \cdot N_{ij}, \tag{7.3}$$

where α_j are the weight coefficients ($\Sigma\alpha_j = 1$) that quantify the importance of each property compared to the others. In this way, it is possible to associate each material to a unique γ_i *index of weighed properties*, which depends on the importance given to the various properties at stake. The material with the highest index is that which best responds to requirements.

Table 7.2 shows the values of γ_i, for the five materials examined, calculated for a specific set of weight coefficients ($\alpha_1 = 0.25$, $\alpha_2 = 0.15$, $\alpha_3 = 0.15$, $\alpha_4 = 0.10$, $\alpha_5 = 0.15$, $\alpha_6 = 0.20$). In this case, too, the carbon steel solution appears the most interesting.

The index of weighed properties can also be used in the following form:

$$\gamma'_i = \gamma_i/C\rho, \tag{7.4}$$

relating γ_i to cost and density product, as previously described. It is a simple way to emphasize the economic factor, which is no longer considered as an element to evaluate when the efficiency of

TABLE 7.2
Properties Method: Materials Comparison

| Materials | Material Properties | | | | | Performance Metrics | | | | |
	Yield Strength [MPa]	Young Modulus [GPa]	Toughness [MPa.m½]	Thermal Expansion [μstrain/°C]	Density [kg/m³]	Cost [€/kg]	γ	ω	ρ/σ_s [kg/MPa.m³]	Cm ρ/σ_s [[€//MPa.m³]
Carbon steel	290	210	60	12	7700	0.60	**74.9**	0.52	26.5	**15.9**
Stainless steel	310	200	95	10	7400	1.50	70.3	0.57	23.9	35.8
Aluminum alloy	240	70	35	24	2500	1.60	48.5	**0.43**	10.4	16.6
GFRP	150	20	16	21	1600	3.10	36.4	0.46	10.7	33.2
CFRP	250	110	18	5	1500	16.00	56.6	0.99	**6.0**	96.0

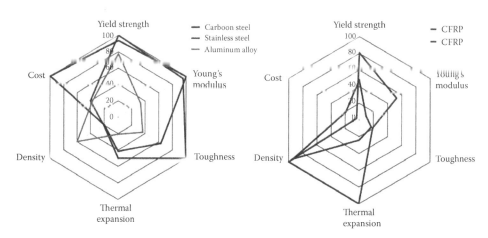

FIGURE 7.2 Properties method: graphic materials comparison.

materials is assessed, but as a primary factor. The cost of materials is therefore ruled out here from the preliminary calculation of γ_i in accordance with Equation 7.3.

This estimate and classification procedure of components can be used even when dealing with non-numerically quantifiable properties, provided they are not quality properties and are correctly standardized numerically.

7.5.2 METHOD OF LIMITS AND OBJECTIVES ON PROPERTIES

The main shortcoming of the standard and weighted properties method is that classification of materials is not made just based on a criterion of efficiency, as most efficient materials indiscriminately maximize or minimize the properties to be detected.

A more effective classification is feasible in the second case, that is, when requirements and constraints are expressible by maximum or minimum limits and also by the specific values these properties have come closest to.

Going back to the pressure tank of Figure 7.1, consider now the case when some project variables have already been set, in particular the tank volume ($V = 30$ L) and the $Q = 2R/L$ aspect ratio (set equal to 0.5). As

$$V = 2\pi \frac{R^3}{Q}\left(1 + \frac{2}{3}Q\right),\tag{7.5}$$

it follows that

$$R = \sqrt[3]{\frac{VQ}{2\pi\left(1 + \frac{2}{3}Q\right)}} = 0.12 \text{ m}\ \ \left(\Rightarrow L = 2R/Q = 0.48 \text{ m}\right).\tag{7.6}$$

The tensile stress the tank is subject to in its cylindrical section, in case of a thin wall ($t < R/4$), can be assessed by simple expressions of σ_r, σ_t, σ_z, radial, tangential, and longitudinal tensions as in the note in Figure 7.1 (see Pipes and Vessels, Chapter 17):

$$\sigma_r = \frac{p}{2} \quad \sigma_t = \frac{pR}{t} \quad \sigma_z = \frac{pR}{2t}. \tag{7.7}$$

Tensile stress on the spherical caps is given by the same σ_r of the cyclindrical section and by tangential tensions equal to σ_t in the same area. Most critical stress conditions occur in the cylindrical section of the tank, and depend on the wall thickness. Assuming $t = 3$ mm, a peak of tension $\sigma_t = 60$ MPa results, as well as an ideal tension according to von Mises of $\sigma_{id} = 52$ MPa (see Failure Theories, Chapter 8).

With proper safety coefficients, a yield point of 140 MPa is set. This property is then considered as an objective, in accordance with a criterion of efficiency aimed at finding the closest solution to the pre-set value so as to ensure the primary function (resistance to internal pressure), exploiting the properties of a single material without using others at too high mechanical characteristics. This is possible only after ensuring that all materials guarantee conditions of structural resistance, with yield point values greater than their target value (it is advisable to insert such values among the primary constraints during the screening phase).

The other $\{PM_j\}$ properties are handled with maximum or minimum limit conditions:

- Young's modulus ≥ 20 GPa
- Resistance to fracture ≥ 10 MPa.m$^{1/2}$
- Thermal expansion $\leq 50 \times 10^{-6}$ 1/°C
- Density ≤ 7000 kg/m^3
- Cost ≤ 5.00 €/kg

These conditions could be used for the initial screening of materials, to rule out those that do not satisfy the requirements. Notice, however, that screenings should be done on the basis of primary constraints only, as this is a rigid selection compared to limit values.

In this case, though limits are used to find excellent materials, they are not to be handled as strict constraints, but as guidelines to support the classification of materials that have passed the screening test.

After defining objective values and maximum and minimum limits, a ω_i index, called a *merit parameter*, can be calculated for each material, from the expression:

$$\omega = \sum_{r=1}^{n_{OBT}} \alpha_r \cdot \left| \frac{V_r}{O_r} - 1 \right| + \sum_{s=1}^{n_{INF}} \alpha_s \cdot \frac{L_s}{V_s} + \sum_{t=1}^{n_{SUP}} \alpha_t \cdot \frac{V_t}{L_t}, \tag{7.8}$$

where n_{OBT}, n_{INF}, and n_{SUP} are, respectively, the number of objective properties, of inferior and superior limits; α_r, α_s, and α_t are the weight coefficients of objective-properties, of those subject to inferior limits, other and superior limits; V_r, V_s, and V_t are the values of tested materials compared to the objective properties, to those subject to inferior limits, and other to superior limits; O_r are the values fixed for objective properties, L_s and L_t are the values of inferior and superior limits.

The material with the lowest merit parameter turns out to be the most effective. In detail, giving more importance to objective properties than to those subject to limits, defines that material as the one with the objective properties closest to fixed values.

In Table 7.2 are the ω_i, values for the five materials tested, calculated for the same set of weight coefficients used for calculating γ_i. As can be seen, the aluminum alloy solution represents the best compromise between objective and imposed limits. The Glass Fiber Reinforced Polymer composite solution, according to γ_i parameters the least effective, reveals itself the second most efficient, as out of all the others it best satisfies the objective property (yield point target value).

7.6   INTEGRATED SELECTION TOOLS

To support design actions aimed at finding solutions to satisfy a wide range of requirements, it is necessary to resort to a different definition of the material selection problem, which may work integrally on all primary design variables, including materials, relevant geometric parameters, and shape features. In fact, in not particularly complex cases, these also affect the materials' properties so that they support primary functions, and so are strictly bound to the choice of materials.

A procedure which, although tending to simplify, can be used for complex cases, looks at a material selection development working on one or more performance functions, deriving from the problem objectives, expressible generically as

$$p = f(F,G,M),$$

where F is the function parameters, G is the geometric parameters, and M is the parameter properties.

When, in these functions, parameters can be de-coupled, they can be expressed as

$$p = f_1(F) \cdot f_2(G) \cdot f_3(M),$$

in which case it is possible to single out all properties of the material under test, defining a *material performance index* where its properties are combined so that they characterize its yield compared to its specific application.

7.6.1   ANALYTICAL FORMULATION

As in any design topic, this approach to integrate selection derives from the formulation of a design problem. In particular, the following are required here:

- Functionality (a component's task)
- Objectives (the factors to be maximized or minimized)
- Constraints (essential requirements to satisfy)
- Free variables (parameters not subject to specific constraints, which can be selected)

The analysis of functionalities deals with the function to solve and the definition of functional parameters to ensure which ones make up the project specifications. Targets tend to direct choices along definite ways, following various efficiency criteria, depending on the case. Constraints most correctly govern this selection process, so that it can provide solutions fully respondent to all constraint-implying requirements. At this level it is presumed that screening primary constraints has been dealt with, and the focus is on significant constraints that are not referable to single properties of materials.

Once the design topic is defined in terms of functionality, targets, constraints, and free variables, the procedure to generate performance indexes is summarized below:

- Definition of objective-expressing equations, which are functions of working, geometric (fixed or variable) parameters, of materials' properties and shape (*objective functions*)
- Definition of constraint-expressing equations, again functions of various project parameters, namely of free variables (*constraint functions*)
- Development of objective functions by the ruling out of free variables, replaced by their analytical expression derived from constraint equations connected to them
- Reformulation of objective functions in a de-coupled form peculiar to functional, geometric parameters, material properties, and definition of the metric that groups all material properties (material performance index)

This index, related to physical–mechanical properties, marks each material for each performance required and, if optimized, maximizes some performance aspects of the application under test. This happens because the form of performance indexes changes with alteration of any of the requirement factors: function, constraints, or objectives.

Consider the simple case of a square-section beam undergoing bending, like that in Figure 7.3a. The component has to perform the function of a leaning-leaning beam undergoing an F midpoint stress. The constraint imposed is the limit to deformations of the beam, which can bend up to a certain angle of the midpoint arrow. This means imposing a limit to its S rigidity (which must be greater than the ratio between the F and the maximum admissible bending arrow f_{max} in the force application point)

Let the objective be to minimize the beam weight, a major efficiency criterion.

As to other design parameters, let us assume the length (L) is fixed in advance, whereas the thickness (b) is not, this being the free problem variable.

Following the same procedure as before, formulate the objective function expression (mass minimization):

$$m = AL\rho = b^2 L\rho, \tag{7.9}$$

and the constraint function expression:

$$S = \frac{48EI}{L^3} = \frac{4Eb^4}{L^3}. \tag{7.10}$$

Deducing the b free variable from the constraint function (Equation 7.10) it follows:

$$b^2 = \left(\frac{SL^3}{4E} \right)^{1/2}, \tag{7.11}$$

which, substituted in the objective function (Equation 7.9), results in

$$m = \left(\frac{SL^5}{4} \right)^{1/2} \cdot \left(\frac{\rho}{E^{1/2}} \right). \tag{7.12}$$

(a)

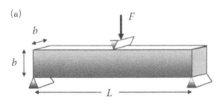

(b)

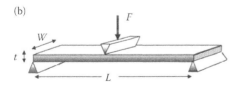

m =	Mass
A =	Section area
L =	Length
b =	Length section side (beam)
W =	Width (panel)
t =	Thickness (panel)
I =	Inertial moment of section
f =	Load
S =	Stiffness
ρ =	Material density
E =	Young's modulus
σ_s =	Yield strength

FIGURE 7.3 Bending load models.

As is clear, the objective function is expressed as $p = f_1(F) \cdot f_2(F) \cdot f_3(M)$ and so it allows the properties of the material to be isolated, and the $\rho / E^{1/2}$ performance index to be found, a combination of ρ density and Young's E modulus. On the basis of Equation 7.12, it can be maintained that, apart from the levels of function (S) and geometric parameters (L), the problem, marked only by the objective and constraint function, can be best solved choosing the material with the $\rho / E^{1/2}$ performance index minimum value.

This index depends only on the function type. Considering the case where resistance constraint is at play instead of rigidity, which prevents the beam from reaching the material yield point, the constraint function (Equation 7.10) is replaced by

$$\frac{M}{2I/b} = \frac{3FL}{b^3} = \sigma_s. \tag{7.13}$$

Deducing the b free variable from this constraint function, a different expression of the same objective function is obtained:

$$m = (3F)^{2/3} \cdot (L)^{5/3} \cdot \left(\frac{\rho}{\sigma_s^{2/3}} \right). \tag{7.14}$$

In this case, the performance index results in $\rho / \sigma_s^{2/3}$, a combination of ρ density and of σ_s yield point.

Considering the case where the function also varies, the index will again change. If, instead of a beam, a panel was undergoing the same stress and constraint condition, such as in Figure 7.3b, the objective function would be

$$m = AL\rho = twL\rho. \tag{7.15}$$

Suppose here that w is another fixed geometric parameter, and the thickness is the free variable. The constraint equation (Equation 7.14) becomes:

$$\frac{M}{2I/t} = \frac{3FL}{wt^2} = \sigma_s. \tag{7.16}$$

Deducing the t free variable out of this new expression and substituting it into Equation 7.15, the following function is obtained:

$$m = (3FwL^3)^{1/2} \cdot \left(\frac{\rho}{\sigma_s^{1/2}} \right). \tag{7.17}$$

The result is a new $\rho / \sigma_s^{1/2}$ performance index, this too is a combination of ρ density and of σ_s yield point, although different in its analytical form.

Changing the objective, with function and constraint being equal, would produce different indexes. Minimization of costs is an immediate case. As costs of materials used for components result from the C_m unit cost times the component weight, cost minimization performance indexes are directly derivable from mass minimization costs, provided that, in their expression, density is multiplied by C_m.

Table 7.3 shows a few performance indexes (in reverse form, to be maximized), if functions vary (tie-rod, bending beam, bending panel, torsion bar, column) and if constraints vary (rigidity, resistance); all are meant to minimize the component mass.

The above indexes refer to most simple stress conditions, representing elementary functionalities of mechanical parts. In practice, many components are referable to these, sometimes mixed, functionalities. In all cases, it is possible to define indexes for components with non-strictly structural functionalities, provided function typology, objectives, and constraints are from time to time defined, and the above procedure is followed.

This approach, then, permits one to spot optimal selection methods of materials, whose properties can be combined to measure their performance potentials, depending on the problem under examination. Performance indexes therefore allow one to look at properties of materials not from a material science viewpoint, which deals with single properties, but from the perspective of the designer, who works with combined rather than single properties, and ranks the potentiality of materials according to the functionality examined. This is the difference between this selection approach and that dealt with in the previous paragraph, concerned with single properties.

Ranking materials on the basis of performance indexes, optimal solutions are found related to functions, objectives, and fixed constraints, independently of all other variables. In particular, once the best materials are spotted, with constraint functions, free variables can be calculated. These values will implicitly be optimal compared to the same objectives that originated performance indexes, thus obtaining a selection that combines with another design variable.

7.6.2 Graphic Selection

The above indices can be used with the help of graphic supports, called *selection charts*, which make it possible to show relations between properties of materials and specific engineering requirements expressed by performance indices.

Consider three distinct cases—the tie-rod, the square-section bending beam, and a bending panel—all undergoing rigidity constraint. From Table 7.3, notice that mass minimization performance indexes are, respectively, E/ρ, $E^{1/2}/\rho$, $E^{1/3}/\rho$ (to maximize). These indexes can be

TABLE 7.3
Performance Indices of Materials

Functions and Specifics on Geometric Parameters	Performance Indices	
	Stiffness Constraint	Strength Constraint
Tie-rod (traction)		
Length fixed, section area free	E/ρ	σ_s/ρ
Beam (bending)		
Length and shape fixed, section area free	$E^{1/2}/\rho$	$\sigma_s^{2/3}/\rho$
Length and height fixed, width free	E/ρ	σ_s/ρ
Length and width fixed, height free	$E^{1/3}/\rho$	$\sigma_s^{1/2}/\rho$
Panel (bending)		
Length and width fixed, thickness free	$E^{1/3}/\rho$	$\sigma_s^{1/2}/\rho$
Torsion bar (torsion)		
Length and shape fixed, section area free	$G^{1/2}/\rho$	$\sigma_s^{2/3}/\rho$
Length and external radius fixed, wall thickness free	G/ρ	σ_s/ρ
Length and wall thickness fixed, external radius free	$G^{1/3}/\rho$	$\sigma_s^{1/2}/\rho$
Column (compression)		
Length and shape fixed, section area free	$E^{1/2}/\rho$	σ_s/ρ

represented on a selection chart showing, logarithmically, Young's modulus and density. On these charts, all potentially usable materials stand out, each in relation to the values that mark them with regard to the two properties (each material is represented by a bubble rather than a dot, as it is assumed to be marked by a range of values rather than exact ones).

Considering then the E–ρ selection chart (Figure 7.4), there is an overall picture of how materials differ compared to the two properties under examination. Moreover, as the scale is logarithmic, performance indexes are graphically representable.

In the case of a tie-rod, the E–ρ performance index is symbolized on the chart by a beam of parallel 1-gradient lines. This occurs because, calling P the generic value taken by the index, it is possible to write

$$P = \frac{E}{\rho} \Rightarrow E = \rho \cdot P, \tag{7.18}$$

which, passing to logarithms, becomes

$$\log E = \log \rho + \log P. \tag{7.19}$$

In the ($y = \log E$, $x = \log \rho$) logarithmic graph, then, to each P value assumed by the index, there corresponds a unitary gradient line that intersects the y-axis on the log P value. It follows that all materials lying on the same unitary gradient line are marked by the same E/ρ value, and so are equivalent as regards the satisfaction of the problem under examination (building a tie-rod that under stress elongates within certain limits, minimizing its weight). Working on lines higher and

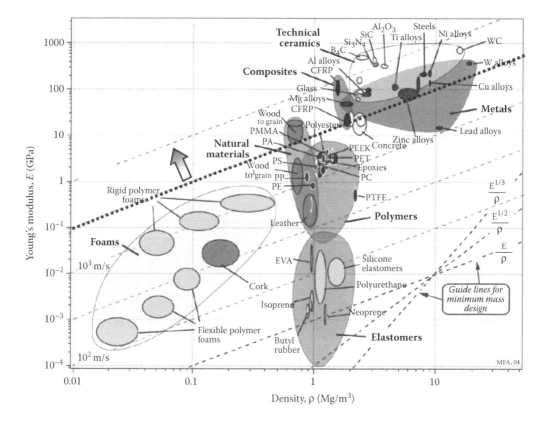

FIGURE 7.4 Young's modulus–density chart.

higher along the y-axis, the index is maximized, and the selection is optimized. For the two other cases, marked by $E^{1/2}/\rho$ and $E^{1/3}/\rho$ indexes, the same selection chart is used in the same way. The variation regards only the gradient of the lines expressing the indexes, which will be two and three (as in Figure 7.4, where the gradient lines are shown, varying depending on the performance index for mass minimization).

Selection charts, connecting physical–mechanical properties of materials suitably combined in relation to the application required, can be variously drawn, also working with groups of more properties on each axis, so as to handle the most diverse design matters.

7.6.3 Application

Going back to the pressure tank introduced in Section 7.5, and drawn in Figure 7.1, this approach to the selection of material first of all needs the problem to be specified in the required form:

- Function—to contain air and resist internal pressure
- Objective—to minimize mass and costs
- Constraint—yield point
- Free variable—wall thickness

All other constraints, apart from what is put into a constraint equation (resistance), are here handled as primary constraints to be screened. It follows that the properties of the material at stake in the problem are handled as follows:

- Service temperature, water resistance, Young's modulus, resistance to rupture, thermal expansion, all become primary constraints for screening, each of them linked to limit values, as before
- The yield point changes into the constraint equation
- Density and costs are handled as objectives (weight and cost minimization of component)

Applying the procedure to calculate the performance index in relation to the first target (mass minimization), first of all the objective equation is obtained:

$$m = 2\pi RLt(1 + Q)\rho. \tag{7.20}$$

It is expressed in this way as the mass to be minimized refers to the material needed to realize the tank, then is equal to the product of ρ density by the volume of the material the tank wall will be made of.

Considering that the maximum stress on the tank wall is tangential to its cylindrical section, and is expressed by the second of Equation 7.7, the constraint equation will be

$$\frac{pR}{t} = \sigma_s. \tag{7.21}$$

Deducing the t free variable from the constraint function and replacing it in Equation 7.20, the objective function is obtained in the form

$$m = 2\pi R^2 L(1 + Q)\cdot(p)\cdot\left(\frac{\rho}{\sigma_s}\right), \tag{7.22}$$

which enables spotting the ρ/σ_s performance index (to minimize), combination of ρ density and σ_s yield point.

Table 7.2 shows the ρ/σ_s values for the five previously examined materials. As can be seen, with this type of approach the Carbon Fiber Reinforced Polymer composite results in the best solution, thanks to its excellent density/strength ratio. The aluminum alloy and the GFRP composite solutions are also quite efficient, unlike the two steel solutions that imply a heavy weight of the tank.

With the constraint equation (Equation 7.21) it is possible to assess the minimum thicknesses a tank wall must have to ensure resistance, depending on the material. Remembering that the maximum p service pressure = 1.5 MPa, and $R = 0.12$ m, minimum thickness values are: 0.6 mm for steels, 0.75 mm for aluminum alloy, 1.2 mm for GFRP, and 0.72 mm for CFRP. In design practice, these estimates are carried out considering a suitable safety coefficient in Equation 7.21.

Charting the σ_s/ρ performance index (reversed form to maximize), from the strength/density selection chart it is possible to have an overview of all potential materials in relation to the performance required. As in Figure 7.5, in the chart the index is symbolized by unitary gradient lines. Focusing on a line in the upper section of the graph (where σ_s/ρ values are high) and excluding all materials below this line, not only is it possible to find confirmation of what was maintained before about the materials considered, but also to spot alternatives, equally efficient, if not more so.

As to the second objective (cost minimization), remembering that these performance indexes are directly derivable from mass minimization indexes, multiplying density by material cost, the performance index will be $C_m\rho/\sigma_s$ combination of ρ density, σ_s yield point, and C_m material unitary cost.

Table 7.2 shows the $C_m\rho/\sigma_s$ values for the five previously examined materials. Considering cost minimization as our objective, the best solution is no longer the CFRP composite, which, owing to its high costs, becomes the worst, but rather the carbon steel solution, and the aluminum alloy also looks almost as efficient.

7.6.4 RESEARCH DEVELOPMENT

In the example case, the objective has been diversified considering two separate aspects, which can provide contrasting indications. Besides, instead of reducing the rigidity constraint to a minimum

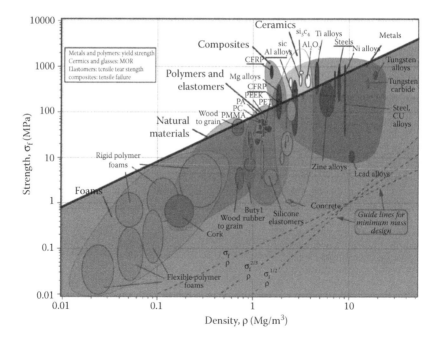

FIGURE 7.5   Strength–density chart.

limit of Young's modulus to be used in the screening phase, it could have been handled like the resistance constraint, writing a second constraint equation into a further performance index. It is easily understood how this approach to selection can in practice be suitably developed, separating objectives and constraints, so as to look into different aspects. This means working with different performance indexes, which express the efficiency of a material compared to the various objectives and constraints imposed. To manage a multifarious, wide set of such indexes, which often gives clashing indications, a designer can always resort to multi-objective analysis techniques, such as that described for the normalized properties method (Section 7.5). In such a case, the mathematical modeling is applied on the performance indexes considered, rather than on individual properties. According to the value attributed to each of them through weight coefficients, the material that best harmonizes all requirements expressed by the above indexes will be spotted and selected.

Part III

Design of Mechanical Components and Systems

Part III of the book is the classic part for a mechanical designer. It treats the arguments that lead to the design of a mechanical component. In several chapters the arguments follow the approach and the treatment sequence as used in the volumes of Giovannozzi that were, and are still, the adopted textbooks. The comparison of Giovannozzi volumes and other books and the results of various studies have suggested only a few integrations. And even if the same calculation models were considered worth mentioning in this collection, we did deliberately avoid using and referring to figures of mechanical systems that would impose the use of Giovannozzi volumes as handbooks.

8 Failure Theories

8.1 MATERIAL STRENGTH AND DESIGN

A material will fail when a certain point is subjected to unfavorable conditions. Breakage occurs when the "load" exerted on that point reaches a particular level. The state of stress that brings about the collapse of the material is known as "stress state limit" and is defined by a group of three main stresses, σ_1, σ_2, and σ_3, and by direction cosines that identify either the planes in which they operate or are from the six stress components σ_x, σ_y, σ_z, τ_{xy}, τ_{yz}, and τ_{zx}. It is well recognized in machine component design that it is indispensible to know about the danger of breakage of any given component. Defining such a parameter is therefore very useful for indicating the degree of safety of the structure being designed.

Evaluating this parameter necessitates different calculation phases once the loads (constant or variable) applied to the structure over time are known.

Initially, the mathematical model should be defined, that is, a move from the real to the conceptual, in which the most important parameters defining the structure are nevertheless vital. The forces operating on the structure are also defined at this stage. In the second phase, once the model has been created, the loads at every point of the structure must be defined, either by traditional methods (Saint-Venant), or by more current design methods as Finite Element Method (FEM) etc.

The next phase requires defining the points of maximum load and calculating the ideal mono-axial load that mimics the danger of breakage at that point. The final phase compares strength data in conditions analogous to reality.

Once the model on which to carry out the calculation has been defined, the loads acting on the structure must be determined. Varying in difficulty, there are a number of methods, from the traditional to the modern (finite elements, boundary elements, etc.), that are relatively simple. What must be determined are all the parameters defining stresses and strains at a particular point.

8.2 STRESSES IN A THREE-DIMENSIONAL OBJECT

A solid body is cut in two and A_2 is removed. To find the equilibrium of A_1, a system of forces must be applied to every single part of the newly created surface S (Figure 8.1).

The aliquot of these forces that acts on dS of the surface could be represented (unless infinitesimally larger) by one single force dF that could be broken down into a normal force dN and a force dT, parallel or at a tangent to dS (Figure 8.2).

Considering dF infinitesimal and having a limited ability to resist any force exerted on it, just as dS is equally infinitesimal, then

$$\sigma = \frac{dN}{dS} \quad \tau = \frac{dT}{dS},$$

where σ and τ are normal and tangential stresses expressed in N/mm^2.

The loads exerted on dS are distinguished by an index to indicate the perpendicular, so that σ_x and τ_x act on dS, whose perpendicular is parallel to the x-axis.

σ_x needs only one index for identification, whereas τ_x, which can relate to any direction of dS plane, needs an extra component depending on y- and z-axes: τ_{xy}, τ_{xz}.

At one single point:

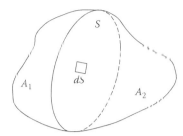

FIGURE 8.1   A solid body is cut in two part.

$$\tau_{xy} = \tau_{yx}, \quad \tau_{yz} = \tau_{zy} \quad \tau_{zx} = \tau_{xz}.$$

$$\frac{\partial \sigma_x}{\partial x} + \frac{\partial \tau_{xy}}{\partial y} + \frac{\partial \tau_{xz}}{\partial z} + \rho_x = 0,$$

$$\frac{\partial \tau_{yx}}{\partial x} + \frac{\partial \sigma_y}{\partial y} + \frac{\partial \tau_{yz}}{\partial z} + \rho_y = 0,$$

$$\frac{\partial \tau_{zx}}{\partial x} + \frac{\partial \tau_{zy}}{\partial y} + \frac{\partial \sigma_z}{\partial z} + \rho_z = 0.$$

ρ_x, ρ_y, and ρ_z are the axis forces exerted on the mass ρdv of minimal unit volume including the one single point with ρ the density. As mentioned above, the load state can also be defined by the main stresses and by directional cosines that indicate the planes' attitudes, that is, those three orthogonal planes between which only perpendicular loads exist ($\tau = 0$).

Let us consider a small straight prism with a right-angle triangular base and height dh in a set direction σ_3 (Figure 8.3). It is possible to work out the stress components (σ_x, σ_y, σ_z, τ_{xy}, τ_{xz}, τ_{yz}) and the main stresses (σ_1, σ_2, σ_3).

Just to remind ourselves of ground already covered in the construction science course, let us consider the simplest case—one plane. The equilibrium equations for the prism in directions σ_φ and τ_φ (Figure 8.3) are:

$$\begin{cases} \tau_\varphi \, ds \, dh = \sigma_x \, dy \, dh \cdot \cos\varphi - \tau \, dy \, dh \cdot \sin\varphi - \sigma_y \, dx \, dh \cdot \sin\varphi + \tau \, dx \, dh \cdot \cos\varphi \\ \sigma_\varphi \, ds \, dh = \sigma_x \, dy \, dh \cdot \sin\varphi + \tau \, dy \, dh \cdot \cos\varphi + \sigma_y \, dx \, dh \cdot \cos\varphi + \tau \, dx \, dh \cdot \sin\varphi \end{cases}. \qquad (8.1)$$

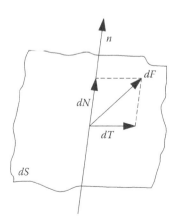

FIGURE 8.2   Elementary surface and elementary forces.

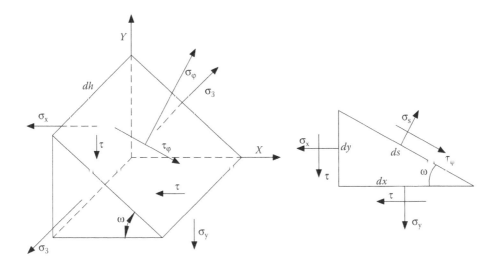

FIGURE 8.3 Stresses components on a prism.

Since

$$\frac{dx}{ds} = \cos\varphi \qquad \frac{dy}{ds} = \sin\varphi,$$

remembering

$$2\sin\varphi\cos\varphi = \sin 2\varphi \quad \cos^2\varphi - \sin^2\varphi = \cos 2\varphi,$$

and dividing Equation 8.1 by $ds\,dh$ is

$$\begin{cases} \tau_\varphi = \sigma_x \dfrac{dy}{ds}\cos\varphi - \tau\dfrac{dy}{ds}\sin\varphi - \sigma_y\dfrac{dx}{ds}\sin\varphi + \tau\dfrac{dx}{ds}\cos\varphi \\ \sigma_\varphi = \sigma_x \dfrac{dy}{ds}\sin\varphi + \tau\dfrac{dy}{ds}\cos\varphi + \sigma_y\dfrac{dx}{ds}\cos\varphi + \tau\dfrac{dx}{ds}\sin\varphi \end{cases}$$

and then

$$\begin{cases} \tau_\varphi = \sigma_x \sin\varphi\cos\varphi - \tau\cos\varphi\sin\varphi - \sigma_y\sin\varphi\sin\varphi + \tau\cos\varphi\cos\varphi \\ \sigma_\varphi = \sigma_x \sin\varphi\sin\varphi + \tau\sin\varphi\cos\varphi + \sigma_y\cos\varphi\cos\varphi + \tau\cos\varphi\sin\varphi \end{cases},$$

leading to

$$\begin{cases} \tau_\varphi = \dfrac{1}{2}(\sigma_x - \sigma_y)\sin 2\varphi + \tau\cos 2\varphi \\ \sigma_\varphi = \sigma_x \cos^2\varphi + \sigma_y\sin^2\varphi + \tau\sin 2\varphi \end{cases} \tag{8.2}$$

The equations above express σ and τ as functions of stresses on the two perpendiculars at x and y in relation to a generic element with variable angle φ. For two elements at angles φ and $\varphi-90°$:

$$\sigma_a = \sigma_x \cos^2\varphi + \sigma_y\sin^2\varphi + \tau\sin 2\varphi,$$

$$\sigma_b = \sigma_x \sin^2\varphi + \sigma_y\cos^2\varphi + \tau\sin 2\varphi,$$

their sum being

$$\sigma_a + \sigma_b = \sigma_x + \sigma_y = \text{cost},$$

defining the sum of perpendicular stresses on two orthogonal elements as constant.

8.3 MAIN STRESSES

From the equations above it is possible to go back to the main stresses, remembering that by definition, in the case of planes, they have a maximum and minimum. So

$$\frac{d\sigma}{d\varphi} = 0.$$

For the classroom, it is better to relate to the main stresses with equations referring to x and y and so for τ zero, the main stresses with τ zero are

$$\begin{cases} \sigma_x \sin\varphi + \tau_{yx} \cos\varphi - \sigma \sin\varphi = 0 \\ \sigma_y \cos\varphi + \tau_{xy} \sin\varphi - \sigma \cos\varphi = 0 \end{cases}.$$

An homogeneous equation system with unknown $\sin\varphi$ and $\cos\varphi$ leads to the main stresses. To provide solutions beyond the banal, the determinant must be zero. For a three-dimensional case (the most common), the determinant is

$$\begin{vmatrix} \sigma_x - \sigma & \tau_{xy} & \tau_{xz} \\ \tau_{yx} & \sigma_y - \sigma & \tau_{yz} \\ \tau_{zx} & \tau_{zy} & \sigma_z - \sigma \end{vmatrix} = 0.$$

The three equation roots are the main stresses ordered as $\sigma_1 > \sigma_2 > \sigma_3$.

For plane case, it is easy to verify that between the stress components and the two main stresses is

$$\begin{cases} \sigma_1 = \frac{1}{2}(\sigma_x + \sigma_y) + \frac{1}{2}\sqrt{(\sigma_x + \sigma_y)^2 + 4\tau^2} \\ \sigma_2 = \frac{1}{2}(\sigma_x + \sigma_y) - \frac{1}{2}\sqrt{(\sigma_x + \sigma_y)^2 + 4\tau^2} \end{cases}.$$

Supposing orthogonal dx and dy to σ_1 and σ_2 and therefore $\tau = 0$, Equations 8.2 are:

$$\tau_\phi = \frac{\sigma_1 - \sigma_2}{2} \sin 2\phi,$$

$$\sigma_\phi = \frac{\sigma_1 + \sigma_2}{2} + \frac{\sigma_1 - \sigma_2}{2} \cos 2\phi.$$

For $\varphi = 45°$, the first gives

$$\tau_{max} = \frac{\sigma_1 - \sigma_2}{2}.$$

8.4 MOHR'S CIRCLES

Mohr used the preceding equations to graphically represent the state of stresses. For plane case on two orthogonal axes, the perpendicular σ and tangential τ stresses on $ds\ dh$ as the angle φ varies, are reported (Figure 8.4):

With

$$OA = \sigma_2,$$

$$OB = \sigma_1,$$

$$AB = \sigma_1 - \sigma_2,$$

$$OC = OA + \frac{AB}{2} = \frac{\sigma_1 + \sigma_2}{2}.$$

For a circle with diameter $AB = \sigma_1 - \sigma_2$ and radius CR at angle 2φ with axis σ:

$$CD = R\cos 2\phi = \frac{\sigma_1 - \sigma_2}{2}\cos 2\phi,$$

$$OD = OC - DC = \frac{\sigma_1 + \sigma_2}{2} - \frac{\sigma_1 - \sigma_2}{2}\cos 2\phi,$$

$$DR = RC\sin 2\phi = \frac{\sigma_1 - \sigma_2}{2}\sin 2\phi.$$

So

OD is the normal stress σ_φ

DR is the tangential load τ_φ inclined at φ to σ_1

The full state of stresses is given by three Mohr's circles, each referring to the three main planes when the three main stresses σ_1, σ_2, and σ_3 are known (Figure 8.5).

As already stated, the main stresses act on the main planes or those planes in which there are only normal stresses (tangential stresses $\tau = 0$).

For plane case, the determinant that links the stress components with the main stresses is

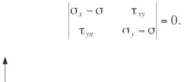

$$\begin{vmatrix} \sigma_x - \sigma & \tau_{xy} \\ \tau_{yx} & \sigma_y - \sigma \end{vmatrix} = 0.$$

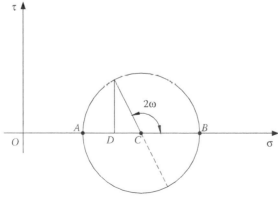

FIGURE 8.4 Mohr's circle.

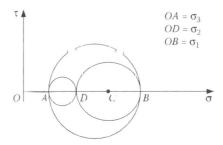

FIGURE 8.5   Mohr's circles for full state of stresses.

For simpler cases, Mohr's circles for each case, and the equations derived from the determinant above, are reported.

8.4.1   SIMPLE TENSILE COMPRESSION STRESS COMPRESSION (FIGURE 8.6)

$$\sigma_1 = 0 \quad \sigma_2 = 0 \quad \sigma_3 = -\sigma \quad \tau_{max} = \sigma / 2.$$

8.4.2   BENDING AND SHEARING (FIGURE 8.7)

$$\sigma_1 = \sigma_3 = \frac{1}{2}\left[\sigma \pm (\sigma^2 + 4\tau^2)^{\frac{1}{2}}\right]; \quad \sigma_2 = 0 \cdot \quad \tau_{max} = \frac{1}{2}\left[(\sigma^2 + 4\tau^2)^{\frac{1}{2}}\right].$$

8.4.3   PURE TORSION (FIGURE 8.8)

$$\sigma_1 = \sigma_3 = \pm\tau \cdot \quad \sigma_2 = 0 \cdot \quad \tau_{max} = \tau.$$

8.5   FAILURE HYPOTHESIS

As regards defining a simple parameter that reflects the degree of safety of a designed structure, the third phase mentioned in Section 8.1 derives from the necessity to evaluate with a mono-axial stress the danger of failure generated by actual loads defined by pluri-axial stresses.

The need for simple failure models in the design phase has led several researchers to different hypotheses depending on the data available. In practice it is quite simple to verify the behavior of

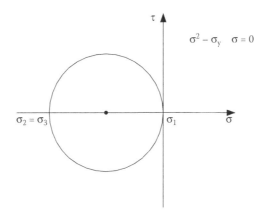

FIGURE 8.6   Mohr's circle for simple tensile stress (compression).

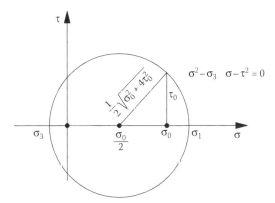

FIGURE 8.7 Mohr's circle for bending and shearing.

a material with mono-directional loads. Through tensile stress or compression tests, data can be obtained that will reveal failure or yield stresses.

As seen before, stress state is generally defined by three main stresses that might cause failure if they work simultaneously. It is indispensable to define an ideal single-axis stress σ_i (ideal stress) that could equal the same failure risk of the three real stresses. This ideal stress σ_i should compare to the single-axis stress (ultimate stress) typical of a material subjected to a single direction stress.

If ever-greater forces are applied to an unloaded object, after a time, some plastic deformation is likely to occur. Each tensile state generating a small deformation corresponds to a set of points defining a limit of elastic state that can be written as

$$f(\sigma_1, \sigma_2, \sigma_3) = 0.$$

This equation provides solutions to problems related to plastic deformation.

The hypothesis of *maximum normal stress* set forth by Rankine, which is inapplicable and cited only for the record, was based on the assumption that plastic deformation only starts when the greatest of the main stresses reaches a value characteristic of the material under stress; or that two states of stress, one on a single axis, the other tri-axis, are equally dangerous when one of the three main stresses reaches the single axis stress (K value) characterizing the material's behavior.

Under such an hypothesis:

$$\sigma_i = \sigma_1 \leq K \ \sigma_i = \sigma_2 \leq -K \ \sigma_3 = 0,$$

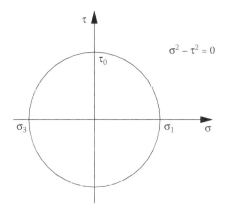

FIGURE 8.8 Mohr's circle for pure torsion.

corresponding to the response of a cylindrical body to simple tensile stress.

Rankine's hypothesis proved inadequate and at odds with experience, as it disregarded the recip-rocal influence of the three stresses in generating plastic deformation. If we consider the cube in Figure 8.9 and want to produce a slight plastic deformation with unloaded lateral sides, load σ_1 ∂K must be applied to the two lined sides. However, if loads $\sigma_2 = \sigma_3$ are applied to the lateral sides, the value of σ_1 that produces equal plastic deformation is definitely greater.

8.5.1 MAXIMUM LINEAR STRAIN HYPOTHESIS

In the case of a cube where $\sigma_1 = \sigma_2 = \sigma_3$, corresponding to a cube surrounded by a pressurized fluid, no plastic deformation will ever occur whatever the load value. Saint-Venant hypothesized that the *maximum linear strain* signals the start of plastic deformation. In other words, two stress states, one single-axis, the other tri-axis, are equally dangerous if they generate identical maximum linear deformation.

Consider the cube (Figure 8.9) again, with Young's modulus of elasticity E and Poisson's coef-ficient $1/m$ (lateral contraction ratio).

The strain components in the three main directions are

$$\varepsilon_1 = \frac{1}{E}\left(\sigma_1 - \frac{\sigma_2 + \sigma_3}{m}\right) \quad \varepsilon_2 = \frac{1}{E}\left(\sigma_2 - \frac{\sigma_1 + \sigma_3}{m}\right) \quad \varepsilon_3 = \frac{1}{E}\left(\sigma_3 - \frac{\sigma_2 + \sigma_1}{m}\right)$$

A single-axis stress (ideal) would produce, for the three directions, the same dilations $\varepsilon_i = \sigma_i/E$ brought about by the actual state of stress, so

$$\sigma_i = \left(\sigma_1 - \frac{\sigma_2 + \sigma_3}{m}\right) \quad \sigma_i = \left(\sigma_2 - \frac{\sigma_1 + \sigma_3}{m}\right) \quad \sigma_i = \left(\sigma_3 - \frac{\sigma_1 + \sigma_2}{m}\right)$$

According to Saint-Venant's hypothesis, the difference between the yield traction stress and the yield compression stress should be around three, this being the assumed m for different metallic materials. Instead, experience shows that the two limits, without interference factors, are practically identical. Saint-Venant's hypothesis was a step forward from the overly simplistic previous one; the danger of failure also depends on the other stresses exerted on the cube.

8.5.2 MAXIMUM TANGENTIAL STRESS HYPOTHESIS

This is a special case of Mohr's hypothesis, according to which failure not only occurs because of normal or tangential loads, but also because of a combination of the two according to a precise law peculiar to each material. In other words, at each normal stress there corresponds a tangential stress

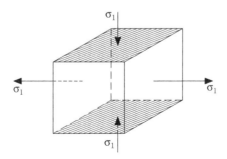

FIGURE 8.9 Cube with unloaded lateral sides.

that causes failure. The curve defining the link between the normal and tangential stresses that cause failure can be obtained from the test results of materials where normal and tangential stresses combine in different ways. This could lead to further complications. Remember that easily obtainable data for material behavior relates to single-axis stresses (traction stress, compression stress).

Mohr's hypothesis refers to the Mohr circle limits for these two states of stress and shows that the curve connecting normal stresses σ and tangential stresses τ is comparable to the dashed line formed by the two tangents to Mohr's circles that represent failure as a consequence of single-axis traction stress or compression stress. The maximum tangential stress hypothesis defines the law between normal and tangential stresses by accounting for the fact that the structure or loaded component yields when the greatest tangential stress reaches a particular limit value and the yield occurs by sliding along 45° planes countering the main stress axes. This means that failure values for traction stress and compression stress are identical and so, with reference to Mohr, the dashed line that replaced the curve connecting normal and tangential stresses degenerates into two parallel lines (meeting at infinity). The formulae defining ideal stress assume that the two stresses, a single-axis and another multi-axis, are equally dangerous when they generate the same maximum tangential stress. Resistance is expressed by

$$-k < \tau_{max} < k.$$

In a two-dimensional state of stress, the maximum tangential stress acting in the plane of the two main stresses equals their semi-difference (see Mohr's circles of figure 8.10). When σ_1, σ_2, and σ_3 are at play in a state of spatial stress, material resistance is assured only if the following ratios apply:

$$-\tau_L < \pm\frac{(\sigma_1 - \sigma_2)}{2} < \tau_L \quad -\tau_L < \pm\frac{(\sigma_2 - \sigma_3)}{2} < \tau_L \quad -\tau_L < \pm\frac{(\sigma_3 - \sigma_1)}{2} < \tau_L.$$

For single-axis stress:

$$\tau_{max} = \frac{\sigma_1}{2},$$

and so the ideal stress σ_i is

$$-\sigma_i < \pm(\sigma_1 - \sigma_2) < \sigma_i \quad -\sigma_i < \pm(\sigma_2 - \sigma_3) < \sigma_i \quad -\sigma_i < \pm(\sigma_3 - \sigma_1) < \sigma_i$$

When one of the main stresses is null, i.c., σ_3 ($\sigma_i = \sigma_1 - \sigma_2$), the limit in the $\sigma_1\sigma_2$ plane is given by the dashed line (1) in Figure 8.11, since $\sigma_i \le k$ always.

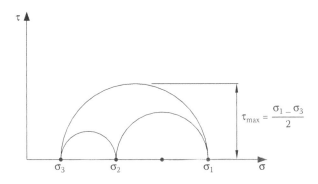

FIGURE 8.10   Mohr's circles for a state of spatial stress.

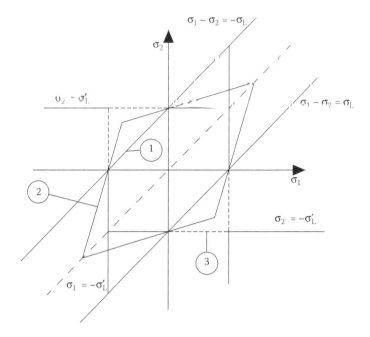

FIGURE 8.11 Comparative stress plane dominion.

This figure shows for a stress plane, for comparative purposes only, the dominions referring to the hypothesis of *maximum tangential stress*, *maximum deformation* (2) and *maximum normal stress* (3). These dominions were obtained with one of the three stresses being null and with materials of k resistance to traction and $-k$ to compression. The figure shows that the safety domain of the hypothesis is totally within the safety domains of the other two cases. This means that hypothesis 1 is more precautionary than the other two. However, remember that symmetrical behavior of a material is at the core of our hypothesis and applies to ductile materials for which $m \approx 3$. Figure 8.12 shows a Mohr representation.

Figure 8.12 shows how effective it is to limit the maximum tangential stress between τ_L and $-\tau_L$ for any state of stress.

For generic plane states of stress ($\sigma_3 = 0$) and stress components σ_x, σ_y, and τ, then

$$\sigma_i = \pm [(\sigma_x - \sigma_y)^2 + 4\tau^2)]^{1/2},$$

and if

$$\sigma_x = \sigma \neq 0 \text{ and } \sigma_y = 0,$$

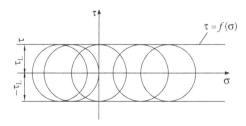

FIGURE 8.12 Mohr's circle for maximum tangential stress hypothesis.

then

$$\sigma_i = \pm [\sigma^2 + 4\tau^2)]^{1/2}.$$

From the equations above for sheer sliding, we have

$$\sigma_i = \pm 2\tau.$$

The result obtained is extremely precautionary as regards states of sheer sliding compared to those derived from other hypotheses, combining caution and appreciable structural simplicity of its formulae, but it only considers two of the three stresses.

Finally, this hypothesis imposes too severe a ratio of τ_s/σ_s (yield values), since it conflicts with reality; however, it does apply to ductile materials. From Mohr's reasoning, it is not hard to find formulae, even when tensile stress and compression failure are different for a given material. These formulae are omitted here since this case best suits materials used in mechanical constructions.

8.5.3 DEFORMATION ENERGY (BELTRAMI'S) HYPOTHESIS

Formulating hypotheses based purely on energy considerations, all three main stresses are taken into account.

The deformation energy hypothesis, or Beltrami's hypothesis, states that the elastic limit is reached when the elastic energy accumulated in a body reaches the maximum value peculiar to that material. Thus, two states of stress, single or multi-axis, are equally dangerous when the strain energy per unit volume reaches the same value.

The energy stored due to the elastic state of stress around one point per unit volume is

$$\Phi = \frac{1}{2}(\sigma_1\varepsilon_1 + \sigma_2\varepsilon_2 + \sigma_3\varepsilon_3),$$

where σ_1, σ_2, and σ_3 and are the main stresses and ε_1, ε_2, and ε_3 the strains in their direction.

By replacing the deformation expressions:

$$\varepsilon_1 = \frac{1}{E}\left(\sigma_1 - \frac{\sigma_2 + \sigma_3}{m}\right) \quad \varepsilon_2 = \frac{1}{E}\left(\sigma_2 - \frac{\sigma_1 + \sigma_3}{m}\right) \quad \varepsilon_3 = \frac{1}{E}\left(\sigma_3 - \frac{\sigma_2 + \sigma_1}{m}\right)$$

it follows that

$$2E\Phi = \sigma_1^2 + \sigma_2^2 + \sigma_3^2 - \frac{2}{m}(\sigma_1\sigma_2 + \sigma_2\sigma_3 + \sigma_3\sigma_1).$$

In the case of single axis stress for deformation energy, we have

$$2E\Phi = \sigma_i^2,$$

leading to

$$\sigma_i^2 = \sigma_1^2 + \sigma_2^2 + \sigma_3^2 - \frac{2}{m}(\sigma_1\sigma_2 + \sigma_2\sigma_3 + \sigma_3\sigma_1).$$

In the case of hydrostatic stress ($\sigma = \cos t$):

$$\sigma_1^2 = 3\sigma^2(m-2)/m.$$

Beltrami's hypothesis is not confirmed in reality since experiments prove that materials can withstand high pressures without breaking. Though insufficient, it has led to formulating other major hypotheses like Von Mises's.

8.5.4 Distortion Energy (Von Mises's) Hypothesis

This hypothesis holds that the yielding of materials does not depend entirely on deformation, but only in the part defined as "twisting" or "shape variation."

According to this hypothesis, the deformation work consists of two parts:

- The former, called Φ', is the energy required for deformation, whereby even if the volume changes, the deformed structure remains geometrically similar to the original.
- The latter, called Φ'', is the energy exerted without any change in volume activating a shape variation; starting from that geometrically similar to the original, then the deformation work is

$$\Phi = \Phi' + \Phi''.$$

This hypothesis, then, clarifies that impending danger to materials depends on "shape variation energy" that must remain below a certain value. So, two states of stress are equally dangerous in terms of failure if they generate the same distortion energy.

The stress state defined by the three main stresses can be thought of as the sum of a state whose volume varies geometrically, and a state that varies geometrically without modifying the volume.

Physically, stable deformation (all geometrical parameters equal) is only obtainable when σ_1, σ_2, and σ_3 are identical. This happens when along each of the main directions there is an average stress equal to

$$\sigma_m = \frac{\sigma_1 + \sigma_2 + \sigma_3}{3}.$$

So, strain along each main axis becomes:

$$\varepsilon_m = \frac{1}{E}\sigma_m\left(1 - \frac{2}{m}\right).$$

Remembering that $\Phi = 1/2(\sigma_1\varepsilon_1 + \sigma_2\varepsilon_2 + \sigma_3\varepsilon_3)$, we can go back to Φ work that is given by

$$\Phi' = \frac{3}{2}\sigma_m\varepsilon_m = \frac{3}{2}\sigma_m^2\left(1 - \frac{2}{m}\right).$$

By substituting σ_m from the previous expression:

$$\Phi' = (\sigma_1 + \sigma_2 + \sigma_3)^2(m-2)/6Em.$$

The state of stress that generates shape variation is defined by the three main stresses $(\sigma_1 - \sigma_m)$, $(\sigma_2 - \sigma_m)$, and $(\sigma_3 - \sigma_m)$. The corresponding energy can be easily calculated with the equation $\Phi = \Phi' + \Phi''$. Then, it follows

$$\Phi'' = \Phi - \Phi'.$$

Substituting the expressions of Φ and Φ', we get

$$\Phi'' = \frac{1}{3E}\frac{m+1}{m}[\sigma_1^2 + \sigma_2^2 + \sigma_3^2 - (\sigma_1\sigma_2 + \sigma_2\sigma_3 + \sigma_1\sigma_3)]$$

For single axis stresses $(\sigma_1 = \sigma, \sigma_2 = \sigma_3 = 0)$, we get

$$\Phi'' = \frac{1}{3E}\left(\frac{m+1}{m}\right)\sigma^2.$$

so σ_{eq} is expressed by the following ratio:

$$\sigma_i = \pm\sqrt{\sigma_1^2 + \sigma_2^2 + \sigma_3^2 - (\sigma_1\sigma_2 + \sigma_2\sigma_3 + \sigma_3\sigma_1)}$$

or

$$\sigma_i = \pm\sqrt{\frac{1}{2}[(\sigma_1 - \sigma_2)^2 + (\sigma_2 - \sigma_3)^2 + (\sigma_3 - \sigma_1)^2]} \tag{8.4}$$

This equation is quite similar to Beltrami's except that it is independent of Poisson's ratio m and allows symmetrical material behavior. If $m = 2$, the two expressions would coincide. In practice, construction materials have a Poisson coefficient greater than two (for steels it is $m = 10/3$), which shows how unreliable Beltrami's hypothesis is and why it is never applied.

For $\sigma_3 = 0$ (plane stress state), σ_i becomes:

$$\sigma_i = \pm\sqrt{\sigma_1^2 + \sigma_2^2 - \sigma_1\sigma_2},$$

and referring to stress components σ_x, σ_y, τ, considering the link between the main stresses, we obtain:

$$\sigma_i = \pm(\sigma_x^2 + \sigma_y^2 - \sigma_x\sigma_y + 3\tau^2)^{1/2}.$$

From this equation, for $\sigma_x = \sigma_y = 0$ (pure shear), it is possible to find the relation between the limit values of normal stress σ_l and tangential stress τ_l.

$$\sigma_L = \tau_L\sqrt{3},$$

leading to

$$\tau_L = \frac{\sigma_L}{\sqrt{3}}. \tag{8.5}$$

Referring to yield point stress, we have

$$\tau_s = 0.58\sigma_s$$

For steels, Φ'' work in a state of plane stress is

$$\Phi'' = \frac{1 + \dfrac{1}{m}}{3E}(\sigma_1^2 + \sigma_2^2 - \sigma_1\sigma_2),$$

also confirmed by experimental tests.

For the case of bending and shearing (Figure 8.7), comparing this work with the analogous single axis strain, we have

$$\sigma_i = \sqrt{\sigma^2 + 3\tau^2}.$$

According to Von Mises, to prevent the material from giving way in the case of plane state stress, it is necessary to comply with the following condition:

$$\sigma_i = \sqrt{\sigma^2 + 3\tau^2} \le k,$$

k being the yield stress of the material. Von Mises's hypothesis is generally reliable for any tensile state and especially for states of pure sliding; therefore it is often used in structural sizing of mechanical components that are made of rather ductile materials.

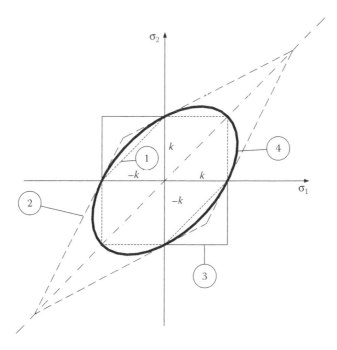

FIGURE 8.13 Comparing of the various theories.

8.6 FAILURE HYPOTHESES: A COMPARISON

A quick survey of the above hypotheses shows there is no single theory that suits all materials types. For example, in two of them, Mohr's and maximum tangential stress, failure depends on two main stresses, but it is easy to see that in practice failure depends on all three stresses. The deformation energy hypothesis itself, using the squares of the main stresses, ignores the sign of the stress and therefore the variety of failure due to traction stress and compression of the many materials used in practice.

Comparing the various theories based on only two non-zero stresses and for materials with equal ultimate traction and compression stress, brings us to a situation like that in Figure 8.13.

Note that Mohr's hypothesis (3) is more cautious, being a hexahedron within an ellipse that corresponds to the distortion energy (Von Mises's) hypothesis (4), thus lending its application to ductile materials (which often have equal ultimate strength values for traction and compression).

A parameter worth considering when choosing the best failure hypothesis is the ratio r between τ_{max} and σ_r, which, for ductile steels, is about 0.55. This ratio suggests the Von Mises's hypothesis as the most suitable, its ratio r being 0.58 (see the pure shear case). For ductile steels and all those alloys that have equal values of ultimate strength to traction and compression, Mohr's theory seems best, as it is simple, easy to apply, and the most frequently used. It is also used in its more general form for materials with different values for failure to tensile stress and compression.

9 Hertz Theory

9.1 INTRODUCTION

The calculation of contact stresses and deformations between two curved elastic bodies is carried out considering the outcome of Hertz theory. The most frequent applications concern the design of ball and roller bearings, and the inspection of gears.

Hertz theory is based on three hypotheses:

- Perfectly elastic behavior of the materials of which the bodies in contact are made
- Lack of friction forces
- Reduced contact surfaces compared to the dimensions of the bodies in contact

If a load is applied to a tennis ball lying on a plane surface in the direction of the sphere axis perpendicular to the plane, the mark in the contact area will have a circular shape. The same happens with a ball bearing, as long as the strain applied does not exceed the elastic limit of the material.

In general, by two elastic bodies pressing against each other, the contact area belongs to a second order surface and has an elliptical shape (Figure 9.1).

The ellipse's semi-axes have the following expressions:

$$a = \mu \cdot q \quad b = v \cdot q$$

where

$$q = \sqrt[3]{\frac{3}{8} P \cdot \frac{\vartheta_1 + \vartheta_2}{\sum \rho}},$$

In this expression, P is the force with which the two bodies are pressed, ϑ_1 and ϑ_2 are two parameters expressed as functions of the elastic moduli E_1 and E_2 and the inverses of Poisson's moduli m_1 and m_2 of the two bodies in contact:

$$\vartheta_1 = 4\frac{m_1^2 - 1}{m_1^2 E_1} \qquad \vartheta_2 = 4\frac{m_2^2 - 1}{m_2^2 E_2}.$$

The term $\sum \rho$ indicates the sum of the principal (maximum and minimum) curvatures of the two bodies at their contact point. By marking the maximum radiuses of curvature without superscript and the minimum ones with superscript for the bodies 1 and 2, it follows that

$$\sum \rho = \rho_1 + \rho_1' + \rho_2 + \rho_2' = \frac{1}{R_1} + \frac{1}{R_1'} + \frac{1}{R_2} + \frac{1}{R_2'}$$

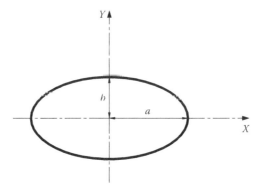

FIGURE 9.1   Elliptical shape of the contact area.

It should be remembered that for a curve:

$$\rho = \frac{\dfrac{d^2 y}{dx^2}}{\left[1 + \left(\dfrac{dy}{dx} \right)^2 \right]^{3/2}}.$$

From a graphic point of view, the curvature center of a curve on a plane at a generic point is obtained by drawing the perpendicular for two points distant δ astride that point; the intersection between the two normals is the center of curvature (Figure 9.2).

The distance $\overline{OP}$ represents exactly the radius of curvature R, whereas $\rho = 1/R$ is the curvature itself.

At a generic point of a surface there are infinite values of ρ, due to the infinitive number of planes with which it is possible to form a sheaf having as axis the line perpendicular to the tangent plane at that point (Figure 9.3). Each plane of such a sheaf intersects the surface along a curve, leading again to a problem on the plane.

Among the infinite planes, two are called principal planes, orthogonal to each other. They contain the minimum and the maximum radius of curvature, respectively. The corresponding curvatures are the principal curvatures.

Take, for instance, a cylinder; consider a point P on any generatrix and draw the tangential plane and the normal line at that point. There are infinite sheaf planes orthogonal to the tangent plane,

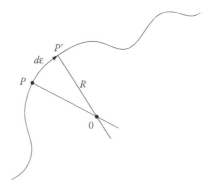

FIGURE 9.2   Graphic curvature center.

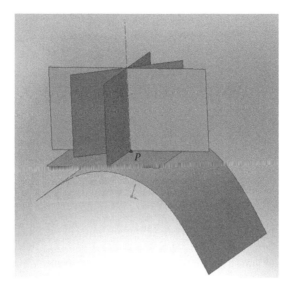

FIGURE 9.3   Sheaf of intersecting planes.

passing through the axis at right angles to the cylindrical surface. Each plane marks a curve in the intersection with the cylinder. The circle obtained as intersection of the sheaf plane, orthogonal to the cylinder axis, with the surface will have a minimum radius of curvature (R), whereas the line obtained as intersection of the sheaf plane containing the cylinder axis with the surface will have the maximum radius (∞) (Figure 9.4).

In a sphere, the radiuses of curvature at any point are all identical. Euler's and Meusnier's theorems are used to determine curvatures on surfaces in revolution solids. Euler's theorem allows the calculation of the radius of curvature in a plane perpendicular to the tangential plane in P, other than the principal plane, known as the principal curvature.

With reference to the sheaf plane forming a generic angle ω with the principal plane (Figure 9.5) having the maximum radius R_1, the curvature $1/R_n$ in P of the intersection curve is calculated by applying Euler's theorem according to

$$\frac{1}{R_n} = \frac{\cos^2 \omega}{R_1} + \frac{\sin^2 \omega}{R_1'}.$$

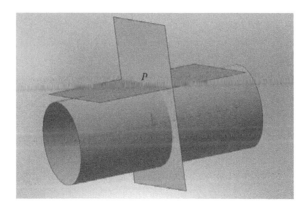

FIGURE 9.4   Intersecting principal planes for a cylindrical surface.

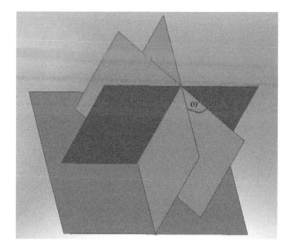

FIGURE 9.5   Sheaf plane forming a generic angle ω with a principal plane.

Once the value R_n is found, it is possible to determine the radius of curvature R in P on a generic plane part of a new sheaf having as axis the track line (P in Figure 9.6b) located by the intersection of the tangent plane with the normal one containing the radius R_n (Figure 9.6b).

Meusnier's theorem allows calculation of the radius of curvature R of the curve obtained as intersection of the surface with an inclined plane of generic angle α with the normal one, where R_n has been previously determined, according to

$$R = R_n \cdot \cos \alpha.$$

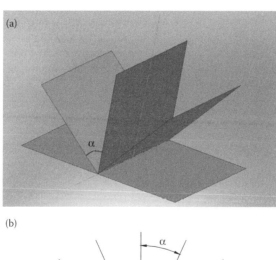

FIGURE 9.6   (a to b) Sheaf of intersection planes forming a generic angle α with the normal principal plane.

Concluding, the curvature related to the normal plane curve can be determined by applying Euler's theorem, and it is possible to calculate the curvature related to the curve located by the plane passing through the track of the former and forming with it the α angle by means of Meusnier's theorem.

When using Hertz theory, curvatures and their radiuses must be assumed as value and sign, positive or negative, depending on the position of their centers.

If the center of curvature is in the "full" section, the radius of curvature is positive; if this center is in the empty section, the radius is negative (Figure 9.7).

Consider μ and ν. They are two coefficients, functions of an auxiliary angle τ, defined by the ratio:

$$\cos\tau = \frac{\sqrt{(\rho_1 - \rho_1')^2 + 2(\rho_1 - \rho_1')(\rho_2 - \rho_2')\cos 2\omega + (\rho_2 - \rho_2')^2}}{\Sigma\rho}$$

where ω is the angle formed by the maximum (or minimum) curvature planes of the two surfaces in contact. The radicand sign always has to be chosen as positive.

As to the special case of rolling bearings, the $\cos\tau$ formula takes on the simplified form:

$$\cos\tau = \frac{(\rho_1 - \rho_1') \pm (\rho_2 - \rho_2')}{\Sigma\rho}$$

being in this case $\omega = 0$.

In this case, $\cos\tau$ is always positive and smaller than 1, as can be verified by observing that the sum of the curvatures obtained by transecting the two bodies in contact with a plane passing through their shared normal is always positive. The dependency of μ and ν on τ remains, as all of them are assigned as functions of a single auxiliary angle ε.

Namely, μ and ν are functions of ε, according to the expressions:

$$\nu = \sqrt[3]{\frac{2E(\varepsilon)\cdot\cos\varepsilon}{\pi}} \qquad \mu = \frac{\nu}{\cos\varepsilon}$$

whereas τ depends on ε according to the formula:

$$\cos\tau = 1 - \frac{2}{tg^2\varepsilon}\cdot\frac{K(\varepsilon) - E(\varepsilon)}{E(\varepsilon)}$$

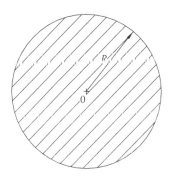
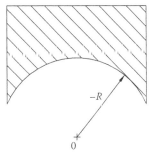

FIGURE 9.7   Radius of curvature positive, and negative.

where

$$K(\varepsilon) = \int_0^{\pi/2} \frac{d\psi}{\sqrt{1 - \sin^2 \varepsilon \cdot \sin^2 \varphi}} \qquad \bar{E}(\varepsilon) = \int_0^{\pi/2} \sqrt{1 - \sin^2 \varepsilon \cdot \sin^2 \psi} \, d\psi$$

are the complete elliptic integrals of first and second kind of modulus $\sin \varepsilon$, respectively. The values μ, ν, in the product $\mu\nu$ as well as the quantity $2K(\varepsilon)/\pi\mu$ are reported in a table (see Giovannozzi's, Vol. I, pp 588–599), directly as functions of $\cos \tau$ with a 1% approximation. When lesser precision is enough, these quantities can also be obtained from the curves in Figure 9.8. in the same text.

The calculation procedure is as follows:

After determining the angle formed by the planes of the principal, either maximum or minimum, curvatures (in case of contact with sphere, which has ∞ principal planes, $\omega = 0$), $\cos \tau$ can be calculated by the previously mentioned formula.

- $\cos \tau$ being known, μ and ν can be inferred from the diagram.

Once μ and ν are known, it is possible to calculate the two semi-axes:

$$a = \mu \cdot q \quad b = \nu \cdot q.$$

Considering the contact mark as elliptic, the average stress value is obviously:

$$\sigma_m = \frac{P}{\pi \cdot a \cdot b}.$$

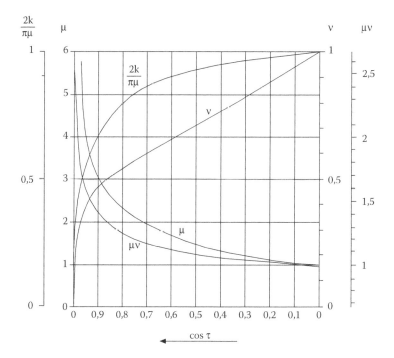

FIGURE 9.8 Curves of Hertz's parameters versus $\cos \tau$.

Of course, stresses are not constant on the entire ellipse: they are at their lowest on the rims, at their top in the center. The shape of the surface obtained by drawing all the points of the vectors representing the σ_s is an ellipsoid (Figure 9.9).

The value c of the semi-axis of the ellipsoid along the z-axis is σ_{max}.

The ellipsoid's generic equation is

$$\frac{X^2}{a^2} + \frac{Y^2}{b^2} + \frac{Z^2}{c^2} = 1$$

whence

$$\frac{X^2}{a^2} + \frac{Y^2}{b^2} = \left(\frac{Z^2}{c^2} - 1\right) = \lambda^2.$$

Then, in the equation

$$\frac{x^2}{(\lambda a)^2} + \frac{y^2}{(\lambda b)^2} = 1,$$

λa and λb are the ellipses' semi-axes obtained intersecting the ellipsoid with the $z = \text{cost}$ planes. The ellipse will reduce to one point for $z = c$ and will be equal to the contact mark for $z = 0$.

To find the ratio existing between σ_{max} and $\sigma_m = P / \pi \cdot a \cdot b$, it is necessary to calculate the ellipsoid volume, which will be

$$V = \int_0^c (\pi \cdot \lambda a \cdot \lambda b)\, dz = \int_0^c \pi ab\left(1 - \frac{z^2}{c^2}\right) dz = \pi ab\left(c - \frac{c3}{3c^2}\right) = \frac{2}{3}\pi abc.$$

The volume V found defines the load P:

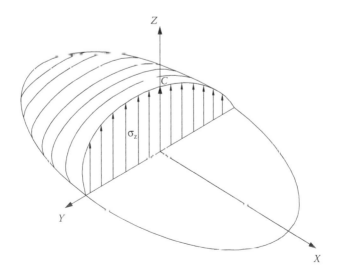

FIGURE 9.9 Qualitative general σ_z stress distribution.

$$P = \frac{2}{3}\pi abc,$$

whence

$$\sigma_{max} = c = \frac{3}{2}\frac{P}{\pi ab} = \frac{3}{2}\sigma_m.$$

Notice that such value is obtained not exactly on the surface, but below it. This explains why, when Hertzian fracture occurs, little craters open on the surface; as a matter of fact, below the surface, micro-cracks are generated through which strong pressures push the lubricant outside, causing the material to break.

In the same way, the stress in one point of the ellipse with x, y coordinates is

$$\sigma = \frac{3}{2}\sigma_m\sqrt{1 - \frac{x^2}{a^2} - \frac{y^2}{b^2}}.$$

In the particular case of contact between two cylindrical surfaces along one of their generatrices $\rho_1^1 = \rho_2^1 = \varpi = 0 \cdot \varpi$ ($\rho_1 = \rho_2 = \omega = 0$), the contact mark shape is a rectangle (a sort of central section of an infinite elongation ellipse). If l is the length of the cylinders, the semi-width of contact rectangle (Figure 9.10) turns out to be

$$b = \sqrt{\frac{P}{l\pi}\frac{\vartheta_1 + \vartheta_2}{\sum \rho}}.$$

The average stress will be equal to

$$\sigma_m = \frac{P}{4bl}.$$

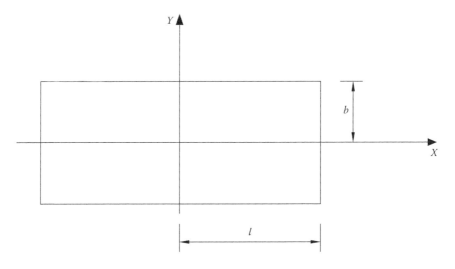

FIGURE 9.10 Rectangular contact area for cylinders.

In this case the surface obtained by joining all the vector points representing the stresses on each point is a semi-cylinder of elliptical cross-section (Figure 9.11).

The ellipse equation is

$$\frac{y^2}{a^2} + \frac{z^2}{b^2} = 1,$$

where c stands for the value of σ_{max}.

The semi-cylinder volume is

$$V = \pi bcl,$$

with b and c as semi-axes of the ellipse into which the ellipsoid has turned.

The volume also represents the vertical load P and then it follows that

$$c = \sigma_{max} = \frac{P}{\pi bl},$$

leading to

$$\sigma_{max} = \frac{4}{\pi} \cdot \sigma_m = 1.272\,\sigma_{max}.$$

Going back to the general case, it can be said that for reasons of symmetry, the most highly stressed point has to be on the z-axis, which is on the axis at right angles to the surface and passing through the ellipse center.

The stress components along the three axes x, y, z are expressed as follows:

$$\sigma_x = \frac{P}{\pi ab} \sum x \quad \sigma_y = \frac{P}{\pi ab} \sum y \quad \sigma_z = \frac{P}{\pi ab} \sum z,$$

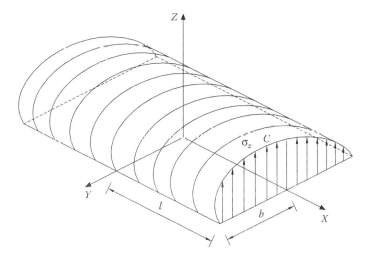

FIGURE 9.11   Qualitative σ_z stress distribution for cylinders.

with

$$\sum x = I_x - i_x \qquad \sum y = I_y - i_y \qquad \sum \quad \frac{3}{2} \sin \varphi \cos \varphi_1$$

where

$$I_x = \frac{3}{2}\xi \cos^2 \varepsilon \left(I_1 + \frac{I_2}{m} \right) \qquad I_y = \frac{3}{2}\xi \cos^2 \varepsilon \left(I_2 + \frac{I_1}{m} \right).$$

$$i_x = \frac{3}{m}\left[\sin \varphi \cos \varphi_1 + \frac{m-2}{2tg^2\varepsilon}\left(\frac{1}{\cos \varepsilon} - \frac{\sin \varphi}{\cos \varphi_1} \right) \right],$$

$$i_y = \frac{3}{m}\left[\sin \varphi \cos \varphi_1 + \frac{m-2}{2\sin^2 \varepsilon}\left(\frac{\cos \varphi_1}{\sin \varphi} - \cos \varepsilon \right) \right],$$

with

$$\cos \varepsilon = \frac{b}{a} = \frac{\nu}{\mu},$$

$$\xi = \frac{z}{b} = tg\varphi_1 \qquad\qquad \xi \cos \varepsilon = \cos \varphi,$$

$$I_1 = \frac{2}{\sin^2 \varepsilon}[F(\varepsilon,\varphi) - E(\varepsilon,\varphi)],$$

$$I_2 = \frac{2}{\sin^2 \varepsilon \cos^2 \varepsilon}\{E(\varepsilon) - E(\varepsilon,\varphi_1) - \cos^2 \varepsilon[K(\varepsilon) - F(\varepsilon,\varphi_1)]\},$$

$$F(\varepsilon,\varphi) = \int_0^{\varphi} \frac{d\varphi}{\sqrt{1 - \sin^2 \varepsilon \sin^2 \varphi}},$$

$$E(\varepsilon,\varphi) = \int_0^{\varphi} \sqrt{1 - \sin^2 \varepsilon \sin^2 \varphi} \cdot d\varphi.$$

These last two expressions are the elliptic integrals of the first and second kind of modulus $\sin \varepsilon$, respectively.

In these two cases $\varepsilon = 0$ (circular contact area) and $\varepsilon = \pi/2$ (elliptic contact area with infinite elongation), the previous expressions take on the following, simple form:
for $\varepsilon = 0$

$$
\begin{cases}
\sum_x \sum_y = \dfrac{3}{2}\left[\dfrac{1}{2}\dfrac{1}{1+\xi^2} - \dfrac{m+1}{m}(1-\xi ar\cot\xi)\right] \\[2ex]
\sum_z = \dfrac{3}{2}\dfrac{1}{1+\xi^2}
\end{cases},
$$

for $\varepsilon = \pi/2$

$$
\begin{cases}
\sum_x = -\dfrac{3}{m}\left(\sqrt{1+\xi^2}-5\right) \\[2ex]
\sum_y = -3\left(\dfrac{\xi^2-0.5}{\sqrt{1+\xi^2}} - \xi\right). \\[2ex]
\sum_z = -\dfrac{3}{2}\dfrac{1}{1+\xi^2}
\end{cases}
$$

From the previous general formulae, other simpler ones are obtained, case by case, referring to particular circumstances and specified values of μ, ν, $2K(\varepsilon)/\pi\mu$.

If the bodies in contact are made of the same material, i.e., steel, it is possible to write

$$
E_1 = E_2 = E \quad m_1 = m_2 = \frac{10}{3},
$$

so

$$
\vartheta_1 = \vartheta_2 = 4\frac{m_1^2-1}{m_1^2 E} = 4\frac{\left(\dfrac{10}{3}\right)^2-1}{\left(\dfrac{10}{3}\right)^2 E} = \frac{3.64}{E}.
$$

Then q is

$$
q = \sqrt[3]{\frac{3}{8}P\cdot\frac{\vartheta_1+\vartheta_2}{\displaystyle\sum\rho}} = \sqrt[3]{\frac{3}{4}P\cdot\frac{3.64}{E}\cdot\frac{1}{\displaystyle\sum\rho}} = 1.397\sqrt[3]{\frac{P}{E\cdot\displaystyle\sum\rho}},
$$

no matter the shape of the surfaces in contact.

Two most frequent cases of contact are sphere/ball versus sphere/ball and cylinder versus cylinder.

If R_1 and R_2 are the radiuses of sphere/cylinders 1 and 2 in contact with each other, the $\Sigma\rho$ is equal to

$$
\sum\rho = \frac{1}{R_1} + \frac{1}{R_1'} + \frac{1}{R_2} + \frac{1}{R_2'}.
$$

9.1.1 Sphere versus Sphere Contact

It obviously follows that $R_1 = R'_1$ and $R_2 = R'_2$, then

$$\sum \mu = 2\left(\frac{1}{R_1} + \frac{1}{R_2}\right).$$

Assuming

$$\frac{1}{R_1} + \frac{1}{R_2} = \rho,$$

it is possible to write

$$\sum \rho = 2\rho,$$

and then

$$q = 1.397 \sqrt[3]{\frac{P}{E \cdot 2\rho}} = 1.11 \sqrt[3]{\frac{P}{E \cdot \rho}}.$$

To determine the values of the semi-axes, it is necessary to know μ and ν values.

As regards rolling bearings, the simplified formula can be used:

$$\cos \tau = \frac{(\rho_1 - \rho'_1) \pm (\rho_2 - \rho'_2)}{\sum \rho}.$$

In this case, it is clear that $\cos \tau = 0$. Setting the abacus to $\cos \tau = 0$, $\mu = \nu = 1$ is found and so

$$a = b = q = 1.11 \sqrt[3]{\frac{P}{E \cdot \rho}}.$$

The contact mark is a circle with a diameter so much the greater, the greater is the P force and the smaller is ρ.

The σ_{max} likewise is equal to

$$\sigma_{max} = 1.5\sigma_m = 1.5\frac{P}{\pi a^2},$$

making a explicit, the maximum stress becomes

$$\sigma_{max} = 0.388 \sqrt{PE^2\rho^2}.$$

9.1.2 Cylinder versus Cylinder Contact

Consider the contact along a generatrix and proceed as in the case of contact between spheres. It follows that

$$b = 1.524\sqrt{\frac{P}{El\rho}},$$

$$\sigma_{max} = 0.418\sqrt{\frac{PE\rho}{l}}.$$

The formula expressing the semi-width b is obtained by taking the limit of general formulae, to which students should refer for further study.

From the previous formulae it may be maintained that, the load being equal, stresses are so much greater, the greater is the relative curvature. In the case of the contact of a sphere versus an equal sphere, sphere versus plane and sphere versus double-radius cave sphere, maximum stresses are in the ratio 1:0.63:0.40.

9.2 APPLICATION TO ROLLING CONTACT BEARINGS

As regards roller bearings, coefficients are usually calculated by abacus. One of these, the coefficient c, multiplied by a parameter, function of the load, and the diameter of spheres or rollers, simultaneously allows calculation of maximum stress.

Both spheres and rollers have curvatures defined by a single quantity (diameter). Rings, instead, have principal curvatures defined by two further quantities (line section diameter, throat diameter, or similar).

In Figure 9.12, d stands for the sphere diameter, d' for the ring cross-section diameter, and D for the diameter of the whole bearing.

In general, D is the diameter along the line passing through the contact point at the right angle to the bearing axis; a principal radius of curvature related to the contact point is equal to $D/\cos\alpha$.

In these cases, the $\cos\tau$ expression can be considered a function of two ratios among the three above radiuses or of two functions of these ratios.

As to ball bearings, the ratio $C = d/d'$ related to the ball-race fitting and the ratio $(D/\cos\alpha)/d$ can be taken as variables on which $\cos\tau$ depends.

Since μ and ν are functions of $\cos\tau$, for given values of E_1, E_2, m_1, m_2, the σ_{max} can be written most simply as

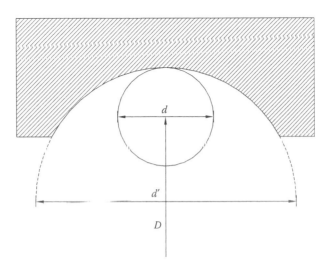

FIGURE 9.12 Curvatures for rolling bearings.

$$\sigma_{max} = c \cdot \sqrt{\frac{P}{d^2}}.$$

This expression is similar to that of sphere-plane contact, as the coefficient c is obtainable from an abacus like that in Figure 9.13 (Giovanozzi, Vol. I).

Similarly, when two cylinders are in contact along a generatrix, the maximum stress can be expressed as follows

$$\sigma_0 = c \cdot \sqrt{\frac{P}{l \cdot d}},$$

c being, in this case also, (for materials assigned), a quantity only dependent on the principal radiuses of curvature under consideration (see classic manuals on bearings for diagrams of c and for the ratio of geometric parameters).

9.3 APPROACH

The reports described thus far refer mainly to the calculation of local stress and deformation.

Consider two points, Q and R, on two bodies in contact, at a certain (big) d distance from each other, measured without load (Figure 9.14). Applying a load P, the distance between Q and R will change. Approach (α) is the variation of the distance between these points:

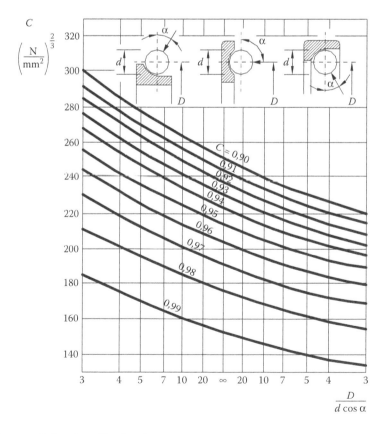

FIGURE 9.13 c coefficient for rolling bearings.

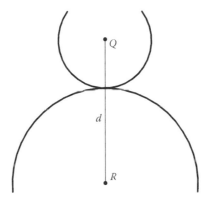

FIGURE 9.14 Contact between solids.

$$\alpha = \frac{3}{2} \frac{2K(\varepsilon)}{\pi\mu} \frac{P(\vartheta_1 + \vartheta_2)}{8q}.$$

For $m_1 = m_2 = m$ and $E_1 = E_2$ this expression takes on the form:

$$\alpha = \frac{3}{2} \frac{2K(\varepsilon)}{\pi\mu} \frac{m_2 - 1}{m_2} \frac{P}{q}.$$

The quantity $2K(\varepsilon)/\pi\mu$ in the formula is tabled or diagrammed as a function of $\cos \tau$.

The distance variation should not be influenced by local deformation, so Q and R have to be distant enough. Indeed, the α approach values change depending on the points chosen. For Q' to R' closer to the contact area, α will be different and affected by local deformation. The quality variation of α, depending on d distance, is shown in Figure 9.15.

As the distance d between the two points increases, the value of α becomes stable. This is because the closer the contact area, the greater the deformation. In general, when parameters such

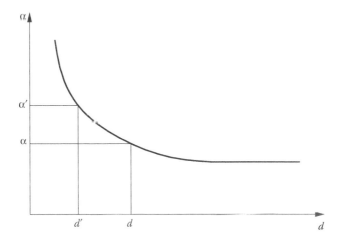

FIGURE 9.15 Qualitative α versus d.

as the approach have to be read, it is important to place reference points in areas not directly affected by deformation.

In the specific case of two cylindrical surfaces in contact along one of their generatrixes, ($\rho_1 = \rho_2 = \omega = 0$), the approach between two points of the bodies located on the common norm does not tend to a finite limit if the distance between the points increases (as in all other cases). Its value, then, will depend on the distance of the two bodies to be measured.

Hertz theory regarding approach leads to values dependent on reference points. Several theoretical and experimental researches have been done to solve this case. However, the most commonly used is the empirical one by H. Bochmann, based on measurements of steel cylinders pressed between two steel planes.

The elastic steel modulus on which the approach depends is inserted in the value of the constant:

$$\alpha = 4.62 \cdot 10^{-4} \frac{P}{\sqrt{dl}} \,.$$

As a first approximation, this expression can also be applied in contact between tracks and cylinders.

The above empirical formula can also be used for cylinders in contact, with materials other than steel, by changing the values of the constant in accordance with the general criterion inferable from Hertz theory.

Therefore, if neither material is steel, the formula can still be used, provided, as in Hertz theory, the approach is proportional to the sum ($\vartheta_1 + \vartheta_2$). Moreover, the obtained numeric coefficient has to be multiplied by the ratio w between the sum ($\vartheta_1 + \vartheta_2$) related to the materials in contact and the sum ($\vartheta_1 + \vartheta_2$) for steel.

The correction factor can be found in Giovannozzi. Vol. I, p. 596.

According to Hertz theory, and taking into account roughly the curvature difference from $2/d$ existing in the case of tracks and cylinders in contact, $1/d$ could be replaced with the sum $1/d + 1/d'$ in the formula, d' being the curvature diameter of the cross-section of the roller track in value and sign.

Finally, applying the general formula to the sphere-to-sphere case, the approach is

$$\alpha = 1.23 \sqrt[3]{\frac{P^2 \rho}{E^2}} \,.$$

9.4   EQUIVALENT IDEAL STRESS IN HERTZIAN CONTACT

In order to define the load limits and stresses that might lead to failure, it is commonly necessary to refer to a proper yield criterion that takes into account the quantities of the three main stresses at their most stressed points.

In particular, note that the σ_{max} stress value at the center of the contact ellipse is not enough to define any risk of failure, both because the stress field is tri-dimensional and because it is not given that the maximum stress will be at that point.

Using four different yield criteria, F. Karas repeatedly tried to find the point of maximum stress and the value of the maximum ideal stress $\sigma'_{id\,max}$ computed at that point. He found that the maximum stress occurs at a point on the z-axis and not on the bodies' surfaces, but a little below it, at a certain coordinate ξ_0. Moreover, he defined $\sigma'_{id\,max}$ as not only smaller than σ_{max}, but also smaller than average stress σ_m. That is due to the fact that, apart from σ_{max}, the other stresses σ are also at work to prevent lateral deformation and reduce σ_m.

Karas obtained the following values:

- Circular contact surface ($\varepsilon = 0$)

$$\xi_0 = 0.429 \quad \frac{\sigma'_{id\,max}}{\sigma_m} = 0.982 \quad \frac{\sigma'_{id\,max}}{\sigma_{max}} = 0.655$$

- Rectangular contact surface ($\varepsilon = \pi 2$)

$$\xi_0 = 0.489 \quad \frac{\sigma'_{id\,max}}{\sigma_m} = 0.801 \quad \frac{\sigma'_{id\,max}}{\sigma_m} = 0.629$$

$\sigma'_{id\,max}$ is then on average only 63–65% of σ_{max}.

This is why in bearings σ_{max} values are higher than breaking stress σ_R of the material of which they are made. So, although the bearing might be expected to break, this does not happen because of the stresses in the other directions that reduce the σ_{id}.

In practice, the stresses actually generated are equivalent to fatigue failure loads, although they are higher than these, as is proven by the fact that the strength of roller bearings is not unlimited, but time dependent, so failure occurs after a certain number of cycles.

9.5 LOADS ON THE BALL BEARINGS

According to Hertz theory, Stribeck deduced that in main bearings the maximum load P_0 sustained by the most stressed ball is equal to

$$P_0 = 4.37 \frac{P}{z}$$

P being the total load on the bearing, z the number of balls.

In practice, the coefficient 4.37 is usually rounded up to the precautionary value of 5, whence the traditional rule to compute P_0, dividing P by the fifth of the total number of balls. This formula, experimentally obtained by Stribeck, can be proved in theory.

In general, the angle between two adjacent spheres is

$$\gamma = \frac{2\pi}{z},$$

so the position direction of the i-th sphere forms an angle with the midpoint vertical axis defined as

$$\gamma_i = i\frac{2\pi}{z}, \quad i = 0,1,...,\frac{z}{2}.$$

With reference to Figure 9.16, extending the sum to all the loads located in the lower half of the bearing, the equilibrium equation along load direction is

$$P = P_0 + 2P_1 \cdot \cos\gamma_1 + 2P_2 \cdot \cos\gamma_2 + \cdots$$

If the approaches are compared with the load P according to the formula:

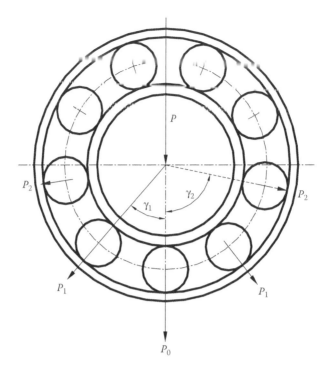

FIGURE 9.16   Load distribution in a ball bearings.

$$\alpha = 1.23 \sqrt[3]{\frac{P^2}{E^2}} \rho \rightarrow \alpha^3 = \frac{1.23 P^2 \rho}{E^2},$$

then P will be proportional to the approach raised to power $3/2$ in accordance with coefficient K, that is

$$P_0 = K \cdot \alpha_0^{3/2} \quad P_1 = K \cdot \alpha_1^{3/2} \quad P_2 = K \cdot \alpha_2^{3/2}.$$

The term K is always the same so substituting it in the equilibrium equation it follows that

$$P = K \cdot \alpha_0^{3/2} + 2K \cdot \alpha_1^{3/2} \cdot \cos \gamma_1 + 2K \cdot \alpha_2^{3/2} \cdot \cos \gamma_2 + \cdots$$

Assuming that the shaft and the outer ring housing are infinitely rigid:

$$\alpha_1 = \alpha_0 \cos \gamma_1 \quad \alpha_2 = \alpha_0 \cos \gamma_2,$$

and substituting them in the equilibrium equation it follows

$$P = K \cdot \alpha_0^{3/2} + 2K \cdot \alpha_0^{3/2} \cdot \cos^{5/2} \gamma_1 + 2K \cdot \alpha_0^{3/2} \cdot \cos^{5/2} \gamma_2 + \cdots$$

whence, observing that $\alpha^{3/2}_0 = P_0 / K$, the expression becomes

$$P = P_0 \cdot (1 + 2\cos^{5/2} \gamma_1 + 2\cos^{5/2} \gamma_2 + \cdots).$$

In general, it is possible to write

$$P = P_0 \left(\sum_{i=-(z/4)}^{z/4} \cos^{5/2} \gamma_i \right).$$

The summation in this expression can be calculated dividing the equality:

$$\int_{-(\pi/2)}^{\pi/2} \cos^{5/2} \gamma \, d\gamma = 0.92 \frac{\pi}{2},$$

by π and multiplying the result by the width of integration interval $(-z/4) - (+ z/4)$. So, it follows

$$P = P_0 \frac{z}{2} \cdot \frac{0.92}{2} = P_0 \frac{z \cdot 0.92}{4},$$

whence maximum load P_0 is obtained

$$P_0 = \frac{4}{0.92} \frac{P}{z} = 4.37 \frac{P}{z}.$$

As stated before, this value is prudently rounded up to 5.

As regards roller bearings, with a similar computation it is possible to get a similar formula and round off the coefficient to 4, so it is assumed that

$$P_0 = 4 \frac{P}{z}.$$

10 Lubrication

10.1 INTRODUCTION

In applied mechanics courses, descriptions of lubrication theory are usually limited to the flat case (all sections in direction perpendicular to the velocity behave in the same way); this means infinite length. In practice, both for thrust Michell bearings and journal bearings, we deal with finite, indeed rather small lengths, that is, with dimensions perpendicular to the velocity comparable to either of the remaining ones (i.e., the length of the journal is comparable to its diameter).

This chapter will present some approximated methods and refer to their results, either graphically or as formulae, that make it possible to calculate the parameters typical of prismatic joint or revolute joint (Michell and journal bearings) with infinite and finite length, within the hypothesis of perfect lubrication.

The lubrication theory formulae we deal with always refer to the classic hypothesis of viscosity μ = cost. Namely, a single viscosity value, corresponding to the mean lubricant temperature in the clearance between reciprocally moving surfaces while in operation, is adopted at all times. In reality, viscosity also varies with pressure, as well as with temperature. However, in classical theory these changes are not considered. In the same way, the elastic deformation of component is not considered; this would entail a shape variation of the clearance. It is not difficult, though, to find studies in the literature where variations in these parameters are taken into due account. For these, refer to specialized texts and works.

The difference between *contour* and *hydrodynamic* lubrication is highlighted in the chart in Figure 10.1, where the coefficient of friction f is shown as a function of parameters peculiar to lubrication, such as μ (dynamic viscosity of the lubricant in pascal per second), N (rotations per second of the shaft), P (load in Newtons).

The chart shows that, when the values of the typical parameter $\mu N/P$ are lower than B, lubrication is unstable, as the lubricating film is too thin and the surface roughness irregularities are higher than the lubricant thickness; there is direct contact between surfaces and the coefficient of friction rises consequently. Point B marks the changeover from film lubrication (the lubricant is interposed) and boundary lubrication. For values greater than B, film thickness is greater than the roughness, and lubrication is stable. In these conditions, friction can be approximated by Petroff's Law that describes the case of a journal with radius r immersed in oil with no eccentricity, and clearance thickness equal to c.

$$f = 2\pi^2(\mu N/P)(r/c).$$

Furthermore, when abscissa values are greater than B, increasing heat decreases viscosity, which lowers the coefficient of friction with a consequent temperature reduction. The phenomenon tends to reach a new status of stability. If abscissa values are lower than B, raising the temperature reduces viscosity, which causes the coefficient of friction to rise with a further rise in temperature. Subsequently, there is an additional reduction in the coefficient of friction and consequent exaltation of the phenomenon.

From Petroff's formula, the type of lubrication is a function of the ratio of rotation velocity and load. The greater the load, the greater the velocity to have mediated lubrication.

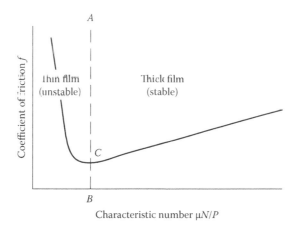

FIGURE 10.1 Qualitative friction coefficient f versus $\mu N/P$.

Many researchers have dealt with hydrodynamic lubrication. At first, it was studied experimentally by Beauchamp Tower. The systematic nature of the experimental results obtained by Tower urged Osborne Reynolds to search for a theory that could describe them.

Reynolds hypothesized that the lubricant adhered to the two faces of the journal and the bearing. He disregarded the curvatures involved, since the thickness of the lubricant film is very small compared to the curvature radius of the journal. Other hypotheses were:

- Newtonian fluid and incompressible lubricants
- Null inertial forces of the lubricant
- Constant viscosity throughout the film (independent on temperature and pressure)

According to the previous hypotheses, equilibrium and continuity equations were applied in the case of the lubricated journal, for which further hypotheses were made:

- Infinite length, that is, in the reference system in Figure 10.2, constant pressure p along the z-axis, zero flow in direction z
- Constant pressure along the y-axis and so only dependent on x
- Velocity of lubricant variable only along the y-axis

The equilibrium equation for an infinitesimal lubricant cube, according to the previous hypotheses, leads to

$$dp/dx = d\tau/dy,$$

and, remembering that Newton's law applies, it follows that

$$dp/dx = \mu \cdot d^2v/dy^2.$$

By integrating twice with respect to y, and assuming p constant along y, we get

$$v = \frac{1}{2\mu} y^2 \frac{dp}{dx} + C_1 y + C_2.$$

The constants C_1 and C_2 are defined by the following contour conditions (the fluid adheres to the fixed wall and to the one in motion):

$$\begin{cases} y = 0 & v = 0 \\ y = h & v = -V \end{cases}.$$

The velocity equation becomes

$$v = \frac{1}{2\mu}\frac{dp}{dx}y(y - h) + V\left(\frac{y}{h} - 1\right).$$

As can be seen, velocity results from the sum of a parabolic and linear term. The conditions of continuity (constant lubricant flow),

$$Q = \int_0^h V dy,$$

and incompressibility of liquids,

$$dQ/dx = 0,$$

lead to Reynolds' equation:

$$\frac{d}{dx}\left(h^3\frac{dp}{dx}\right) = -6\mu V\left(\frac{dh}{dx}\right).$$

In the case of even flow along the z-axis (bearing with finite length), and so a non-constant pressure along this axis, we obtain

$$\frac{\partial}{\partial x}\left(h^3\frac{\partial p}{\partial x}\right) + \frac{\partial}{\partial z}\left(h^3\frac{\partial p}{\partial z}\right) = -6\mu V\left(\frac{dh}{dx}\right).$$

It should be noted that when the journal and bearing axes are not aligned (due to imperfect shaft alignment in the assembly phase or to the effect of bending stresses) we get $h = h(x,z)$, so the derivative of the second member is partial.

Analytical approximate and numerical solutions of the previous equations have been found. For the flat case (journal with infinite length), Arnold Sommerfeld in 1904 came to a twofold solution. one with a journal totally immersed, the other, only a half-immersed:

$$\left(\frac{r}{c}\right)f = \Phi\left[\left(\frac{r}{c}\right)^2\mu\frac{N}{P}\right],$$

with Φ function characterized by the two above situations.

The square bracketed figure is called *Sommerfeld number S*. It is an important project parameter, as it is dimensionless and a function of well-known project parameters. Indeed, charts and diagrams are drawn with it to determine *a priori* unknown parameters.

More generally, when calculating lubricated pairs, load *P*, viscosity μ, rotations *N*, and the geometry of the bearing are considered as design parameters. It follows that we need to estimate:

- The friction coefficient *f* and power loss
- Temperature rise ΔT
- Lubricant flow rate *Q*
- Minimum clearance thickness h_0, which must not be less than the maximum value of unevenness

So, lubricant viscosity decreases appreciably and non-linearly as temperature rises (charts feature non-constant height declines, so they are not semi-logarithmic). As temperature under operating conditions is unknown beforehand, the calculation is always iterative unless, obviously, the viscosity value chosen for it coincides with that of the operating temperature.

So, all the following cases aim to find values for the four parameters (f, ΔT, Q, h_0) above, provided that the geometry and input parameters P, μ, and N are known.

10.2 PRISMATIC PAIR OF INFINITE AND FINITE LENGTH

Consider the sliding block in Figure 10.2. If h_1 and h_2 ($h_2 > h_1$) are the heights of the clearance ends, and *l* and *b* the dimensions of the sliding block in the direction of velocity *V*, then b/l is the length.

P is the normal load on the sliding block and $T = fP$ the resistance. The eccentricity *e* is the distance over which *P* operates from the plate center and $\varepsilon = e/l$ is the relative eccentricity. We have already noted that Reynolds' equation in the two-dimensional case is

$$\frac{\partial}{\partial x}\left(h^3 \frac{\partial p}{\partial x}\right) + \frac{\partial}{\partial z}\left(h^3 \frac{\partial p}{\partial z}\right) = -6\mu V\left(\frac{dh}{dx}\right).$$

If $dh/dx = 0$ (that is $h = $ cost) or $V = 0$, it turns into a harmonic equation like the following:

$$\frac{\partial^2 p}{\partial x^2} + \frac{\partial^2 p}{\partial z^2} = 0 \quad \left(\frac{dh}{dx} = 0 \rightarrow h = \text{cost}\right).$$

From Cauchy integral theorem, *p* will assume all boundary values in all points. As the external pressure is atmospheric, there is no pressure gradient in the clearance and so there is no support.

To ensure hydrodynamic lubrication it is necessary that

$$\frac{dh}{dx} \uparrow 0 \quad \text{and} \quad V \uparrow 0.$$

In practical applications, once the system geometry is defined, the only requirement is to determine load *P*, resistance *T* (and so *f* to define the rise in temperature), lubricant flow rate, and the minimum clearance height h_0.

The resisting force *T* is given by the resulting tangential actions:

$$T = \int \tau dA = \int \mu \frac{dV}{dh} dx dz.$$

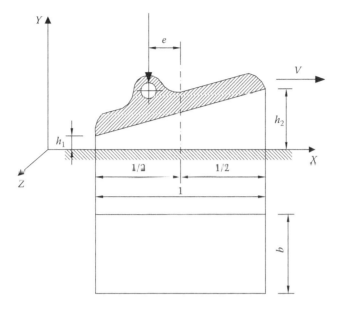

FIGURE 10.2 Scheme for a sliding block.

To understand qualitatively how much T depends on geometrical parameters, suppose by rough approximation, that the velocity gradient is constant. Therefore

$$T = \bar{K}\mu b \frac{l}{h_1} V.$$

Coefficient $\bar{K}$ would depend on the geometrical characteristics (clearance shape), and with the symbols in Figure 10.2:

$$m = \frac{h_2 - h_1}{h_1} = \frac{h_2}{h_1} - 1,$$

or $\bar{K} = f(m)$.

In classical theory, load P for the case of infinite length is

$$P = k\mu V \left(\frac{l}{h_1}\right)^2 b \quad \text{with} \quad k = \frac{6}{m^2}\left[\ln\left(m+1\right) - \frac{2m}{2+m}\right],$$

(P/b represents, in this case, the load per unit length in direction z) and for the coefficient of friction f, we get

$$f = \frac{T}{P} = k' \frac{h_1}{l} \quad \text{with} \quad k' = \frac{1}{3}\left(2m + \frac{6}{2+m}\frac{1}{K}\right).$$

Finally, it should be pointed out that relative eccentricity e/l is a function of m and k:

$$e/l = f(m,k),$$

and more precisely:

$$e/l = \frac{2+m}{2m} + \frac{m+1}{m(2+m)} - \frac{3}{(2+m)^2}\frac{1}{k}$$

The above formula reveals that, for a fixed geometry, both load and resistance are proportional to velocity and viscosity. In comparing the coefficient of friction to f to the load P, it is proportional to velocity and its square root is inversely proportional to the square root of load P. This is true as long as P does not give clearance minimum thickness to less than surface unevenness; that would lead to dry friction and none of the above formulae would apply.

In 1942, W. Frössel found the exact solution for sliding block finite length. His results were plotted (Figure 10.3), the abscissas being:

$$a = 1 + \frac{1}{m} = \frac{h_2 - h_1}{h_1},$$

and the ordinates $mk^2/6$ and mkk'.

Having fixed the ratio h_2/h_1 (clearance shape), these charts can determine the coefficients k and k' and so load and resistance as well as the relative eccentricity e/l of the load in the sliding block finite length.

For practical applications, such as Michell supports with an oscillating sliding block, it is quicker to refer to relative eccentricity e/l, a function of m. The support on which the sliding block oscillates is defined, and so is the position of the sum of the forces, as it necessarily passes through that point. Relative eccentricity e/l is fixed in this way. So, by processing Frössel's results, Funaioli produced an abacus containing the values of k and k' as functions of e/l (Figure 10.4) for various lengths. Thus, k and k' can be obtained directly, if e/l is known.

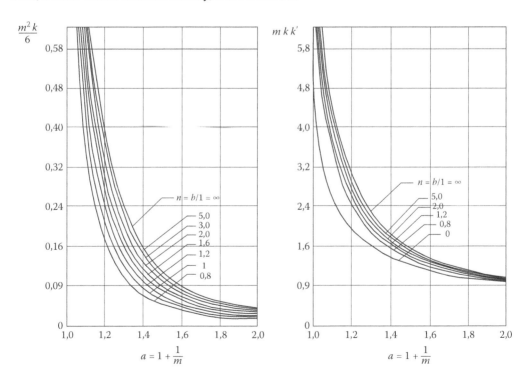

FIGURE 10.3   W. Frössel results for sliding block finite length.

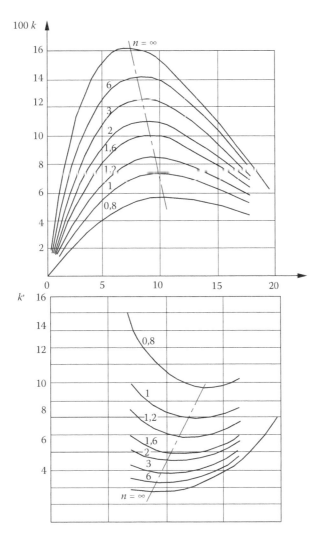

FIGURE 10.4 Funaioli's abacus.

Load variations are a function of e/l, and the maximum load occurs when e/l is around 7.5% for infinite lengths, whereas it reaches greater values for finite lengths (up to around 10% if length is equal to 1). Observing the diagrams, the load is always applied to the sliding block half where thickness is lower. Generally, given the same load and V and h_1 are constant, two solutions are possible. They will be the same when the eccentricity is the one of the maximum load. Minimum resistance values occur roughly for eccentricities to which maximum load values correspond.

In the k' diagram, the intersections between the curves and the dashed line, very close to minimum k', are those required for practical applications.

10.3 CALCULATION OF MICHELL THRUST BEARINGS

Figure 10.5 shows a Michell thrust bearing. The calculation will refer to the symbols below.

Assuming that peripheral velocity variation from internal to external diameter is negligible, if N is the shaft angular velocity in rpm and d is the average sliding block diameter, the average reference velocity will be $V = \pi d N / 60$.

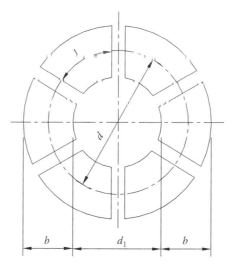

FIGURE 10.5 Scheme of a Michell thrust bearing.

By way of approximation, the real segments are replaced with rectangular surfaces, b width and l length. Having fixed the n ratio (between 1 and 2), optimum k (k_0) is determined from the above diagram in Figure 10.4, that is eccentricity e/l, giving the minimum k'.

If i is the number of segments (sliding blocks), the load is

$$P = i k_0 \mu V b \left(\frac{l}{h_1} \right)^2 .$$

For lengths between 1 and 2, with less than 0.5% error, optimum k_0 coefficients:

$$k_0 = \frac{0.162}{1 + 1.22 \left(\dfrac{l}{b} \right)^{1.45}} ,$$

(as seen above, k_0 coincides with the intersections of the k curves with the dashed line, Figure 10.4) and so load becomes

$$P = i \frac{0.162}{1 + 1.22 \left(\dfrac{l}{b} \right)^{1.45}} \frac{\mu V}{h_1^2} b l^2 .$$

From the above formulae, to obtain the maximum load from the fluid layer, for a given average rotation velocity and geometry (sliding block shape), the h_1 values of clearance minimum thickness have to be as small as possible. However, h_1 cannot have values below a certain limit, as this could cause the surface roughness to be greater than h_1 and break the lubricating film (lubrication is no longer hydrodynamic).

Currently, surface roughness is defined by the average of the local unevenness values (Figure 10.6). This average value can be chosen and defined in various ways. The most common definition

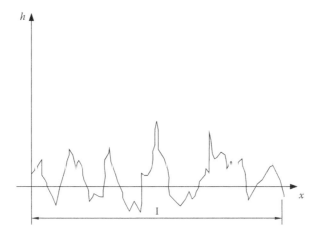

FIGURE 10.6   Qualitative surface roughness.

is the average arithmetic value h_m of the absolute protrusion values measured above (and below) a straight line along the surface profile:

$$h_m = \frac{1}{l} \int_0^l |h(x)| \, dx.$$

Alternatively, the root mean square is used:

$$h_m = \frac{1}{l} \sqrt{\int_0^l h^2(x) \, dx}.$$

To avoid contact between surface protrusions, a minimum thickness reference of $h_1 = 2 \div 3 h_m$ can be used as for high quality finish. For lower quality, values may be up to $10 \, \overline{h}_m$.

The filling ratio φ

$$\varphi = \frac{il}{\pi d},$$

is the ratio between the length of segments and the entire circumference of the bearing, measured on the average diameter d. In practice, φ is comprised between 0.85 and 0.95.

By expressing l as a function of φ and substituting it in P:

$$P = ik_0 \frac{\pi V}{h_1^2} b \frac{\pi^2 d^2 \varphi^2}{i^2} = \frac{k_0 \cdot \mu \cdot V \cdot b \cdot \pi^2 \cdot d^2 \cdot \varphi^2}{h_1^2 \cdot i},$$

where k_0, $n = b/l$, φ and h_m are known. The only unknown is average diameter d, obtained as follows

$$d = \frac{1}{\pi} \sqrt[3]{\frac{60 P h_1^2 i}{\mu k_0 b N \varphi^2}},$$

μ is known, so after selecting the lubricant and defining the oil's operating temperature T_1, the coefficient of friction f expressed as a function of P is

$$f = \frac{T_l}{P} = k^1 \frac{h_1}{l}$$

where k' can be obtained from Funaioli's charts (Figure 10.4) or directly from the formula on p. 221. By plotting f as a function of V, we obtain the result shown in Figure 10.7.

It is known that when $V = 0$, no lubrication occurs, and the coefficient of friction is the static one.

Only when the curves intersect in V^* the film of lubricant gives a load. In V^* the clearance height exceeds the unevenness, that is $h_1 > 2 \div 3h_m$ (in practice it is never lower than 5 μm) and the coefficient of friction varies with the velocity square root. So, a system is needed to put the oil under pressure and, at the same time, set the journal velocity higher than V^*. In other words, when starting, lubrication is hydrostatic and the load is obtained by pressure oiling the bearings through an independent hydraulic circuit.

It should be pointed out that this oil has to have a certain viscosity as, in transition phases, when the velocity is low compared to the design velocity (less than V^*), the minimum height of the clearance decreases, leaving hydrodynamic lubrication to become viscous lubrication (the lubricant behaves like grease).

Once the coefficient of friction is found, power in terms of heat dissipated is $W = fPV$ and from this, through the mechanical equivalent of calories A, the amount of heat Q produced per unit time is calculated.

It is then necessary that the heat produced has to be dispersed, which thing can be done either by air-cooled (for low heat flow) or water-cooled heat exchangers (high heat flow). In other words, the temperature of the lubricating oil must be stable at T_1 to guarantee correct operation of the bearing.

If, for any reason, it not possible to dissipate heat until the oil temperature returns to T_1, the computation must be recalculated with the coefficient of viscosity corresponding to the oil temperature obtained with the chosen heat exchanger.

10.4 JOURNAL-BEARING PAIR: SETTING REYNOLDS' EQUATION AND CLEARANCE HEIGHT

Reynolds' equation is made dimensionless, so that its integration provides results valid for all similar bearings with the same geometrical characteristics.

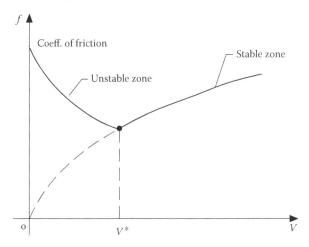

FIGURE 10.7 Qualitative friction coefficient f versus velocity V.

The circumference coordinate x is divided by the journal radius. The axial coordinate z is divided by half the length of the journal, the fluid film height is divided by the clearance:

$$\varphi = x/R \quad \xi = \frac{z}{L/2} \quad H = h/c .$$

Making pressure dimensionless,

$$\bar{p} = \frac{p}{12\pi\mu N}\left(\frac{c}{R}\right)^2 .$$

Reynolds' equation becomes:

$$\frac{\partial}{\partial\varphi}\left(H^3\frac{\partial\bar{p}}{\partial\varphi}\right) + \left(\frac{D}{L}\right)^2\frac{\partial}{\partial\xi}\left(H^3\frac{\partial\bar{p}}{\partial\xi}\right) = \frac{dH}{d\varphi} .$$

If R and r are the bearing and journal radii, φ the angle corresponding to the sections of minimum height h_0, e the eccentricity (distance between journal and bearing axes during operation), and c the clearance in radial direction, we should define relative eccentricity ε and relative clearance ψ as:

$$\varepsilon = \frac{e}{R-r} = \frac{e}{c} \quad \psi = \frac{R-r}{r} = \frac{c}{r} .$$

Solving the equation requires an expression for the clearance height. Referring to Figure 10.9:

$$AO' = h + r = R\cos\xi + e\cos(\pi - \varphi),$$

since $\cos\xi \cong 1$,

$$h = R - r - e\cos\varphi,$$

then

$$h = (R-r)(1 - \varepsilon\cos\varphi).$$

10.5   INFINITE LENGTH JOURNAL-BEARING PAIR

The study of the infinite length lubricated journal bearing (figure 10.8 and 10.9) is carried out by integrating:

$$\frac{d\bar{p}}{d\varphi} = \frac{6\mu V(H_m - h)}{h^3} ,$$

(H_m is clearance height at the maximum pressure point), in the experimentally verified hypothesis of detachment of the lubricating layer in the section where $dp/d\varphi = 0$, of abscissa $-\varphi_2$, symmetrical,

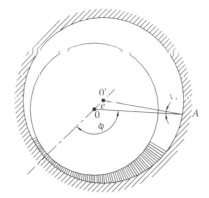

FIGURE 10.8   Scheme of a journal bearing.

compared to the reduced section of abscissa $\varphi = 0$, at the maximum pressure point of abscissa $+ \varphi_2$. This hypothesis is equivalent, when Reynolds' equation is integrated, to imposing a particular condition on the boundary values, which generates further complications in the computation; it is not opportune to provide the solution here.[*]

Therefore, the maximum and minimum pressure points are symmetrical at the point where the clearance has minimal thickness h_0.

In Figure 10.10, the chart shows the pressure trend as function of φ.

In practice, the oil inlet is where the hollow starts, to prevent any cavitation. The operational pressure range is defined by $\varphi_1 + \varphi_2$. By computation, for which we refer to applied mechanics courses, the load P per unit length, directed toward the journal axis is

$$P = A\mu V \frac{r^2}{\left(R - r\right)^2} \left(\varphi_1 + \varphi_2\right)^2.$$

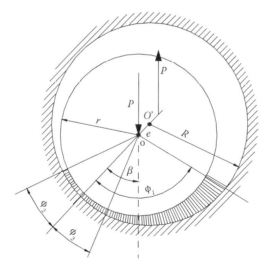

FIGURE 10.9   Load on a journal bearing.

---

[*] The solution is dealt with in detail in. Pinkus, O. and Sternlicht, B. *Theory of Hydrodynamic Lubrication*, McGraw-Hill Book Company Inc., 1961.

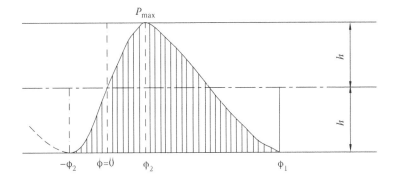

FIGUER 10.10   Pressure trend as function of φ.

A depends on the clearance geometric characteristics.

Combining $(\varphi_1 + \varphi_2)^2$ in a constant term and replacing *y*:

$$P = k\frac{\mu V}{\psi^2}.$$

The value of *k* (dimensionless *load coefficient*, an expression quite similar to Sommerfeld's number, $k = 1/\pi S$) is shown in the charts and with $\varphi_1 - \beta$ (angle formed by the oil inlet point with the direction of load *P*), is one of the input parameters for the calculation. For the coefficient of friction, the expression is

$$f = c\psi = c\sqrt{k}\sqrt{\frac{\mu V}{P}} = K\sqrt{\frac{\mu V}{P}},$$

k, *C*, and *K* are functions of the angle φ_1 (for oil inlet) and the relative eccentricity ε.

ε as well as φ_1 are unknown, so the position of the clearance minimal thickness turns out to be unknown. After processing and taking into account of the relationship between φ_1, φ_2, and ε, the quantities necessary for the calculation are obtained as a function of

$$k = \frac{P\psi^2}{\mu V} \quad \text{and} \quad (\varphi_1 - \beta).$$

The first parameter is known because vertical load *P* that the bearing must withstand is known, as well as its relative play ψ, rotation speed and lubricant viscosity. The second parameter is also known because the oil inlet position is defined.

Ten Bosch thought of presenting the results of these computations in graphic form. The first of these (Figure 10.11), shows the *k* values in the abscissa, and the various classes of curves with angle $(\varphi_1 - \beta)$ as ordinates.

In particular, the coordinates are

- The quantities $h_0/(R - r)$, which help determine the minimal thickness $h_0 (\geq 2 - 3\ h_m)$ of the lubricating layer and relative eccentricity $\varepsilon = 1 - (h_0/R - r)$.
- The coefficients *c* and *K* which, with $f = K\sqrt{\Box V/P}$, help calculate the journal's coefficient of friction.

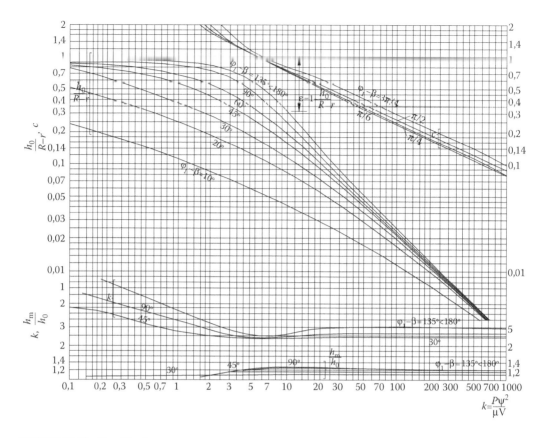

FIGURE 10.11 Bosch's abacus.

- The h_0/h_m ratios that help calculate h_m and the oil flow rate q_x in peripheral areas, by unit length. This flow rate, which is constant throughout all sections according to the principle of continuity, in the maximum pressure section is

$$q_x = \frac{V \cdot h_m}{2}.$$

For practical use of the graphs, given the load P, first the load coefficient k must be found in order to access the abscissa axis. According to the parameter $(\varphi_1 - \beta)$, the ratio h_0/h_m is found. Referring to k, $h_0/(R - r)$ is found on the third set of curves.

As R and r are known and h_0/h_m has just been identified, h_m must be calculated.

To determine φ_1, φ_2, and β, refer to the graph in Figure 10.12, where two abacuses have been included with different abscissas and ordinate scales, divided by the dashed diagonal from the bottom left corner to the top right one.

The first chart gives ε and $(\varphi_1 - \beta)$. Then, we have to refer to the top left side of Figure 10.13; where ε (continuous line) and $(\varphi_1 - \beta)$ (dashed lines) intersect φ_1 and β can be read.

Then, since φ_1 and ε are known, the bottom right side of Figure 10.12 shows the intersection point where, in the ordinates, we read φ_2. In the same way, the length $r(\varphi_1 + \varphi_2)$ of the lubricated arc can also be determined; it will be used, as we shall see later on, to approximate the effect of finite length.

So, all the data have been found to characterize an infinite length journal. After obtaining ε with the first diagram, β and, consequently, φ_1 can be found with the polar diagram in Figure 10.13. The

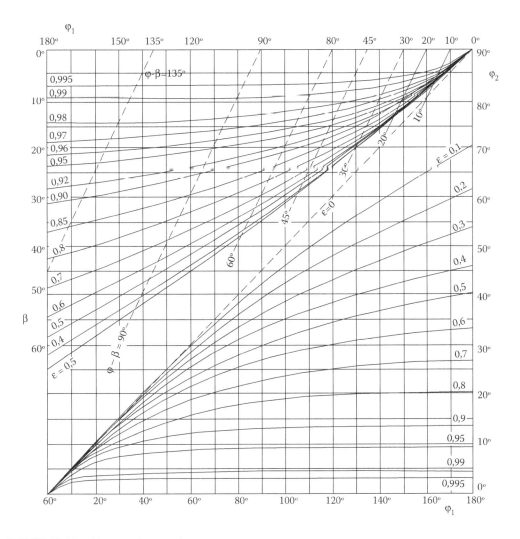

FIGURE 10.12   Abacuses for φ_1 and φ_2.

journal center is easily identified at the intersection of curve $(\varphi_1 - \beta)$ with the radius circumference ε, as shown in Figure 10.14 for the case $(\varphi_1 - \beta) = 90°$, $\varepsilon = 0.67$.

In Figure 10.13, the dashed-dotted semicircle refers to the case $(\varphi_1 - \beta) = 180°$.

From the graph in Figure 10.11, k should not be lower than 2, to prevent high friction torque; high values of k (greater than 100) are also undesirable as they result in too thin a lubricant layer.

Note that for values of k ranging between 2 and 400, K is roughly constant and equal to $2.6 \div 3$. The coefficient of friction is

$$f = \left(2.6 \div 3\right)\sqrt{a}\sqrt{\frac{\mu V}{P}}.$$

Now, all the reasoning behind the calculation for Michell bearings is applicable, in particular as regards determining the lubricant temperature if f is known. Final temperatures, which differ from the hypothetical ones, whose coefficient of viscosity was used to calculate Sommerfeld's number, have to be recalculated.

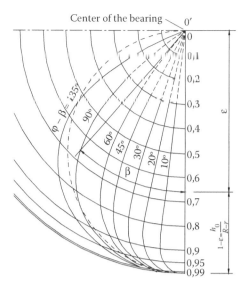

FIGURE 10.13   Identification of the bearing's center.

10.6   JOURNAL-BEARING PAIR OF FINITE LENGTH: SCHIEBEL'S FORMULAE

An approximate solution of the fundamental equation of lubrication for the journal-bearing pair of finite length was developed by A. Schiebel:

$$\frac{\partial}{\partial x}\left(h^3 \frac{\partial p}{\partial x}\right) + \frac{\partial}{\partial z}\left(h^3 \frac{\partial p}{\partial z}\right) = -6\mu V \frac{dh}{dx}.$$

Schiebel managed to integrate the fundamental equation of lubrication theory by admitting *a priori* a transverse parabolic distribution of pressure (in the z direction).

In fact, instead of integrating, he used the variational approach, finding the minimum function integral that satisfies the Euler–Lagrange equation. This method is used when it is easier to find the minimum of a function than to integrate it.

A classic application case referred to in all mathematical texts regards the solution of finding the trajectory that minimizes the time needed to reach point $B(b, \beta)$ from point $A(a, \alpha)$ in a gravitational field (g acceleration of gravity). The time is

$$t = \frac{1}{\sqrt{2g}}\int_a^b \frac{\sqrt{1 + y'^2}}{\sqrt{a - y}}\, dx.$$

The problem can be solved by defining the function $F(y, y', x)$ and drawing the corresponding Euler–Lagrange equation:

$$\frac{\partial F}{\partial y} - \frac{d}{dx}\frac{\partial F}{\partial y'} = 0.$$

In this case, the function is

$$F = \frac{\sqrt{1 + y'^2}}{\sqrt{a - y}}.$$

From the Euler–Lagrange equation, we obtain y, the minimum time trajectory.

Analogously, Schiebel realized that the fundamental equation of the lubrication theory could be interpreted as Euler's minimum variation condition of the integral, extended over the entire lubricated surface:

$$F = \frac{h^3}{2}\left(\frac{\partial p}{\partial x}\right)^2 + \frac{h^3}{2}\left(\frac{\partial p}{\partial z}\right)^2 - 6\mu V \frac{dh}{dx}\, p.$$

First, we verify that this function, replaced in the Euler–Lagrange equation, corresponds to the starting equation.

Considering that $F = F(p, \partial P/\partial x, \partial P/\partial z)$, which allows first order derivatives, the Euler–Lagrange equation is expressed as follows:

$$\frac{\partial F}{\partial p} - \frac{\partial}{\partial x}\left[\frac{\partial F}{\left(\frac{\partial p}{\partial x}\right)}\right] - \frac{\partial}{\partial z}\left[\frac{\partial F}{\left(\frac{\partial p}{\partial z}\right)}\right] = 0.$$

We define equation terms using

$$\frac{\partial F}{\partial p} = -6\mu V \frac{dh}{dx} \qquad \frac{\partial F}{\left(\frac{\partial p}{\partial x}\right)} = \frac{1}{2} 2 \cdot h^3 \frac{\partial p}{\partial x} \qquad \frac{\partial F}{\left(\frac{\partial p}{\partial z}\right)} = \frac{1}{2} 2 \cdot h^3 \frac{\partial p}{\partial z}.$$

By substituting these terms into the Euler–Lagrange equation we obtain:

$$6\mu V \frac{dh}{dx} - \frac{\partial}{\partial x}\left(h^3 \frac{\partial p}{\partial x}\right) - \frac{\partial}{\partial z}\left(h^3 \frac{\partial p}{\partial z}\right) = 0.$$

This is Reynolds' fundamental equation, which we have met before. So, integrating the equation is equivalent to finding Euler–Lagrange's minimum variation condition of the integral extended to the entire domain, in other words:

$$\int F\left(P, \frac{\partial p}{\partial x}, \frac{\partial p}{\partial z}\right) dx\,dz = \min.$$

For an approximate solution to this problem, Schiebel exploited Ritz's (also called relaxation) method. Writing F as the sum of the functions of the same variables through the a_i coefficients leads to:

$$F = a_1 F_1 + a_2 F_2 + a_3 F_3.$$

By substituting, we obtain

$$\int \left(a_1 F_1 + a_2 F_2 + \dots \right) dx dz = \min$$

The previous integral becomes

$$G = a_1 \int F_1 dx dz + a_2 \int F_2 dx dz + \dots = \min.$$

By implementing the derivatives of the a_i coefficients $\partial G / \partial a_1$, $\partial G / \partial a_2$, $\partial G / \partial a_{3i}$ and equalizing them to zero to obtain the minimum, a linear system will result of n equations in n unknowns. F_1 contains other coefficients that initially are not defined but are made clear as the system is solved.

The results of Schiebel's complex calculations refer to two main cases: $(\varphi_1 - \beta) = 180°$ and $(\varphi_1 - \beta) = 90°$.

The first case is the one in which roughly the entire bearing is operative. In the second more frequent case (which is desirable because it gives lower friction), only about half the bearing is at work. Here follows Schiebel's final formula.

Entire bearing operative $((\varphi_1 - \beta) = 180°)$

$$P = 0.366 \frac{\mu N}{\psi^2} \frac{d}{1 + 8 \left(\dfrac{d}{l} \right)^2} \sqrt{\frac{R - r}{h_0} - 1},$$

$$f = 3.44 \sqrt{\frac{\mu V}{P} \left[1 + 8 \left(\frac{d}{l} \right)^2 \right]},$$

$$Q_1 = \frac{10 \dfrac{d}{l}}{1 + 8 \left(\dfrac{d}{l} \right)^2} Vde.$$

Half bearing operative $((\varphi_1 - \beta) = 90°)$

$$P = 0.183 \frac{\mu N}{\psi^2} \frac{d}{1 + 2 \left(\dfrac{d}{l} \right)^2} \sqrt{\frac{R - r}{h_0} - 1},$$

$$f = 2.81 \sqrt{\frac{\mu V}{P} \left[1 + 2 \left(\frac{d}{l} \right)^2 \right]},$$

$$Q_1 = \frac{1.5 \dfrac{d}{l}}{1 + 2 \left(\dfrac{d}{l} \right)^2} Vde.$$

The coefficient values in the above formula are precautionary, that is, they show the maximum coefficient of friction and minimal bearing load capacity.

In these formulae, Q_1 is the amount of oil flowing laterally because of finite length, to which the circumferential flow $q_x = Vh_m/2$ of the infinite length journal must be added.

d is journal diameter, N the revolutions per minute of the shaft, l the axial journal length, and Pl the total load in N.

When only the lower half of the bearing is operative, its upper half has a non-pressured film of oil that hampers motion and generates an increase Δf of the friction coefficient already calculated.

If the journal and bearing were concentric and the clearance had a constant value $h = R - r = \psi\, r$, the tangential stress τ would be equal to $\mu V/h$ everywhere.

As a matter of fact, the journal has an eccentric position and the clearance thickness is not constant. Any oil distribution channels or lubrication rings diminish friction surfaces. Any additional friction, assuming $\tau - \lambda(\mu V/h)$ operates everywhere, λ being lower than 1 (experience suggests $\lambda = 0.5$), is

$$\Delta f = \lambda \pi \frac{\mu V b}{P\psi},$$

b being the axial length of the journal.

The total drag force is given by

$$T_{\mathrm{TOT}} = T + \Delta f \cdot P \quad \text{with } T = f \cdot P.$$

The values of viscosity for the formula are those obtained for the lubricant's average temperature in the clearance. In the case of power train shaft journals, the reference is room temperature increased by ΔT due to friction and equal to the power lost to friction:

$$\left(f + \Delta f\right)PVl = \left(f + \Delta f\right)P\frac{\pi N}{30}\frac{d}{2}b,$$

and equalized to the heat generated by the journal box which, if no oil coolers are available, can be expressed as $a \cdot \pi\, d\, b\, \Delta T$, that is to say, proportional to the operational surface of the bearing as well as to the thermal increment ΔT. So

$$\Delta T = \frac{\left(f + \Delta f\right)PN}{60a}.$$

For ordinary journal boxes without oil coolers, Lasche suggests the following values for a:

- $a = 5$ for light journal boxes
- $a = 8$ for ordinary journal boxes
- $a = 14$ for heavy journal boxes, with large metal masses

The previous formula provides the maximum temperature allowed for a bearing under normal operation.

The lost power $W = T_{\mathrm{TOT}}\,\omega r$ is the mechanical power dispersed as heat. To determine lubricant temperature, refer to the Michell bearing calculation.

As above, the finite length journal calculation helps to determine the lubricant flow rate from the bearing sides. Total flow rate is

$$Q = Q_1 + \frac{V \cdot h_m}{2}.$$

with h_m determined from Ten Bosch's diagrams above. Knowing the flow rate leads to oil temperature that must equal the hypothesized one at the beginning of the calculation, for which oil type and viscosity were chosen.

If the calculation reveals higher oil temperatures than those expected, it needs recalculating for natural oil cooling, or forced cooling systems might be necessary to bring the oil temperature down to the computational one.

This is usually the case when the difference between the hypothesized starting temperature and the final temperature is over 20°. During the design phase, it is also possible to hypothesize parameters values that impact on a coefficient of friction adapted to limited temperatures, such as radius play (increasing it lowers the coefficient of friction), but at the expense of other factors such as load, which is one of the design input data.

In order to guarantee the design value of this factor, it is necessary to work on other parameters (diameter, length, viscosity), which in turn affect the power lost. So, repeated calculations for the lubricated journal (generally of any lubricated component) are required to optimize its operation.

10.7 JOURNAL-BEARING WITH FINITE LENGTH: BOSCH'S APPROXIMATE COMPUTATION

With an approximate calculation, Ten Bosch worked out the effect of finite length, based on three, easily understandable, fundamental points:

- Reduced load capacity compared to the infinite journal
- Reduced clearance minimal thickness due to pressure variations along the diagram axis
- Increased coefficient of friction due to decreasing minimal thickness compared to the case of infinite length

His computation used the formulae for the infinite length journal, adjusted for the above points.

For the first point, assuming that the length of the lubricated pair is defined by the ratio $n = l/r(\varphi_1 + \varphi_2)$, the reduction in load capacity respect to the infinite length case can be obtained by considering a k value reduced of the ratio $1/\alpha$, where α, for lengths from 0.8–3, can be calculated from the formula:

$$\alpha = 1 + 0.84 \left(\frac{l}{b} + 0.2 \right)^{1.83}. \tag{10.1}$$

For the second point, assuming the shaft takes on the position it would adopt if the length was infinite and the load is α times the real one $P(\alpha P)$, $h_0/(R - r)$, ε, β are determined using Ten Bosch's diagrams, until Sommerfeld number $k = \alpha P \psi^2/\mu V$ is obtained, for a given load direction and oil inlet position.

For the third point, as clearance shape and size are equal, the peripheral force of friction T per unit length remains the same whether the length is infinite or finite. In order to have, with infinite length, the same clearance shape of a pair of finite lengths with P load, the "equivalent" infinite length journal should have αP load.

In the previous hypothesis, this journal peripheral force of friction is identical to that of the real journal:

$$T = \alpha c \psi P \text{ instead of } T = c\psi P,$$

c corresponding to $k = \alpha P \psi^2/\mu V$.

For the real finite length pair, loaded with real load P, we obtain the coefficient of friction:

$$f = \frac{T}{P} = \alpha C \psi = K \sqrt{a} \sqrt{\frac{\mu V}{P}},$$

c and K corresponding to $k = \alpha P \psi^2 / \mu V$.

So, knowing the geometric parameters, load P, journal velocity, and oil inlet point, the calculation procedure for $(\varphi_1 - \beta)$ is:

1. Calculation of $k = P \psi^2 / \mu V$
2. Calculation of φ_1 and φ_2 with the charts of Figures 10.11 and 10.12.
3. Calculation of $n = l / r(\varphi_1 + \varphi_2)$ and α from the lubricated sliding block chart

$$\alpha = \frac{k_\infty}{k_{b/l}} = \frac{k_\infty}{k\left(\dfrac{b}{r\left(\varphi_1 + \varphi_2\right)}\right)},$$

or through Equation 10.1
4. Evaluation of h_m, h_0, ε, as far as the coefficient of friction f, by substituting the new value of $(\varphi_1 - \beta)$ and $k = \alpha P \psi^2 / \mu V$. into Figure 10.11.
5. Use the chart in Figure 10.12 to recalculate angles φ_1 and φ_2.

An example of computation that follows the cited method of Giovannozzi, Vol. I, p. 574.

The above procedure is then completed following the general indications above, and more precisely, by ensuring that the coefficient of friction f does not generate excessive temperature variations compared to those conjectured at the start.

10.8 JOURNAL-NARROW BEARING PAIR: OCVIRK'S HYPOTHESIS

The idea of removing $\partial p / \partial x$ from Reynolds' equation derives from the fact that, in bearings with small length l/b, the axis-oriented pressure gradient is predominant on the circumference-oriented one ($\partial p / \partial x \ll \partial p / \partial z$). Therefore, Reynolds' equation can easily be integrated.

The narrow bearing approximation was first introduced by A.G.M. Michell (who patented the thrust bearing above). After a cool reception from the scientific community, his work was resumed several years later by F.W. Ocvirk (1955). It looks more topical compared to the infinite length one, since technological development has moved toward narrower bearings.

According to this hypothesis, Reynolds' equation becomes

$$\frac{\partial}{\partial z}\left(h^3 \frac{\partial p}{\partial z}\right) = -6\mu V \frac{dh}{dx}.$$

If misalignment $h \neq f(z)$ is ignored, we can further simplify

$$\frac{d^2 p}{dz^2} = \frac{6V\mu}{h^3} \frac{dh}{dx}.$$

Integrating twice

$$p = \frac{6V\mu}{h^3} \frac{dh}{dx} \frac{y^2}{2} + C_1 y + C_2.$$

The boundary conditions are the so-called Gumbel conditions: cavitation is not considered and negative pressures which are generated for angles $x > \pi/2$ are eliminated.

For $z = \pm L/2$:

$$p = 0 \qquad \text{for } z = \pm L/2$$
$$dp/dz = 0 \qquad \text{for } z = 0 \text{ (because of symmetry respect to the central plane)}$$

Replacing them in the equation, we find

$$C_1 = 0 \text{ and } C_2 = \frac{-3V\mu}{h^3} \frac{dh}{dx} \frac{l^2}{4}.$$

Whence, the pressure distribution in the narrow bearing is given by the formula:

$$p = \frac{3V\mu}{h^3} \frac{dh}{dx} \left(y^2 - \frac{l^2}{4} \right),$$

which leads to the load, coefficient of friction, and all the other characteristic parameters. Comparing this with numerical integration, results shows a certain similarity of results when length values are lower than ¼.

10.9 LUBRICATED JOURNAL WITH FINITE/INFINITE LENGTH: RAIMONDI–BOYD CALCULATION

Apart from the study and results above that solved the problem of the lubricated journal with infinite and finite length, this chapter deals with the work of A.A. Raimondi and J. Boyd in 1958, of Westinghouse Research Labs., who in a three-fold project, obtained results for various cases of lubricated journals, by numerically integrating Reynolds' two-dimensional equation with the finite differences method.

To reduce the number of variables, the equation is expressed in a dimensionless way, so that the results are a function of dimensionless geometric parameters ($\varepsilon = e/c$) and the length l/d. This facilitates applying the results to the real bearing, once the geometric and other parameters are known (rotating speed, viscosity, etc.). The equation is then numerically integrated by means of finite differences.

The integration domain is divided into a mesh of finite points n_i and n_j, each given a pressure $p_{i,j}$, which is a function of the pressure of adjacent nodes. In this way, a system of $n_i \cdot n_j$ algebraic linearly independent equations is obtained (plus boundary conditions) that can be solved in several ways (e.g., by direct inversion of the coefficient matrix or by relaxation). A pressure distribution is then obtained, like the one in Figure 10.14, from which all bearing parameters can be derived.

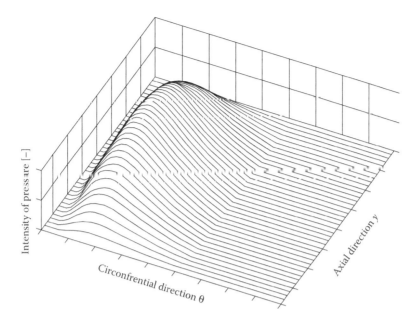

FIGURE 10.14 Pressure distribution along a journal axis.

The cases dealt with by the two authors refer to l/d values of $1/4$, $1/2$ and 1, with β values ranging from $60°$–$360°$. In the third part of the original work, thanks to newly imposed boundary conditions, cavitation is taken into account in those areas of the film in which pressure is less than the atmospheric pressure.

Nowadays, the results of this work are the most commonly used in any general survey of lubricated journals, so they are found in almost all American machines design texts. This is also why there is little information about how to read the diagrams, and students are advised to consult the original work as well as various texts of mechanical engineering design texts.[*]

We would like to stress that results of case studies on steam turbine journals closely coincided with results obtained using Ten Bosch's calculation and with Raimondi and Boyd's formulae and charts.

Figure 10.14 shows that the pressure distribution along the journal axis is not constant at all, and increases as length reduces. This gives some idea of the error made when considering constant axial pressure distribution, that is, if axial lubricant flow is not taken into account.

In this text, the diagrams and results relating to $1/6$–3 lengths are shown (figures from 10.15– 10.21), using the same symbols adopted in the original work,[†] which students will have to refer to when they apply different cases. These diagrams were obtained according to Raimondi–Boyd, integrating the pressure distribution in the clearance. They show how, as length diminishes, the results of the finite length pattern progressively differ from the infinite length one approaching the "narrow" case.

Before applying Raimondi and Boyd's results, one should bear in mind that these results are based on viscosity being constant during operation. However, we do know that lubricant input and output temperatures differ, making viscosity not constant. Here we suggest inserting the average of the input and output temperatures in the temperature formula.

[*] Shigley, J.E. and Mischke, C.R. *Mechanical Engineering Design*, 5th ed. McGraw-Hill, 1989; Juvinall, R.C. and Marshek, K.M. *Fundamentals of Mechanical Design*, Edizioni ETS, 1993.

[†] Raimondi, A.A. and Boyd, J. A solution for the finite journal bearing and its application to analysis and design, parts, I, II, III, Trans. ASLE vol.1, in *Lubrication Science and Technology*, Pergamon, New York, 1958, pp 159–209.

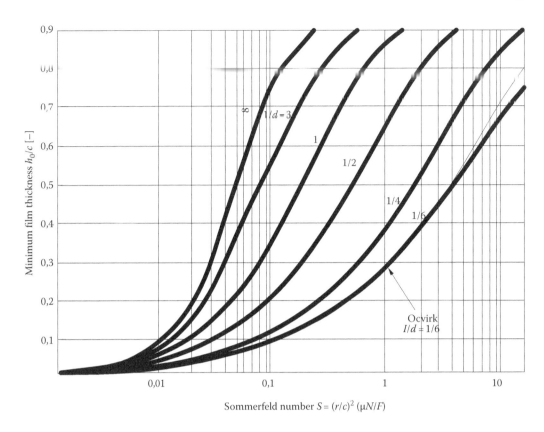

FIGURE 10.15 Minimum film vs. Sommerfeld number S.

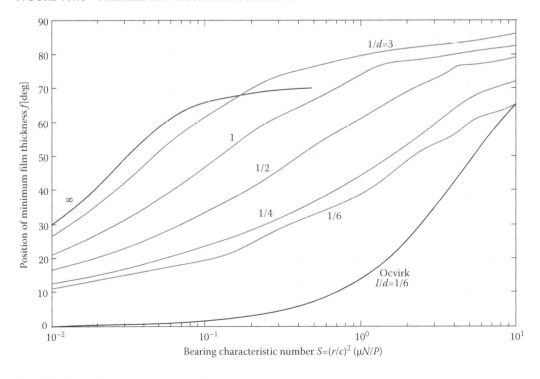

FIGURE 10.16 Position of minimum film vs. bearing characteristic number S.

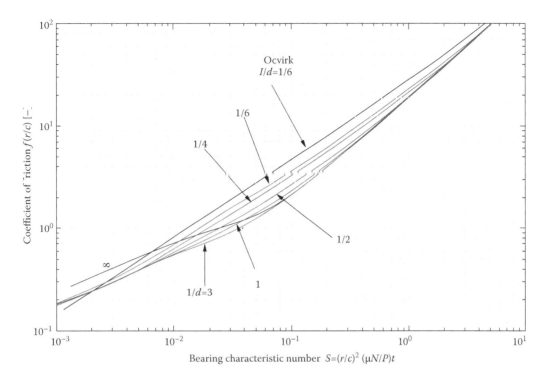

FIGURE 10.17   Coefficient of friction vs. bearing characteristic number S.

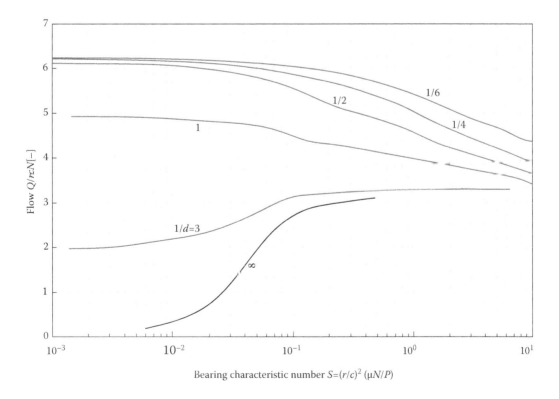

FIGURE 10.18   Flow Q vs. bearing characteristic number S.

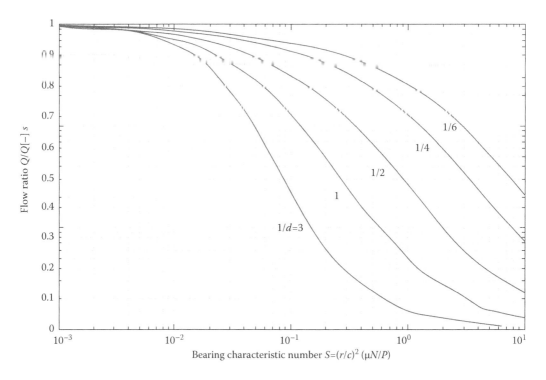

FIGURE 10.19   Flow ratio vs. bearing characteristic number S.

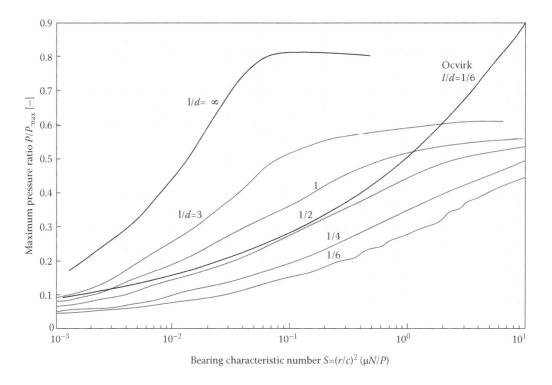

FIGURE 10.20   Maximum pressure ratio vs. bearing characteristic number S.

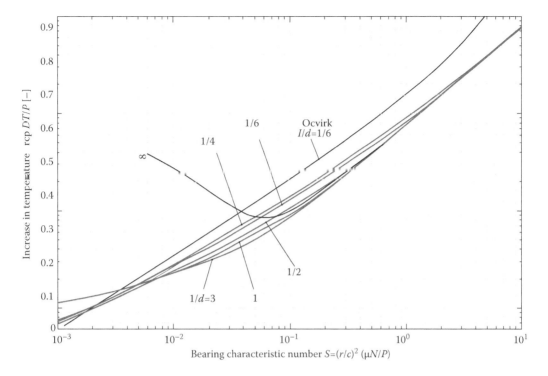

FIGURE 10.21 Increase in temperature vs. bearing characteristic number S.

As an example, here is how to determine the parameters in a typical case, where the lubricant's outflow temperature is unknown (see Figure 10.22).

Data:

- Complete bearing $\beta = 360°$
- Journal length $l = 200$ mm
- Diameter $d = 200$ mm
- Radius play $c = 0.25$ mm
- Rotation velocity $N = 30$ rpm
- Load $W = 10,000$ N
- Load per unit surface $P = W/ld = 250,000$ N/m^2
- Input temperature $T_{in} = 45°$C
- SAE20 oil, density $\rho = 800$ kg/m^3, specific heat $c_p = 1700$ J/kgK

The calculation starts by assuming an average outflow temperature, so

$$T_{media} = T_{in} + \frac{T_{out} + T_{in}}{2} = T_{in} + \frac{\Delta T}{2}.$$

For example, one could conjecture that $\Delta T_1 = 30°$C: $T_{mean1} = 45 + 15 = 60°$C. From the chart in Figure 10.23, the viscosity (in millipascal second) for an SAE20 oil at the given temperature is about:

$$\mu_{SAE20}\left(T_{medial}\right) \cong 20 \cdot 10^{-3} \text{ Pa·s}.$$

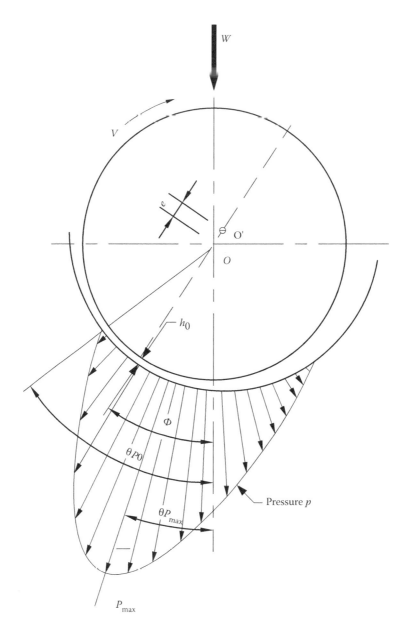

FIGURE 10.22   Polar diagram of film pressure distribution (Raimondi and Boyd).

With this figure, Sommerfeld number can be obtained at the first attempt:

$$S_1 = \frac{\mu N}{P}\left(\frac{R}{c}\right)^2 = \frac{3.5 \cdot 10^{-3} \cdot 30}{250,000}\left(\frac{0.100}{0.00025}\right)^2 = 0.384.$$

This S_1 is then entered into the temperature rise chart: the value found is $\Delta T_2 \approx 35P/\rho c_p = 6°C$. The ΔT thermal difference calculated is then compared to the conjectured calculation at the beginning; if the two values differ, the calculation has to be repeated using the value of ΔT obtained from the diagrams until they coincide. Since $\Delta T_2 \neq \Delta T_1$, we calculate Sommerfeld number S_2 again, finding a new viscosity corresponding with ΔT_2. The calculation should be repeated until $\Delta T_n - \Delta T_{n-1}$.

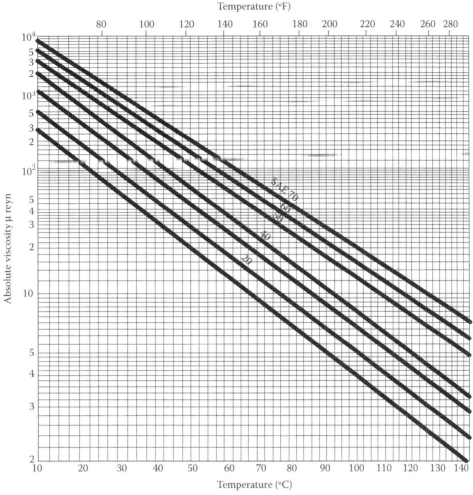

FIGURE 10.23   Viscosity–temperature chart for typical oil SAE.

The chart showing the dimensionless rise of temperature is obtained by using the tank as a control volume and writing an equilibrium equation of enthalpy lost to friction that will be equal to that of the liquid flowing into the tank when the system is fully operative.

The liquid flowing out laterally (Q_s) is believed to undergo a temperature rise equal to $\Delta T/2$ (the average between input and outflow), whereas the liquid flowing over the entire bearing ($Q - Q_s$) increases in temperature by ΔT.

$$H_{\text{lost}} = \rho C_p Q_s \Delta T/2 + \rho C_p (Q - Q_s) \Delta T = \rho C_p Q_s \Delta T \left(1 - \frac{1}{2} \frac{Q_s}{Q} \right).$$

H_{lost} is equal to the mechanical work the journal carries out on the film.

$$H_{\text{lost}} = 2\pi N \cdot fWr = 2\pi N \left(f \frac{r}{c} \right) (Pl2r) c.$$

By equalizing, we obtain a dimensionless expression that is a function of known parameters.

$$\mu c_p \Delta l / F = 4\pi \frac{f(r/c)}{\left(1 - \frac{1}{2}\frac{Q_s}{Q}\right)\frac{Q}{rcNl}} .$$

Once the correct viscosity is determined and consequently S, we refer to the diagrams and by l/d length ratio we obtain clearance minimal thickness h_0, relative eccentricity ε, the load angle φ that sets the position of the clearance minimal thickness, the coefficient of friction $f(r/c)$, load $Q/rcNl$, and the lateral outflow capacity Q_s/Q.

For length values other than those in the diagrams, the parameters (generally called y) can be computed using the values $y_{l/d}$ from the charts (shown in the diagrams for the four values of l/d) by the expression:

$$y = \frac{1}{(l/d)^3}\left[-\frac{1}{8}\left(1 - \frac{l}{d}\right)\left(1 - 2\frac{l}{d}\right)\left(1 - 4\frac{l}{d}\right)y_\infty + \frac{1}{3}\left(1 - 2\frac{l}{d}\right)\left(1 - 4\frac{l}{d}\right)y_1\right.$$

$$\left. - \frac{1}{4}\left(1 - \frac{l}{d}\right)\left(1 - 4\frac{l}{d}\right)y_{1/2} + \frac{1}{24}\left(1 - \frac{l}{d}\right)\left(1 - 2\frac{l}{d}\right)y_{1/4}\right].$$

As stated earlier, the design of a lubricated journal that responds adequately to stress requirements and imposed external conditions, therefore depending on several variables, requires experience. The various calculation methods, if used correctly, achieve quite similar results. On the basis of long experience, different computing methods are recommended to validate the dependability of the results.

10.10   NOTES ON LUBRICANTS

Before ending the chapter, here are some notes on lubricants and their importance to the durability of transmission components.

An effective lubricant must not only generate suitable layers, but must also reduce friction, transfer heat and protect components from wear and tear. An important feature of any good lubricant is its capacity to keep its characteristics stable over time, even in changing conditions. A lubricant can reach target areas by splashing or, in more severe cases, by being pumped through channels.

The choice of lubricant also depends on its maximum operating temperature. According to application, there are different average operating temperatures. These range from around 50°–60° for the bearings of a big steam turbine, to around 100°–120° for automotive gears. A lubricant should also have a detergent capacity for wear-and-tear products.

As for automotive oils, multigrade oils are employed, which remain effective and maintain good viscosity in changing conditions. As touched on earlier in this chapter, viscosity is not constant with temperature. Figure 10.23 reports the behavior of an SAE (Society of Automotive Engineers) oil.

Viscosity is assessed with a viscometer which, broadly speaking, is a vessel with an opening of known size at its bottom. A given amount of liquid at a given temperature is poured into it and the time the liquid takes to flow out is measured. The viscosity (cinematic or dynamic) is assessed by diagrams or by interpolating curves. An SAE 10 oil, for instance at 130°F, has a viscosity ranging from 160–200 centipoises [cP] (1×10^{-2} dine s/cm^2); for an SAE 20 it ranges from 230–300 cP.

In the international system (SI), viscosity is expressed in pascal second (Pa s) and is converted thus: 1 cP $= 10^{-3}$ Pa s.

Apart from dynamic viscosity, there is also kinematic viscosity, which is measured in Stokes (St) or centiStokes (cSt). It equals the dynamic viscosity divided by lubricant density, so $cSt = \rho cP$ with ρ as density.

The diagram showing the temperature-related variation of viscosity of the lubricant highlights how computations of lubricated components have to be carried out several times. Indeed, assuming an operating temperature at the start of the calculation, a viscosity value is selected that is applied over the entire process until the final oil temperature is determined; if it coincides with the temperature assumed at the start, all well and good, otherwise it is necessary to start from this last temperature value and repeat the calculation.

11 Shafts and Bearings

11.1 SHAFTS

Shafts are long machine components that in operation either rotate or swing around a rectilinear axis. Shafts and axles can be distinguished from gudgeons according to the stress involved:

- Gudgeons are quite compact, so the bending moment stress is unimportant (leverage is small) compared to shearing and torque stress.
- Axles and shafts are more similar to very long beams, so shearing stress is negligible and there are bending and twisting stresses, respectively.

The distinction between axles and shafts is purely conventional, as in practice, bending and torsional stresses are involved. In practice, shafts are subject to dynamic stress, so fatigue checks are necessary. These are rarely subject to static loads.

For computation, a shaft can be roughly described as a rectilinear elastic beam with a certain number of supports corresponding to bearings, with the forces applied by rotating components (gears, pulleys, joints, or clutches). Generally, shafts are isostatically constrained and so there are two bearings, one as a hinge, the other as a carriage. More generally, a shaft is exposed to bending, shearing stress, and torsional stress due to the application of moment.

Normally, bending stress is symmetrical and due only to shaft rotation. Thus, fatigue stress fluctuates when the average tension is zero $\sigma_m = 0$, and fatigue checks must compare real stress to the material's resistance to fluctuation. As regards torsional stress, there are obviously several cases.

If, for example, the shaft is linked to a piston engine, the result is a variable pulsating twisting torque with a tension lower than zero, $\tau_i = 0$. The dynamic component of comparison will be the fatigue limit for pulsating stress, or its lifetime durability.

After specifying shaft stress, it must be checked for fatigue; the Smith-Goodman's diagram (or similar) is generally used, depending on the material and load type (axial, bending, or torsional), proceeding according to fatigue procedure, and bearing in mind that these components are almost invariably subject to compound stress. Therefore, it is necessary to assess the point of maximum stress and adopt such criteria as will consistently deal with varying compound stress.

The analysis should always take into account the notching effects to which components like shafts are usually subject. As we saw in Chapter 6, notching effects can account for stress increases equal to β_i times those computed.

More generally, connecting keyways, or diameter variations with connectors of various sizes, or interference-forced components, give rise to local stress increases that can lead to possible breakage.

11.2 SHAFT MEASURING

Shafts are first roughly measured through a static bend-twisting check that applies less stress than used in the traditional static check. This stress is obtained from Bach's empirical criterion that is equal to

$$\sigma_{am} = \frac{\sigma_s}{B\beta},$$

where B is Bach's coefficient, which equals 3, 2, or 1 depending on the type of stress (swinging, pulsating, static), and β is the effective notching factor.

As mentioned earlier, in the computation, a shaft is considered as a rectilinear elastic beam, isostatically constrained with supports corresponding to bearings, and the forces are applied by rotating components (wheels, pulleys, flywheels).

If junctions and clutches that link various components of a shaft convey only torque (e.g., leather mesh joints), they resemble a hinge. If they ensure perfect stability of the components connected (flange unions, disk joints) they can be considered as directly belonging to the beam, although with a different bending rigidity.

In general, the data used to measure a shaft are the following:

- Input power W
- Angular shaft speed $\omega = 2\pi n/60$

From these values, the input torque can be assessed, given by

$$M_t = \frac{750\,W}{\omega} \quad \text{if } [W] = \text{horsepower,}$$

$$M_t = \frac{1020\,W}{\omega} \quad \text{if } [W] = \text{kilowatt,}$$

where in both cases the torque is expressed in N·mm.

In general, knowing how motion is conveyed, it is not difficult to deduce the forces working on a shaft and therefore make a simple model of the stress components. In consequence, if transmission is by means of cogwheels, the torque operating on the various shafts of a gearing can be worked out if the incoming torque and transmission gear ratios are known. If transmission occurs by means of spur gears (Chapter 15, Spur Gears), the force exchanged $\bar{F}$ is inclined by the angle of pressure against the tangent to the pitch, and so it is easy to understand which components (radial and tangential) exert power on the shaft. In this case, there is no axial component to the shaft, which is the case for spiral or cylindrical bevel gears. With spiral gears, the shaft supports can use radial bearings, but with cylindrical bevel gears, bearings are needed (at least one) that can also withstand axial loads.

Rotating shafts (rotors), therefore, rest on supports (bearings) that are subject to static or dynamic forces. At each of these sections (gudgeons) the shaft must be fitted into its bearings, so they must be designed in such a way as to facilitate this fitting (usually of high quality). Positioning, bearing, and consequently gudgeon type depends on systems with critical speeds possibly much lower than operating speeds. Therefore, shaft dimensions (and mass) are also components that affect possible resonance effects.

As an example, Figure 11.1 shows a three-stage steam turbine rotor where the five gudgeons are numbered, and through which, forces are applied to the crankcase F.

11.2.1 Two-Bearing Shafts (Isostatic)

As before, shafts are generally isostatically constrained by two bearings, one a carriage, the other a hinge. The latter ensures axial adjustment of the shaft. Anything that might skew the shaft is calculated by equilibrium equations related to the two perpendicular planes passing through the shaft axis.

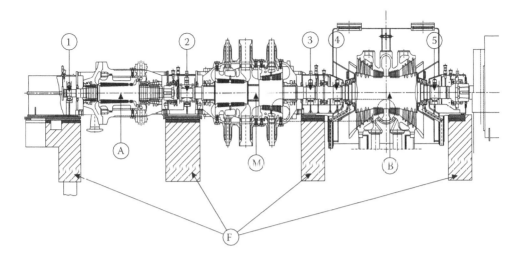

FIGURE 11.1   Three-stage steam turbine rotor.

In order to calculate constraint reactions, it is necessary to know the distance of the forces from the bearings. However, this depends on shaft diameter which is not known 'a priori'. By considering the mid section of the bearing and neglecting shaft weight, initially it is determined the bending moment in the two perpendicular planes, then the overall moment (principle of superposition of effects) and at end will be carried out a bending-twisting check at the points of peak stress.

Shaft sizing is carried out as follows:

1. The first sizing is based only on the torque moment to find a provisional shaft diameter.
2. The shaft and components fitted to it are designed and the forces exerted on it are calculated, having defined the constraint positions.
3. Knowing the forces exerted, the component weight, and the distance between the various bearings, the bending moment diagrams are drawn in the two perpendicular planes.
4. The total bending moment is $M_f = \sqrt{M_x^2 + M_y^2}$. Note that, in general, whereas the bending moment varies along the axis, the torque remains constant. Therefore, generally, maximum stress occurs in the section where the total bending moment is at a peak.
5. Having calculated the stress parameters, the corresponding tensions σ and τ in the most stressed section are evaluated.

$$\sigma = \frac{M_F}{I} \cdot R = \frac{4M_F}{\pi R^3} \quad \tau = \frac{M_t}{I_z} \cdot R = \frac{2M_t}{\pi R^3}.$$

In some cases, though, it may be necessary to evaluate axial or shear stress. The former adds to bending stress, the latter to torque stress.
6. Since pluriaxial stress is being dealt with, it is necessary to take into account ideal tension σ_{id} which alone is as dangerous as all the tensions working together. By using the distortion energy hypothesis, we obtain:

$$\sigma_{id} = \sqrt{\sigma^2 + 3\tau^2}.$$

Different expressions are produced when other breakage hypotheses are adopted.

7. Knowing the ideal tension, a fatigue check can be carried out. The easiest way is to use the Bach criterion (above). Safety conditions are when $\sigma_{id} \le \sigma_{am}$, where s_{am} is given by (using the symbols most commonly used)

$$\sigma_{am} = \frac{\sigma_s}{B\beta_l} c_1 c_2 c_3.$$

8. By equalizing with the previous expression, the minimum shaft diameter able to withstand compound bending-torque stress can be calculated. By making $\sigma_{id} = \sigma_{am}$, we get

$$\sigma_{id} = \sqrt{\sigma^2 + 3\tau^2} = \sqrt{\left(\frac{4M_F}{\pi R^3}\right)^2 + 3\left(\frac{2M_t}{\pi R^3}\right)^2} = \sigma_{am},$$

and

$$d_{min} = \sqrt[3]{\frac{16}{\pi\sigma_{am}} \sqrt{4M_F^2 + 3M_t^2}}.$$

This minimum diameter corrects the preliminary diameter that only accounted for torque.

9. This procedure is repeated as the bearing location varies, so as to determine its best position.

If the shaft cross-section varies, it must be taken into account in the previous formulae when calculating moments of polar inertia $J = \pi R^4 / 2$.

β_i in the σ_{am} expression obviously must account for any variations in cross-section or keyways, or other types of machining that might raise tension due to notching.

In general, in the construction of machine components likely to have notching, just as shafts have, it is not advisable to use noble steels with relatively high breakage points, because they are most notch-sensitive and, all things being equal, they might be overly costly. Indeed, comparing production and work environment, the reduction coefficient C_1 acceptable stress decreases as tensile stress R rises, as shown in the general fatigue diagram.

11.2.2 Shafts with More than Two Bearings (Hyperstatic)

Sometimes a shaft can be constrained hyperstatically, or it can be considered as a beam on several supports in a computational model.

In this case, Casigliano's theorem is applied, according to which the partial derivative of elastic deformation compared to a constraint reaction is equal to the shift δ of that reaction.

From constructional sciences courses, the elastic deformation is given by

$$L = \frac{1}{2} \int \frac{M_F^2}{EJ} ds \quad \text{given} \quad \frac{dy}{dx} = -\int \frac{M}{EJ} dx,$$

and deriving from the hyperstatic unknowns X_i:

$$\frac{\delta L}{\delta X_i} = \frac{\delta}{\delta X_i} \int \frac{M_F^2}{2EJ} ds = \int \frac{\delta}{\delta X_i} \frac{M_F^2}{2EJ} ds = \int \frac{M_F}{EJ} \frac{\delta M_F}{\delta X_i} ds = \delta_i.$$

The shifts corresponding to the constraints are zero or proportional to the reaction if they are elastically pliable constraints.

If the law of moment or the inertial moments vary according to not easily integrable functions, it is necessary to "split" the integral and discretize the equation to solve it.

11.3 DEFORMATION LIMITS AND DETERMINATION OF ELASTIC DEFLECTIONS

Apart from setting a limit to maximum stress, it is often advisable to set limits to the maximum deformation shafts can be liable to. In particular, bending deformations must be such that the ratio between the maximum deflection of the shaft and the distance between the supports is less than 1/3000, and the tangential slope (*tan α*) to the elastic line in proximity to the bearings must be less than 1/1000:

$$\frac{f_{max}}{l} < \frac{1}{3000} \quad tg\alpha < \frac{1}{1000}.$$

Particular attention must be paid to gear shafts, since vectors and slopes in excess can compromise correct meshing, increasing wear and reducing service life.

As regards torsional deformation, unit rotation must be less than ¼ degree per meter:

$$\vartheta_t = \frac{\tau}{GR} = \frac{M_t}{GRW_t} = \frac{2M_t}{GR\pi R^3} = \frac{2M_t}{G\pi R^4} < \frac{1}{4} \frac{grado}{metro} = \frac{1}{4} \frac{\pi}{180}.$$

For low torsional rotation, $\vartheta = \tau/GR$, the material should work little (low τ) and not be too small (big R).

As a general rule, to limit deformation, less expensive materials with lower breaking points should be exploited, so as to oversize the shafts and ensure greater rigidity.

11.3.1 ASSESSMENT OF SHAFT ELASTICITY

Computing the elasticity of shafts with varying cross-sections, with a variable inertia moment along the shaft axis, is carried out by Mohr's method following the numerical procedure proposed by Giovannozzi (Vol. 1, page 666).

This method is based on integrating the equation of the elastic line of inflected beams according to the hypothesis of negligible shear deformation:

$$\frac{\partial^2 y}{\partial x^2} = -\frac{M}{EJ}.$$

In the diagram of the bending moment within a plane, the beam is divided into all those sections where the following conditions are true:

- The moment of inertia is constant ($J = const$);
- The bending moment M changes linearly.

If the moment varies non-linearly (as in the case of a distributed load), the beam is divided into many small sections where the non-linear trend is close to a linear one. According to this hypothesis the moment of the nth section is expressed by:

$$M_n - M_{n-1} + \Delta M,$$

that is, moment M_n in the nth section equals the moment M_{n-1} in the previous section increased by the variation ΔM between the two sections.

With reference to Figure 11.2 and according to the hypothesis of linearity, the moment in the generic section of abscissa z can be expressed as

$$M(z) = M_{i-1} + \frac{z}{l_i} \Delta M.$$

Owing to the rotation of the generic section of abscissa z, we have

$$\alpha(z) = \frac{\partial y}{\partial z} = -\int_0^z \frac{M(z)}{EJ} dz.$$

And the absolute rotation of the abscissa z section is

$$\alpha(z) = \int_0^z \frac{M(z)}{EJ} dz = \frac{1}{EJ} \left(M_{i-1} \cdot z + \frac{\Delta M}{2li} z^2 \right).$$

To calculate the deflection in section z it is sufficient to integrate the above equation:

$$f(z) = \int_0^z \alpha(z) dz = \frac{1}{EJ} \left(M_{i-1} \frac{z^2}{2} + \frac{\Delta M_i}{6li} z^3 \right).$$

This represents the deflection of a generic section compared to the elastic line tangent in section $z = 0$.

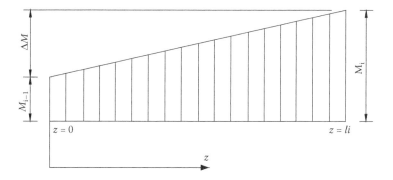

FIGURE 11.2 Qualitative linear bending moment.

Rotation and deflection variations between end sections of an nth small section are

$$\Delta\alpha_i = \alpha(li) - \alpha(0) = \frac{1}{EJ_i}\left(M_{i-1}li + \frac{\Delta M}{li}\frac{li^2}{2}\right) - 0 = \frac{1}{EJ_i}li\left(\frac{M_{i-1} + M_i}{2}\right),$$

$$\Delta f_i = f(li) - f(0) = \frac{1}{EJ_i}\left(M_{i-1}\frac{li^2}{2} + \frac{\Delta M}{li}\frac{li^2}{6}\right) - 0 = \frac{1}{EJ_i}li^2\left(\frac{2M_{i-1} + M_i}{6}\right).$$

Rotation and deflection values are not absolute but relative so we assume $\alpha(0) = 0$ and $f(0) = 0$. To obtain absolute values, the actual rotation and deflection values in the first section must be known.

The overall values compared to the first shaft section are given in Figure 11.3:

$$\alpha_i = \sum_{i=0}^{i} \Delta\alpha_i \quad f_i = f_{i-1} + \alpha_{i-1}li + \Delta f_i,$$

nth section rotation is equal to the sum of all the small section rotations located before the nth section. The nth section vector equals the sum of vectors in the first sub-section of the nth small section plus the curvature effect of the elastic line plus the bending effect of the nth small section.

The real deflection y_i, however, are given by the distance in the various sections between the elastic line and the line joining the two supports (non-deformed axis). From the similarity of the triangles it follows that

$$\frac{y_i + f_i}{x_i} = \frac{f_d}{l} \quad \text{whence} \quad y_i = f_d\frac{x_i}{l} - f_i,$$

where f_d is the final section ($z = l$) vector against the tangent in the initial section ($z = 0$).

If, apart from the M_y moment in a plane, there is an M_x moment in the perpendicular plane, the total vector is

$$f = \sqrt{f_x^2 + f_y^2},$$

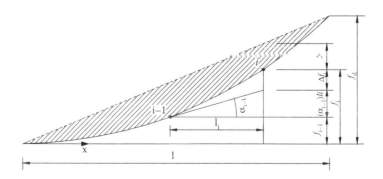

FIGURE 11.3 Rotation and deflection in a finite part.

and it is located in a plane defined by the angle:

$$\tan \gamma = \frac{f_x}{f_y},$$

where γ is the angle of the plane containing the shaft elasticity against y.

Had the shaft been hyperstatically constrained, to calculate the vector and the rotations, the fourth order differential equation would have had to be integrated:

$$\frac{d^2 y}{dx^4} = q(x),$$

where $q(x)$ is the perpendicular distributed load imposing the constraint-created conditions as boundary conditions so as to generate the four integration constants.

Summarizing, these are the procedural steps:

1. Computation of the bending moment M.
2. Subdivision of the beam into several small sections with J = cost and M that varies linearly.
3. For each small section, calculation of the relative rotation and vector variations between the two end sections $\Delta\alpha_i$, Δf_i.
4. For the first section of the shaft, calculation of the overall values of the rotations and deformations relative to α_i, f_i.
5. Calculation of actual vectors y_i bearing in mind that:

$$\alpha_1 = \Delta\alpha_1 \qquad\qquad\qquad\qquad f_1 = \Delta f_1$$

$$\alpha_2 = \Delta\alpha_1 + \Delta\alpha_2 \qquad\qquad\qquad f_2 = f_1 + \alpha_1 l_2 + \Delta f_2$$

$$\alpha_3 = \Delta\alpha_1 + \Delta\alpha_2 + \Delta\alpha_3 \qquad\qquad f_3 = f_2 + \alpha_2 l_3 + \Delta f_3$$

$$\alpha_4 = \Delta\alpha_1 + \Delta\alpha_2 + \Delta\alpha_3 + \Delta\alpha_4 \qquad f_4 = f_3 + \alpha_3 l_4 + \Delta f_4$$

...   .

Once the calculation is made, it is necessary to verify any deformations or calculate shaft vibration frequencies or critical velocities of rotating shafts.

As mentioned several times, both the static and dynamic calculations for shafts can be carried out by finite element method (FEM) numeric simulations, rather than with the traditional method above. The first step is to define the solid model (Figure 11.4 shows the solid model of the shaft in Figure 11.1), schematizing it into appropriately shaped components and for the subsequent definition of stress and deformation due to external loads, masses, thermal stress, etc.

It should be remembered that using analytical solutions, even simpler schematized solutions, helps analyze parameters that might influence the issue and can point to a more suitable model on which to apply appropriate numerical methods.

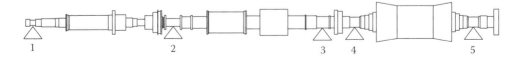

1   2   3   4   5

FIGURE 11.4   Model of the shaft of Figure 11.1.

In Figure 11.4 there is a solid model of a steam turbine shaft (power 70 MW). In the model, the fulcrum points of the shaft have been numbered (1, 2, 3, 4, 5) where the bearing housings are fitted. The previous outline, relating to a large system, shows how all power transmission shafts use bearings albeit with different models. Depending on the fulcrum point, anti-friction or friction bearings are fitted.

11.4   ROLLING-CONTACT BEARINGS (ANTI-FRICTION BEARINGS)

For this topic, refer directly to Giovannozzi, vol. I, which contains several examples of shaft fittings, apart from describing various types of rolling-contact bearings. Here, only the general concepts to bear in mind while studying and examining the designs in the text are dealt with. For detailed descriptions and illustrations of various types of bearings please refer to, besides the textbook, the catalogues and manuals of the manufacturing companies (e.g., RIV-SKF, FAG).

11.4.1   CLASSIFICATION OF BEARINGS

Bearings are designed to facilitate the rotation of inter-connected components, guaranteeing their relative positions and withstanding certain loads. They can be classified on the basis of their characteristics.

11.4.2   CLASSIFICATION BY ROLLING BODY SHAPE

A first classification can be the shape of the rolling body:

- Ball bearings
- Roller bearings (conical, barrel/tumbler, cylindrical rollers)

11.4.3   CLASSIFICATION BY ASSEMBLY METHOD

A further classification can be on the basis of assembly method, which depends on whether the bearing can be dismantled:

- Non-dismantlable bearings (radial ball bearings, and flanged roller bearings)
- Dismantlable bearings (roller bearings and angular-contact bearings in general), in which the two pieces can be mounted separately)

11.4.4   CLASSIFICATION BY THRUST DIRECTION

According to thrust direction, the following distinctions can be made:

- Radial bearings (loads perpendicular to the axis of rotation)
- Axial bearings (loads along the axis of rotation)
- Mixed bearings (force direction oblique to the axis)

Radial bearings   (ball and roller bearings) can withstand small axial loads via the lateral supports (rimlets).

Axial bearings   (thrust bearings) are rolling components (balls or rolls) held between two plates. They are not engineered to rotate at high speeds, as the lubricant tends to flow out of the bearing due to centrifugal forces. To prevent this, a radial ball bearing can be used with four faces of the two rings blocked, which ensures the axial adjustment of the shaft together with the thrust bearing that takes the whole axial load.

Mixed bearings such as barrel or roller bearings withstand remarkable radial and axial loads. Bear in mind that the load on these bearings, whether radial or axial, also generates axial (or radial) forces. When a mixed bearing (either angular-contact, ball, or conical bearing) is subjected to a load, reactions are generated along the contact perpendicular between the ball race and rolling bodies and so both axial A and radial R components are present. If the load applied is only axial, the radial components are canceled out, because the external radial component is equal to zero, $R = 0$.

If a ball bearing is subjected to a radial load, for instance directed upward, the resulting force between the various balls must be such as to balance the load itself. The radial force R tends to compress the balls of the top half of the bearing and free those of the bottom half, therefore only some of the balls react to force R. Furthermore, this force tends to rigidly translate the inner ring compared to the external one, compressing the balls even more. The central ball is the most compressed, whereas the side balls are less compressed. (see Stribeck's hypothesis, Paragraph 9.4).

Rigid translation means a non-uniform load on the balls: each sphere reacts with a strength N_i different for each sphere, both radial and axial components being different.

So, applying radial load R causes the bearing to react with an induced axial force generated by the bearing itself.

With reference to Figures 12.7, 12.8, 12.9 (conical roller), there are two separate contact perpendiculars, because the contact generators are not parallel. The force directions are $n - n$ and $n' - n'$, producing a small force on each roller along the cone axis that tends to eject it from its outer race.

The cone is held inside the bearing due to the shaft located behind it (on the inner side). The shaft generates a third reaction such as to prevent the cone unthreading itself. So, the conical roller is held in equilibrium by three forces.

The reasoning for a roller bearing also applies to ball bearings. Conical roller bearings withstand axial thrust in one direction only. For bi-directional axial thrust, these bearings are mounted in pairs.

Conical bearings can disassemble into three components: outer ring, rollers, and inner ring. For this reason, they are pre-loaded in assembly so no axial shifts occur in operation.

Usually, these bearings are mounted by putting two identical bearings back-to-back so that one reacts against the other. Similarly, a simple thrust bearing can be coupled to a conical roller bearing to oppose the axial load induced by a radial load.

In assembly, bearings are generally pre-loaded, so the distance between opposing bearings is small to reduce any clearance between the chassis and shaft components (even owing to different materials or temperatures). It should be pointed out that any elongation caused by thermal dilation is greater the longer the length involved. If temperature changes by ΔT, the corresponding length variation is as follows:

$$\Delta l = l - l_0 = l_0 \alpha \Delta T.$$

If the Δl dilation is prevented, compression stress is generated on the shaft equal to

$$\sigma = E\varepsilon = E\frac{\Delta l}{l_0},$$

corresponding to forces on the bearing rings equal to

$$F = \sigma A = AE\varepsilon = AE\frac{\Delta l}{l_0} = AE\alpha\Delta T,$$

as can be seen, proportional to α and independent of l_0. The force F then generates a system of forces on the bearing rings that depend on the extent of the Δl shift.

11.5   GENERAL ASSEMBLY RULES

During the bearing assembly, some rules must be observed to make the whole system work efficiently. These rules can be summarized as follows:

1. It must always be possible to assemble and disassemble every mechanical part, including the bearings.
2. For optimum reliability, the number of mechanical system components must be as small as possible. This is also valid for bearings (generally two bearings are enough to constrain a shaft).
3. The system must not be hyperstatic, otherwise stress is generated both on the shaft and bearings. Ideally, one bearing works as a hinge and the other as a carriage.
4. Each bearing has one function only: one bearing (the hinge) ensures the axial position (coordinate x) of the shaft and its position along the z-axis; the other bearing (the carriage) guarantees axis centering and its position along the z-axis.
5. When the distance between supports is too long, any shaft deformation must not strain the bearing. However, orientable bearings are used to adjust for any small misalignments.
6. If one section of the shaft needs constraining more rigidly, either a couple of bearings close to each other or one very big one is used. Thus, a fixed or a semi-fixed end is produced, which must be accounted for in the calculation.
7. The shaft bearing inner-ring couplings with outer-ring chassis are the shaft-hole type and tolerances are defined by UNI guidelines, as reported in the bearing manufacturers' catalogues. Furthermore, the chassis section where the outer or inner ring of the bearing moves axially needs to be rectified. If a roller bearing is used without its inner ring, it is necessary to rectify the shaft section where the bearing is mounted.
8. The junction radiuses of the rounded bearing housings are fixed by UNI norms and reported in the manufacturers' manuals. If there is any variation in shaft bore due to notching, it is common practice to use shaft junctions with radiuses not less than $d/10$, where d is the diameter of the smallest bore. This rule might prove impossible to follow should bore variation be due to the "unballing" of a bearing's inner ring. In such a case, the radius r of the shaft throat junction must be shorter than the radius r of the bearing edge. If they were equal, there would be a coming together. So, the following must be ensured:

$$\frac{d}{10} < \rho < r.$$

   If this is not possible, a spacer should be fitted between unballing and bearing that can be rounded to suit.
9. The transmission of forces between the rotating part (shaft) and the fixed part (chassis) must only take place through the rolling components; there must be no other points of contact or in-between components.
10. The bearing ring that rotates in the direction of the load must be mounted sufficiently tightly in accordance with the manufacturer's instructions.

11.6   BEARING ADJUSTMENT

In most current applications, bearings enable shafts to rotate at certain speeds while maintaining their position and withstanding specific loads. Bearings must be mounted to ensure the durability of reciprocal positioning of the interconnected components (shaft, bearing, chassis).

As mentioned before, it is advisable that each bearing performs one task only. In order to guarantee the correct position of components over time, a bearing's axis must coincide with the shaft axis as well as with the housing axis. To do this, the bearing is mounted rigidly in the pocket. Another bearing must ensure the reciprocal position of the three components in the plane defined by the front section of the bearing itself, in other words, the shaft position has to be defined in relation to the system (e.g., the chassis).

To fix the shaft axis with respect to the chassis, one point of it should always be fixed and, if there is any clearance, adjustments must be made to offset this clearance and restore the initial setup.

The bearings that ensure the shaft's axial adjustment are usually either radial ball bearings, with both their outer and inner rings blocked axially on the chassis and the shaft respectively, or thrust bearings.

Radial ball bearings that ensure free shaft dilation that have their non-rotating ring against the load, must be mounted with a free coupling or a push fit, so they run axially in their housings with little force, and can therefore follow the axial shifting of the shaft almost freely.

These bearings, though, can have both rings blocked axially on the shaft and chassis when they are open roller-type bearings or small roller bearings. In this case axial shifting is generated between the rollers and the edgeless ring.

The simultaneous axial adjustment of two radial bearings is carried out only for angular-contact bearings that are mounted in very close pairs so that thermal dilation is limited.

Therefore, for the shafts supported by two bearings, the following applies:

- A bearing (*A*) adjusts its shaft axially and withstands radial as well axial loads. For this reason, radial ball bearings or orientable roller bearings are used. In the case of thick shafts, angular-contact or conical bearings can be used, mounted in pairs and pre-loaded. The expression "assembly in opposition" is used to refer to a condition in which each of the two conical bearings can take an axial load and prevent any axial shift in one direction. Being modular, conical bearings are mounted pre-loaded. The amount of pre-loading in conical bearings is regulated by moving one of the two rings axially against the other to the desired force. The rings of the two conical bearings are fixed axially on one side only.
- A bearing (*B*) adjusts the shaft radially, enabling axial movement and withstanding only radial loads. For this reason flangeless bearings are used that are constrained on four faces (movement occurs within the bearing). For radial ball bearings or roller bearings with a free ring (the unloaded ring, or the one not coupled to the rotating part), the regulator takes up the clearance that occurs during operation, thus setting the pre-load. Conical bearings for mixed loads are set axially for both bearings, each bearing ring being blocked on one side only; in addition, being modular, conical bearings are mounted in opposition, the pre-load being set by the regulator.

11.7   ASSEMBLY MODELS

The instructions above for correctly assembling and adjusting bearings are summarized in the descriptions of some typical bearing assemblies.

Bearing assembly for unidirectional axial loads (Figure 11.5)
Mechanical components
Outer chassis
Fifth wheel
Orientable ring of the axial thrust bearing
Flat ring of the axial thrust bearing
Spacer
Orientable ball bearing
Shaft

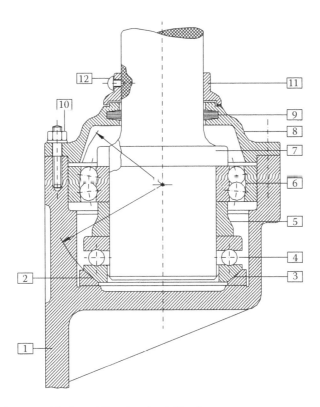

FIGURE 11.5   Bearing assembly for unidirectional axial loads.

Cover
Felt grease seal
Screw
Outer cover
Screw

Assembly Order
1-(2-3)-(4-5-6-7)-(8-9)-10-11-12

Loading Model
The only function of the orientable radial ball bearing (6) is to orientate the shaft, since in this assembly there are no radial loads.

The orientable thrust bearing (2-3-4) only works axially, and absorbs the load unidirectionally; it can also follow the shaft orientation. This type of bearing is not suitable for high-speed rotation because lubricant would be lost.

The thrust bearing's orientable ring is not directly mounted on the shaft but it has a radial clearance.

The centers of oscillation of both bearings coincide.

Adjustment
Radial adjustment is by the orientable bearing whereas the axial adjustment is provided by the thrust bearing and axial load; there is no axial adjustment for the orientable ball bearing (6). The adjustment is provided by the axial load.

The surface supporting the outer ring of the orientable ball bearing needs to be refaced.

Component 11 works as an axial blocker for cover 8.

Bearing assembly for bi-directional axial loads (Figure 11.6)

Mechanical components
Outer chassis
Fifth wheel with axial thrust bearing orientable ring
Flat ring of axial thrust bearing
Shaft
Axial thrust bearing with fifth wheel
Spacer
Orientable ball bearing
Spacer
Blocking disk
Screw
Metal ring
Felt grease seals
Cover
Screw

Assembly Order
1-(2-3)-(4-(5-6)-7-8-9-10)-11-12-13-14

Loading Model
An orientable radial ball bearing (7) works only as a shaft orientator, since in this type of assembly there are no radial loads.

Orientable thrust bearings (3–5) work only axially and absorb the load unidirectionally. This type of bearing is not suitable for high-speed rotation, because lubricant would be lost. The thrust bearing orientable ring is not directly mounted on the shaft, but it has radial clearance.

The oscillation centers of the three bearings coincide.

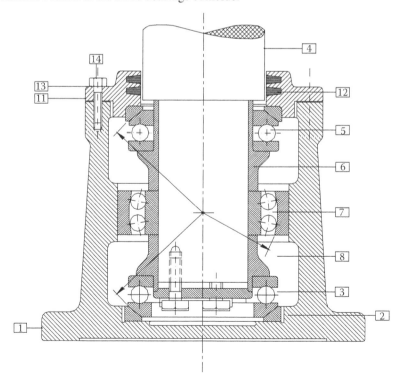

FIGURE 11.6   Bearing assembly for bi-directional axial loads.

Adjustment

Axial adjustment is provided by a metal screw system (11–14) or by a vacuum screw; in orientable ball bearings there is no adjustment (7).

Radial adjustment is provided by the orientable bearing, axial adjustment by the two thrust bearings.

The area where the outer ring of the orientable ball bearing is placed needs refacing. The screw (10) is sized for a load turned upward.

Convergent conical bearings with inner-ring adjustment (Figure 11.7)

Mechanical components

Circular components

Threaded grain

Conical bearing inner ring

Conical bearing outer ring

External case

Conical bearing inner ring

Conical bearing outer ring

Threaded ferrule with blocking washer

Assembly Order

1-2-(3-4-5)-(6-7)-8

Loading Model

By screwing the ferrule (8), the conical bearing (6–7) is pushed against the external case (5) and at the same time component 1 gets pulled right pushing the left conical bearing (3–4) against the case (5), thus generating a closed loop of axial forces. While these forces are exerted the case must be in equilibrium.

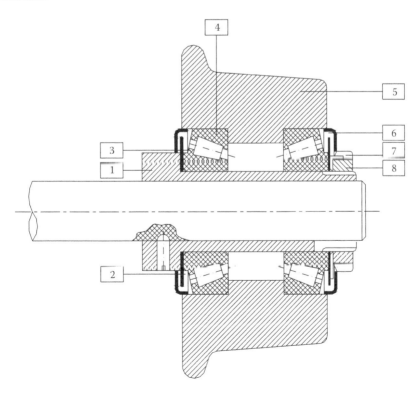

FIGURE 11.7 Convergent conical bearings with inner-ring adjustment.

Adjustment

The threaded grain (2) helps re-locate the circular component (1) to the right place after the system has been dismantled. The circular component (1) can be obtained from the shaft.

In conical bearings axial adjustment is carried out on both bearings, as per thrust bearings. Radial adjustment is carried out on both bearings

Conical bearings are modular and must be pre-loaded; this pre-loading is provided at the assembly phase by an adjustment component, the ferrule (8) and is carried out on the inner ring. As conical bearings withstand axial loads unidirectionally (like thrust bearings), it is necessary to mount them in opposition, resulting in the geometrical layout of an "X."

Convergent conical bearings with outer-ring adjustment (Figure 11.8)

Mechanical system components
Circular component
Conical bearing inner ring
Conical bearing outer ring
Blocking component
Spacer
Conical bearing inner ring
Conical bearing outer ring
Blocking component
Ferrule with washer
External case
Screw
Adjustment ferrule
Threaded grain

Assembly Order
1-(2-3-4-5)-(6-7-8)-9-(10-11)-12-13

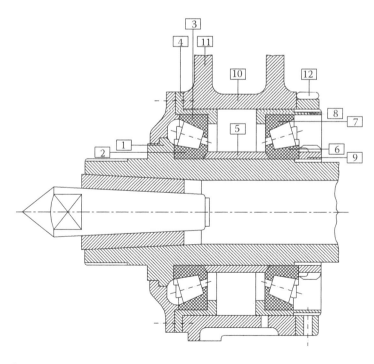

FIGURE 11.8 Convergent conical bearings with outer-ring adjustment.

Loading Model

By screwing the ferrule (12), component 8 is pushed against the right conical bearing (7–6), which pushes against the ferrule (9), which in turn pulls component 1 right, pushing the left conical bearing (2–3) against component 4.

Adjustment

In conical bearings both bearings have axial and radial adjustment.

Conical bearings are modular and must be pre-loaded; this pre-loading is provided to them at the assembly phase by an adjustment component, the ferrule (12).

As conical bearings withstand axial loads unidirectionally (like thrust bearings), it is necessary to assemble them in opposition, resulting in the geometrical layout of an "X."

Adjustment is via ferrule 12 and is carried out on the outer ring. Ferrule 9 works as an unballing, not as an adjustment component.

The spacer could be eliminated, not being an axial adjustment component, but it does keep the bearing inner rings at the desired distance during dismantling; a spacer transmits no forces.

The threaded grain (13) prevents the outer ferrule (12) from unscrewing, given that it is not fixed with a safety washer.

Divergent conical bearing with outer-ring adjustment (Figure 11.9)

Mechanical Components

Cover

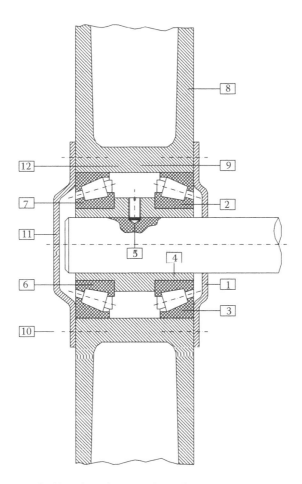

FIGURE 11.9   Divergent conical bearing with outer-ring adjustment.

Conical bearing outer ring
Conical bearing inner ring
Circular component
Threaded grain
Conical bearing inner ring
Conical bearing outer ring
External case
Screw
Metal ring
Cover
Screw

Assembly Order
1-(2-3-4-5)-(6-7)-(8-9)-10-11-12

Loading Model
By turning the screw (12), component 11 gets pushed onto the left conical bearing (7–6) which, through component 5, conveys the load to the right conical bearing (2–3), pushing it against cover 1, which is screwed to external box 8.

Adjustment
The adjustment component is made up of a metal screw-ring device. Both bearings have axial as well radial adjustment.

11.8   FRICTION BEARINGS

As previously maintained, the shafts must be accurately machined in the support areas as they rest on components or release forces (friction bearings, bushes), and they rotate with non-zero, relative speed. These components have to be shaped and produced so they respond to operational requirements, having some clearance to ensure rotation between shaft and bearing, and must be produced with materials that reduce friction, and withstand the specific stresses generated even at high temperatures.

In Figure 11.10 we see a classic one-piece bush. Depending on assembly type, bushes can also be made of two semi-shells. They always have lubrication holes.

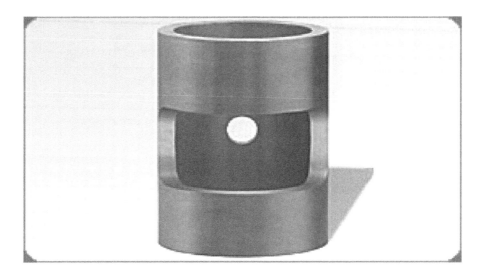

FIGURE 11.10   Example of a one-piece bush.

The life of mechanical components depends, as is often the case, on correct assembly. For bushes operating with friction between two mechanical components, the need for correct assembly is all the more important, though they are not at all costly and have a great impact on machine reliability.

Needless to say, the off-project use of components can be lethal for the machine. Usually, these phenomena are connected to temperature rise of all machine parts.

When bushes are mounted in series (see shaft on page 257) perfect shaft axiality must be ensured.

When the distance between supports is too long (15 times gudgeon diameter), the end supports should use orientable bushes that can adapt to the shaft line. In any case, the main idea is to ensure minimal, more regular wear.

As mentioned before, friction bearings must be mounted with some clearance (clearance to gudgeon diameter ratio ranges from 0.7% to 3.5% of the diameter). This parameter should be checked frequently, especially when the shaft is not rigid. Here, variations in clearance can alter support rigidity and significantly modify the whole system's rigidity. Often, as pointed out in Chapter 10, lubrication dictates the values for clearance.

Under normal wear, bushes are subject to more or less advanced ovalization, which often depends on non-symmetrical axial operation, bringing about non-uniform load distribution over the entire circumference over time.

In the case of large machines, such as high-powered steam turbines, clearance values are given in vertical instead of horizontal directions of the load.

So, bush design consists of:

1. Checking all the stresses involved
2. Choosing the material
3. Assessing the operational temperature
4. Assessing the product pv (average pressure × tangential speed)

As far as checking specific pressure is concerned, the average pressure value can be calculated once the forces on the supports in static and dynamic equilibrium are known (reactions in static condition, lift in operating condition). If F is the force on the support, the average pressure equals:

$$p = \frac{F}{Dl},$$

with D the bearing diameter and l its axial length.

Maximum specific pressure can be calculated assuming a pressure trend such as $p = p_0 \cos^2 \varphi$ for the semicylinder on which the shaft lies, and then calculating p_0 when the load-directed resultant of pressures is equal to force F. This produces a value of p_0 equal to about four times the computed pressure above.

It should also be pointed out that, especially for end bearings, the pressure trend along the bush axis is discontinuous, which overloads the bush ends all the more, the shorter is the bearings axial length. An approximation can be made hypothesizing an off-center bearing reaction compared to the mid-part of the bearing itself, hypothesizing a linear distribution of pressure along the axis such as to zero the pressure at one end. More simply, given the uncertainty of the calculation, this effect is allowed for by assuming $5\,p_0$ as the maximum admissible pressure of the material.

The maximum pressure to which these components can be subjected must not exceed 15 N/mm^2.

After determining the maximum specific pressure, go to point 2, the choice of material.

Bushes are made from materials that must withstand specific stress, be very durable, have relatively low friction coefficients with the material the shaft is made of (generally steel), have low thermal expansion coefficients similar to those of the materials involved with the coupling, be plasticly deformable, and have a low melting point.

To meet these requirements, the materials should derive from the following:

- Bronze alloys
- Tin alloys (white metals)
- Copper alloys
- Lead alloys
- Self-lubricating materials (materials obtained through sintering with steel powders and silver or tin powders)

The trend today is to make bushes with an outer surface of strain resistant material (steel) and with an inner shell of soft material.

Figures 11.11 and 11.12 show examples of bushes made with different technologies (steel/white metal bearings in the former, steel/pink metal in the latter).

Numerous types exist, with different shapes and different hole locations and webbing. Thin-walled bushes are available on the market, whose diameters go up to 350 mm.

Figures 11.13 and 11.14 show examples of anti-friction rings and axial thrust bearings. The rings are used for low axial loads and work as friction reducers for contiguous components mounted on the same shaft that can rotate independently at different speeds. The thrust bearings can withstand high thrust and are mounted on large turbine shafts.

As for the temperature calculation of friction bearing material in operation, it begins with computing the power lost to friction, which equals:

$$W = fF\omega D/2.$$

As seen in Chapter 10 on lubrication, if the bearing is hydrodynamically oil-lubricated, the coefficient f is quite low and the power lost is absorbed by the oil. If the bearing is poorly lubricated (either at the start or owing to low operating speeds), the power lost is absorbed by the bearing surface sometimes generating a high enough temperature to damage or even melt the bearing.

FIGURE 11.11   Examples of anti-friction rings.

FIGURE 11.12 Examples of anti-friction rings.

The quantity of heat discharged per unit time just by the effect of conduction, in the case of i shells, can be approximated as

$$W = \frac{2\pi R L \Delta T}{\Sigma S_i / k_i},$$

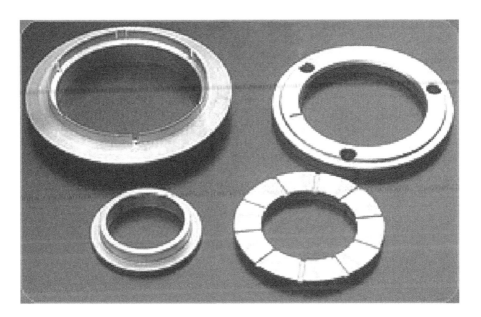

FIGURE 11.13 Example of anti-friction rings.

FIGURE 11.14   Example of axial thrust bearings.

R being the bush's mean radius, L the axial length, ΔT the thermal difference between the inner and outer surface of the bush, S the shell thickness, and k the coefficient of thermal conductivity of the material. For k, we can assume:

- Bronze $k = 380$ W/(m K)
- Aluminum $k = 220$ W/(m K)
- Steel $k = 30{-}40$ W/(m K)

As these are the materials used to manufacture bushes, they only melt if temperatures of about $1000°$ are reached at the contact surface.

One parameter which limits temperatures is the product pv, the result of average pressure and tangential velocity proportional to lost power, which must not exceed $(1.1 - 1.7) \times 10^6$ W/m for the materials of which the bushes are made.

The coefficient values provided in this paragraph are only approximate. Whenever a bush is selected or designed, it is best to refer to the manufacturers who can provide accurate coefficients and specific features of their products.

12 Splined Couplings, Splines and Keys

12.1 SPLINED COUPLINGS

As already seen in the previous chapter, a shaft consists of components with different functions and diameters. When coupling shafts or different components (flywheel, gearings, etc.), the ends must match so the coupling is as perfect as possible, which means ensuring co-axiality of both parts. Splined couplings are used when velocities are high and vibrations must be minimized.

There are two categories of splined coupling, straight-sided and curved-sided; the latter, in particular, have an involute curved profile. The splined part is usually machined by disc cutter for straight sides. Shaped disc cutters or rack cutters (see Chapter 15, Gear Wheel Manufacture) or rolling methods are used for curved sides with an involute curved profile. Higher pressure angles than those typically required in geared wheels (Generally 30°) and shorter teeth (Total height equal to a Modulus) are used in such a case.

Curved-sided splined shafts are only used in key mechanical systems in view of their high costs compared with other linkages. The centering of the splined couplings can be made either on the shaft (male) or on the hub (female) or directly on the sides. Norms specify standardized profiles both for rectilinear and involute types.

Calculations for the resistance of splined profiles refer to tooth resistance (whether rectilinear or involute) in relation to shaft resistance. In the case of rectilinear sides, the calculation is made assuming constant pressure p_a on the whole contact area.

With reference to Figure 12.1, where l indicates the active axial length in the coupling, the maximum tangential strength T that can be endured by each tooth is

$$T = p_a l \left(\frac{D-d}{2} - 2c \right),$$

c being the radial height of the bevel.

In order to account for non-uniform pressure over the whole active surface due to manufacturing imperfections, elastic deformation due to various stresses during operation, and all the other dynamic factors, the transmittable moment of each single tooth is multiplied by coefficient $\psi < 1$.

If all the teeth were to contribute equally, with $(D+d)/4$ being the arm of z teeth, the transmittable moment by all the teeth of the spline would be equal to the transmittable moment of shaft diameter d, and for an allowable stress, τ_a, the result would be Figure 12.1 and Figure 12.2

$$\psi p_a l \left(\frac{D-d}{2} - 2c \right) \frac{D+d}{4} z = \frac{\pi d^3}{16} \tau_a.$$

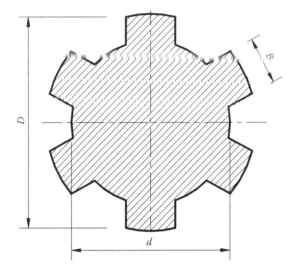

FIGURE 12.1   Straight-sided spline coupling.

Since

$$m = \frac{\pi}{2\psi} \quad \Omega = \frac{d^2}{z(D+d)(D-d-4c)} \quad K = \frac{p_a}{\tau_a},$$

then

$$\frac{l}{d} = \frac{m\Omega}{K}.$$

The values of m and K and all the other parameters can be found in Giovannozzi's tables, vol. I, p. 343.

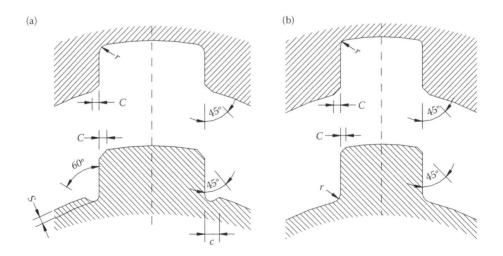

FIGURE 12.2   Geometric values for straight-sided spline coupling.

In the calculation above, we assumed that all z teeth contributed equally to the transmission of moment. This is close to reality when the shaft is under maximum load, meaning that deformation can involve all the faces of the teeth. In reality, because of production imperfections, not all the teeth work in the same way, which means the starting hypothesis is only partly true. Taking this into consideration, the ratio l/d calculated above can be increased by dividing by coefficient y as a function of the number of teeth, type of material, and quality of manufacture. In the design, only the number of teeth is considered and so the ratio can be approximated to 0.98 when z (teeth) is less than 10, and 0.96 when z is more than 10.

In the case of involute teeth, the calculation is similar to that for involute geared wheels (Chapter 15). However, following the previous straight tooth procedure, using the symbols in Figure 12.1:

$$\psi p_a lh\, d_p z = 2\,\frac{\pi d^3}{16}\tau_a.$$

The working height h is equal to a fraction of the modulus, the total height being equal to a modulus, as already mentioned:

$$h = \lambda d_p /z.$$

For d_p up to 50 mm, λ is assumed to be 0.7 and over 70 mm, λ is 0.95 interpolating 50–70 mm linearly. To obtain a formula similar to that for rectilinear sides, we can write

$$\frac{l}{d} = \frac{m\Omega'}{K},$$

where $\Omega' = d^2 /(4\lambda\, d_p^2)$.

The straight-sided tooth procedure can be used again for non-uniform loads on teeth as a result of manufacturing imperfections. However, in the case of involute splined profiles, a more precise performance is required, so the y coefficient can be assumed as 1 for fewer than six teeth and 0.98 for more. Furthermore, even more so than with straight teeth, gauging depends on the level of surface finishing and the experience of manufacturers, who in special applications (aeronautical construction), slightly twist the surface along its axis to take elastic deformation into account during operation to optimize load distribution over the whole tooth surface.

As already mentioned, shaft testing should be carried out at the most critical section, taking into account all the factors that affect fatigue resistance. The most critical section is often the one closest to the spline because, in practice, it is affected by the "shock" of machining and therefore shape factors at the tooth base. Once provisional calculations have been made, this part must be fatigue tested. Experience shows that shafts at a first approximation seem to be oversized (static safety coefficients about 10), and are at their fatigue limit when the test is carried out on the parts closest to the splined sections where breakage is most probable.

12.2   SPLINES AND TONGUES

Splined profiles and tongues are mobile linkage parts. They are used to render a shaft integral with another component flush fitted to it (pulley, gear, etc.), allowing transmission of torque. Splined and tongue linkages can vibrate as they spin owing to even the slightest asymmetry. So, this type of linkage is only used when velocities are low. The simplest spline is a rectangular parallelepiped that partly fits into a key-way in the shaft and partly into another in the linked part. The difference between a spline and a tongue depends on how they work when torque is applied.

Splines are generally used when the main stress is generated by the bottom of the splines inter-acting with the upper and lower sides of the connecting element. Tongues are used when the main stress is due to the pressure caused by the side surfaces of the key way and the side surfaces of the connecting element. In reality, if the spline is not mounted with side clearance, as torque increases it functions as a tongue and not as a spline. The operation and torque of an embedded spline with medium side precision is different to that of one with lower precision.

Consider the case of an embedded spline of l length with a b section width and h height connect-ing a shaft, diameter D, to a hub (see Figure 12.3). It works as a spline if the key-way is slightly wider than the width b of the spline itself. Because of upper and lower side contact, the linkage only works until the applied load produces sliding. The sides do not transmit torque, being produced by normal operation and friction that develop on the sides opposite the contact.

On account of a certain play between the side surfaces of the spline and that of the key-way, as the load increases, the contacting surfaces slide and the load transmitted decreases compared to what it might have been. So, the surface pressure diagram changes from its initial rectangular shape to a trapezium and finally at its limit to a triangle, when the ends of the spline break off. Further load increase can produce contact with the sides of the spline. In this case, its operation is mixed (the spline works partly as a tongue) and load transmission occurs because of the side action. The pressure on the splined edges produces plasticization.

Let us examine the various phases. At the beginning, with no sliding (constant p_0), the splined surface pressure is equal to

$$P = p_0 bl.$$

The same force is being generated opposite the shaft.

When the load induces sliding, pressure on the splined surfaces is not constant but varies, and according to the hypothesis of the conservation of plane sections, the pressure varies linearly. One end of the spline is more loaded than the other end. This produces an eccentricity of force, P, there-fore loading the shaft.

To balance the spline in such conditions (see Figure 12.4):

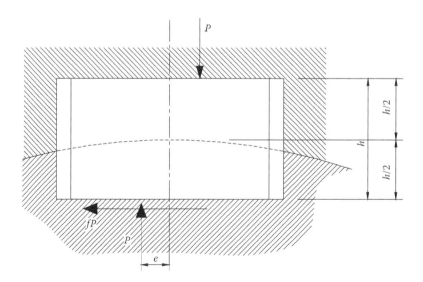

FIGURE 12.3 Spline in work conditions.

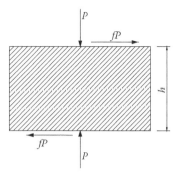

FIGURE 12.4 Loads on the surfaces of a spline.

$$P2e = fPh.$$

At the limit when sliding starts, the following will occur to the shaft:

- A normal force P with eccentricity $e = fh/2$ etc.
- A tangential force fP applied at distance $(D - h)/2$ from the shaft axis
- A tangential force fP on the shaft generated opposite the spline at distance $D/2$ from the shaft center

Balancing the above-mentioned forces leads to a load equal to

$$M = fPD = fp_0blD \qquad (12.1)$$

Therefore, the higher the initial shrinkage pressure p_0—which must be compatible with the resistance characteristics of the three components' materials (shaft, spline, hub)—the higher the transmittable load.

The extra pressure at the ends of the spline, in the hypothesis of the conservation of the plane sections, is equal to

$$p' = \frac{Mb}{2I} = \frac{6fPh}{2b^2l} = 3p_0f\frac{h}{b}.$$

The overall strain will therefore be $p_0 + p'$ and it must be lower or, at most, equal to the safety load at compression K, that is

$$p_0 + p' = p_0\left(1 + 3f\frac{h}{b}\right) = K,$$

and the maximum allowable value, p_0, compatible with the material's resistance is equal to

$$p_0 = \frac{K}{1 + 3f\dfrac{h}{b}}.$$

By substituting a definite value in (12.1), we get

$$M = fUlD \frac{K}{1 + 3f \dfrac{h}{b}},$$

If we exert the maximum tangential strain on the shaft $t = M / W_t$, we have

$$\frac{K}{\tau} = \frac{\pi D^2}{16} \left(1 + 3f \frac{h}{b}\right) \frac{1}{fbl},$$

and assuming

$$A = \frac{\pi}{16} \left(1 + 3f \frac{h}{b}\right) \frac{1}{f} \frac{D}{b},$$

we have

$$\frac{K}{\tau} = A \frac{D}{l}.$$

The coefficient A is related to system geometry and the friction coefficient between the contacting parts. In the field of the most commonly used geometric values, f being equal, it remains almost constant. Consequently, the previous formula can provide immediate data on splined length in relation to steel type. It can also identify steel type in relation to the chosen geometry and optimize the way of manufacturing the material in relation to the size of the whole system.

The calculation above refers to the case of a certain initial play between the key-way and spline for non-lowered splines (high hs).

In the case of "precise" shaft-splined couplings (embedded splines), even if the upper and lower surfaces of the spline initially need forcing, load transmission is produced by side actions generated when the external load is applied. In fact, high pressures would be necessary to transmit the great loads caused just by friction, pressures that would not be compatible with the resistance characteristics of the materials. So, by balancing the transmittable moment by friction ($M = fPD$) with shaft resistance ($M = \tau \pi D^3 / 16$), we get

$$P = \frac{\pi}{16f} \tau D^2,$$

which, applied in practice, shows that the force generated is incompatible with forced assembly.

So, it is realistic to suppose that load transmission is due to forces on the splined sides (Figures 12.5 and 12.6) that makes the key works like a tongue, as does of shaft deformation.

In that case, normal pressures of intensity p, act on the two active halves of the splined sides and therefore total force is given by: $phl/2$

The moment generated by the pressures on the splined sides equals $1/4ph^2l$ and must be balanced by the moment of pressures generated between the splined side bases and the key-way.

To avoid any contacting parts separating, and supposing a linear distribution of pressure along width b of the key-way, the initial force p'_0 between the upper and lower surfaces of the spline must be equal to

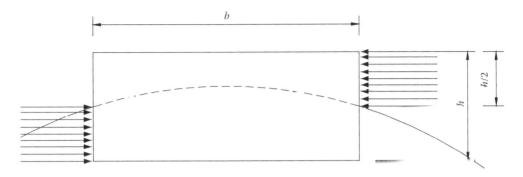

FIGURE 12.5   Pressure on the lateral surfaces of a spline.

$$2p_0' = \frac{3}{2} p \left(\frac{h}{b}\right)^2 .$$

This obviously must be lower or, at most, equal to the unit compression tension σ' of the materials in contact, that is

$$\sigma' = 2p_0' = \frac{3}{2} p \left(\frac{h}{b}\right)^2 .$$

The previous ratio defines the maximum pressure on surface width b as a function of pressure p and as a function of the ratio between the two dimensions of the straight cross-section.

Equalizing the pressure moment in relation to the shaft center with the transmitted torque (Figure 12.7) helps define the ratio between the pressures on the splined sides and the shaft shearing stress τ, so it is equal to

$$\frac{1}{4} phlD = \tau\pi \frac{D^3}{16},$$

leading to

$$\frac{p}{\tau} = \frac{\pi}{4} \frac{D^2}{hl}.$$

If the most commonly used values are added to the formula, the ratio is equal to

$$\frac{p}{\tau} = 1.6 \div 4.7.$$

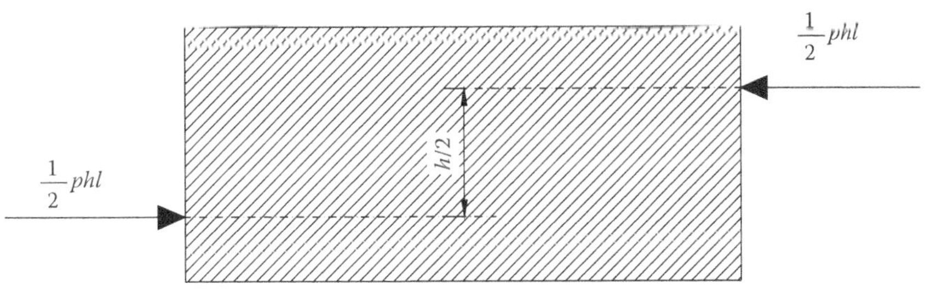

$$\frac{1}{2} phl$$

$$\frac{1}{2} phl$$

FIGURE 12.6   Forces on the lateral surfaces of a spline.

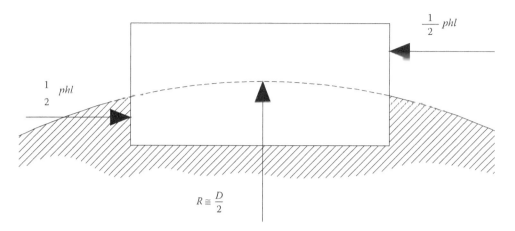

FIGURE 12.7   Forces transmitted from the shaft.

Taking into account the allowable strain values τ that are used in practice with steel shafts, the maximum pressure values are about 40 kg/mm².

Considering that there are always extra loads on the splined edges, the compression strains calculated in the two previous hypotheses suggest using high carbon steels for this type of component. This would limit plasticization that, with time, can cause slackening and then disengagement.

As regards shearing stress, for a rectangular cross section, the maximum value is equal to

$$\tau = \frac{3}{2}\frac{phl}{2bl} \cong \frac{3M}{blD},$$

and remembering that pressure is

$$p = \frac{4M}{hlD},$$

we have

$$\frac{p}{\tau} = \frac{4b}{3h}.$$

With normal splined proportions, such a ratio is always satisfied.

With reference to tongues, as already mentioned, they transmit the moment exclusively through side action. For this reason, they have to be mounted with no play between the side surfaces. In practice and also because of the material's elastic deformability, as the moment increases, the upper and lower surfaces are involved, as occurs with splines.

From the hypothesis that the upper and lower surfaces are not involved, assuming that the moment is transmitted because of the pressures (lateral distribution) on the four side surfaces (two for the two slots), maximum pressure values on the edges equal three times the pressure defined by $p = 4M/hlD$, higher than that for splines. For this reason, they are used when moments are low. In some cases, to avoid high edge pressures, they are screwed onto the shaft, which frees tongues from the effect of moment.

As already mentioned at the beginning, the previous calculations only give general approximate data. The results obtained through the various calculations assume the validity of applying

Saint-Venant's theory for solids, which does not respect the hypothesis for which the theory itself is valid.

In the previous calculations, we referred to mean shaft strain as an effect of torque moment, without considering strain increases due to the shaft key-ways that embed the spline or tongue.

The fatigue calculation for a shaft that is mostly subject to torsion must also consider the notch effect β equal to about 3 in the case of key-ways made using button milling cutters (Figure 12.8a), and not lower than 2.2 for key-ways made using disc milling cutters (Figure 12.8b). Lower values can be used in the case of American tongue key-ways (Figure 12.8c), where β could equal 2.

The calculations give an idea of the necessary splined proportions to avoid excessively high local pressures on the spline and key-way.

Analysis of system stress shows that, due to the initial force, non-symmetrical radial pressures are produced on the shaft that cause local compression stress and that must be taken into account when the shaft is tested.

Evaluating local pressure can be done by assuming a pressure trend of type $p = p_0 \cos^2 \phi$ for an arc equal to 180° or 240°, depending on the manufacture type and type of forced spline mounting.

So, it is possible to define the force P concentrated at the point diametrically opposite the spline. The value of p_0 is easily defined considering that the resultant of the pressure components in P direction must be equal to the P force.

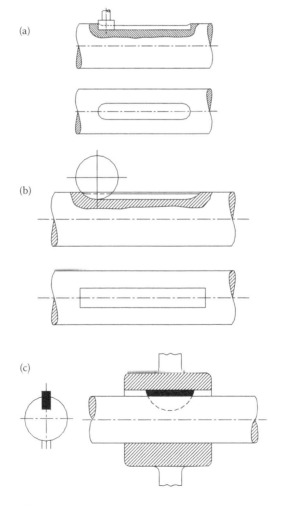

FIGURE 12.8 Examples of key-ways.

12.3   TRANSVERSE SPLINES

These are machine components used to connect rods to sleeves or other components. They are mounted transversally to the rod axis in a key-way in the rod itself (Figure 12.9).

As shown below, transverse splines require high quality manufacture and materials given that they are precision mounted to prevent any shock from load variation.

In some applications (e.g., connecting a stem to a ring), to prevent load variation, they are mounted with interference to produce pre-loading that precludes load inversions and limits $\Delta\sigma$ on the stem section weakened by the splined key-ways, thus helping system durability.

The calculation considers the spline and ring as a single piece and proceeds as per forced linkages, explained in Chapter 16. In particular, referring to Figure 12.9 coupling, stem rigidity is defined first and then the force-deformation link is expressed as follows:

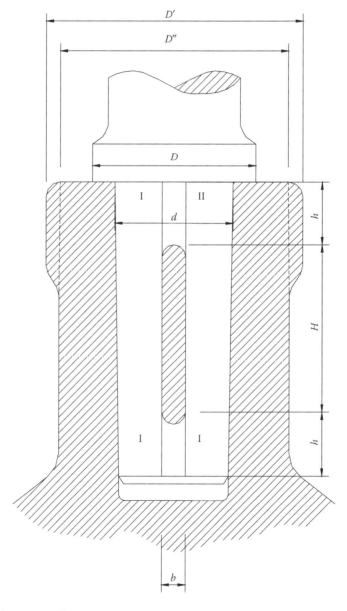

FIGURE 12.9   Transverse spline.

$$\lambda = \frac{Fl}{ES},$$

E being the material's elastic modulus and S is the area of the stem cross section.

For the corresponding ring

$$\lambda' = \frac{Fl}{E'S'},$$

(E' being the elastic modulus and S' is the ring cross-sectional area).

The calculation will also have to be made graphically, as in Chapter 16 (Forced Linkages), drawing an appropriate diagram after defining the specific pressure between the crown of the stem and ring and after defining F_0, remembering that

$$i = \lambda + \lambda' = \frac{F_0 l}{ES} + \frac{F_0 l}{E'S'}.$$

The values of F'_0 and T are easy to define, being effects of an external force.

The previous calculation left out spline bending and shearing. However, it is not difficult to include these when they are comparable to other elements and add them, with the correct sign, to the ring deformation in the λ' calculation.

Once the maximum and minimum forces and their corresponding stresses are known, the resulting fatigue can be calculated.

With an easier calculation, the stress variation on the stem is left out and everything is made relative to a maximum value of T opportunely corrected by the coefficient m.

With reference to Figure 12.10, if K and τ indicate allowable tensile and shear stresses on the stem, with K' the allowable bending of the spline, with $\bar{K}$ and $\bar{\tau}$ the allowable tensile stress and shearing on the ring, and with p the allowable pressure between the stem and the ring, the following ratios will be produced.

For stem resistance to tensile stress:

$$K = \frac{mF}{\dfrac{\pi d^2}{4} - db},$$

db being the area of the key-way.

The joint pressure of stem and ring is produced by the force mF acting on the circular ring whose area is $\pi(D^2 - d^2)/4$ and equals:

$$p = \frac{4mF}{\pi(D^2 - d^2)}.$$

The external diameter of the ring can be defined assuming that the stem and ring operate at the maximum allowable stress:

$$\frac{\pi}{4}(D''^2 - d^2)\bar{K} \cong \frac{\pi d^2}{4}K,$$

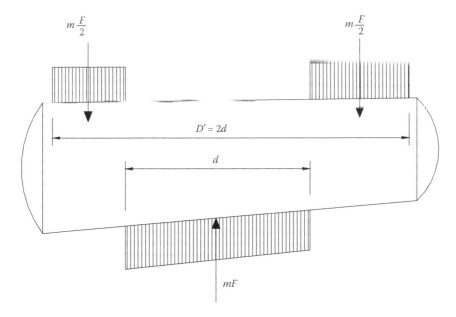

FIGURE 12.10   Loads on the spline.

In practice, data suggest that the width of the key-way is

$$b = \frac{d}{3} \div \frac{d}{4},$$

which suggests $D' \cong 2d$ as the maximum stem diameter.

Splined shearing stress is

$$\tau_{max} = \frac{3}{2}\frac{T}{bh}.$$

Shearing stress for the rectangular sections I and II indicated in Figure 12.9 must be calculated as follows:

$$\tau = \frac{3}{2}\frac{mF}{2d \cdot h'} \quad \bar{\tau} = \frac{3}{2}\frac{mF}{2(D'-d)h''}.$$

The diagram in Figure 12.9 is used to define splined bending, where the load should be distributed uniformly on the supports and the central part. In such an hypothesis, the following occurs:

$$K' = \frac{M}{W} = \frac{3mFd}{2bh^2}.$$

As already mentioned, it is not difficult to define values for shearing and bending by calculation if the linkage is a forced one.

The fatigue calculation, referring to the simple model above, supposes that the stem is completely free of any external force. This means that stress varies from a maximum calculated stress (considering shearing and bending) to zero. For mean stress in the previously defined conditions, the safety coefficient can be calculated after taking into account all the factors affecting it.

The previous calculations refer to a cylindrical stem but, by applying the same logic, can be used for a conical stem (Figure 12.11).

In this case, because of the axial force A, the inclination α of the stem, produces compression on the stem and internal pressures on the external collar with radial and tangential components which tend to widen it. The resultant of the normal components to a diametric plane of the collar, equals T_0, and the straight section of the collar undergoes normal stresses σ_t (see internally fortified pipes).

It $f = tan\varphi$ is the friction coefficient between stem and ring, the axial force per unit length of stem in contact with the collar equals:

$$A = 2\pi Rp(\sin\alpha + f\cos\alpha) = 2\pi R \frac{p}{\cos\varphi}\sin(\alpha + \varphi),$$

and consequently (see Chapter 13, compression springs) T_0 equals:

$$T_0 = R \cdot p(\cos\alpha - f\sin\alpha) = R\frac{p}{\cos\varphi}\cos(\alpha + \varphi),$$

If the two equations are combined, we get:

$$T_0 = \frac{A}{2\pi\tan(\alpha + \varphi)},$$

and the mean normal stress in the ring is equal to

$$\sigma_t = \frac{T_0}{S},$$

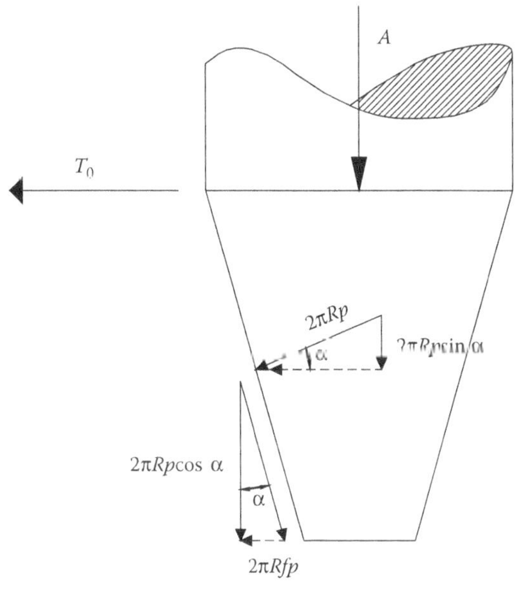

FIGURE 12.11 Conical stem.

S being the area of the cross-section meridian of the ring.

It is not difficult to apply the above to calculating the loads on various components to obtain the fatigue limit

12.4   SPUR GEARS

Before closing this chapter, we should touch on another less common inter-shaft linkage known as Hirth's spur gears.

At the ends of the shafts, spur gears transmit the power in play. The linkage is by way of engaging two gears with equal toothing but different pitch, or by a double-end stay bolt locked by an end nut (see scheme of Figure 12.12 for the calculation).

The flat-sided teeth are specially cut so that when interlocked they guarantee perfectly centered co-axiality of the shafts. Generally, the teeth faces are thermally treated and have special finishes.

Since spur gears are relatively expensive, they are only used when the requirements of other components (bearings, wheels, pulleys) demand a break in a shaft. Details on manufacture and angles can be found in Giovannozzi, vol. I. Specific data in tables can be found in company catalogues.

It should be noted that test calculations and sizing are largely approximate. To obtain an indicator of pressure, reference is made to a calculation chart that treats teeth as Saint-Venant's beams even though the tooth's form is not at all close to that for which the theory is valid.

Referring to Figure 12.12, the tangential force P applied to the centroid (barycenter) of a tooth cross section at distance R from the axis is

$$P = \frac{M_t}{R},$$

where M_t is the torque.

The axial force A required to transmit the moment neglecting friction is

$$A = P \tan \beta / 2,$$

with β the angle of mean semi-aperture between two teeth.

To guarantee continuous interlocking of teeth, the bolt is sized for a load m times that first defined. The coefficient m generally varies between 1.8 and 2.2, depending on application and torque variation. So, the pressure on the bolt core S is

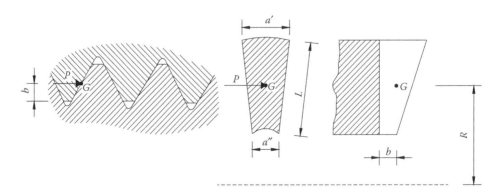

FIGURE 12.12   Scheme for the stresses calculation in spur gears.

$$\sigma = \frac{N}{S} = \frac{mP \tan \beta/2}{S}.$$

The specific pressure on the teeth must be tested.

Bearing in mind that because of defects in manufacture it is assumed that only four teeth transmit the load P and that all the teeth n are subject to the axial force produced by locking the bolt or from another thrust that keeps the teeth interlocked:

$$p = \frac{1}{bl}\left(\frac{P\cos\beta/2}{4} + \frac{A\sin\beta/2}{2n}\right),$$

For the maximum specific pressures, reference may be made to a fraction (about 1/3) of the Brinnel hardness of the material.

Just as for teeth, the tooth is sized as though it were a mortised beam with a rectangular base and width equal to the mean of widths a' and a'' of the largest and smallest base of the trapezium.

In that case, and supposing that of all the teeth only half transmit the moment, the unit bending stress is

$$\sigma = \frac{M}{W} = \frac{6\left(\dfrac{2P}{n}\right)b}{L\left(\dfrac{a'+a''}{2}\right)^2}.$$

At the tooth base there are also shearing stresses that, according to the previous hypothesis, are

$$\tau_m = \frac{\left(\dfrac{2P}{n}\right)}{L\left(\dfrac{a'+a''}{2}\right)},$$

with a τ_{max} equal to 1.5 times the value τ_m calculated earlier.

For practical data on maximum pressures and geometry as well as on construction designs, refer to Chapter 8 of Giovannozzi, vol. I.

13 Springs

13.1 INTRODUCTION

Springs are mechanical components that are easily deformed by relatively small loads. The basic potential energy stored in a spring is expressed by

$$dL = Fdf,$$

which as a linear characteristic (loads proportional to deformations) takes the form of

$$L = \frac{1}{2} F \cdot f.$$

Springs are mechanical components that can store a certain energy and are used to prevent impacts (sudden thrusts) that may damage or abnormally stress any mechanical components. Properly adjusted, they can be used in measurement instruments such as dynamometers. Linear springs generate well-defined oscillation frequency systems that must be duly considered in certain applications.

The materials commonly used to build springs are carbon or alloy steels (AISI 1085, AISI 1065, AISI 1066) in minor applications. In applications with higher requirements, alloy steels are exploited where silicon is often a component (AISI 9254). Chromium–vanadium steels are also employed (AISI 6150). There are, moreover, springs made of non-ferrous material like bronze or copper alloys. In applications of some significance, springs, which are made by high material deformation, a heat treatment of tempering and extension is necessary. Further enhancements can be obtained with procedures like presetting or shot blasting. The values of linear and tangential elastic coefficients refer, as a general rule, to steels:

- $E = 210,000$ N/mm^2 for tensile stress
- $E = 215,000$ N/mm^2 for bending
- $G = 83,000$ N/mm^2 for shearing stress and torsion

For bronze springs, the E values are 108,000 N/mm^2 and the G values are 45,000 N/mm^2. For brass springs, the E values are 96,000 N/mm^2 and the G values are 35,000 N/mm^2.

Acceptable tangential stress is on average 0.6 of acceptable normal stress.

In most applications, springs operate under fatigue. Good design means determining the parameters peculiar to the operating conditions: σ_s and σ_i and their derivatives. For suitable design and complying with the principles mentioned in fatigue, one must be familiar with the steel fatigue diagrams (Smith's diagrams, Figure 13.1) and then enter the average stress and oscillation width values.

Apart from the fatigue resistance factors dealt with in Chapter 6, other factors such as operating conditions should not be overlooked (temperature, thorough lubrication of leaf springs) as well as the consistency of calculation and operating conditions (i.e., perfect axiality of the load in the case of helical springs), or any eccentricities of actual loads compared to calculation hypotheses.

Generally speaking, the steels employed for variable load springs are steels with lower values of static tensile stress but able to withstand greater oscillation widths (Smith's "thick" diagrams). For rational calculation, it is always advisable to use materials with easy access data peculiar to the material itself. Tensile stress and torsion fatigue diagram data related to the variation of

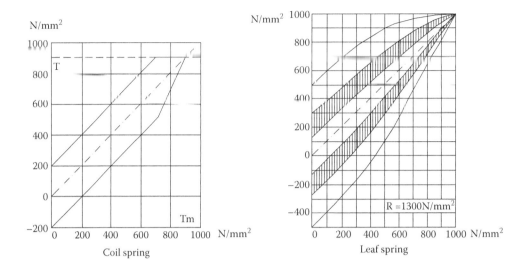

FIGURE 13.1 Fatigue diagrams for springs.

characteristics after heat treatment should be consulted and so too any data that help take into account the individual features of the designed spring.

13.2   SPRING COMPUTATION

Springs are generally classified according to the type of load they undergo; thus, there are bending, torsion and tensile stress springs. Spring calculations involve determining the loads applied to them (static or dynamic), assessing their geometric features for the given material, determining the spring characteristics (load deflections), and defining clearance in extreme operating conditions.

In spring studies, the "utilization coefficient" m takes into account the characteristics of the spring's material. It is defined as the ratio between the energy stored by a spring under load and the energy the spring would store if the entire spring were subjected to the same maximum stress caused by that load. So,

$$m = \frac{\frac{1}{2}F \cdot f}{\frac{1}{2}\frac{\sigma^2}{E} \cdot V} \quad \text{for bending or tensile stress springs,}$$

$$m = \frac{\frac{1}{2}F \cdot f}{\frac{1}{2}\frac{\tau^2}{G} \cdot V} \quad \text{for torsion springs.}$$

When there is both σ and τ, coefficient m is given by

$$m = \frac{\frac{1}{2}F \cdot f}{\left(\frac{1}{2}\frac{\sigma^2}{E} + \frac{1}{2}\frac{\tau^2}{G}\right)V}.$$

The optimum condition would be if $m = 1$, meaning all points in the spring were equally stressed.

Generally speaking, in mechanical applications of springs, $m = 1$ never occurs, except in just one system (at odds with the definition of spring), in which a constant-cross-section shaft is stressed by force F applied to its free end. In this case, the normal stress in all points of the rod is obtained by

$$\sigma = \frac{F}{S},$$

all points being equally stressed, whereas the elastic deflection is given by

$$f = \frac{Fl}{ES}.$$

Deformation turns out to be

$$L = \frac{1}{2} F \cdot f = \frac{1}{2} F \cdot \frac{Fl}{ES} = \frac{1}{2} \frac{F^2 l}{ES}.$$

The elastic energy stored in all points subject to the maximum perpendicular stress is given by

$$\varepsilon = \frac{1}{2} \frac{\sigma^2}{E} \cdot V = \frac{1}{2} \frac{F^2}{S^2 E} \cdot Sl = \frac{1}{2} \frac{F^2 l}{ES}.$$

The utilization coefficient is therefore:

$$m = \frac{\dfrac{1}{2} \dfrac{F^2 l}{ES}}{\dfrac{1}{2} \dfrac{F^2 l}{ES}} = 1.$$

Here, the shaft does not have the pre-requisites of a spring since its deformation f is extremely small compared to the load applied.

Usually, the "characteristic" spring in a Cartesian diagram with force F on the y-axis and f on the x-axis (Figure 13.2), is a line passing through the origin, that is, F load and corresponding f deflection are directly proportional according to a constant coefficient:

$$C = \frac{F}{f},$$

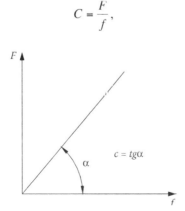

FIGURE 13.2 Linear spring "characteristic".

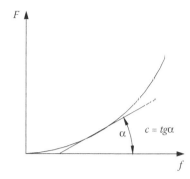

FIGURE 13.3 Nonlinear spring "characteristic".

called "stiffness" or "elastic constant" of the spring, uniquely dependent on the spring's geometric dimensions, on the elasticity coefficient of its material, and on constraint conditions.

Generally, stiffness is expressed by the ratio:

$$C = \frac{dF}{df},$$

and represents the slope on the f axis of the tangent to the stress/deformation curve at one point being variable if the load changes.

The most common case is when spring or spring system stiffness grows as the load increases (Figure 13.3).

Before continuing this study of springs, the elasticity ellipse theory will come in handy for assessing deformation.

Suppose a beam has a force F applied (Figure 13.4), the deflection in whatever direction is

$$f = F\int_0^1 \frac{ds}{EJ} \cdot x \cdot y,$$

which is the product of force F multiplied by the centrifugal moment of elastic weight (ds/EJ) compared to force direction and the deflection direction required.

Analogously, the deflection pointing in the same direction as the force can be calculated using the same expression but substituting product $x \cdot y$ with x^2:

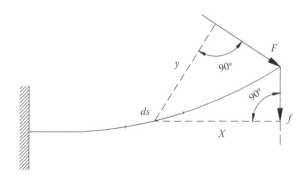

FIGURE 13.4 Scheme of a beam.

$$f = F \int_0^l \frac{ds}{EJ} x^2,$$

which is the product of the force F multiplied by the inertia moment of elastic weights compared to the force direction.

The theory says the rotation of the beam ends due to force F is given by

$$\Delta\varphi = F \int_0^l \frac{ds}{EJ} y,$$

which is the product of the force F multiplied by the static moment of elastic weights.

Apart from the force deflection f, it is also possible to calculate the deflection f and the rotation $\Delta\varphi$ produced by a moment M with the following formula:

$$f = M \int_0^l \frac{ds}{EJ} x,$$

which is the product of M multiplied by the static moment of elastic weights compared to the deflection direction.

$$\Delta\varphi = M \int_0^l \frac{ds}{EJ},$$

which is the product of the M moment multiplied by the beam's elastic weight.

13.3 BENDING SPRINGS

The simplest case of bending springs is provided by a rectangular cross-section beam fixed at one end and free at the other where load F is applied (Figure 13.5).

When the beam size is such that the ratio between beam length l and inertial radius ρ of the cross section does not exceed the approximate value of 200 (it is still possible to refer to it as a massive beam) the beam theory applies.

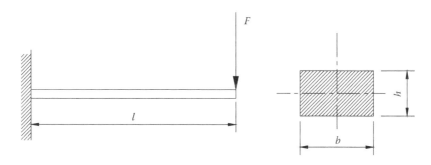

FIGURE 13.5   Rectangular bending spring.

By applying the elasticity theory, the deflection is equal to

$$f = F \int_0^l \frac{dA}{EJ} \cdot x^2 = \frac{F}{EJ} \left[\frac{x^3}{3} \right]_0^l = \frac{Fl^3}{3EJ}.$$

Maximum stress occurs at the fixed-end section where at its extreme point (having $y = h/2$ as coordinate) it equals:

$$\sigma_{max} = \frac{M}{J} y = \frac{Fl \cdot 12 \cdot h}{bh^3 2} = \frac{6Fl}{bh^2}.$$

If this beam is compared to a spring, its various components do not all operate in the same way as several stresses are at play.

Indeed, the bending moment varies from section to section and even in a single section, stress is not uniform (Sporer's law on Figure 13.6).

Therefore, there is a double variation: all the points are stressed differently. The material works at its best only at the fixed end, then the utilization coefficient is definitely different than 1.

In order to determine the m value, the external work is to be assessed first, which results from:

$$L = \frac{1}{2} F \cdot f = \frac{1}{2} F \frac{4Fl^3}{Ebh^3} = \frac{2F^2 l^3}{Ebh^3},$$

then the energy the beam can store is given by the ratio:

$$\varepsilon = \frac{1 \sigma_{max}^2}{2E} V = \frac{1}{2E} \frac{36 F^2 l^2}{b^2 h^4} bhl = \frac{18 F^2 l^3}{Ebh^3},$$

The utilization coefficient becomes:

$$m = \frac{L}{\varepsilon} = \frac{1}{9}.$$

Compared to other springs, it is quite low. To improve the m utilization coefficient for the same material volume, to increase the quantity of storable potential energy, the material volume must be greater right where the moment is stronger (i.e., at the fixed end), and lesser where the moment is weaker. So, springs must be made to form a solid that can withstand uniform bending resistance.

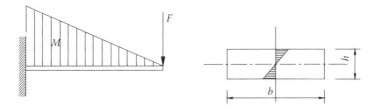

FIGURE 13.6   Qualitative stress in a rectangular bending spring.

Given distance x of a generic end cross-section, the value of σ_{max} section by section will be

$$\sigma(x) = \frac{6Fx}{b(x)h^2(x)}.$$

To obtain a constant σ there are two possibilities:

- Keep h constant and change b
- Keep b constant and change h

If $h = \cos t$, we get

$$\sigma(x) \cdot b(x) = \frac{6Fx}{h^2},$$

or, it should be

$$b(x) = K_1 \cdot x = \frac{b_0}{l} x,$$

in other words, b must vary by triangular law.

From above, the profile in Figure 13.7 appears. We shall, therefore, have a constant thickness triangular spring.

If $b = \cos t$, we get

$$\sigma(x)h^2(x) = \frac{6Fx}{b},$$

or, it should be

$$h^2(x) = k_2 \cdot x,$$

in other terms, height h varies according to a parabolic law. From one side, the profile in Figure 13.8 would be visible.

Defining the deflection, in the case of the $b(x) = b_0 \cdot x/l$ triangular profile, can be done using the elasticity ellipse theorem:

$$f = F\int_0^l \frac{dx}{EJ} x^2 = F\int_0^l \frac{12x^2 dx}{Eb(x)h^3}.$$

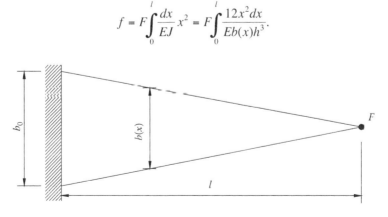

FIGURE 13.7 Triangular bending spring.

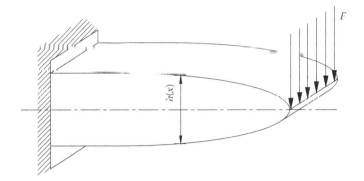

FIGURE 13.8 Parabolic bending spring.

Replacing $b(x) = b_0 \cdot x / l$, we obtain

$$f = F \int_0^l \frac{12lx^2}{Eh^3 b_0 x} dx = \frac{12F}{Eh^3} \frac{l}{b_0} \frac{l^2}{2} = \frac{6Fl^3}{Eb_0 h^3},$$

σ_{max} will then be

$$\sigma_{max} = \frac{6Fxl}{b_0 x h^2} = \frac{6Fl}{b_0 h^2}.$$

The utilization coefficient is

$$m = \frac{\dfrac{1}{2} F \dfrac{6Fl^3}{Eb_0 h^3}}{\dfrac{1}{2E} \dfrac{36F^2 l^2}{b_0^2 h^4} \dfrac{b_0 l}{2} h} = \frac{1}{3},$$

equal to triple m for constant cross-section springs.

Triangular springs loaded on the side opposite the fixed end have an elastic deformation (elastic line) that is a curved arc. This curvature is given by the expression:

$$\frac{d^2 y}{dx^2} = \frac{M}{EJ} = \frac{Fx}{E \dfrac{b_0 h^3 x}{12l}} = const$$

During manufacture, the theoretical triangular shape is always replaced, in practical applications, by a trapezoid with base b at the fixed end and base b' at the free end, so as to let the loads be applied (Figure 13.9).

In the case of a trapezoid plate, the deflection can be determined with the usual formula:

$$f = F \int_0^l \frac{12x^2 dx}{Eb(x)h^3}.$$

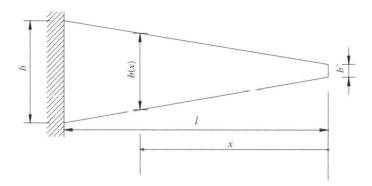

FIGURE 13.9 Trapezoid bending spring.

Substituting $b(x)$ with $b(x) = b' + x \cdot (b - b')/l$, it equals:

$$f = \frac{12F}{Eh^3} \int_0^l \frac{x^2 dx}{b' + \frac{b - b'}{l} x}.$$

Assuming $\beta = b'/b$, $A = x \cdot (1 + b)/l$, it equals

$$f = \frac{12F}{Ebh^3} \int_0^l \frac{x^2 dx}{\beta + Ax}.$$

Resolving the integral remembering that for Ruffini,

$$\frac{x^2}{\beta + Ax} = \frac{x}{A} - \frac{x}{\beta + Ax},$$

the following expression is obtained:

$$f = 4K \frac{Fl^3}{Ebh^3} \quad \text{with } K = \frac{3}{2} \frac{1}{1-\beta} \left\{ 1 - \frac{2}{1-\beta} \left[\beta + \frac{\beta^2}{1-\beta} \ln \beta \right] \right\}.$$

Table 13.1 shows the value of K as a function of β

TABLE 13.1
Shows the Values of K as a Function of β

$\beta = \dfrac{b'}{b} =$	0.0	0.1	0.2	0.3	0.4	0.5	0.6	0.7	0.8	0.9	1.0
K =	1.500	1.390	1.316	1.250	1.202	1.160	1.121	1.085	1.654	1.025	1.000
K' =	0.657	0.609	0.569	0.539	0.513	0.492	0.473	0.457	0.442	0.429	0.417

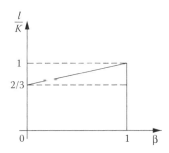

FIGURE 13.10 Approximate coefficient l/K for the calculation of the trapezoid spring.

K can be approximated with a maximum error not exceeding 4%, with the formula $K = 3/(2 + \beta)$, considering the stiffness of the trapezoid spring ratio β is like that of two independent in-parallel springs, one rectangular with a width b', the other triangular $(b - b')$ wide.

According to these hypotheses, l/K, proportional to stiffness, varies linearly from $2/3$ for triangular springs to 1 for rectangular ones (see Figure 13.10) and K takes on the above shape.

For practical reasons, the trapezoid spring replaces the constant-resistance, triangular spring because it makes the best use of the material.

If the ratio $\beta = b'/b$ is quite small, the elastic deformation of the trapezoid spring takes on an almost circular configuration.

The reasoning above applies when the L/D ratio of the solid is quite low and we are not dealing with "rods," which when forces still compatible with the material's resistance are applied, may produce deformations that affect the values of the stress components that generated them. In this case, the deformations must be studied using successive approximations or, in simple cases, with parametric techniques beyond the remit of this course.

13.3.1 LEAF SPRINGS

The previous study also applies to the calculation for leaf springs. The triangular plate springs above are not normally used, owing to their bulkiness. To reduce this, springs made up of strips (leaves) are employed, each strip being designed as if derived from joining two half strips symmetrically to the midpoint of the triangular leaf from which they derive (Figure 13.11).

For the constraint mode adopted during operation, all the leaves are "forced," using u-bolts, so they remain in contact and have similar curvature all the time. If the leaves are sufficiently lubricated, so they can slide against each other, they operate as though they were a single triangular plate. Each leaflet absorbs its own part of the moment. Past experience confirms that when the leaves can move freely, they behave as though they were one.

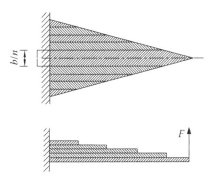

FIGURE 13.11 Triangular spring divided in strips.

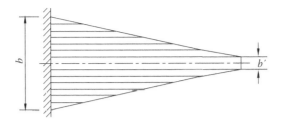

FIGURE 13.12 Trapezoid spring divided in strips.

Due to the impracticality of making the longest leaf "pointed," the leaf spring is more similar to a trapezoid spring than a triangular plate (Figure 13.12). Furthermore, it is at this extremity that the load is applied through an eyelet.

When the leaf curvature is relatively small, the formulae obtained for the trapezoid leaf fully satisfy the design requirements, especially when any friction between the leaves is eliminated (limited).

For special application springs, the leaves may also be bent differently, so as to create pre-deformations and pre-load single components. Here the behavior of the leaf spring differs even more from the trapezoid spring, owing to the resulting friction. The deflection load curve under the best conditions approaches that reported in Figure 13.13, for a leaf spring with constant-curvature leaves but covered with rust.

By way of experiment, it has been verified that the friction effect, as an absolute value, is identical both for increasing and decreasing loads. The ideal characteristic of a spring with no friction, for which calculations are made, is given by the mean line between the two lines of increasing or decreasing load. It is this line that must be compared to the ideal. For such springs 6SiCr8 UNI 3545 steel can be used ($R = 1450–1700$ N/mm^2) or 52SiCr5 UNI 3545 steel ($R = 1400–1650$ N/mm^2).

13.3.2 LEAF SPRINGS WITH SPRING SHACKLES

In most practical applications, leaf springs are held together by components called spring shackles (Figure 13.14) and generally only loaded axially. In this way, deformations may be increased letting the suspended mass oscillate slightly longitudinally.

Referring to Figure 13.15, notice the M and N points do not shift vertically but follow a curve (φ is the spring semi-aperture angle).

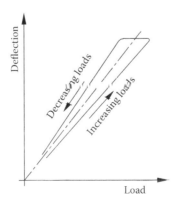

FIGURE 13.13 Qualitative real "characteristic" for a leaf spring.

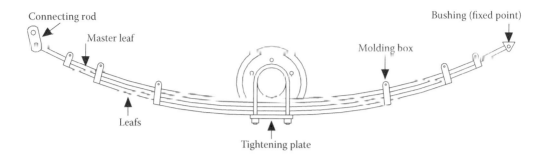

FIGURE 13.14   Spring with spring shackles.

Spring shackles are hinged shafts, so their reaction will always be axial. From the notes in Figure 13.15:

$$\frac{F_0}{F} = tg\alpha.$$

Therefore, while the vertical reaction is always F, the horizontal one always has different values depending on the position of the shackle.

Both vertical forces F and the horizontal $F_0 = Ftg\alpha$ generate a vertical shift. The total vertically deflection will be

$$f = f_v + f_0,$$

where f_v is the vertical deflection generated by F and f_0 is the vertical deflection generated by F_0 (horizontal force).

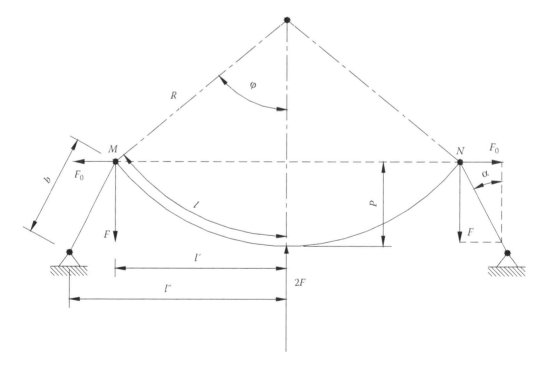

FIGURE 13.15   Scheme for a spring shackles calculation.

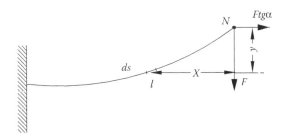

FIGURE 13.16   Scheme for the deflection's calculation

For reasons of symmetry, during deformation the tangent at the midpoint remains horizontal and in the calculation model we consider the end as fixed (Figures 13.16 and 13.17).

Applying the ellipse of elasticity theory we obtain:

$$f_v = F \int \frac{ds}{EJ} x^2 \quad f_0 = Ftg\alpha \int \frac{ds}{EJ} xy.$$

Since it is a leaf spring, the N shift generated by F, and disregarding any curvature effect, is given by

$$f_v = K \frac{Fl^3}{3EJ}.$$

The vertical shift generated by $Ftg\alpha$ is calculated by referring again to the ellipse elasticity theorem. With reference to the diagram in Figure 13.17, the horizontal force-generated vertical shift is given by

$$f_0 = Ftg\alpha \int \frac{ds}{EJx} f(x)(l' - x),$$

where Jx, defined above, is the inertial moment of the trapezoid spring.

With an unloaded spring, ds is a curved arc (see below), whereas when the spring is loaded, it is a generic curve. For small deformations, behavior is linear-elastic, so to calculate length, we refer to initial shape that can be simply approximated to a curved arc.

In practice, we refer to an unloaded beam, whose function characterizing the axis line (curved arc) is known. In the calculation, we suppose that by applying a load, deformation configures as a parabola with the radius of the osculating circle that is a function of the load itself. As radius R varies (Figure 13.15), there is a corresponding variation of angle a and so too of the two load components.

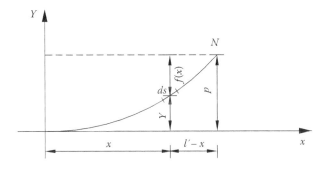

FIGURE 13.17   Scheme of the elastic deflection.

In the proposed simplified hypothesis there is

$$ds = \sqrt{dx^2 + dy^2}$$

The generic circumference with radius R is assimilated into the corresponding arc of the parabola ($y = ax^2$) with equal-curvature.
Since

$$\frac{d^2y}{dx^2} = 2a = \frac{1}{R} \Rightarrow a = \frac{1}{2R},$$

it follows:

$$y = \frac{1}{2R}x^2,$$

and dy is given by

$$dy = \frac{1}{2R}2xdx.$$

Substituting in the ds expression we obtain:

$$ds = dx\sqrt{1 - \left(\frac{dy}{dx}\right)^2}.$$

Substituting in the integral and solving it, it is finally possible to provide a form similar to that generated by the vertical force component (above), which is

$$f_0 = \frac{K'Ftg\alpha l^2}{EJ}p,$$

In conclusion, the total deflection $f = f_v + f_0$ will be

$$f = K\frac{Fl^3}{3EJ}\left[1 + \frac{3K'}{K}\frac{p}{l}tg\alpha\right].$$

In the deflection formula, K is the coefficient that is a function of β relating to the trapezoid spring, whereas K' is another coefficient, also a function of β, related to the horizontal force effect.
With a similar reasoning to determining K for the trapezoid plate, the coefficient K' can be given the following form:

$$K' = (2 - \beta)/3.$$

This result is obtained by assuming the horizontal force applied at $2/3p$ from the horizontal (not all basic components of l length operate in the same way in forming the deflection) and calculating K' as a linear interpolation of the two extreme cases (rectangular, triangular spring).

However, the term K'/K varies little: from 0.445 if $\beta = 0$ (triangular spring) to 0.417 if $\beta = 1$ (rectangular spring) and, since the estimate is approximate, the average can be used, always assuming $3K'/K = 1.3$.

Assuming

$$\frac{KFl^2}{3EJ} = \frac{1}{A},$$

f becomes

$$f = p_0 - p = \frac{1}{A}(l + 1.3 p t g\alpha)$$

having called p_0 the p deflection with an unloaded spring.

Substituting p in the third member of the previous formula with $p_0 - f$ and solving for f, we obtain:

$$f = \frac{l + 1.3 p_0 t g\alpha}{A + 1.3 t g\alpha}$$

According to the previous hypotheses (spring elastic deformation is always circular), $p = p_0 - fl$ is linked to distance l between the spring midpoint and its end. Since

$$p = R(1 - \cos\varphi) \quad l = R\varphi,$$

it follows:

$$\frac{1 - \cos\varphi}{\varphi} = \frac{p}{l},$$

which provides φ as a function of p, and so l' from the equation:

$$l' = R\sin\varphi = l\frac{\sin\varphi}{\varphi},$$

and the α angle from (Figure 13.15)

$$\sin\alpha = \frac{l'' - l'}{b}.$$

In practice, for reasons of space, full loads are normally referred to when p and α correspond to $Q = 2F$ maximum. Therefore, p', α', and A' become input data for that load condition. So, f becomes

$$f = \frac{1}{A'}(l + 1.3 p' t g\alpha')$$

and so the unloaded deflection $p_0 = p' - f$.

The system characteristic is determined for each load value as follows: the f deflection, which corresponds to a different $Q = 2F$ load, is assessed by an iterative calculation. First p is calculated with

$$p = p_0 - \frac{1}{A}(l + 1.3 p t g\alpha)$$

in which, to calculate f, A is approximated to the new Q, and α can be approximated to the one used to assess p_0. The φ value is then determined from

$$\frac{1-\cos\varphi}{\varphi} = \frac{p}{l},$$

and l' through:

$$l' = l\frac{\sin\varphi}{\varphi},$$

to obtain a new (closer) value of α with

$$\sin\alpha = \frac{l''-l'}{b}.$$

The calculation is then repeated with the previously found α re-calculating p and, as before, a new α value is recalculated, and so on, until the two consecutive α quantities coincide. Thus for Q, the desired p value has been found (the last in the iteration). The calculation is repeated as before with a new Q value, then A, starting with the computation of p from α's latest value.

For each Q load, a certain number of iterations have to be done to find the corresponding p deflection. Next, the vertical load shift due to the variation of α must be considered which, passing from a' to a'', would generate a vertical shift of the whole system equal to:

$$s = b(\cos\alpha'' - \cos\alpha').$$

which should be added to f for the total vertical shift.

Finally, the misalignment of the connecting rod's center should be considered, as well as the elastic deformation of the curved tract due to the eyelet's size, which is feasible by observing the system geometry (Figure 13.18).

Leaf springs are mostly employed for jeeps, trucks, buses, and rail wagons.

The maximum stress is

$$\sigma = \frac{6Fl'}{bnh^2},$$

where h stands for the thickness of each spring leaf plate that is generally identical for all plates.

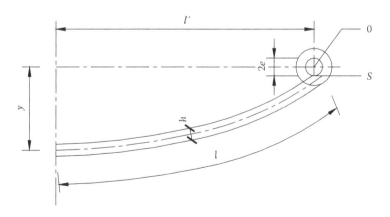

FIGURE 13.18 Geometric form of a leaf spring.

13.3.3 CHARACTERISTICS OF LEAF SPRING SYSTEMS

Given the applications of leaf springs (rail wagons, buses), one issue defines the most suitable characteristic to prevent any annoying frequencies that may negatively impact the structure.

As regards the physiological issue, consider a bus, at first empty and then loaded with people. If annoying oscillations are to be avoided, it is necessary for the oscillation frequency to remain almost constant. Either in the front or in the rear suspension system, the frequency is set at around 2 Hz.

For the study, the mass-spring system (Figure 13.19) with elastic constant c is the reference on which a Q load is applied (some of the load is suspended).

Since c is the spring elastic constant, the ω pulsation, peculiar to such a system, is

$$\omega^2 = \frac{c}{Q/g}.$$

Thus, being a function of the suspended mass.
Since it is desirable that ω remains constant, then

$$\frac{c}{Q} = \frac{1}{B} = \text{cons.}$$

Bearing in mind that $Q = c \cdot f$ and $dQ = cdf$ (f is the deflection), we have

$$\frac{dQ}{Q} = \frac{c}{Q}df.$$

Replacing the $c/Q = 1/B$ condition, it follows that

$$\frac{dQ}{Q} = \frac{1}{B}df.$$

The equation shows how the deflection must vary as load varies. Integrating, we get

$$f = B \ln Q + C_1.$$

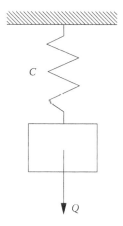

FIGURE 13.19 Mass-spring system.

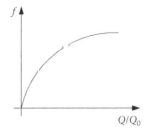

FIGURE 13.20   Logarithmic characteristic.

The C_1 constant can be defined on the basis of the initial conditions.
If $f = 0$ when $Q = Q_0$ (Q_0 is the load at 0 weight), we obtain

$$C_1 = -B \ln Q_0,$$

and then

$$f = B \ln \frac{Q}{Q_0}.$$

The ideal variation of the deflection f with load Q is logarithmic (Figure 13.20).

However, in reality a leaf spring with this characteristic cannot be created, so we try and come close to it with a polygonal curve, and in reality by arranging more springs in parallel according to the pattern in Figure 13.21. In this way, when the second spring touches the floor of the chassis, $c = c_1 + c_2$, and with the third $c = c_1 + c_2 + c_3$.

13.3.4   COIL SPRINGS

Coil springs operate by bending and are used in delicate mechanical systems (instruments, watches, etc.). Being of circular or rectangular cross-section, their axis obeys the law of Archimedes' spiral. In practical applications there are two cases of constraint other than the extremes shown in Figures 13.22 and 13.23. The conditions for these two cases involve different systems of stresses and deformations.

Archimedes' spiral equation is $\rho = K\alpha$. It states that, owing to an anomaly that increases by a round angle, there is a radius variation equal to pitch a. So, constant K is equal to $a / 2\pi$ and the spiral is in general:

$$\rho = \frac{a}{2\pi}\alpha.$$

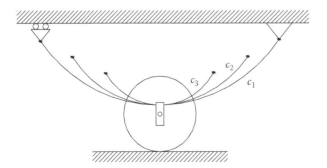
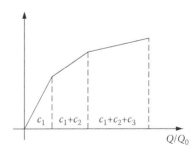

FIGURE 13.21   Spring in parallel and qualitative characteristic.

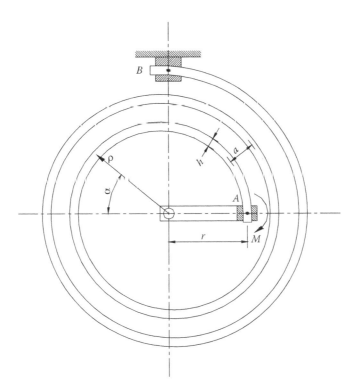

FIGURE 13.22 Coil spring with a applied moment.

Analyzing the constraint system reported in Figure 13.22, it may be observed that for any force applied from outside the lever, the spring is subject to an M moment that is constant along the entire spring, which is not directly loaded. A survey of deformations applying the elasticity ellipse theorem concludes the following:

- The $\Delta\varphi$ rotation of the element itself around the elastic center of gravity is

$$\Delta\varphi = M \int_0^l \frac{ds}{EJ} = \frac{Ml}{EJ},$$

l/EJ being the total elastic weight of the spring, l long.

- The M moment generates an f shift of A in a direction perpendicular to the joining line A with the center of gravity of the elastic weights, which is

$$f = M \int \frac{dS}{EJ} x.$$

If a is small, the elastic center of gravity definitely coincides with the spring center, so

$$f = \frac{Ml}{EJ} r,$$

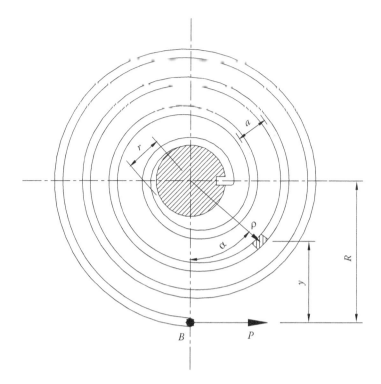

FIGURE 13.23   Coil spring with an applied force.

leading to

$$f = \Delta\varphi \cdot r,$$

that is, the deflection is equal to the rotation distance of the elastic weights' center of gravity from the extreme end. Only if the moment is constant along the entire length of a beam, does this result apply. In fact, this can be easily verified, for example, in the case of rectilinear-axis beams.

Deformation work in the spring is

$$L = \frac{1}{2} M \cdot \Delta\varphi,$$

from which, since $\Delta\varphi = Ml / EJ$, it follows that

$$L = \frac{M^2 l}{2EJ}.$$

If the cross-section is rectangular with sides b and h, there is

$$L = \frac{M^2 l}{2E \dfrac{bh^3}{12}} = \frac{6M^2 l}{Ebh^3}.$$

σ is given by

$$\sigma = \frac{M}{W} = \frac{M}{J}\frac{h}{2} = \frac{M \cdot 12 \cdot h}{bh^3 2} = \frac{6M}{bh^2},$$

and the utilization coefficient is

$$m = \frac{L}{\varepsilon} = \frac{\frac{\sigma^2}{6E}V}{\frac{1}{2}\frac{\sigma^2}{E}V} = \frac{1}{3}.$$

For circular cross-section springs with diameter d similarly:

$$\sigma = \frac{32M}{\pi d^3} \quad L = \frac{1}{8}\frac{\sigma^2}{E}V,$$

and the utilization coefficient is

$$m = \frac{L}{\varepsilon} = \frac{\frac{1}{8}\frac{\sigma^2}{E}V}{\frac{1}{2}\frac{\sigma^2}{E}V} = \frac{1}{4}.$$

Now let's analyze the second constraint pattern, shown in Figure 13.23. The relative rotation between points A and B, with y the distance from force line P to the ds component, is

$$\Delta\varphi = \int \frac{P}{EJ}yds.$$

Also in this case, if a is small, it is possible to consider the center of gravity of the entire spring as coinciding with its center, so

$$\int yds = lR,$$

substituting, we obtain

$$\Delta\varphi = \frac{PR}{EJ}l,$$

identical to that found in the case with PR instead of M.

Note that in this case the value obtained is approximate, not having taken into account the fact that the end B is not fixed. With the approximation defined earlier, work is expressed as

$$L = \frac{1}{2}P \cdot R\Delta\varphi.$$

As $\Delta\varphi = PRl/EJ$, work becomes

$$L = \frac{1}{2}\frac{P^2R^2l}{EJ}.$$

For rectangular cross-sections of sides b and h:

$$L = \frac{6P^2R^2l}{Ebh^3}.$$

The maximum moment in the cross section diametrically opposite to B is $M_{max} = 2PR$, then

$$\sigma = \frac{M}{W} = \frac{M}{J}\frac{h}{2} = \frac{12PR}{bh^2}.$$

Leading to

$$L = \frac{1}{24}\frac{\sigma^2}{E}V.$$

The utilization coefficient is equal to

$$m = \frac{L}{\varepsilon} = \frac{\dfrac{1}{24}\dfrac{\sigma^2}{E}V}{\dfrac{1}{2}\dfrac{\sigma^2}{E}V} = \frac{1}{12}.$$

For circular cross-section springs with diameter d, there is similarly:

$$\sigma = \frac{PRd}{J} \qquad L = \frac{1}{32}\frac{\sigma^2}{E}V,$$

and the utilization coefficient is equal to

$$m = \frac{L}{\varepsilon} = \frac{\dfrac{1}{32}\dfrac{\sigma^2}{E}V}{\dfrac{1}{2}\dfrac{\sigma^2}{E}V} = \frac{1}{16}.$$

As for the use of the material, the first constraint proves more suitable.

For a more exact calculation, σ bending expressed by $\sigma = 12PR/bh^2$ and $\sigma = PRd/J$, should be added, respectively, to P/bh (tensile stress) and $4P/\pi d^2$ (compression force P), each with its own sign, to the cross section diametrically opposite to B for which σ has been calculated.

If a is large compared to the thread diameter, $\int yds$ must be calculated more precisely. Starting from the spiral equation $\rho = (a/2\pi) \cdot \alpha$, see Figure 13.23, assuming $R = (a/2\pi) \cdot \alpha_2$, for a ρ radius cross-section and α anomaly, we get

$$y = R - \rho\cos(\alpha_2 - \alpha),$$

and still

$$ds = \sqrt{\rho^2 + \left(\frac{a}{2\pi}\right)^2}.d\alpha$$

Now the integral $\int y \, ds$ can easily be carried out.

The previously found ratios require a high number of coils, and require that at the extreme end of the spring only a tangential force be applied according to the direction of the force applied.

For springs with a small number of coils, the calculation turns out to be more complicated, since the elastic center of gravity can no longer be considered as coinciding with the spiral's center, and the shifts are such that stress parameters can vary owing to deformation.

The arc element ds must be determined exactly, with no approximation, and the same done for the cross section of the maximum bending moment that is to be assessed by equalling to zero the derivative of the general bending moment expression. Besides, by not considering the extreme end constrained to move in the direction of the force, the shift of point B would generate in it a radial component that would tend to make the spring shrink further, changing its configuration.

Things get even more complicated when one of the thread dimensions is very small compared to its length, so we would be dealing with a "rod" rather than a "beam." So, a calculation would have to be carried out from an elastic, non-linear point of view or by successive approximations, or with parametric procedures.

13.3.5 Cylindrical Helical Springs

The previous calculations and formulae apply to cylindrical helical springs (Figure 13.24) that operate by bending, where at the spring ends a pure moment M is applied (similar to what happens with spiral springs like in Figure 13.22) or a force F perpendicular to the thread axis. These springs are used in several mechanical systems such as car starter motors, door hinges, etc.

For a better idea of how they work, think of the "double body" springs used in mousetraps.

Using the adopted symbols, let's consider a half spring (single body). The force F, perpendicular to the thread axis, is applied by a rectilinear terminal, which is R from the spring's axis of radius $\rho = r = \cos t$. In the layout, the spring is represented only by one of the n coils ($a = 0$).

Given the perfectly symmetrical conditions, consider the central coil of the single body as fixed with a fixing moment equal to M.

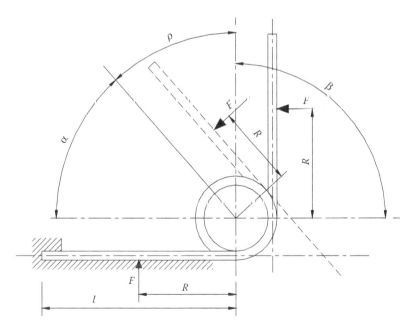

FIGURE 13.24 Bending helical spring.

The shift in force F consists of two components: one is due to the bending of the rectilinear terminal, the other is caused by the spiral-shaped part being subject to an M moment applied to the ends of the first and last coil.

The above formulae can then be used and more precisely:
for the rectilinear part the deflection

$$f_1 = \frac{FR^3}{3EJ},$$

for the spiral-shaped part the deflection

$$f_2 = \frac{MlR}{EJ},$$

that is, the spiral spring on Figure 13.22.

If the thread cross-section is circular with diameter d, the total theoretical deflection for the whole, single-body, n-coiled spring with terminal length R, is

$$f = 2\left(\frac{64MR^2}{3E\pi d^4} + \frac{64MRrn}{Ed^4} \right).$$

The second formula member represents the rotation effect of all the coils on the deflection.

When the number of coils becomes too high, the springs tend to deform along their axis, so an axis must be inserted to prevent any deformation phenomena. However, this axis must have a diameter that does not interfere with spring deformation. For this reason, in the maximum deformation layout, its diameter must be a few tenths less than the inner minimum diameter of the deformed spring.

Furthermore, it is also advisable that coil pitch is not too small, to prevent any coil contact with another coil during deformation.

The diameter, when the rotation of the spring ends, is equal to

$$\varphi = 2\frac{64Mrn}{Ed^4},$$

can be calculated seeing that the total spring length does not change; it follows that the $2r^1$ diameter of the spring after deformation turns out to be

$$2r^1 = \frac{2r}{1 + \varphi / \pi}.$$

The pitch variation of the winding spiral after total rotation must be such that the coils, at maximum deformation, do not touch each other, and so it must be greater than nd.

The values calculated up to now are theoretical and neither coil-to-coil nor coil-axis interference occurs. In fact, mainly in long-axis springs, it is difficult to avoid interference with the support axis resulting in a deflection value around 4%–6% less than the above calculated one, also shown experimentally.

As for the calculation of resistance, unlike coil springs where the L/d ratios do not require taking curvature into account depending on use, here this effect must also be considered.

These springs are built by winding on cylinders of relatively small diameter ($2r$), which generates residual axial stress.

If loads are applied to the spring according to coiling, the maximum bending stress generated is partly compensated by the residual ones. The maximum bending stress occurs at the fixed end (maximum distance of the force) and so for circular cross-sections:

$$\sigma = \frac{M}{W} = \frac{32M}{\pi d^3}.$$

The stress calculated this way must again be adjusted, the $d/2r = c$ ratio being high in most applications (greater than 0.2). This entails non-coincidence between the neutral and center of gravity axes of the cross section (see bending calculation for flange joints). If the cross sections remain flat before and after deformation by moment M, the maximum bending stress occurs in the inner rim section. For any evaluation, refer to rectilinear beam tensions and "adjust" with Wahl's K_1 coefficient for high curvature solids subject to bending. So, in the most stressed point, the maximum tension is

$$\sigma_{max} = k_i \frac{32M}{\pi d^3} \quad \text{with} \quad k_i = \frac{4c^2 - c - 1}{4c(c - 1)}.$$

Calculations similar to those above can be made for cross sections other than circular ones, which, however, are not used significantly. If deformation and stress values are needed, it is easy to substitute the geometric parameters of the cross section into the general formulae. The manuals report the k_1 coefficients relative to various cross-section layouts.

13.3.6 CONICAL DISC SPRINGS

Conical disc springs are used in several applications. Depending on geometry, they may have non-linear characteristics and be used in specific fields. They are conical (see Figure 13.25) with an inner diameter d and an outer D. The height h of the cone is generally short and its semi-opening is close to 90°. Thickness b, other conditions being equal, makes the spring more or less rigid.

Variations in stiffness can be obtained by arranging different components according to different models (Figure 13.26). At a first approximate calculation, stiffness approximates to that of the components arranged in series without considering friction.

To calculate stress and deformation, we refer to an approximate theory (Almen and Lazlo) that, although less accurate than those obtainable with the conical shell theory, is definitely easier to apply. The approximate theory is based on the hypothesis of midsection rigid rotation around point O at radius $r_0 = (R - r)/\ln(R/r)$ from the cone axis (Figure 13.27).

According to this hypothesis, loads and deflection s are linked by the following ratio:

$$P = A \frac{E}{1 - v^2} \frac{b^4}{R^2} \frac{f}{b} \left[\left(\frac{h}{b} - \frac{f}{b} \right) \left(\frac{h}{b} - \frac{f}{2b} \right) + 1 \right].$$

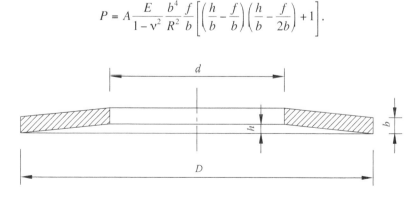

FIGURE 13.25 Conical disc spring.

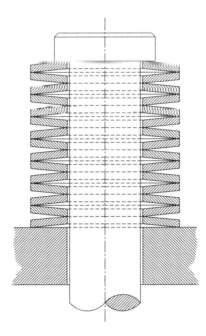

FIGURE 13.26   Conical disc spring in working condition.

The maximum peripheral compression tension on the r radius is

$$\sigma = A\frac{E}{1-\nu^2}\frac{b^2}{R^2}\frac{f}{b}\left[B\left(\frac{h}{b}-\frac{f}{2b}\right)+C\right].$$

E being the material's elasticity coefficient, υ the reciprocal of Poisson's modulus, and A, B, and C constants of the ratio $\gamma = R/r$ in accordance with the formulae:

$$A = \pi\frac{\gamma}{(\gamma-1)^2}\left(\frac{\gamma+1}{\gamma-1}-\frac{2}{\ln\gamma}\right),$$

$$B = \frac{\delta}{\pi\ln\gamma}\left(\frac{\gamma-1}{\ln\gamma}-1\right),$$

$$C = \frac{3}{\pi}\frac{\gamma-1}{\ln\gamma}.$$

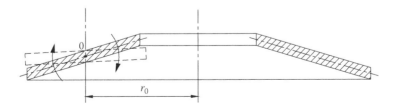

FIGURE 13.27   Scheme for the conical disc spring calculation.

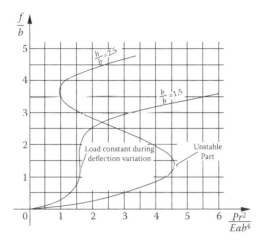

FIGURE 13.28 Qualitative conical disc spring characteristic.

From the formulae for P and σ, the link between stress and deformation is not linear over the whole field, and so while the characteristic is continuous, it does have some variably sloping sections, depending on the geometry of the component.

In Figure 13.28, for γ equal to 4, several spring characteristics are shown (h/b), the slope variation of the various curves under the load, and, in particular, the diagram shows how for the spring with h/b equal to 1.5, there is a 90° slope, with deflection variations caused by the smallest (practically absent) load variations.

Due to this peculiarity, these springs are used in oleo-dynamic valves or magneto-thermal switches. By exploiting geometry, the vertical tract characteristic can be made to be the greatest obtainable.

13.3.7 VARIABLE THICKNESS DISC SPRINGS

Another parameter that can provide wider deflection fields at constant load is the form of the plate. To calculate tensions and deflection s in these springs, refer to the formulae below taken from a work by this author et al.*

The thickness variation law (see Figure 13.29) is defined by

$$t(x) = T_0 + T_1 \cdot x,$$

where

$$T_0 = t_{c'} \left(1 + \tau \frac{a+b}{a-b} - \frac{2 \cdot c \cdot \tau}{a-b} \right),$$

$$T_1 = \frac{2t_{c'}}{a-b} \cdot \tau,$$

$$\tau = \frac{t_b - t_a}{2t_{c'}},$$

---

* La Rosa, G., Messina M., and Risitano A., Stiffness of variable thickness Belleville springs, *Transaction of ASME: Journal of Mechanical Design*, 123, 94–299, 2001.

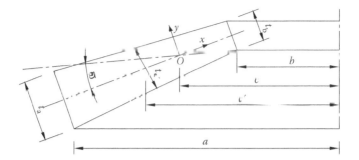

FIGURE 13.29 Variable thickness conical disc spring.

where t_b is the thickness at the inner radius, t_a is the thickness at the outer radius, t_c is the thickness at $c' = (a + b)/2$ point.

As for τ, it is an a-dimensional parameter ranging between −1 and 1, which defines the spring's tapering (Figure 13.30).

From Almen's and Laszlo's hypotheses, followed by Curti et al.'s guidelines, and calculating according to what Almen and Laszlo said, the following expressions apply:

- Rotation center—rigid

$$c = \frac{(a - b)^2}{-2\tau(a - b) + (a - b)\ln\dfrac{a}{b} + \tau(a + b)\ln\dfrac{a}{b}}.$$

- Characteristic:

$$P = \frac{2\pi \cdot E \cdot \varphi}{a - b}\left\{\begin{array}{l}(\beta - \varphi)(\beta - \varphi/2)\left[T_0(I) + T_1(II)\right]\\[2mm] +\dfrac{1}{12}[T_0^3(III) + 3T_0^2 T_1(IV) + 3T_0 T_1^2(I) + T_1^3(II)]\end{array}\right\},$$

with

$$I = \frac{1}{2}\cdot(a^2 - b^2) - 2\cdot c\cdot(a - b) + c^2\cdot\ln\frac{a}{b}$$

$$II = -\frac{1}{3}\cdot(a^3 - b^3) + \frac{3}{2}\cdot c\cdot(a^2 - b^2) - 3\cdot c^2\cdot(a - b) + c^3\cdot\ln\frac{a}{b}$$

$$III = \ln\frac{a}{b}.$$

$$IV = -(a - b) + c\cdot\ln\frac{a}{b}.$$

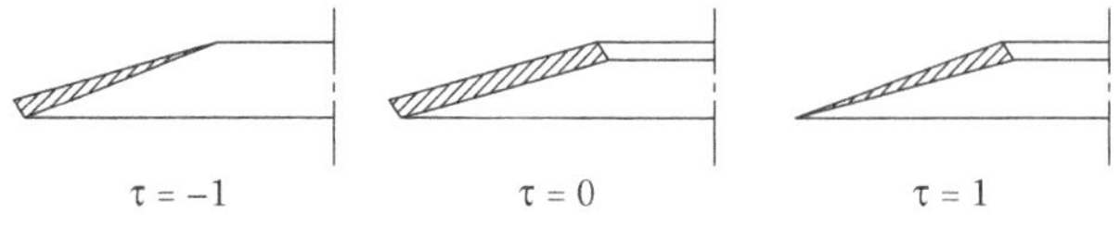

$$\tau = -1 \qquad\qquad \tau = 0 \qquad\qquad \tau = 1$$

FIGURE 13.30 Extreme variable thickness.

Tangential tensions:

$$\sigma_t(x,y) = \frac{E\varphi}{c - x}[x(\beta - \varphi/2) + y],$$

with

$$-\frac{t(x)}{2} \le y \le \frac{t(x)}{2}.$$

The tapering (variation of thickness with radius), widens the field of constant load for different deflection s.

For some geometrical parameters, a spring loses its essential capacity for constant deflection s at variable loads both for constant and variable thickness springs. As load decreases, the spring deflection increases (see Figure 13.28 the characteristic curve for $h/b = 1.5$). In this case, geometrical instability is at work: the spring suddenly passes from an upward to a downward concavity.

With the previously reported formulae (constant and variable thickness), peak stress values are obtained for the smaller-radius contact points of the various components.

In these points, theoretical stress is almost always greater than the material can withstand, so there is plastic subsidence of the material and a re-adjustment of the stress at these points. This is exactly why in some experimental examinations of springs tested for static stress, it was possible to verify reaching maximum theoretical stress (calculated with the given formula) of 1400–1500 MPa, with materials that had a 900 MPa yield point.

To approximate average stress, refer to the supported plate formulae (conical aperture 90°) or, with a quite simple calculation, to the average tension generated assuming the resultant of the contour-distributed load is applied to the center of gravity of the semi-circumferences (Figure 13.31).

With reference to Figure 13.32, the bending moment in the midpoint cross-section is given by

$$M = \frac{P}{2}\left(\frac{2R}{\pi} - \frac{2r}{\pi}\right) = P\frac{(R - r)}{\pi},$$

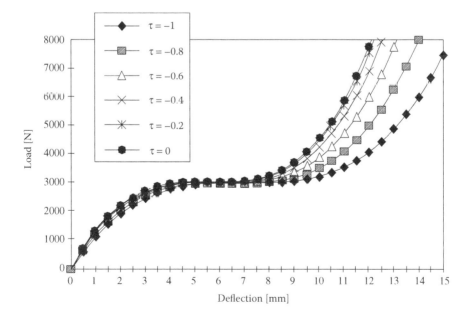

FIGURE 13.31 "Characteristic" of variable thickness conical disc spring.

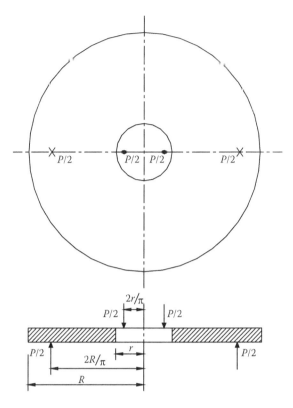

FIGURE 13.32   Scheme for approximate stresses calculation.

since the modulus of resistance is expressed by

$$W = \frac{1}{6}2(R-r)b^2 = \frac{1}{3}(R-r)b^2,$$

the stress is

$$\sigma_m = \frac{3P}{\pi b^2},$$

which is around 40%–50% of that calculated by Almen and Laszlo for the contact points.

If springs are subjected to dynamic loads, their construction and the heat treatments that follow must be precise. In any case, springs need to be sized using the general criteria shown for component fatigue calculation (Chapter 6).

13.3.8   DIAPHRAGM SPRING

Another type of spring used in automobile clutches is that in Figure 13.33 (diaphragm spring).

This spring is made up of a conical steel disc of radius R, cut along generatrixes. The incisions that start from a smaller-diameter circumference ($2\,r_i$), end up near the greater-diameter circumference ($2\,r$), with a ring thus providing a series of leaves, as many as the incisions, with width a at the base and a^1 at the tip ($\beta = a/a^1$ tapering ratio).

In operating conditions, the spring lies on the side of the larger ring, each leaf being loaded at the opposite end where all together they form a seat for the load ring.

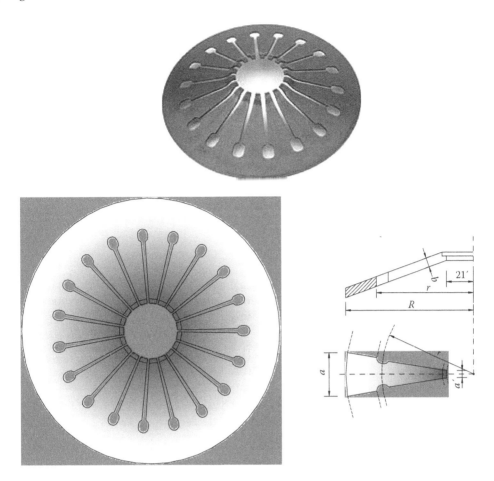

FIGURE 13.33 Diaphragm spring.

The maximum stress at the leaf base can easily be calculated with the simple rectangular leaf formulae and so the maximum stress in the ring section, which can be assumed as fixed, is

$$\sigma_{max} = \frac{6F(r - r_i)}{ab^2}$$

where F is the force exerted on each leaf, a the width at the leaf base, and b leaf thickness (the same across the whole disc) (Figure 13.34).

At the design phase, an important characteristic because it determines probable disc wear up to malfunction due to pressure loss between the disc and friction material, can be approximately and

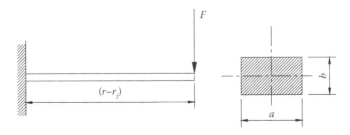

FIGURE 13.34 Scheme for approximate deformations calculation.

simply calculated by adding Almen and Laszlo's component (due to rigid rotation of the fully conical part) to the component deflection (due to leaf bending).

Considering that the fillet effect "masks" the effect of the less material closest to the base, the
vertical deflection of the trapezoid leaves can be approximated to that of a trapezoid leaf of width a
at the fillet end closest to the solid cone and a^1 at its tip ($\beta - a^1/a$), fixed at its base, b high, loaded
at its end by vertical load F/n and so it is

$$f_L = K \frac{4F(r - r_i)^3}{nEab^3} \cos^2 \varphi,$$

since

$$K = \frac{3}{(2 + \beta)}.$$

E the elastic modulus of the material, φ the angle formed by the cone generatrix with the horizontal (tan $g\ \varphi = H/R$). If the bending effect on the rotation is neglected. The deflection due to the
rigid rotation of the solid conical part can be approximated to

$$f_R = \frac{f(r - r_i)}{(r_0 - r)},$$

since $r_0 = (R - r)/ln(R/r)$ the distance of the point around which, in Almen and Laszlo's theory,
the rigid rotation of the solid part occurs and f being the deflection linked to the load which, according to this theory, is

$$F = A \frac{E}{1 - v^2} \frac{b^4}{R^2} \frac{f}{b} \left[\left(\frac{h}{b} - \frac{f}{b} \right) \left(\frac{h}{b} - \frac{f}{2b} \right) + 1 \right].$$

Since

$$A - \pi \frac{\gamma}{(\gamma - 1)^2} \left(\frac{\gamma + 1}{\gamma - 1} - \frac{2}{\ln \gamma} \right).$$

Alongside the previously defined shift components, the rotation generated by the bending moment of
transport distributed along the two edges of the solid conical part should be considered. This calculation
could follow the suggestion of Curti & Orlando* or by determining the coefficients of the conical shell.

Given the size of the diaphragm springs used in mechanical applications, as also shown in*, a
shift component does not exceed a tenth of the other two components.

Its impact can be approximated by considering the cone-shaped disc component as a rectangular
plate ($R - r$) of unit length at whose end the moment by length unit is applied, equal to: $F(r-r_i)/2\pi r$
(Figure 13.34). Under these conditions, the effect of the moment generates a vertical deflection of
the fin end equal to

$$f_M = \frac{6F(R - r)(r - r_i)^2}{\pi r Eb^3} + \frac{3P(R - r)^2(r - r_i)}{\pi r Eb^3}.$$

* Curti G. and Orland M., Ein neues Berechnungsverfahren fur Tellerfedern, *DRAHT* 30-1, 17–22, 1979.

The geometric form with incisions makes diaphragm springs more flexible compared to disc springs (Belleville) with similar geometries. Diaphragm spring deformation is closer to that previously seen in tapered disc springs. The leaves amplify the peculiarity of traditional disc springs (constant thickness), increasing the variability of deflection s with constant load.

As an example, the diagram shows the diaphragm spring of a production clutch, with the following characteristics:

Outer radius $R = 60$ mm   Inner radius $r = 15$ mm
Thickness $b = 1.3$ mm   Leaf width (base) $a = 8$ mm $\beta = 1$
Elastic modulus $E = 210,000$ N/mm^2 Poisson coefficient $\upsilon = 0.33$ F [N]

Along with the Belleville spring curves, the diagram of Figure 13.35 shows the lines of the various rigidities at the start phase. The Belleville spring behavior most resembling diaphragm spring behavior is that marked by a "rhombus" line (stiffness $K = 181.62$ N/mm); since it has an h/b equal to 0.5, it represents the spring equivalent of a diaphragm spring as regards deformation.

If it was necessary to have a constant force over a wider operational field, radial dimension being equal, thickness b could be varied equaling the diaphragm spring of the "triangle" line.

With this procedure, and with no complex calculations for the characteristics of diaphragm springs, it is easy to design them so that load remains constant throughout for the deflection s affected by probable wear of the clutch disc.

13.4 TORSION SPRINGS

Torsion springs are made of steel and their major stress is torsion. The most used materials are alloyed steels such as C48Si7 UNI 3545 ($R = 1300$–1550 N/mm^2 purged), C55Si7 UNI 3545 ($R = 1350$–1600 N/mm^2 purged) and 60Si7 UNI 3545 ($R = 1380$–1730 N/mm^2 purged).

Calculations for these can be done by referring to Saint-Venant torsion theory formulae. Even when a spring has no rectilinear axis, the calculation can adjust for deformation and torsion with coefficients that take operating geometries into account.

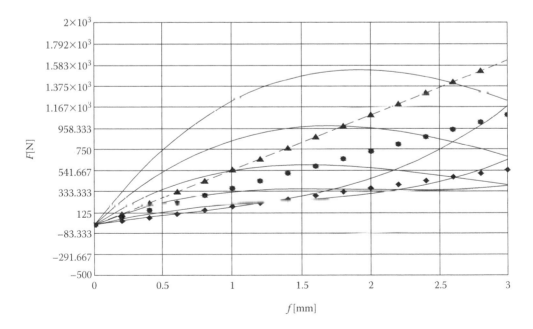

FIGURE 13.35 Qualitative characteristic curves for diaphragm spring.

When rectilinear axis components are not being dealt with, it is normally necessary to consider the helical structure of the spring axis as well as the marked curvature of a basic cross-section. Clearly the adjustment values are all the greater, the smaller the coil radius is, compared to the inertia radius of the spring cross-section and the steeper the helix inclination.

We know that If M_t is the twisting moment, the relative rotation $d\vartheta$ between the extreme ends of a basic cross-section ds, the coefficient of elastic shearing stress G, the maximum shear stress τ, then:

$$\tau = \frac{M_t}{W_t} \quad \frac{d\vartheta}{ds} = \frac{M_t}{GJ_p}$$

for a circular cross-section diameter d, since

$$W_t = \frac{\pi d^3}{16} \quad J_p = \frac{\pi d^4}{32}$$

it follows that

$$\tau = \frac{16M_t}{\pi d^3} \quad \frac{d\vartheta}{dS} = \frac{32M_t}{G\pi d^4}$$

Instead, for a rectangular cross-section with dimensions b and h ($h \geq b$, $A = bh$)

$$\tau = \frac{M_t}{\mu h b^2} \quad \frac{d\vartheta}{ds} = \frac{M_t}{\nu G h b^3}$$

μ and ν being a-dimensional quantities dependent on the h/b cross-section elongation shown in Figure 13.36.

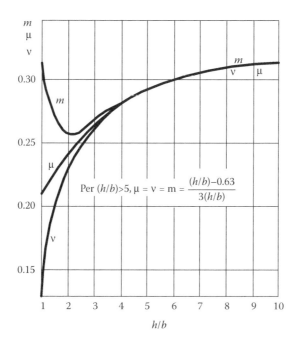

FIGURE 13.36 μ and ν a-dimensional quantities dependent on the h/b cross-section elongation.

In this case deformation is

$$L = \frac{1}{2} M_t \vartheta,$$

ϑ being the relative rotation. The utilization coefficient will be

$$m = \frac{\dfrac{1}{2} M_t \vartheta}{\dfrac{1}{2} \dfrac{\tau^2}{G} V}$$

If M_t is constant for the entire length of the spring, the utilization coefficient for the circular cross-section will be

$$m = \frac{M_t \vartheta}{V \dfrac{\tau^2}{G}} = \frac{M_t \cdot \dfrac{32 M_t l}{G \pi d^4}}{\dfrac{\pi d^2}{4} l \dfrac{256 M_t^2}{\pi^2 d^6}} = \frac{1}{2}.$$

Similarly, for rectangular cross-sections, we find

$$m = \frac{\mu^2}{\nu}.$$

In circular cross-sections, m reaches the value of 1/2, greater than that which could be obtained under the best conditions for bending springs (1/3, triangular plate).

Note that in the case of circular cross-section torsion springs, not only is the moment constant throughout the whole spring length, but tension is equal to zero in one point only and reaches values that remain constant with the coordinate defining boundary values, which never occurs when there is bending stress. If, then, for metallic materials, the energy storable is compared in both cases, assuming conditions would permit the entire spring to operate at maximum levels of bending and torsion, τ being a fraction of σ and the modulus of elastic shearing stress G being an even lower fraction of the modulus of elasticity E, it can be seen that, m and V being equal, torsion springs can store greater energy (1.15–1.3 times) than the energy storable by bending springs.

13.4.1 Torsion Bars

The simplest torsion bar consists of a solid circular cross-section bar. The type of constraint is always with one end fixed and the calculation model is shown in Figure 13.37. This type of spring is generally used in vehicle suspension. It is normally a round cross-section bar, one end of which is fixed to the bodywork, the other to the fulcrum of the suspension's oscillating arm; therefore, it operates under torsion. The pre-loading of this spring (quite significant for its limited size) can easily be changed by modifying the mounting of either end (for this reason, a grooved coupling or a flange with several positioning modes) is frequently used).

With reference to Figure 13.36, generally the deflection is the sum of bending and torsion:

$$f = f_f + f_t.$$

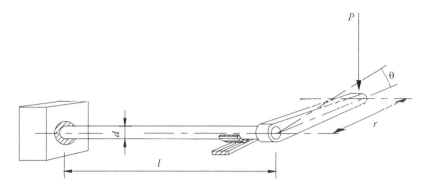

FIGURE 13.37 Torsion bar.

The bending deflection is given by

$$f_{\mathrm{f}} = \frac{Pr^3}{3EJ}.$$

In practical applications, given the high stiffness of cross-section length r, f_{f} is negligible compared to f_{t}, which is given by

$$f_{\mathrm{t}} = r \cdot \Delta\vartheta = r\frac{Pr}{GJ_{\mathrm{p}}}l = r\frac{32Pr}{G\pi d^4}l = \frac{32Pr^2l}{G\pi d^4},$$

the maximum τ generated by the torque is

$$\tau = \frac{16Pr}{\pi d^3},$$

so

$$f_{\mathrm{t}} = \frac{2rl}{Gd}\tau.$$

The utilization coefficient is $m = 1/2$, as previously seen, M being constant throughout the entire shaft.

13.4.2 Cylindrical Helix Torsion Springs

These consist of a thread (with circular or rectangular cross-section) wound helically around a cylinder with $D = 2r$ diameter (Figures 13.38a and 13.38b).

The load is applied to the last coil and the reaction is on the first one. Because of this, these coils are manufactured so that the top and bottom coils have a flat surface. The resultant of the pressure applied to the last coil is in line with its axis and generates a constant moment throughout the spring.

These springs usually have linear characteristics. However, non-linear characteristics can be obtained with variable-radius springs. Here, coils come into contact at different moments and the contribution of each to the total deflection is obviously different.

The axial force P generates a constant torque given by

$$M_{\mathrm{t}} = Pr,$$

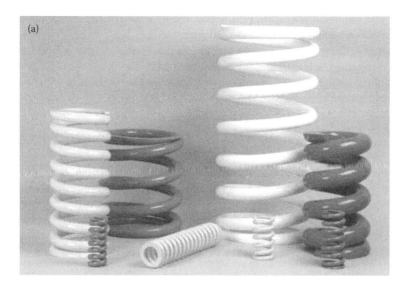

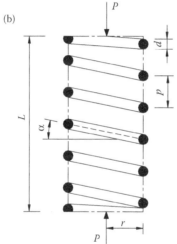

FIGURE 13.38 (a) Torsion springs, (b) Cylindrical helix.

for the entire length of the spring, which is equal to

$$l = \frac{i \cdot 2\pi r}{\cos \alpha},$$

where i is the number of effective coils.

For $\alpha \to 0$ and $r \to \infty$, the spring becomes a torsion bar.

Therefore, under these conditions, the f_t and τ formulae (used for bars) can be applied by simply replacing $l = i \cdot 2\pi r / \cos \alpha$:

$$f_t = \frac{32Pr^2}{G\pi d^4} \cdot \frac{i \cdot 2\pi r}{\cos \alpha} \quad \tau = \frac{16Pr}{\pi d^3}.$$

Generally, when the $2r/d = 1/c$ ratio is high, the case approaches that of a torsion bar. In practice, c values range between 0.07 and 0.17.

When α is substantial and, above all, r decreases, these formulae cannot be applied as the conditions are other than those for Saint-Venant's rectilinear beam.

For these springs, the theory of strong curvature solids applies. These effects can be approximated by observing that the length of the external fiber has a ratio of $(r + d/2)/r$ compared to the average radius of the spring. This raises the tension compared to the average value equal to $(1 + c)$, which approximates only to the twisting moment, so

$$\tau_{max} = \tau(1 + c),$$

which does not produce substantial errors for practical values of c (0.07–0.17).

A theoretical study (Göhner) that accounts more correctly for the curvature effect of the spring's helical structure, as well as the tensions due to shearing stress and made with successive approximations, produces increases in maximum tangential stress just short of those provided by the previous relation.

Göhner's exact theory also shows that curvature has a small effect on deformation. Using his results for tension and deformation, correction coefficients can be used:

$$f_t = \lambda'' \frac{32Pr^2}{G\pi d^4} \cdot \frac{i \cdot 2\pi r}{\cos\alpha} \quad \tau_t = \lambda' \frac{16Pr}{\pi d^3}.$$

The λ' and λ'' coefficients are functions of the $c = d/2r$ ratio (see Figure 13.39).
λ' is analytic expression is as follows

$$\lambda' = 1 + \frac{5}{4}c + \frac{7}{8}c^2 + c^3,$$

which highlights that with the earlier, simple formula, stress quantities are obtained that are very close to those computable with the exact theory. Indeed, for the borderline case $c = 0$, in both

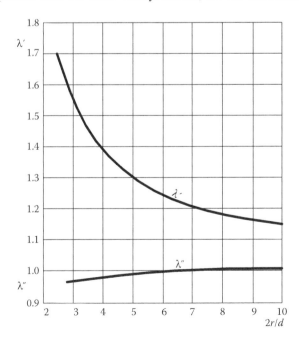

FIGURE 13.39 Diaphragm spring.

expressions the result is 1 and if c tends to 1 (if $c = 1$ the coil is in contact) a value of about 2.6 is obtained in both cases.

In addition, Wahl's theory:

$$\tau_{max} = \left(\frac{1 - 0.25c}{1 - c} + 0.615c \right) \tau,$$

gives the stress value at the point closest to the spring axis and is based on an approximate theory similar to that usually applied to the bending of great curvature solids. This theory also considers shearing stress and its values accord well with the exact theory.

Göhner's theory is important not only because it confirms other simpler theories, but also because, highlighting the small impact of different factors on the deflection, it even reappraises applying the others. For rectangular cross-section springs, the stress increasing formula according to the $(1 + c)$ coefficient or Wahl's coefficient can be approximated, in which $c = \rho / 2r$ is assumed, ρ being the cross-section inertia radius ($\rho^2 = J_p / A$) compared to the cross-section symmetry perpendicular to the spring axis.

For a more accurate computation, Liesecke's formulae and diagrams exploit Göhner's results and include the μ and ν coefficients that feature in the torsion formulae. The expressions are

$$\tau = \varphi \frac{2r}{\left(\dfrac{h}{b} \right)^{\frac{3}{2}} b^3} P \qquad f = \chi \frac{8ir^3}{Gh^2 b^2} P = \frac{\chi}{\varphi} \frac{4}{G} \frac{ir^2}{\sqrt{hb}} \tau$$

The φ and χ coefficients can be inferred from the abacus in Figure 13.40.
The utilization coefficient in its general layout is expressed by

$$m = \frac{\dfrac{1}{2} Pf}{\dfrac{1}{2} V \dfrac{\tau^2}{G}}.$$

Neglecting the axes helical structure, its volume is

$$m = \frac{\chi}{\pi \varphi^2}.$$

13.4.3 CONICAL HELIX TORSION SPRINGS

General, these consist of a thread wound around a revolving solid, most usually a cone (Figure 13.41). The reference for the calculation is of a thread winding on a cone with constant pace over the generatrixes of the cone itself. In this case, the horizontal projection of the thread axis is an Archimedes' spiral. Here, the moment varies along the entire spring, and is equal to $M_t - P \cdot \rho$, with minimum and maximum values given by

$$M_{min} = Pr \qquad M_{max} = PR.$$

The basic rotation is expressed by

$$d\vartheta = \frac{P \cdot \rho}{GJ_p} ds.$$

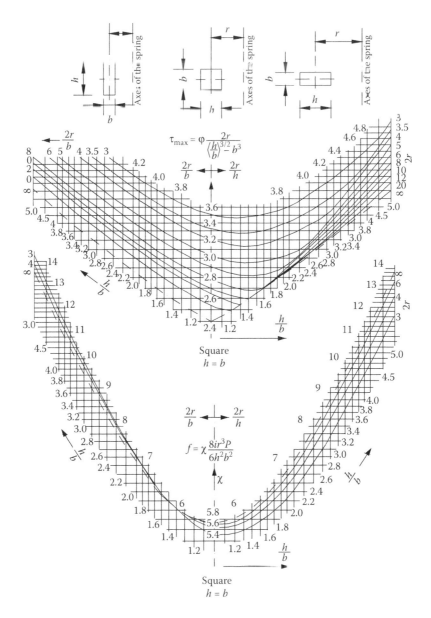

FIGURE 13.40 ϕ and χ coefficients as function of h/b ratio

The basic elastic deflection df is given by

$$df = \rho \, d\vartheta = \frac{P\rho^2 ds}{GJ_p}.$$

By making df a function of ρ, and assuming the spiral pace a is small, we get $ds \cong \rho \, d\alpha$. Since the anomaly α is connected to the radius by

$$d\alpha = \frac{2\pi}{a} d\rho,$$

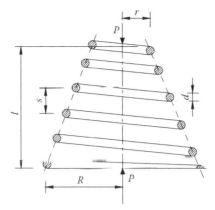

FIGURE 13.41 Conical helix torsion spring.

then

$$df = \frac{P\rho^3 d\rho 2\pi}{GJ_p \cdot a},$$

and as the number of coils is $i = (R - r)/a$, we obtain

$$df = \frac{2\pi Pi\rho^3 d\rho}{GJ_p(R - r)}.$$

Integrating, the total deflection is

$$f = \frac{\pi Pi(R^2 + r^2)(R + r)}{2GJ_p},$$

which for the circular cross-section becomes

$$f = 16\frac{i(R^2 + r^2)(R + r)}{Gd^4}P = \frac{16l(R^2 + r^2)}{\pi Gd^4}P.$$

Based on the maximum tension $\tau = 16/p \cdot (R/d^3) \cdot P$, we have

$$f = \frac{l(R^2 + r^2)}{GRd}\tau.$$

The "utilization coefficient" m becomes

$$m = \frac{1}{4}\left(1 + \frac{r^2}{R^2}\right).$$

For a rectangular cross-section (spiral spring):

$$f = \frac{\pi}{2} \frac{i(R^2 + r^2)(R + r)}{vGhb^3} P = \frac{1}{2} \frac{l(R^2 + r^2)}{vGhb^3} P,$$

and as $\tau = (R/\mu hb^2) \cdot P$, we obtain

$$f = \frac{1}{2} \frac{\mu}{v} \frac{l(R^2 + r^2)}{GRb} \tau,$$

and the utilization coefficient becomes

$$m = \frac{1}{2} \frac{\mu^2}{v} \left(1 + \frac{r^2}{R^2} \right).$$

In order to account for the effect of curvature on stress, the reasoning for cylindrical springs is valid by referring to the c values calculated for coil diameter whose stress increase must be assessed.

Since a basic spring cross-section generates a basic deflection proportional to $\rho^2 ds$, an entire coil with 2ρ average diameter will generate a partial deflection proportional to ρ^3.

As regards the effect of each turn on the entire deflection, it is different from turn to turn, varying the basic contribution by ρ^3. So, if the geometry is such that successive turns under load can come into contact (a smaller than d or b), the initial contribution comes from the largest coil and then from those with smaller average diameters. This gives rise, in conclusion, to a deflection that is not proportional to the load and therefore to a non-linear characteristic of the spring (Figure 13.42).

A spring may be designed in such a way that at maximum load it has a deflection such that between each coil there is a certain gap so the coils do not touch. This is done by giving the cone on which the coils are wound a great angular aperture, in other words with $R - r > id$.

This means that if the spring is totally compressed, the coils are limited one to another, the total fully loaded spring height being equal to diameter d if it is a circular cross-section spring and to height h if it is a rectangular one. In conclusion, with the previous sizing, if coils are overly compressed, they start touching and, as load increases, we have springs with i-1 coils. This does not happen in the last solution where the spring coils cannot touch and the characteristic is linear.

Volute springs (rectangular cross-section) are quite useful when there are space restrictions. Indeed, when they are totally compressed, they reduce to the size of the height of a coil.

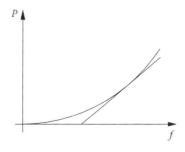

FIGURE 13.42 Possible non-linear characteristic of a conical spring.

13.5 COMPRESSION SPRINGS

We have already seen that to have high utilization coefficients the whole spring must operate with the identical stress. We also saw that a tensile stress stressed shaft has a "utilization coefficient" equal to 1. A thin ring too, used as a spring, which exploits the diameter variation due to constant radial pressure, would have a utilization coefficient equal to 1. In this case, tangential stress over the entire straight cross-section turns out to be practically constant.

Starting from this principle, and to obtain significant deformation generated by a load, E. Kreiszig invented ring springs, in which radial pressures are generated by inter-inserted rings with conical contact surfaces (Figure 13.43).

When an axial load P is applied, conical surfaces ensure that each ring is loaded with radial pressure, with consequent radial deformation. Owing to the conical shape of the surfaces in contact, relative axial shifts are generated (deflection). Therefore, for single rings, there is a tensile stress (external) that causes the average diameter to increase, and compression stress (internal) which reduces the average diameter. At the moment of its release, a spring goes back to its initial position. If surfaces were perfectly lubricated, the deflection in one direction would be equal to that in the other direction at its release.

By referring to the average diameter $d_m = (D + d)/2$, since one ring widens and another shrinks, the average diameter under load will be equal to $d_m + \delta$.

Variation δ entails an axial shift equal to

$$f = \frac{\delta}{tg\beta}.$$

Since generally $\tan g\ \beta = 0.25$, a deflection is generated equal to four times δ. The deflection s generated by deformation of the $n - 1$ ring couples are summed, the total deflection being:

$$f = (n - 1)\frac{\delta}{tg\beta}.$$

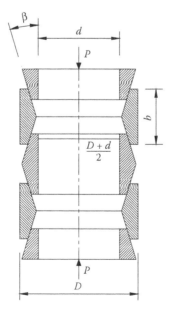

FIGURE 13.43 Compression springs.

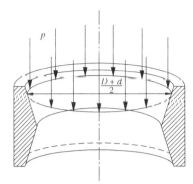

FIGURE 13.44   Ring of the compression rings.

To determine δ, the pressure generated by the axial load can be thought to operate on the circumference of diameter d_m (Figure 13.44).

Pressure p along the axis generated by force P distributed on the circumference of radius d_m is

$$p = \frac{P}{\pi d_m}.$$

Referring to the diagrams in Figure 13.45, the global reaction is inclined at φ to the perpendicular of the surface. So it follows that:

$$\bar{r} = \frac{p}{tg(\beta + \varphi)}.$$

By reversing load R, it finds itself on the opposite friction cone generatrix, since the force of friction has an opposite direction, always being opposite to the direction of motion. So, P becomes P', r is the same, and the angle is β – φ.

In the diagram on the right of Figure 13.41, we have

$$\bar{r} = \frac{p'}{tg(\beta - \varphi)}.$$

Substituting calculated $\bar{r}$, while the load is being applied there, is

$$p' = p\frac{tg(\beta - \varphi)}{tg(\beta - \varphi)}.$$

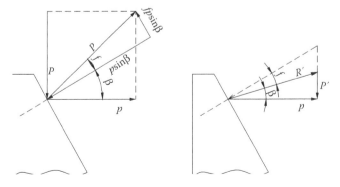

FIGURE 13.45   Pressure on the conical ring face.

It is important to point out that $\bar{r}$ shrinkage is the same both in the loading and release phase.

Mariotte's generic formula tells us that the tangential stress generated in cylindrical tubes undergoing internal pressure is

$$\sigma_t = \frac{p^* \cdot d^*}{2s},$$

adapted to our case $p^* = 2\bar{r}$, $d^* = (D + d)/2$, $s = (D - d)/4$, so:

$$\sigma_t = \frac{2\bar{r}\dfrac{D+d}{2}}{2\dfrac{D-d}{4}b} = \frac{2\bar{r}(D+d)}{(D-d)b}.$$

Anticipating the forced connections in Chapter 16, as a consequence of σ_t, according to Hooke's law there is a unit deformation expressed by

$$\varepsilon_t = \frac{\sigma_t}{E} = \frac{1}{E}\frac{2\bar{r}(D+d)}{(D-d)b},$$

which corresponds to a $\xi = \varepsilon_t \cdot r$ radial shift. Since in our case we have

$$r = \frac{D+d}{4} \quad \bar{r} = \frac{p}{tg(\beta + \varphi)},$$

we obtain

$$\xi = \frac{1}{2 \cdot E}\frac{\bar{r} \cdot (D+d)^2}{(D-d)b}.$$

The shift δ related to the diameter is $\delta = 2\xi$, since ξ is related to the radius. So

$$\delta = \frac{1}{E}\frac{\bar{r} \cdot (D+d)^2}{(D-d)b}.$$

The deflection $f = \delta / tg\beta$, leading to

$$f = \frac{\bar{r} \cdot (D+d)^2}{E \cdot (D-d) \cdot b \cdot tg\beta} = \frac{\sigma_t \cdot (D+d)}{2 \cdot tg\beta \cdot E}.$$

The total deflection is obtained by multiplying by $n - 1$, so

$$f = \frac{n-1}{2tg\beta}(D+d)\frac{\sigma_t}{E}.$$

Similarly, we can express P as a function of σ_t:

$$P = \frac{\sigma_t \pi tg(\beta + \varphi) \cdot b(D-d)}{4}.$$

Finally, the utilization coefficient can be determined by

$$m = \frac{\dfrac{1}{2}Pf}{\dfrac{1}{2}\dfrac{\sigma_t^2}{E}V} = \frac{\dfrac{1}{2}\dfrac{\pi(n-1)}{8}\dfrac{tg(\beta+\varphi)}{tg\beta}h(D^2-d^2)\dfrac{\sigma_t^2}{E}}{\dfrac{1}{2}\dfrac{(n-1)}{2}\pi\dfrac{(D^2-d^2)}{4}b\dfrac{\sigma_t^2}{E}} = \frac{tg(\beta+\psi)}{tg\beta}.$$

The formulae for springs with semi-rings at their ends are exact. In the case of complete solid rings at the ends, undergoing a stress equal to half in confrontation to that which is in the semi-rings at the ends, and consequently subject to diameter variations equal to half of the central rings, the previous formulae should be used replacing $(n-1)$ with $(n-1.5)$.

$m = tg(\beta+\varphi)/tg\beta$ provides a "utilization coefficient" m greater than 1, which is due to one part of the compression being employed to overcome the friction rather than deforming the rings.

The force at the beginning of release

$$P' = P\frac{tg(\beta-\varphi)}{tg(\beta+\varphi)},$$

(with quantities actually equal to $P' \cong P/3$), provides work L', restored to release:

$$L' \cong \frac{1}{3}L,$$

which is to say, $2/3$ of compression L are dissipated to friction.

The angle of friction φ in lubricated springs is assumed as $7°-9°$.

To comply with the $\beta > \varphi$ requirement, which expresses retrograde motion, in order for the compressed spring not to block in any event, β angles between $12°$ and $14°3'$ are adopted. In practice, width $b = (1/5 \div 1/6)D$ values are used.

Even when a spring is compressed to its maximum by deflection f, there must be no contact between the two inner or outer adjacent rings, and between their extremes there must be a distance Δb of around $(D+d)/200$ for rough rings and $(D+d)/400$ for worked rings, so total spring length, l_{tot}, is

$$l_{tot} = \frac{n\mp 1}{2}b + \frac{n-1}{2}\Delta b + f,$$

counting the minus (–) or plus (+) according to whether the spring ends with two half rings or two complete solid rings.

To make both the compressed and taut rings operate in the same way, they are made of different thicknesses (compression fatigue resistance is usually greater than tensile resistance), so $h_m = (D-d)/4$, the average thickness h_i of the smaller inner ring as well as h_e of the outer larger ring being

$$h_i = (0.9 \div 0.85)h_m \quad h_e = (1.1 \div 1.15)h_m,$$

with corresponding stresses expressed by

$$\sigma_i = \sigma_t\frac{h_m}{h_i} \quad \sigma_e = \sigma_t\frac{h_m}{h_e},$$

where σ_t is the value from the previous formula. Here the deflection is

$$f = \frac{n-1}{2\pi tg\beta tg(\beta+\varphi)}\left(\frac{D}{h_e}+\frac{d}{h_i}\right)\frac{P}{bE}.$$

13.6 RESONANT SPRING CALCULATIONS

When designing a spring, resonance phenomena that cause collapse must be prevented. Those most likely to experience resonance are the torsion springs that retract valves in internal combustion engines.

Helical springs have their own relatively low frequencies of longitudinal vibration. The characteristic frequencies tyfigureal of the valve retraction springs in internal combustion engines are of the same order as those of the first harmonics of the periodic movement imposed at the spring end by the camshafts. Therefore, there is the danger that the frequency of a given harmonic equals the spring's own frequency.

For springs employed in this way, a calculation is needed to verify whether any overstress may occur. This computation should comply to the following guidelines:

- First, the system or spring frequency is measured under constraint conditions
- Second, knowing the shape of the cam shafts, the displacement harmonics are worked out
- Third, critical resonance is detected, those dangerous harmonics that have a frequency equal to that of the spring system
- Finally, taking whole system damping into account, the elastic curve under resonance is determined and the maximum corresponding τ are calculated

To calculate the frequency peculiar to a spring system, we call L the spring length from support to support; the actual length of the uncompressed spring is instead equal to

$$l = i2\pi r.$$

From the applied mechanics course, the frequency peculiar to the axial vibrations of a fixed-end shaft is

$$f = \frac{m}{2}\frac{1}{L}\sqrt{\frac{E}{\rho}} \quad m = 1.2,$$

with the coefficient of elasticity E and the specific mass ρ of the spring's material.

The previous expression is inferred by considering a tiny element of length dx and straight cross-section S, and creating an equilibrium equation that provides the variation in normal force:

$$\frac{dT}{dx} = \rho S \frac{d^2\xi}{dt^2}.$$

Since

$$\frac{dT}{dx} = S\frac{d\sigma}{dx} = SE\frac{d\varepsilon}{dx} = SE\frac{d^2\xi}{dx^2},$$

and applying $\xi = \xi_0 \sin\cdot\omega t$, the result is

$$f = \gamma\sqrt{\frac{E}{\rho}},$$

where γ is a coefficient that depends on boundary conditions.

The previously reported f applies for a shaft but not for a helical spring. So, to the two systems (shaft, spring) need to be made equivalent. Multiplying and dividing the root term by S, we get

$$f = \frac{m}{2}\sqrt{\frac{ES}{S\rho L}}.$$

The root term stands for the ratio between shaft stiffness and mass by unit length. Using the same values for the spring by comparing $\Delta l = PL/ES$, the axial deformation of the shaft, and $f = 32Pr^2l/G\pi d^4$, the axial deformation of the spring, so

$$\frac{L}{ES} = \frac{32 \cdot r^2 l}{G\pi d^4}.$$

Therefore, the equivalent ES product is

$$ES = \frac{G\pi d^4 L}{32 \cdot r^2 l}.$$

Then, the equivalent ρS is determined by comparing the masses of the two systems, and the result is

$$\rho S = \rho \frac{\pi d^2}{4} \frac{l}{L}.$$

Substituting the ES equivalent and ρS equivalent into the above formula:

$$f = \frac{m}{2} \frac{d}{rl} \sqrt{\frac{G}{8\rho}},$$

which with $m = 1, 2, 3,...$ expresses the n frequencies peculiar to a helical spring. Note that those frequencies are inversely proportional to the number of coils because $l = i2\pi r$.

The elastic deformation corresponding to these peculiar frequencies is expressed by an equation like the following:

$$\xi = A\cos(m\omega x) + B\sin(m\omega x).$$

In general, as m increases, the number of nodes increases by 1.

From the analysis thus far, we have come to know the "shape" of each eigenfunction that corresponds to the m frequencies (eigenvalues) peculiar to that system, but not the actual deformation amplitude (eigenfunctions are defined by less than one constant). In order to know the unequivocal eigenfunction, it is necessary to refer to the damping.

The second phase requires analyzing the external stimulation forces. The valve is triggered by a camshaft (eccentric) that, depending on engine type, runs at the velocity (two-stroke engine) or half (four-stroke engine) the velocity of the drive shaft. By expressing lift h as a function of time, there is a (qualitative) trend like that in Figure 13.46.

Breaking down $h(t)$ into a Fourier series, we get

$$h(t) = A_0 + \sum_i A_i \sin i\omega t + \sum_i B_i \cos i\omega t,$$

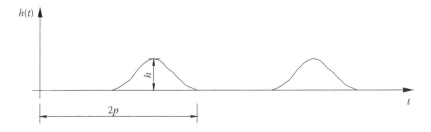

FIGURE 13.46   Qualitative displacement of an eccentric.

where

$$A_0 = \frac{1}{2\pi} \int_0^{2\pi} h(t)d(\omega t),$$

$$A_i = \frac{1}{\pi} \int_0^{2\pi} h(t)\sin i\omega t d(\omega t),$$

$$B_i = \frac{1}{\pi} \int_0^{2\pi} h(t)\cos i\omega t d(\omega t).$$

Remember, we obtained A_0 by integrating the first and second term of the $h(t)$ expression, which is

$$\int_0^{2\pi} h(t)d(\omega t) = \int_0^{2\pi} A_0 d(\omega t)$$

$$+ \int_0^{2\pi} (A_i \sin \omega t + A_2 \sin 2\omega t)d(\omega t)$$

$$+ \int_0^{2\pi} (B_i \cos \omega t + B_2 \cos \omega t)d(\omega t).$$

Recalling that

$$\int_0^{2\pi} \sin(\omega t)d(\omega t) = 0,$$

and that

$$\int_0^{2\pi} \cos(2\omega t)d(\omega t) = 0,$$

we get

$$\int_0^{2\pi} h(t)d(\omega t) = \int_0^{2\pi} A_0 d(\omega t).$$

To obtain A_1 and B_1, both members of the $h(t)$ expression are multiplied by $\sin i\omega t$, so

$$h(t)\sin \omega t = A_0 \sin(\omega t) + \sum_i A_i \sin i\omega t \sin \omega t + \sum B_i \cos i\omega t \sin i\omega t,$$

and by cos $i\omega t$, to get

$$h(t)\cos\omega t = A_0\cos(\omega t) + \sum_i A_i \sin i\omega t \cos \omega t + \sum_i B_i \cos i\omega t \cos i\omega t.$$

Integrating in the period, the integral of the sine and cosine functions equal zero, and so does the integral of the orthogonal function products. Only A_1 and B_1 remain.

With the positions:

$$tg\varphi_i = \frac{A_i}{B_i} \quad M_i = \sqrt{A_i^2 + B_i^2} \quad M_0 = A_0,$$

the series can be expressed as

$$h(t) = M_0 + M_i \sin(i\omega t + \varphi i).$$

So, the $h(t)$ function, which identifies the motion of the small plate, can be expressed as a constant M_0 and rotating vectors M_i, which are just the harmonics the valve is being stimulated by (Figure 13.46).

When $i\omega$ is equal to the frequency peculiar to a spring, resonance phenomena take place.

Let's plot the velocity of the camshaft (eccentric) ω (x-axis) against the velocity of the corresponding rotating deflection $i\omega$ (y-axis) (Figure 13.47).

If $i = 1$, the velocity is the same and so the graph line is at 45°; if $i = 2$ the rotating deflection runs at 2ω, and so on.

When the camshaft runs at a certain velocity ω^*, the first harmonic runs at the same velocity, the second harmonic runs at double velocity, and so on, when $i = 3, 4, 5$, etc.

The diagram shows which harmonics resonate. Indeed, marking on the diagram the frequencies peculiar to the spring, the various camshaft velocities where the system resonates can be found, as well as the harmonics involved in the resonance condition (Figures 13.48 and 13.49).

Generally speaking, the problem must be solved either by preventing resonance conditions by modifying the system to change those peculiar frequencies, or by verifying the maximum stress that can be applied to the regime under such conditions. The lower order harmonics are the most dangerous. For some automobiles, a survey of higher order harmonics should be carried out.

The next calculation phase entails the detection of overstress. To determine the unit stress a longitudinal vibration produces, in conditions of resonance with a given harmonic, the extent of

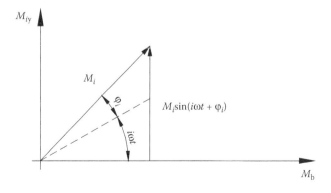

FIGURE 13.47 Harmonic component of the $h(t)$ function.

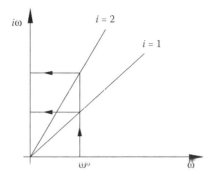

FIGURE 13.48 Harmonic velocity versus shaft velocity.

damping needs assessing, without which stress would be infinite. Usually, damping forces are adjusted proportionally to that particular valve velocity subject to sinusoidal oscillation and to the surface areas that oppose the motion in various ways. This calculation phase can be complicated and unreliable and often the best advisor is experience.

After calculating the stimulation to resonance, it is equalized to damping, which leads to the spring's axial deformation.

To calculate overstress, note that unit stress is proportional to df/dx, $f(x)$ being the deflection, or better, the amplitude of axial shift due to vibration in the section, defined by the x coordinate (Figure 13.50).

The coefficient of proportionality is obtained from the formula for torsion spring deflection s, which is

$$ f = \frac{32Pr^2l}{\pi Gd^4} = \frac{2rl}{Gd}\tau. $$

Note that for static loads, in which $f = 2rl/Gd$ and df/dx are constant throughout the spring:

$$ \frac{f}{l} = \frac{df}{dx}, $$

f being the free end deflection. So, the above coefficient is

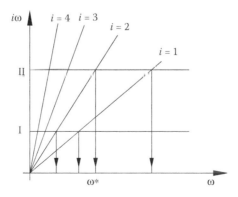

FIGURE 13.49 Resonate conditions.

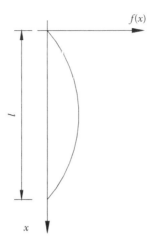

FIGURE 13.50   Qualitative spring displacements.

$$\frac{\tau}{f \,/\, L} = \frac{GdL}{2rl}.$$

Therefore, unit stress for vibrating springs is

$$\tau = \frac{GdL}{2rl} \cdot \frac{df}{dx}.$$

Cam shape must be designed to limit harmonic amplitude so that system damping is sufficient to keep deformation and stress within acceptable limits.

Apart from harmonics at the free end, spring vibrations can also be triggered by the shock of a valve shutting, as well as when it starts to lift. It is evident that the mass and elasticity of valve kinematics also have their own peculiar impact on these vibrations.

14 Flexible Machine Elements

14.1 INTRODUCTION

Machines often use flexible mechanical components to transmit motion between shafts at a distance. The main stress to which they are subject is tensile, but there might also be bending stress, as in the case of steel belts or any material with perfect elasticity in operation.

The most common flexible machine components are chains, belts and hawsers. Their special characteristic is their ability to absorb impact. They are designed with a fixed lifespan, at the end of which they are replaced (aside from servicing). During design, these components are subject to durability testing. Below, we will deal only with belts, referring to Giovannozzi, vol. II, for chains and hawsers, as everything needed to design these components is here.

14.2 BELTS

There are different types of belt, depending on application. The most common are flat (rectangular cross-section), belts with a circular or V-shaped cross-section, and toothed belts.

Flat or circular cross-section belts are able to transmit power between widely spaced shafts (5–6 m) that are not even coplanar. Flat rectangular cross-section (ribbon) belts (Figure 14.1) are often of steel and used where operating temperatures exceed 350°C. Other materials deform and cannot guarantee either technical characteristics or reliability.

Metal belts are often used as towlines, where the transmissible moment depends on friction with the cylinder. When great precision or synchronous systems are required, the belts are perforated with circular or rectangular designs that correspond to the cylinder "teeth."

Generally, the materials used for belts are steel (often stainless) or anodized aluminum with a hard covering. At modest operating temperatures (< 100°C), belts can be Teflon-, neoprene- or silicon-coated, or anodized with no coating.

To transmit motion, they are mounted on pulleys that in turn are fixed to drive and slave shafts. Shafts are nearly always steel and may have surface treatments to improve the friction on which power transmission always depends. For synchronous pulleys, the pivots or teeth guarantee synchrony.

Flat belts are used in various system configurations at variable and continuous velocities (e.g., conical pulleys are for continuously varying velocity). As a rule of thumb, inverted curvature belts are to be avoided, especially if they are steel.

Flat belts can slip, so they are used in low power transmission with some periodic irregularity.

V-shaped belts are used with V-shaped pulley races to increase contact area and limit slippage and jumping, so they are used in applications of greater power and above all for higher velocities.

When constant velocities are required (no periodicity), toothed belts are used (e.g., to drive cam shafts). In the case of ribbon belts, the cylinder has the teeth whereas the belt is perforated. V-shaped belts are mostly used when shaft distances are short (inter-pulley distance < 1 m). These pulleys have teeth, so motion occurs by meshing, which means greater power can be transmitted without slippage. However, due to belt elasticity, their length must be limited or there is the risk of mismeshing and precocious wear. Pre-loading is not generally used in these applications, but for other belt types it can increase the transmissible power.

Belts are made from leather, hemp, nylon, steel, and composites, with reinforced rubber matrices. Timing belts in particular use a composite reinforced along their outer edges (above the pitch line) with fine steel wires.

FIGURE 14.1   Example of a ribbon belt and pulleys.

In some special applications with high diameter pulleys, thin flat belts of steel, coated to prevent slippage, are used (e.g., belts for ribbon brakes). The mechanical efficiency of belts ranges from 0.7 for flat belts to 0.95 for V-shaped and to 0.99 for timing belts. Two of the advantages of belts are that they can absorb impact and limit slave pulley jerking. Here, flat belts are the best. Figure 14.2 shows examples of a timing belt and pulleys.

14.3 SIZING BELTS

Belt transmitted power between one drive shaft and another occurs with a theoretical transmission ratio of

$$\tau = \frac{\omega_2}{\omega_1} = \frac{R_1}{R_2},$$

R_1 (ω_1) and R_2 (ω_2) being the angular velocity of the drive and driven pulley.

This ratio varies when there is slippage that creates irregularities that could damage the drive system. Variations in inertial torque and changes in velocity give rise to instability in the drive system and belts. To limit slippage and to limit tension variation between the two sides of the belt, the belts are pre-loaded. For high-speed belts, the effects of centrifugal force must be accounted for, which in slow applications is negligible.

Belt length calculations take into account the pulley's winding angles for pulley diameters that are different or less than π and greater than π for larger diameters (Figure 14.3). The winding angles θ_1 and θ_2 are

FIGURE 14.2   Examples of a timing belt and pulleys.

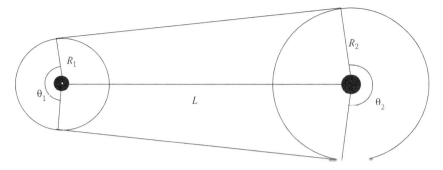

FIGURE 14.3 Scheme of belt and pulleys.

$$\theta_1 = \pi - 2\frac{\arcsin(R_2 - R_1)}{L} \qquad \theta_2 = \pi + 2\frac{\arcsin(R_2 - R_1)}{L}.$$

Belt length l becomes

$$l = 2\left[L^2 - \left(R_2 - R_1\right)^2\right]^{1/2} + \theta_1 \cdot R_1 + \theta_2 \cdot R_2.$$

Transmission moment M, under operating conditions and not taking into account inertia is defined by

$$M = \left(F_1 - F_2\right) \cdot R_1 \qquad \text{where } F_1/F_2 = e^{f\theta_1},$$

and, consequently, power P is

$$P = \left(F_1 - F_2\right) \cdot V,$$

F_1 and F_2 being the belt sides with more and less tension, θ_1 the winding angle of the belt on the drive pulley, f the coefficient of the friction between cylinder and belt, and V the peripheral velocity.

Friction calculations should also take into account dynamic friction that governs any system fluctuation or vibration; these might reduce characteristic friction values.

For steel belts, a practical value of 0.35–0.45 can be assumed; for nylon or similar, assume 0.6–0.8.

Taking inertia into account, the tension ratios (T_1, T_2) for the two sides are

$$\frac{T_1}{T_2} = \frac{(F_1 - m \cdot V^2)}{(F_2 - m \cdot V^2)} = e^{f\theta_1},$$

m being the mass per unit length of the belt.

The effect of centrifugal force may be such as to loosen the belt from the cylinder, gradually losing contact until F_2 equals 0. Total slippage prevents power transmission altogether. To avoid this, the system is pre-loaded with force F_i on both sides, and when v is close to zero (low velocity):

$$F_1 = F_i + \Delta F \qquad F_2 = F_i - \Delta F,$$

where ΔF is the tension variation between both sides and so

$$F_1 + F_2 = 2F_i.$$

In practice, the slave pulley for non-timing belts is loaded with force $2F_i$. Summing the two belt tensions under power transmission, it is:

$$F_1 + F_2 = 2F_i + 2m \cdot V^2.$$

Power transmission does not occur when F_1 equals F_2. For negligible mV^2, the definition of minimum transmittable power is

$$P_{min} = F_i \cdot V.$$

Maximum transmittable power is when $F_2 = 0$:

$$P_{max} = 2F_i \cdot V.$$

The two tension equations show how, in reducing the tension variation percentages compared to average values to increase fatigue resistance, it is advisable to pre-load the belt with higher tensions compatible with the admissible static load:

$$F_1 = F_i + m \cdot V^2 + \Delta F \qquad F_2 = F_i + m \cdot V^2 - \Delta F.$$

In practical applications, basic power is calculated after the formula is corrected with the two coefficients that take into account real conditions.

In particular, the value is multiplied by coefficient cp representing pulley size and cv representing velocity (1–0.7). It is divided by coefficient Ks representing the operating conditions as a function of the uniformity of applied moment and vibrations. So

$$P = \frac{(F_1 - F_2) \cdot V \cdot cp \cdot cv}{Ks}.$$

As regards the coefficient values to use, please refer to the manufacturers' catalogues and norms. The values provided above are only indicative to highlight the issue.

14.4 TIMING BELTS

As previously mentioned, timing belts are used to transmit power at high velocity and when maximum motion regularity is required. They do not permit slippage. Transmission performance is close to that for two intermeshed cogwheels (0.97–0.99). They are made of composites with rubber matrices reinforced with steel wires or gauze along the extremities of the belt (the pitch line).

Transmission takes place as per internal timing belts with a prolonged contact span for the deformability of the belt. Their characteristics are pitch and profile type (toothed) for the toothed pulley. They transmit power between closely spaced shafts because excessive shaft distancing would increase the chances of irregular contact over the teeth sides, which in the long run would increase wear. Generally, tooth number is less than 150, taking into account that average pitch is about 15 mm, making the total length less than 2.5 m for significant applications.

One fairly frequent automobile application is to drive the cam shafts where timing belts have substituted for chains. More recently, they are being used in motorcycles (BMW) where they have replaced the classic rear wheel drive chain.

As previously mentioned, because of the accuracy of the application, they must be thoroughly tested to define their durability so as to provide for their replacement (at the design phase) after a fixed operating span. Estimating their lifespan at the design phase is not easy, which is why manufacturers rely on testing.

It should be emphasized that this accuracy required in automobile engine manufacture is because even the slightest camshaft phase shift would bring about expensive damage throughout the engine. This is why chains are coming back into vogue.

As regards the power and belt tension formulae, in general the above applies.

14.5   PRECAUTIONS FOR TESTING DURABILITY

For the all the above-mentioned belts, calculating durability must take into account maximum tensile stress by referring to the thinnest cross-section. For V-shaped or timing belts, their profiles have a reduced cross-section, which must be multiplied by an average coefficient of 0.9. Acceptable load is a function of the material; manufacturers provide material durability for belts.

For belts reinforced with steel wire or flat belts in sheet steel, bending around the pulley should be considered, because it gives rise to tensions proportional to the inverse of the smallest pulley radius, which are then added to those of tensile stress so

$$\sigma = \frac{E \cdot d}{D},$$

for circular cross-section or steel reinforced belts of diameter d along the pitch line;

$$\sigma = \frac{E \cdot s}{D},$$

for flat sheet steel belts of thickness s.

For non-metallic belts, it is difficult to design for durability under normal operating conditions, so manufacturers' data is used, whereas durability estimates for reinforced belts are always based on the durability of the steel core.

For metallic belts, fatigue design concepts should be applied (see Chapter 6 "Fatigue of Materials"). Generally, the characteristics of steel are well-known, so given the stresses calculated above, one can refer to Smith's diagrams (corrected for durability) and define or verify size.

As previously mentioned, the loads for metallic belts are variable. The load ratio $R = 0$ if belt curvature is not replaced. In this case, at the average tension of the belt is added to the $\Delta\sigma$ bending stress produced by winding around the pulley.

For non-metallic belts, stress is generally more constant, apart from stresses due to operation, because those due to bending are zero. Given their operating conditions, with slippage and velocity variation, environment should also be considered as contributing to fatigue and aging that depreciate their original characteristics.

Steel belts are used when other belts cannot be. Generally, over 350°C, flat sheet steel belts are used because other belts lose their mechanical characteristics. To calculate durability, the perforations must be taken into account, because they can lead to tensions three times higher than nominal tensions.

Part IV

Design of Components and Mechanical Systems

Part IV of this collection deals with a few of the most common systems used in mechanical engineering constructions. The didactic purpose of this book is directed towards the choice to study the systems better suited for the definition of a mechanical component, on which to run calculation of resistance or verification, following the definition of the model.

There has been limited, even short in some cases, description, referring, in the same spirit as adopted in the previous parts, to the two volumes of the maestro.

15 Spur Gears

15.1 INTRODUCTION

As with the machine engineering notes for 2nd and 3rd year students, only the basic foundations of spur gears are dealt with in this chapter, so that later on during their specialization course, students will be able to deal with conical as well as helical toothed gears.

These notes comply with Giovannozzi's guidelines but, for the above reasons, they are made simpler and at the same time, pay greater attention to some analytical points.

Concentrating exclusively on spur gears, in this phase, is a good compromise between not overloading students with work and providing them with the basic concepts for subsequent courses on other types of gears/gears. Users of these notes will complete their training on other types of gears in texts, such as Giovannozzi, Vol. II, which these notes complement.

15.2 INTRODUCTORY IDEAS AND PRELIMINARY DEFINITIONS

In the case of transmission between parallel axes with friction gears, their contact surfaces are made up of two circular cylinders that roll without creeping (Figure 15.1) while conveying motion.

However, friction gears are notorious for not transmitting high power because they are limited by peripheral stress. Providing their pitch circles with teeth, as shown in Figure 15.2, creates gears whose teeth inter-mesh, are always in contact, and so transmit high power.

As the figure shows, gear teeth are radially oriented with respect to the pitch circle, partly externally and partly internally. The tooth top is the part that juts out of the pitch circle; the part below that line is the bottom. The tooth regions of the profiles, corresponding to the top and the base, respectively, are the side and the flank. The top and base are the profile zones.

Almost all cylindrical gears have involute circle profiles, first proposed by Philippe de Lahaire in 1695. There are gears with cycloidal profiles, and arc circle profiles derived from the former with particular modifications. These profiles are on their way out, their application being limited to precision mechanics (watches, meters) or extremely rare cases (e.g., the rotors in Roots compressors).

The main reason for the spread of involute gear teeth is attributable to the greater simplicity of manufacturing the tooth surface compared to cycloidal teeth. Moreover, the easier workability allows for greater precision in tooth dimensions and consequently enhanced transmission.

15.3 DEFINITIONS AND NOMENCLATURE

Without going into the particulars of conjugated profiles (refer to an applied mechanics course for this), the following are just some definitions and properties.

The involute circle is the plain circle generated from point C of a line (generatrix) tangential to a circle (called fundamental or basic circle) that rolls without creeping on the circle itself (Figure 15.3). An involute circle has a peculiarity: the perpendicular to the involute at any point is always tangential to the basic circle. All the perpendiculars of an involute envelop a circle, so there is biunique junction between any point on a given circle and an involute point. So, every circle has two involute groups, a right and left one. Consequently, circle involutes number ∞^2.

The pressure angle ϑ together with the distance r from the basic circle's center univocally locates each point of the involute so each pair of ϑ and r values corresponds to point P of the involute (Figure 15.4).

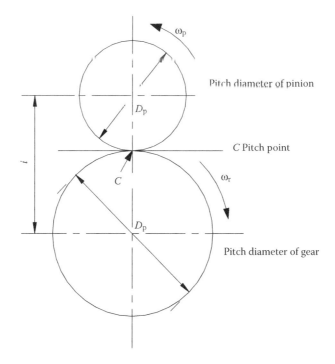

FIGURE 15.1 Spur gear.

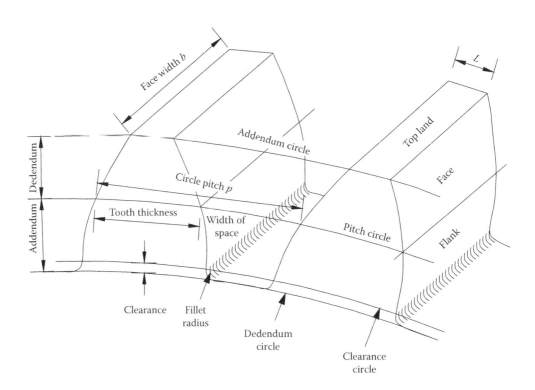

FIGURE 15.2 Nomenclature of spur gears.

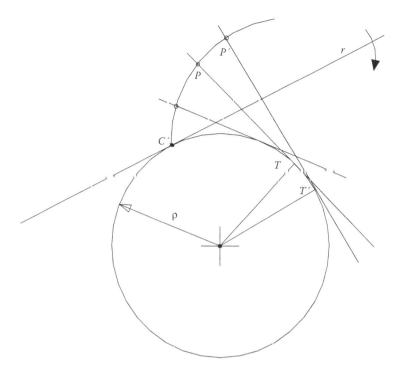

FIGURE 15.3 Involute generation.

Once ϑ is determined, the relation is defined between all the possible circle radii passing through generic P points of the involute and the base circle. With ρ the base circle radius, it follows that

$$r = \rho \cos \vartheta,$$

r being the radius of the circle with P passing through center O.

A few preliminary definitions are essential for analyzing tooth geometry (Figure 15.5):

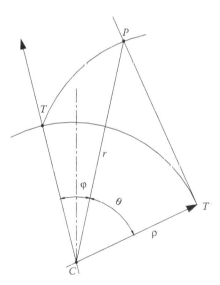

FIGURE 15.4 Coordinate of an involute point P.

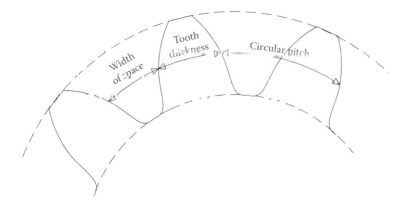

FIGURE 15.5 Circular pitch.

- The *thickness* of a tooth is the arc length between two sides measured along a circle.
- *Span* is the arc distance between opposite sides of two contiguous teeth along a circle.
- *Pitch p* is the distance between two consecutive, homologous points along a circle.

The above parameters usually refer to the cut pitch circle of the gear (the circle linked to the mesh with the cutting tool).

On its cut pitch, the tooth of an ordinary tooth is as thick as its span, so together they equal a half pitch. When meshing with another gear, the cut pitch circle coincides with the pitch working circle (that passes through point C at the same peripheral velocity as in the gear).

Usually, however, to prevent embedding and allow lubrication, a slight clearance is given to the pitch. Of course, for ordinary gears, the closer the thickness approaches $p/2$, the quieter and better the motion is.

As can be seen, the tooth develops across the pitch circle, which is not visible looking at the gear. Clearly visible are the top and the bottom circles. Usually, an involute profile does not intersect the top or bottom circles with a sharp corner, but is radiused to them or is interspersed with suitable chamfers near the circles.

Since the number of teeth n is an entirety, on the pitch circle or radius R the following ratio is

$$pn = 2\pi R,$$

where

$$p = \frac{2\pi R}{n}.$$

Pitch p in this case is an irrational number. Gearings cannot be standardized using this number. To remedy this, a ratio is introduced, called module m, defined as

$$m = \frac{2R}{n},$$

so that, dividing the diameter by parts n they are all m long. This helps define the pitch as

$$p = \pi m.$$

For two gears to suitably engage, they need to have the same pitch and so the same module. Thus, the ratio between the angular velocity of the driven gear and that of the driving gear (gear ratio) can be expressed, as R_1 and R_2, and also as the ratio of number of teeth:

$$\tau = \frac{\omega_2}{\omega_1} = \frac{n_1}{n_2}.$$

The ratio between the number of teeth (or between base and pitch radiuses) of the two gears is often called gear ratio, while the ratio between the angular velocities of the incoming and outgoing elements of a gearing is called transmission ratio.

The axial length of a tooth (face width) is also linked, though rather loosely, to the module. This length b, the ratio $\lambda = b/m$, ranges between 8 and 14; smaller ratios than λ are commonly used in manufacture today, the bigger ones in high power reduction gears.

Finally, the module value has to be chosen from standard values, preferably among these (in millimeters): 1; 1.25; 1.5; 2; 2.5; 3; 4; 5; 6; 7; 8; 10; 12; 16; 20.

The parameters defining the radial development of teeth have a tool for reference. Taking the cut pitch circle of the gear as reference (which passes through point C at the same peripheral velocity as the tool) the tooth profile can be considered as divided into two parts (Figure 15.6):

- Addendum a, the distance between the pitch and the top circle (delimiting the upper tooth)
- Dedendum u, the distance between the pitch and bottom circle (delimiting the lower tooth)

Totalling the two, addendum or projection on the one hand, and indentation or dedendum on the other, is tooth height; while the teeth of two conjugated gears have the same height, the projection and indentation can differ.

Each of the two lateral surfaces of a tooth is called a side. A side is divided in two parts by the pitch circle, the external part being called the addendum side (face), the internal one the dedendum side (flank).

The part of the flank and face where contact takes place between teeth is the active side (effective profile); it falls within the internal and external parting circles. So, normal contact (profiles are circular involutes) develops from the internal parting circle to the external parting circle. Outside them no regular contact can occur since in these segments (fillets), profiles do not conjugate.

The nomenclature cited is not only applied to external but also to internal gears, where the addendum is internal and the dedendum external to the pitch circle.

Summarizing, with reference to Figure 15.7, a tooth is laterally limited by two involute arcs, symmetrically arranged with respect to a radius. Since an involute is totally outside the base circle,

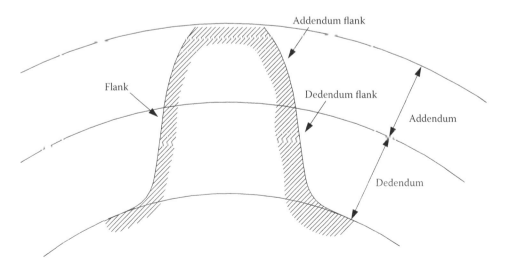

FIGURE 15.6 Nomenclature of the teeth.

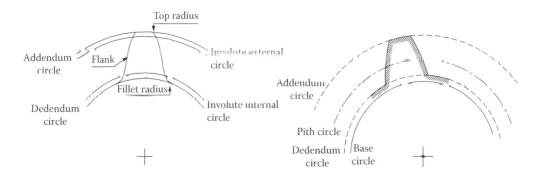

FIGURE 15.7 Nomenclature of the circles.

if a tooth extends inside this circle, its contour in this part is just a fillet that does not touch the other gear profile.

The section of an involute gear tooth is internally interrupted by the concentric internal parting circle, and externally by the external parting circles, also concentric to the pitch circle.

An effective profile, then, is that part of a tooth that comes into contact, or one made following the rules for conjugating profiles.

A tooth extends for a small distance above the external parting circles up to the top circle where a fillet helps initial meshing and prevents any line scoring. An effective profile also has its own internal parting circles beyond which the tooth extends by a suitable fillet as far as the bottom circle, this being a fitting without interference.

In the specific case of involute gears, in an effective profile, the internal parting circle is external (or coincident) to the base one.

An effective profile, therefore (the tooth part where the contact actually occurs), has to be made following the conjugated profile rules.

The shape of top and bottom fillets depends, if the cut occurs by envelope, on the tool profile. If the cut is made with a milling cutter, fitting shape can be made as desired.

Therefore, observing a gear, the bottom and top circles are visible (they delimit a tooth from top to bottom) and so are the flanks. The base circle, instead (this defines the involute profile), the effective profile, and the pitch circle (places of instantaneous rotation, so meshing with the other gear), are not identifiable by sight.

15.4 GEAR SIZING

The number of teeth n in a gear is obviously an integer, so there will be as many full as empty. A certain clearance is needed to allow lubrication and prevent seizure when gears operate. While inter-meshing, radial clearance g is necessary to prevent the top of any tooth from entering the bottom circle of the other gear.

g can range between $0.1 \, m_0$ and $0.45 \, m_0$, where m_0 shows the tool module (tool gear or gear cutter) with which a gear can be cut (generated).

The greater the g/m_0 is, the wider the tooth connecting radiuses can be at their base, that is, near the bottom circle, with a relative improvement in tooth bending resistance.

With the classic value $g = 0.167 \, m_0$, the connection radii between tooth flank and bottom circle are fairly small. The norms provide for a proportional clearance of $g = 0.25 \, m_0$. By these norms, connecting radii are longer compared to those with traditional sizing.

In reality, there are three types of sizing, defined by the ratios expressing a (addendum) and u (dedendum) (that is, projection or indentation as regards the reference line) of the gear cutter as a function of its module m_0.

Radial clearance g, which is the meshing distance of two gears between the top circle of one gear and the bottom circle of the other, is a variable quantity depending on sizing:

$$g = u' - a.$$

With normal sizing the total height of a tooth is equal to $h = 2.25\ m_0$, with

- $a = m_0$
- $u = 1.25\ m_0$
- $g = 0.25\ m_0$

With usual sizing:

- $a = m_0$
- $u = 7/6\ m_0$
- $g = 0.167\ m_0$

With STUB (lowered) sizing, mainly used for highly stressed gears:

- $a = 0.8\ m_0$
- $u = m_0$
- $g = 0.2\ m_0$

For tooth sizing, it is also worth pointing out the following:

- Ordinary gears have identical (a) projection and (u) indentation, compared to the pitch circle for the two gears and standardized values.
- Modified gears (see below) may have different projections and indentations for the two conjugated gears and certainly do not have standardized values.

15.5 CUT AND OPERATING CONDITIONS

The parameters peculiar to gears, as above, unless otherwise specified, refer to pitch circle and, with ordinary gear cutting, they refer to the tool that cut the gear.

Remember that for two gears to engage, the module of the engaging gears must be the same. It follows that if they are cut by envelope (gear cutter) they must be generated by the same tool.

As we shall see later on, if gear cuttings are "modified," the pitches meshing with the tool gear (cut pitch circle) are different from operating pitches (operating pitch circle), defined later during the meshing of two gears cut with the same tool.

Two conditions are then generally taken into account:

- Cut condition, referred to the gear cutter.
- Operating conditions of two engaged gears (cut with the same tool if by envelope), and engaged

So, if n is the number of teeth and p the pitch, for the pitch circle:

$$np = 2\pi R \quad nm = 2R,$$

and with reference to the base circle with radius ρ:

$$np_f = 2\pi\rho \quad nm_f = 2\rho,$$

p_f and m_f being pitch and module measured at the base circle.

If we refer to the gear cutter, the relations below apply:

$$np_0 = 2\pi R_0 \quad nm_0 = 2R_0,$$

(later on, subscript $_0$ will always be used to define cut conditions).

From a geometric viewpoint, two gear cutting tools may differ from each other by pressure angle ϑ_0 and by pitch p_0 (m_0) (Figure 15.8).

15.5.1 ORDINARY SIZING

Given these tool characteristics, all the corresponding gear characteristics can be defined. The geometry of tool fillet defines the geometry of top fillet and the bottom fillet of the gear tooth cut with such a tool.

Gears built with the gear cutter in Figure 15.8 will have a clearance of

$$g = u^1 - a = 1.25m_0 - 1m_0 = 0.25m_0.$$

15.6   INVOLUTE EQUATION

As already seen, an involute circle is the curve described by a point on a line that rolls without sliding on the base circle (Figures 15.3 and 15.4):

$$\rho = r \cos \vartheta.$$

The easiest analytical definition of an involute shape is done by polar coordinates, referred to as an axis passing through the gear center and the involute origin, or in other words, the involute point that belongs to the base circle and so passes through the tangential point of the line that generated it (first point of the involute).

Suppose that the perpendicular to the involute at a generic point P touches T' the base circle with O center and is $\vartheta = P\hat{O}T'$.

In our polar reference frame, referring to line $\overline{OT}$, point P is located by distance $r = \overline{OP}$ and by the φ anomaly function:

$$P = f(r,\varphi).$$

To define an involuted circle, the following equality applies:

$$\hat{T}\hat{T}' = \overline{PT'},$$

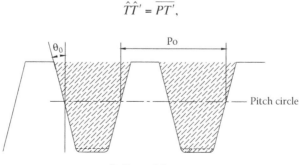

Ordinary sizing

FIGURE 15.8   Profile of the Rack cutter.

which, expressed as function of ρ, ϑ, and φ, becomes

$$\rho(\vartheta + \varphi) = \rho tg\vartheta.$$

Isolating φ:

$$\varphi = tg\vartheta - \vartheta = ev\vartheta.$$

As to the generic radius r that identifies P:

$$r = \frac{\rho}{\cos\vartheta}.$$

Finally in the polar system selected, point P is defined as follows

$$P = f\left(\frac{\rho}{\cos\vartheta}, ev\vartheta\right),$$

where ϑ is variable, $\varphi = ev\theta$ is a generic anomaly and ρ is the constant radius of the base circle defining the involute shape. Now it can be finally said that the parameter (ρ = fundamental radius) does not vary as point P on the involute varies.

Once the gear is built (tooth shape having been defined), radius ρ of the base circle does not vary as operating conditions vary. A variation in the center to center (i.e., while a gear is engaging with another gear) will make no difference to this parameter ρ.

15.7 DETERMINING TOOTH THICKNESS

From the previous paragraph, tooth thickness can be computed at any point P^* of the profile, once the thickness at another point P is known, especially choosing point V where thickness is always equal to 0 (Figure 15.9).

When point P^* coincides with the V vertex, where both tooth involute profiles meet (Figure 15.10), tooth thickness is zero, and we will call γ the angle $\overline{\varphi}$ at that point. Like ρ, $ev\gamma$ does not change once the tooth is made, since it defines the angular opening at the tooth base (on the base circle) and does not depend on the gear's operating conditions, but merely on the tool's structural characteristics. With reference to the semi-thickness corresponding to P^* (Figure 15.9):

$$\frac{s^*}{2} = r \cdot P^*OV.$$

Referring again to the same figure, the P^*OV angle as a function of γ is

$$P^*OV = \frac{s}{2r} - (\varphi^* - \varphi) = \frac{s}{2r} - (ev\vartheta^* - ev\vartheta).$$

So by, substituting into the s^* expression we obtain

$$s^* = r^*\left[\frac{s}{r} + 2(ev\vartheta - ev\vartheta^*)\right].$$

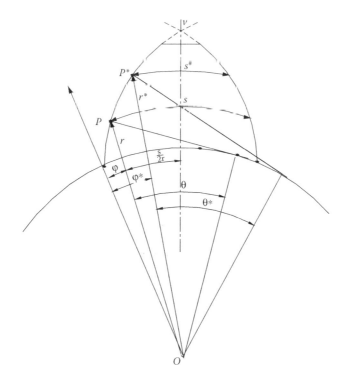

FIGURE 15.9 Thickness of the tooth.

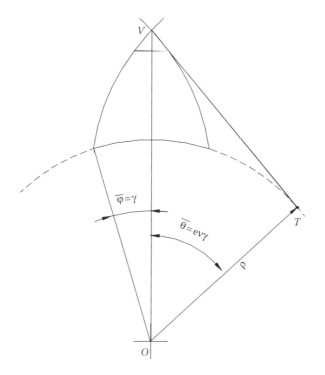

FIGURE 15.10 Angular coordinate of the V point.

This ratio can also be written

$$\frac{s*}{2r*} = \frac{s}{2r} + (ev\vartheta - ev\vartheta*).$$

If $P*$ coincides with point V (Figure 15.10), where the tooth involute profiles meet, the thickness is $s* = 0$ and $ev\vartheta* = ev\gamma$, so

$$0 = \frac{s}{2r} + (ev\vartheta - ev\gamma),$$

leading to

$$s = 2r(ev\gamma - ev\vartheta).$$

While $ev\gamma$ is always the same and depends on the parameters peculiar to the gear cutter, ϑ is in line with point P and r as well.

On the base circle where the involute originates, $\vartheta = 0$, $ev\vartheta = 0$, $r = \rho$, so tooth thickness is

$$s_b = 2\rho ev\gamma.$$

To obtain s_b or the thickness at any other point, refer to the gear cutter with module m_0 (with $p_0 = \pi m_0$ pitch) and pressure angle ϑ_0. The radius of the pitch circle of the gear with n teeth is

$$R_0 = \frac{nm_0}{2},$$

then

$$\rho = R_0 \cos \vartheta_0.$$

The $ev\gamma$ needed to obtain s_b is defined by referring to the gear cutter (i.e., the tool) on whose pitch circle the tooth thickness is half the pitch ($p_0/2$).

So, it is enough to apply the general formula $s = 2r(ev\gamma - ev\vartheta)$, for $r = R_0$ and $ev\vartheta = ev\vartheta_0$, and equalized to $p_0/2$, or

$$\frac{p_0}{2} = 2R_0(ev\gamma - ev\vartheta_0).$$

With module m_0 as reference:

$$\frac{\pi m_0}{2} = \frac{2nm_0}{2}(ev\gamma - ev\vartheta_0).$$

The values in the expression above are all known except $ev\gamma$, which is then

$$ev\gamma = \frac{\pi}{2n} + ev\vartheta_0.$$

Wheel thickness s_b can be obtained on the base circle:

$$s_b = 2R_0 \cos \vartheta_0 \cdot \left(\frac{\pi}{2n} + ev\vartheta_0 \right).$$

Once $ev\gamma$ is known, which depends on the geometry (ϑ_0) and number of teeth to cut, the thickness s_{P_c} can be assessed at any point P_c of the involute, which is distant r_c from the center O of the gear (Figure 15.11).

Applying the tooth thickness formula $s = 2r(ev\gamma - ev\vartheta)$, at the P_c point, we obtain

$$s_{P_c} = 2r_c(ev\gamma - ev\vartheta_c),$$

since

$$\cos \vartheta_c = \frac{\rho}{r_c} = \frac{R_0 \cos \vartheta_0}{r_c}.$$

R_0, ϑ_0, and r_c are known, so it is possible to define ϑ_c and then $ev\vartheta_c$, which is added to the formula to calculate s_{P_c}.

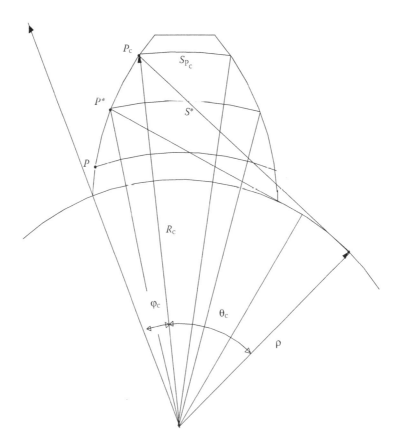

FIGURE 15.11 Thickness in a point of the tooth.

TABLE 15.1

$\vartheta_0 =$	15°	18°	20°	22°	24°
$n \geq$	69	48	39	32	27

15.8 TOOTH BASE FILLETS

As previously stated, a tooth profile in the area nearest to the bottom circle has a fillet with a profile other than the involute one. If, for example, we refer to traditional sizing ($a = m_0$ and $u = 7/6\,m_0$) the active profile will be entirely involute (of course outside the base circle), with the bottom circle outside the base circle only for a number of teeth higher than those defined by the ratio (for each pressure ϑ_0 angle):

$$u = \frac{7}{6}m_0 \leq (R_0 - \rho) = R_0(1 - \cos\vartheta_0),$$

so

$$n \geq \frac{7}{3(1 - \cos\vartheta_0)},$$

or for the tooth numbers, see Table 15.1.

The previous condition is not binding for the construction or running of the tool. Then, if gears are cut with a lower number of teeth than those in Table 15.1, the internal parting circle proves to have a greater diameter than that of the bottom circle, making it necessary to "extend" the profile (fillet) from the base up to the bottom circle (Figure 15.12). If the tooth cut is carried out by envelope, the fillet is defined by the tool tip shape. If the cut is done with a cutter, the tool can be shaped to create the most suitable fillet.

At all events, the curvature radiuses of fillets have upper limits defined by their interference with the motion of the tooth tip that engages with a sprocket. A lower limit is also defined by high notch values.

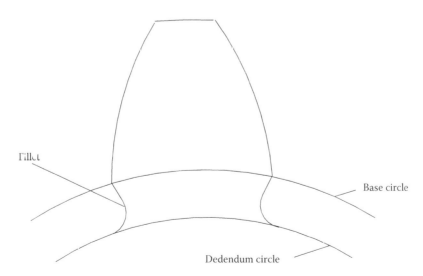

FIGURE 15.12   Fillet between base circle and dedendum circle.

In the case of highly stressed gears, where the rectilinear-prolonged, radially oriented fillet may generate fillets with small radiuses near the bottom circle, special fillets are employed. One of these is defined by the extension of the involute from the base circle to the bottom circle with an elongated epicycloid (Schiebel's fillet, Figure 15.13). So, strong-base teeth are obtained and drawbacks peculiar to circle-flank connections are avoided (manufacturing difficulty, high stress concentration factors).

Using fillets like the one above enhances tooth strength (the notch effect is remarkably reduced and the tooth is thicker at its base), but it is rather costly since suitably adapted cutters must be used, so solutions like this must only be adopted for gears of critical importance (aeronautics, highly stressed velocity gears).

Generally, the fillets recomended for ordinary applications are set by norms. So, for instance, UNI S 3522 suggests a circle fillet with an $r = 0.4 \, m_0$ radius that, in particular cases, can reach the value of $0.45 \, m_0$.

UNI 4503 stipulates, for the reference tools of standardized creators, the values of $r_1 = 0.25 \, m_0$ and $r_2 = 0.37 \, m_0$ for the two cases of creators assigned to cut gears with dendendum equal to $7/6 \, m_0$ and to $1.25 \, m_0$, respectively.

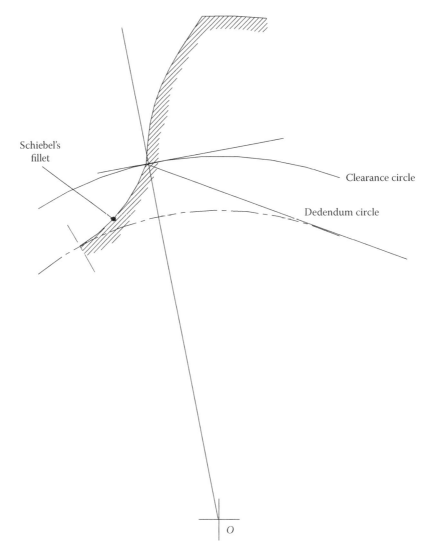

FIGURE 15.13 Qualitative Schiebel's fillet.

As stated before, in major gearings, operating at high velocity and highly stressed, particular geometrically defined fillets are used that depend on the gears they will engage with. Schiebel's fillet is one of these and is dealt with in detail later on.

15.9 EULER–SAVARY THEOREM

To show that a Schiebel fillet works without interference during operation, the Euler–Savary theorem is needed.

t p and a mobile one (mobile polar) p' (see Figure 15.14). While the mobile p' rolls on the fixed p, an M fixed point on the mobile polar traces out a curve (roulette curve of M point)

The Euler–Savary theorem helps determine the radius of curvature of the roulette at the M point. In Figure 15.14, O' is the center of curvature of the mobile polar p', while O is the center of curvature of the fixed polar p. Drawing the perpendicular to the $\overline{CM}$ line at the C point, join M to O' and extend it until intersecting the perpendicular to $\overline{CM}$ at the N point. Then, joining N with O as far as the intersection with $\overline{CM}$, Ω is found, which is the center of curvature of the roulette. Therefore, the radius of curvature of the rollette is $\overline{M\Omega}$. The Euler–Savary theorem states the following:

$$\left(\frac{1}{\overline{C\Omega}} - \frac{1}{\overline{CM}} \right) \cos \beta = \frac{1}{R} - \frac{1}{R'}.$$

Let us demonstrate now that the above construction is defined by the equation before reported. Having chosen an x, y reference system as in the Figure 15.14 write the equations of the two r and r' lines. A generic line has a parametric equation:

$$y = mx + q.$$

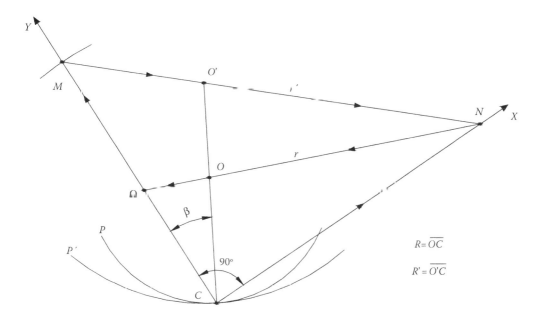

FIGURE 15.14 Curvature center of a roulette.

To determine m and q, it is necessary to impose that the line passes through two points r' passes through the M and N points of coordinates $M \equiv \left(0, \overline{CM}\right)$ and $N \equiv \left(\overline{CN}, 0\right)$. Imposing these two conditions, we have

$$\begin{cases} \overline{CM} = q \\ 0 = \overline{CN}m + q \end{cases},$$

and substituting in $y = mx + q$ and dividing by $\overline{CM}$, the r' equation is

$$\frac{y}{\overline{CM}} + \frac{x}{\overline{CN}} = 1.$$

In the same way, the r equation is determined passing through the Ω and N points:

$$\frac{y}{\overline{C\Omega}} + \frac{x}{\overline{CN}} = 1.$$

By referring to the r' line equation and replacing y and x with the point O' coordinates in the reference system x, y, we obtain

$$\frac{R'\cos\beta}{\overline{CM}} + \frac{R'\sin\beta}{\overline{CN}} = 1.$$

Much in the same way, for the r equation, after replacing y and x with the O point coordinates, we have

$$\frac{R\cos\beta}{\overline{C\Omega}} + \frac{R\sin\beta}{\overline{CN}} = 1.$$

Dividing the former equation by R' and the latter by R, we obtain

$$\frac{\cos\beta}{\overline{CM}} + \frac{\sin\beta}{\overline{CN}} = \frac{1}{R'} \qquad \frac{\cos\beta}{\overline{C\Omega}} + \frac{\sin\beta}{\overline{CN}} = \frac{1}{R}.$$

Subtracting from member to member:

$$\left(\frac{1}{\overline{CM}} - \frac{1}{\overline{C\Omega}}\right)\cos\beta = \frac{1}{R'} - \frac{1}{R}.$$

The graph created corresponds to the ratio found.

15.10   INTERFERENCE AND SCHIEBEL'S FILLET

A type of interference occurs when a gear tooth tip tends to dig inside the bottom circle of the other gear when the top circle of a gear has a diameter which interferes with the bottom circle of the other (Figure 15.15). The condition to prevent such phenomenon is when $u > a'$, that is the dedendum of a gear is greater than that of the other.

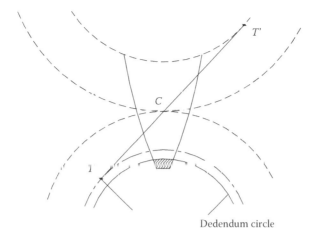

Dedendum circle

FIGURE 15.15 Interference.

Another type of interference is when the external parting of a gear intersects the line of contacts outside the $\overline{TT'}$ segment (Figure 15.16). In this case the tooth's edge would describe a curve, which in the relative motion proves to be internal next to the tooth flank. In practice it would tend to dig into the tooth flank. In this case it would be necessary to arrange for the external parting circle to coincide with the top circle and make sure the tooth tip of one gear has no interference with the other in the connecting area between the base and the bottom circle. If there is a bevel on the external part (external parting circles do not coincide with top circles), it will anyway be necessary to check the curve of the tooth's tip to adjust the bottom fillet of the other gear to prevent any interference.

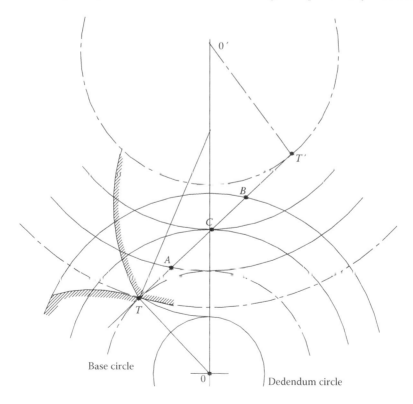

FIGURE 15.16 Possible maximum arc of action.

Exploiting the whole arc of contact but avoiding any interference, is where the Schiebel fillet comes in handy. So, the Schiebel fillet is used when the bottom circle diameter is smaller than that of the base circle. A base fillet is dependent on the characteristics of the gear with which the sprocket wheel will have to engage. This means that for each gear to engage, a different fillet is needed.

The external parting circle of a gear with center O', which engages the gear with center O to avoid interference, at most can reach point T. Similarly, the external parting of the gear with center O can at best reach T'.

In practice, T and T' are unlikely to be reached, but segments near them should be avoided while concentrating on segment AB at the intersection of external parting circle with the line of contacts. However, the borderline case occurs when the external parting circumference passes through T.

By using an epicycloids profile for the fillet—curve drawn from one point in the rolling of a circle (with which the point is integral) onto another (base circle)—according to the procedures below, we get a shape of the fillet itself such that the trajectory of the T point, which also coincides with the edge of the other gear, during meshing proves to be external (Figure 15.16).

This fillet can be created by describing the $\overline{TM}$ segment axis (ends in O^1) and the $\overline{TN}$ segment axis which intersects the $\overline{OO[}$ at the O'' point (Figure 15.17).

By rolling the circle of $\overline{O[}$ radius on the pitch circle of the gear with O center, we obtain the rollett (elongated epicycloid) that connects the involute and the bottom circle. The Euler–Savary theorem makes it possible to demonstrate that this fillet does not interfere with the other gear's teeth. If mobile p' and p'' are made to roll on the fixed polar p, the profile described by point T, thought to belong to p'' (with O' center) must be outside the profile described by the same point T thought to belong to p'' (circle with O''). To demonstrate that the curve described by point T is external to the epicycloids fillet, the above theorem is applied. Following the Euler–Savary theorem steps, we see that

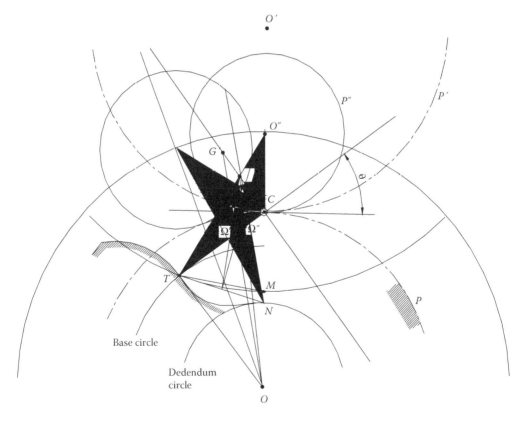

FIGURE 15.17 Curvature of Schiebel's fillet.

$$C \equiv C \quad \overline{CT} \equiv \overline{CM}.$$

To determine the center of curvature of the rollette described by T, thought to belong to p'', draw the n perpendicular to $\overline{CT}$ (passing through C), connect T with O'' and call G'' the intersection with the perpendicular. Then connect G'' to O until it intersects with $\overline{CT}$ at point Ω'', which is the center of curvature of the rollette.

Do the same with p'. C and T are the same points and so is the perpendicular to $\overline{CT}$. Join T to O' and call G' the intersection with the perpendicular; then join G' to O, the intersection with $\overline{CT}$ is Ω', which is the center of curvature of the rollette curve p' traced by T point thought on the p'. As can be seen, the radius of curvature of the former roulette curve (p'') is bigger than the latter (p').

15.11   MESHING

Resuming with involute profiles and considering two meshed gears, the inclined line of angle ϑ is at a tangent common to the two base circles. Tooth profiles are segments of involute, therefore the perpendicular of each tooth tip is at a tangent to the base circle. The perpendicular to the contact point, inclined by angle ϑ, is at a tangent to both base circles. All possible contact points are on this line (called the contact line). It follows that the forces exchanged between two engaging gears are on such a contact line (Figure 15.18).

Angle ϑ is called the pressure angle because the forces exchanged operate along the line inclined at this angle.

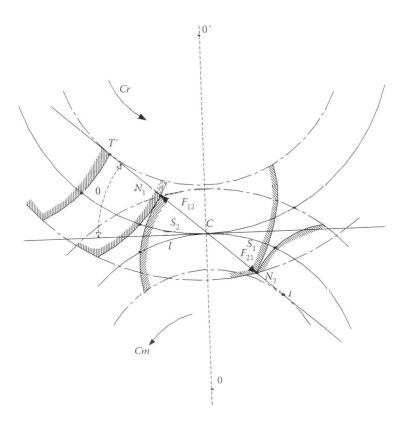

FIGURE 15.18   Action segment $N_1 - N_2$.

If T and T' are the tangential points of the contact line with the base circles, the $\overline{TT'}$ segment is a theoretical (possible) place of contact. The actual place of contact will be defined by the external parting circles of the two gears intersecting the contact line at points N_1 and N_2.

Figure 15.18 shows forces F_{12} and F_{21}, exchanged in zero friction; the distances of forces are ρ, ρ'.

The $\overline{N_1 N_2}$ segment is given the name of contact segment or action segment.

If rotation directions as in Figure 15.18 are considered, contact starts as soon as the tooth of gear 1 touches the tooth tip of gear 2 near its base.

While one tooth is using its tip, the other uses its base. In C, teeth touch one another right in the pitch circle tangential point of each gear (circles with R and R' radii, respectively).

While the pair of meshing teeth passes from N_2 to N_1, the pitch circles roll along each other, describing a circle that is the sum of arc s_1 (from N_2 to C) and s_2 (from C to N_1):

$$S = S_1 + S_2.$$

The arc s is the action arc; S_1 is an access action arc and S_2 is a recess action arc.

The arc length measured along both pitch circles is identical, as they roll without creeping on each other.

For gear 1 to transmit a continuous motion to gear 2, the action arc has to be greater than the pitch:

$$S \geq p.$$

This condition allows at least one pair of teeth to mesh at any given moment; before the teeth separate, another pair starts to mesh.

Now consider the profile passing through N_1 and the same tooth passing through C (Figure 15.19). A profile is usually generated by rolling the base circle line. So, the coresponding arcs and segments are equal:

$$\hat{T}\hat{H} = \overline{TN},$$

and similarly

$$\hat{L}\hat{T} = \overline{CT}.$$

By splitting the following segment:

$$\overline{CN_1} = \overline{TN_1} - \overline{TC},$$

and replacing the two equalities:

$$\overline{CN_1} = \hat{T}\hat{H} - \hat{L}\hat{T} = \hat{H}\hat{L}.$$

Note that points B and C are homologous to H and L. The rotation being rigid, the angle subtended to the two arcs is the same. So, referring to Figure 15.19, it follows that

$$\hat{H}\hat{L} = \rho\alpha \quad \hat{B}\hat{C} = R\alpha.$$

As $\rho_1 = R_1 \cos \vartheta$, so:

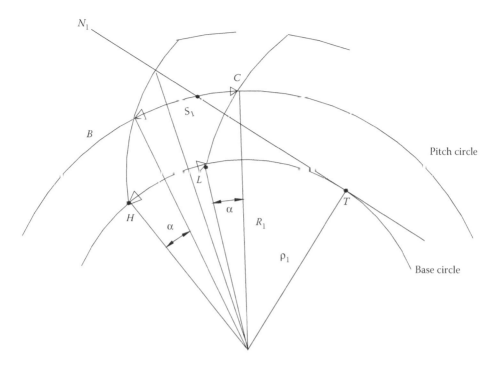

FIGURE 15.19 Relation between base arc and pitch circle arc.

$$\rho_1 = R_1 \cos\vartheta,$$

we have

$$\hat{H}\hat{L} = \hat{B}\hat{C} \cos\vartheta,$$

$$\overline{CN_1} = \hat{B}\hat{C} \cos\vartheta = S_1 \cos\vartheta.$$

Similarly for $\overline{CN_2}$:

$$\overline{CN_2} = S_2 \cos\vartheta.$$

The sum of $\overline{CN_1}$ and $\overline{CN_2}$ is equal to the $\overline{N_1N_2}$ segment, so

$$\overline{N_1N_2} = (S_1 + S_2)\cos\vartheta = S\cos\vartheta,$$

from which

$$S = \frac{\overline{N_1N_2}}{\cos\vartheta}.$$

In conclusion, the $S > p$ condition is expressed as follows

$$p < \frac{\overline{N_1 N_2}}{\cos \vartheta}.$$

This inequality is strictly linked to the ϑ pressure angle.

If smaller and smaller pressure angles are employed, the denominator increases, so, external partings being equal, the pitch decreases or, all conditions being equal, the number of teeth must be greater.

Substituting $p = 2\pi R/n$ and expressing $\overline{N_1 N_2}$ as a function of R, R', a, a', ϑ, the condition takes the form:

$$\frac{2\pi R}{n} < \frac{1}{\cos \vartheta} \left[R \sqrt{\sin^2 \vartheta + \left(2 + \frac{a}{R}\right) \frac{a}{R}} + R' \sqrt{\sin^2 \vartheta \left(2 + \frac{a'}{R'}\right) \frac{a'}{R'}} - (R + R') \right].$$

In the specific case of equal normally proportioned ($R = R' = R$, $a = a' = m$) external gears, the expression becomes

$$\frac{2\pi R}{n} < \frac{2R}{\cos \vartheta} \sqrt{\sin^2 \vartheta + \frac{4}{n}\left(1 + \frac{1}{n}\right)} - \sin \vartheta,$$

which can be written

$$\sqrt{n^2 \sin^2 \vartheta + 4(1 + n)} - n \sin \vartheta > \pi \cos \vartheta.$$

It then follows that

$$n > \frac{\pi^2 \cos^2 \vartheta - 4}{2(2 - \pi \sin \vartheta \cos \vartheta)}.$$

For $\vartheta = 20°$ $n > 3$ is obtained. In fact, the limitation is only theoretical. However, the S/p ratio is significant. The integer of the ratio represents the number of pairs meshing at the same time. If it is two, for instance, there are at least two pairs meshing (or from two to three). The decimal part, instead, is the percentage of the action arc covered by a number of pairs equal to the whole plus 1. If the ratio $S/p = 2.4$, it follows that at any time there are at least two pairs of teeth meshing and for 40% of the time there are three pairs. The tooth pairs meshing help us identify the total force exchanged that is distributed over the number of teeth in contact. Since tooth durability is likened to fatigue-stressing a beam, loading and non-loading times are very important.

15.12   EXPRESSING CONTACT SEGMENT AND ACTION ARC

Let us express the $\overline{AB}$ contacts segment length as a function of gear size (Figure 15.20). Applying Carnot's theorem to the two triangles $O'AC$ and OBC, we have

$$(R' + a')^2 = R'^2 + \overline{AC}^2 - 2R' \cdot \overline{AC} \cdot \cos(90 + \vartheta),$$

$$(R + a)^2 = R^2 + \overline{CB}^2 - 2R \cdot \overline{CB} \cdot \cos(90 + \vartheta).$$

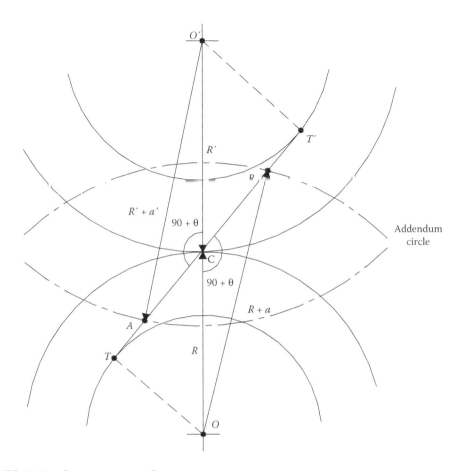

FIGURE 15.20   Contact segment *AB*.

Isolating $\overline{AC}$ and $\overline{CB}$, and applying the resolving formula of the second grade equation, we obtain

$$\overline{AC} = -R' \sin \vartheta + \sqrt{R'^2 \sin^2 \vartheta + a'(a' + 2R')},$$

$$\overline{CB} = -R \sin \vartheta + \sqrt{R^2 \sin^2 \vartheta + a(a + 2R)}.$$

Adding the two access and recess parts, we have

$$\overline{AB} = \overline{AC} + \overline{CB} = R' \sqrt{\sin^2 \vartheta + \left(2 + \frac{a'}{R'}\right)\frac{a'}{R'}} + R \sqrt{\sin^2 \vartheta + \left(2 + \frac{a}{R}\right)\frac{a}{R}} - (R' + R) \sin \vartheta.$$

To determine the action arc S it is sufficient to divide by $\cos \vartheta$, so

$$S = \frac{\overline{AB}}{\cos \vartheta}.$$

15.13   CONDITION OF NON-INTERFERENCE

As already seen, the external parting circles of meshed gears cannot have long radii. The intersection points of these circles with the contact line must be inside the $\overline{TT}$ segment, otherwise contact does not take place correctly. This precludes having few teeth, especially in normal gears (Figure 15.21).

Setting an upper limit to the top radius length means setting an upper limit to projection and so, for normal gears, to the module; (for such gears the addendum is equal to the module, or $a = m$). However, setting an upper limit to the module, given the additional radius, means setting a lower limit to number of teeth.

When two generic gears mesh, non-interference is obtained starting from the relation (Figure 15.22):

$$\overline{TT'} \geq \overline{AB},$$

Calculating $\overline{AB}$ with Carnot's theorem applied to the OCB triangle of which two sides R and $(R + a)$ and an angle $(\vartheta + 90°)$ are known, as well as to the $O'AC$ triangle $R',(R' + a)$ and $(\vartheta + 90°)$ that are known, and knowing that:

$$\overline{TT'} = R\sin\vartheta + R'\sin\vartheta,$$

$$n_{min} = \frac{2K'\tau}{-1 + \sqrt{1 + \tau(2 + \tau)\sin^2\vartheta}}.$$

For a sprocket-rack cutter pair ($\tau = 0$), applying the de l'Hôpital theorem, the expression becomes

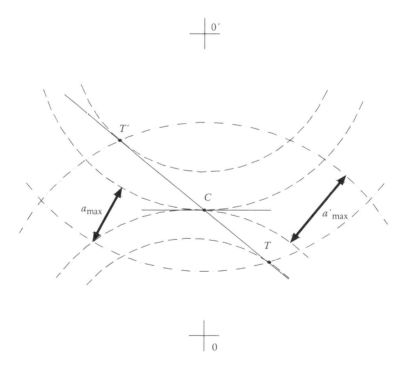

FIGURE 15.21   Maximum value for the addendum.

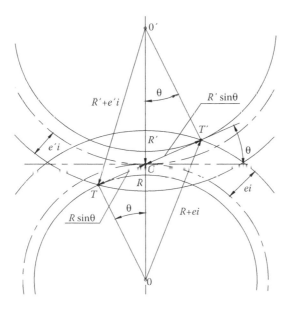

FIGURE 15.22 Extreme situation for non-interference.

$$n_{\min} = \frac{2K'}{\sin^2 \vartheta}.$$

To highlight the reasoning so far, we look at the case of a sprocket-rack mesh.

With reference to Figure 15.23, the point A is defined by the external parting of the driving gear which, in this case, is the tip of the tool.

For non-interference:

$$\overline{TC} \geq \overline{AC},$$

or

$$MC \geq \overline{NC},$$

and since $\overline{TC} = \rho tg \vartheta$, where $\overline{TC} = \rho tg \vartheta$:

$$\overline{MC} = \overline{TC} \sin \vartheta = \rho tg \vartheta \sin \vartheta.$$

Bearing in mind that $\overline{NC} = a' = K'm$, non-interference can be written as follows

$$\rho tg \vartheta \sin \vartheta \geq K'm.$$

Substituting $\rho = R \cos \vartheta$ and making $tg \vartheta$ explicit, we obtain

$$n_{\min} \geq \frac{2K'}{\sin^2 \vartheta}.$$

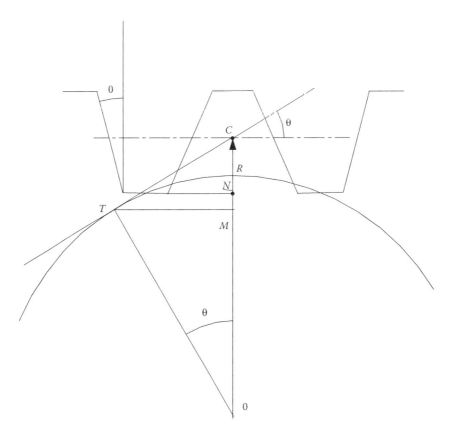

FIGURE 15.23 Pinion and rack.

This expression coincides with the previous formula deduced for $\tau = 0$.

The diagram in Figure 15.24 (Giovannozzi, Vol. II) obtained with the general formula, refers to the case $K' = 1$. It shows the minimum number of teeth as the transmission ratio changes and for different values of ϑ (curve parameter).

Bearing in mind the above expression for minimum number of teeth, for sizing beyond the norm (k' other than 1) minimum teeth number is obtained by multiplying n values in the chart by k'. Norms usually assume $\vartheta = 20°$. Other adopted pressure angles are $14°30'$, $15°$, $22°$, and $30'$. The case $\vartheta = 0°$ is related to friction gears.

Generally, overly high pressure values are not used, to prevent radial loads on the bearings and shafts being excessive. For modified gears (see below), $\vartheta = 15°$ or $\vartheta = 14°30'$ values are used.

The above formulae are also needed to determine the minimum teeth number that may be notched by envelope with a cut gear or a cut rack. To do so, you just need to use for τ the ratio between the tooth number of the gear and the tooth number of the cut gear ($\tau = 0$ in the case of cut rack). If gears are to be cut with as low a number of teeth as desired, a milling cutter is needed.

By referring to the abacus in Figure 15.24, if where $\tau = 0$ we draw the parallel to the abscissa until it intersects the ϑ curves, the abacus will show the minimum number of teeth that can be cut by the rack without any interference (Figure 15.25).

If the number of teeth to be cut is below n_{min}, and the work is done with a cut rack, tooth profile will prove to be engraved in the area next to the base circle (Figure 15.26). Under such conditions, a part of the involute profile is lost, as well as a segment of the action arc, with a decrease in the contact ratio ε.

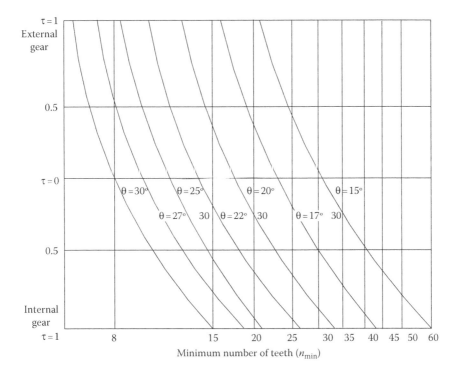

FIGURE 15.24 Transmission ratio versus minimum teeth number.

15.14 REDUCTION OF THE CONTACT LINE'S USABLE SEGMENT DUE TO INTERFERENCE

It has already been stated that cutting gears with fewer teeth than the minimum, generates tooth profiles that have no involute geometry in the part nearest to the base circle. In this case, the rack's external parting goes beyond the T point generating cut interference, thus losing part of the involute profile, and in the nearest part to the base circle the tooth is "hollow" (Figure 15.27).

When such a tooth engages with another involute tooth, part of the contact is lost because in the interference area during manufacture, the two profiles are not conjugated in this part. Consequently, a part of the contact segment will be lost too, with a corresponding reduction the action arc and contact ratio ε. In the design phase, the consequences of a cut under these conditions must be evaluated to guarantee acceptable contact ratio ε values when teeth mesh.

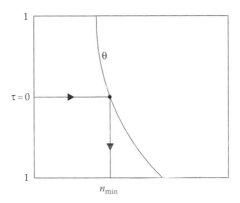

FIGURE 15.25 Minimum number of the pinion in engaging with a rack.

FIGURE 15.26   Qualitative shape teeth when the number of the teeth is below the n_{min}.

With reference to Figure 15.28, it is necessary to find out where on the involute of the tooth profile to start cutting away material. Note that this position is the point where for the first time the tool tip edge touches a possible involute representing the gear tooth profile, which originates at A on the base circle. Thus, the rectilinear profile of the cut rack will be at a tangent to the other involute segment passing through A.

As the line rolls on the base circle in two opposite directions, the tangential point describes two conjugated involutes (Figure 15.29).

So, assuming the rack external parting is beyond $\overline{CT}$, there is a point (P) where the rack touches the involute for the first time. This point is defined by the intersection of the horizontal line (the external parting) and the involute.

With reference to Figure 15.30, having defined λ so that $\overline{HN} = \rho\lambda$ (proportional to the projection of the tool edge compared to the pitch circle), it follows that

$$\overline{BP} = \rho(\lambda - \psi)tg\vartheta.$$

It is evident that the rectilinear rack profile is at a tangent in B to the other segment of the (dotted) involute passing through A. The rack profile must always remain at a tangent to an involute whose center of curvature is perpendicular to the rectilinear profile (the face) of the tool.

At a distance equal to the rack pitch in the base circle, point A' is the origin of a contact gear profile within the usable segment of contacts. Figure 15.30 shows an early phase of interference.

The $\hat{P}\hat{A}$ profile arc is lost. Because the involute's properties are known, the loss of the $\hat{P}\hat{A}$ profile arc corresponds to the loss of a length of contact segment equal to the distance between P and point K of contact of the perpendicular on P to the profile with the base circle:

$$\overline{PK} = \rho\varphi.$$

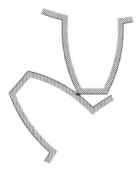

FIGURE 15.27   First contact in interference conditions.

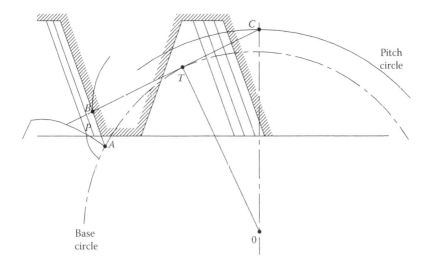

FIGURE 15.28 Relative position between the rack cutter and the pinion.

Again, due to the involute's properties, this length is also equal to $\hat{K}\hat{A}$ arc taken on the base circle.

So, $\rho\varphi$, that is φ, must be found. To this end, project the *OKPBN* straight line segments in directions *CN* and *ON*, to produce the equations:

$$\psi = \sin(\psi + \varphi) - \varphi\cos(\psi + \varphi),\tag{15.1}$$

$$1 = \cos(\psi + \varphi) + \varphi\sin(\psi + \varphi) + (\lambda - \psi)tg\vartheta.\tag{15.2}$$

From the two relations obtained, by removing the additional ψ parameter, it is possible to obtain φ as a function of ϑ and λ.

To describe the $\varphi(\lambda)$ function for a given value of ϑ, it is advisable to assume for simplicity:

$$\psi + \varphi = \gamma,$$

and then, from Equation 15.1, obtain ψ function of γ so as to get

$$\varphi = \frac{\gamma - \sin\gamma}{1 - \cos\gamma},$$

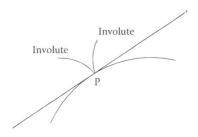

FIGURE 15.29 Conjugated involutes.

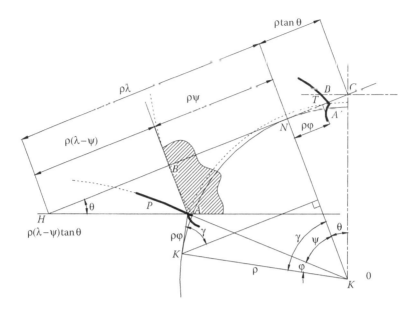

FIGURE 15.30 First contact of the cutter in interference conditions.

and then obtain λ with the equation:

$$(1 - \cos\gamma - \varphi\sin\gamma)\cot g\vartheta + (\gamma - \varphi) = \lambda.$$

In this way the $\varphi(\lambda)$, ϑ-parameterized curves can be described (Figure 15.31).

Knowing λ, we can enter it into the abacus, intersect the curve of the chosen ϑ and then obtain φ, from which the $\rho\varphi$ lost segment is obtained.

λ can be obtained as follows:

$$\rho\lambda = \overline{HN}.$$

Point H shows how much the external parting projects, compared to the tangential spot N; the more the tooth tip projects, compared to the pitch line of the cut rack, the greater the λ will be. The external parting of the tool (rack) juts out of its pitch line as much as u', equal to the dedendum

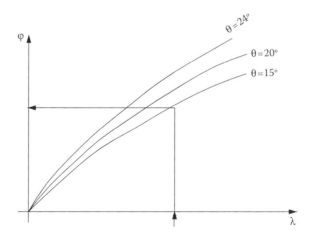

FIGURE 15.31 ϕ versus λ diagram.

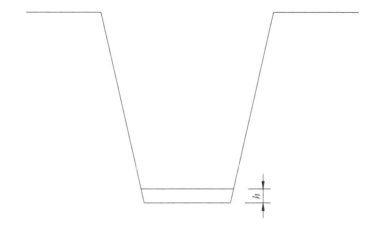

FIGURE 15.32 Cutter's profile.

of the gear being cut. From u', apart from interference effects, projection h must be subtracted, corresponding to the tip tooth fillet, which has no involute profile (Figure 15.32).

Therefore, the projection is

$$u' - h.$$

Expressing λ as function of this quantity, we obtain (Figure 15.33):

$$u' - h = \rho\lambda \sin \vartheta + \rho tg\vartheta \sin \vartheta.$$

therefore,

$$\lambda = \frac{u' - h}{\rho \sin \vartheta} - tg\vartheta,$$

from which, since $\rho = (nm/2)\cos \vartheta$, we obtain

$$\lambda = \frac{4}{n \sin 2\vartheta} \frac{u' - h}{m} - tg\vartheta.$$

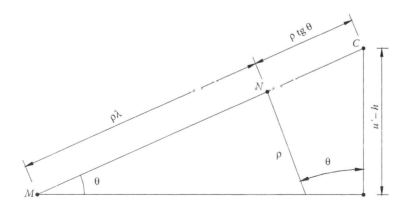

FIGURE 15.33 Relation between λ and the addendum of the cutter.

Knowing λ, we can determine angle φ and so the arc lost due to cut interference. Similar reasoning can be applied when cut is carried out by a cut wheel.

15.15 SPECIFIC CREEP

If two profiles are in contact at point P other than C, and we call ds and ds' the distances covered in infinitesimal time dt from point P of the profile thought to belong to a tooth and from the same point thought to belong to the other tooth, absolute velocity being different, the two infinitesimal spaces will be different.

We call the specific creep of a profile K_s, the ratio between the relative and absolute shift of the point thought to belong to the gear for which creep is being assessed. In other words:

$$K_s = \frac{ds - ds'}{ds} \quad \text{for the first profile,}$$

$$K_s' = \frac{ds' - ds}{ds'} \quad \text{for the second profile.}$$

In the same figure, the absolute velocity vectors of point P (V_p and V_p'), at one time thought to belong to center circle O, at another to the circle O', are perpendicular to segments $\overline{OP}$ and $\overline{O'P}$. Obviously, V_p is perpendicular to $\overline{OP}$, V_p' is perpendicular to $\overline{O'P}$ and the two velocities have the same contact line components equal to

$$V_c / \cos \vartheta \quad (V_c = \omega R = \omega' R'),$$

ω' and ω being the angular velocities of the two gears and V_C being the velocity of the C point.

Different shifts are generated by (different) velocity components on the contact line perpendiculars; so, to calculate (different) shifts and referring only to these components we get

$$ds = V_p \sin \gamma \, dt = \overline{OP}\omega \sin \gamma \, dt,$$

$$ds' = V_p' \sin \gamma' \, dt = \overline{O'P}\omega' \sin \gamma' \, dt.$$

Observe Figure 15.34 and note that

$$\overline{OP} \sin \gamma = \overline{TP} = R \sin \vartheta + \delta,$$

$$\overline{O'P} \sin \gamma' = \overline{T'P} = R' \sin \vartheta + \delta,$$

and then, substituting:

$$ds = (R \sin \vartheta + \delta)\omega \, dt,$$

$$ds' = (R' \sin \vartheta + \delta)\omega' \, dt.$$

Substituting in $K_s = (ds - ds')/ds$, we obtain

$$K_s = \frac{(R \sin \vartheta + \delta)\omega \, dt - (R' \sin \vartheta - \delta)\omega' \, dt}{(R \sin \vartheta + \delta)\omega \, dt}.$$

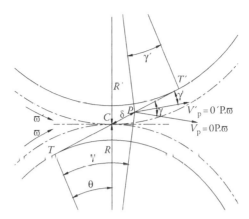

FIGURE 15.34 Velocities of a generic contact point.

Dividing by ω and substituting $\tau = \omega' / \omega$, we obtain

$$K_s = \frac{(R \sin \vartheta + \delta) - (R' \sin \vartheta - \delta)\tau}{(R \sin \vartheta + \delta)},$$

since also $\tau = R / R'$, then

$$K_s = \frac{(1 + \tau)\delta}{R \sin \vartheta + \delta}.$$

Similarly for K'_s, a relation like this will apply:

$$K'_s = \frac{-\left(1 + \dfrac{1}{\tau}\right)\delta}{R' \sin \vartheta - \delta}.$$

In the two former relations, δ must be assumed positive in recess and negative in access.

It may happen that specific creep is continuously variable. Absolute maximum values occur when specific creep is negative, in access for the driving gear, in recess for the driven one.

It might be pointed out that for $\delta = -R \sin \vartheta$ ($\overline{CT}$ segment) $K'_s \to -\infty$. Whereas $K'_s \to -\infty$ for $\delta = R' \sin \vartheta$ ($\overline{CT'}$ segment).

Therefore, if the contact segment is used up to the limit points T and T' where the line intersects the base circles, $K_s = -\infty$ in T and $K'_s = -\infty$ in T'. The opportunity to prevent contact of the involute section closest to the base circle is again confirmed. Indeed, high relative creep means greater wear.

For $\delta = 0$ there is $K_s = 0$ and $K'_s = 0$, as in C the two gears have the same velocity.

For $\delta = R' \sin \vartheta$, corresponding to point T', it follows:

$$K_s = \frac{(1 + \tau)R' \sin \vartheta}{R \sin \vartheta + R' \sin \vartheta} = 1.$$

The diagram in Figure 15.35 shows specific creep for both gears. If creep is to remain below a 4–5 limit, the two external parting circles should not exceed points α and β. Otherwise, the gear will have to be sized down (stubbed).

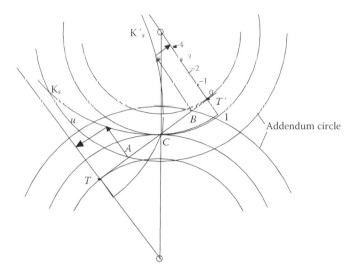

FIGURE 15.35 Specific creep curves.

In fact, visible wear eliminates part of the tooth thickness, generating high temperatures, so the oil becomes contaminated by the heat waste, spoiling the lubricating properties of the oil itself.

The experience of various gear manufacturing companies indicates, as previously maintained, 4–5 for creep values and suggests fairly similar values for the two gearings.

So, specific creep generates material for removal (Figure 15.36):

$$dV = Adh,$$

where dh is the variation of thickness according to the perpendicular of the tooth profile and the area of contact A.

As for maintaining the involute profiles, it is advisable that the wear material is removed uniformly, but this cannot happen. Indeed, according to Reye's hypothesis, the volume V of removed material is proportional to friction work, or

$$dV \propto fApvdt,$$

where f is the coefficient of friction, p is specific pressure, A the contact area, and v the velocity perpendicular to the contact area.

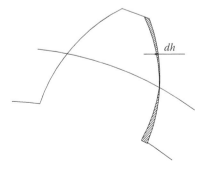

FIGURE 15.36 Qualitative variation of the thickness.

As $dV = Adh$, it follows that

$$dh \propto vdt.$$

In other words, there is proportionality between relative shift, in its turn proportional to K_S and height variation. The heavier specific creep, the further the involute profile is different.

The calculations also help assess output η_m of a pair of gears. The numerator of ks ($k's$) represents relative shift at one point of the contact segment. Transmission ratio τ being equal, it changes linearly with δ.

The friction force results from force N by the coefficient of friction f.

Were there a single tooth meshing ($\epsilon = 1$), the force N would be constant all along the contact arc and elementary friction work would be $dLp = fNd\delta$.

The average lost work would result from the elementary work integral between ($\delta = -R \sin \vartheta$) and ($\delta = R' \sin \vartheta$); in the former case, the lost work would be represented by the area of the two triangles with their vertices in C, and with bases $R \sin \vartheta$ and $R' \sin \vartheta$, and heights $(1 + \tau) R \sin \vartheta$ and $(1 + \tau) R \sin \vartheta$.

The ratio between lost work and motion work, which for gears practically coincides with effective work, represents loss factor which, in this case (arc of action equal to the pitch) is

$$1 - \eta_m = \pi f \left(\frac{1}{n} + \frac{1}{n} \right).$$

When more pairs of teeth mesh, force N is distributed over more teeth and the integral used to determine lost work must consider the segments where there are one or more teeth meshing, as well as the consequent load distribution.

By acting thus, an expression similar to the previous one would be written where the second member would be multiplied by factor K, a function of access and recess contact segments. This factor is quite close to 1, so in a practical assessment of the η_m output of each pair of gears, refer to the above formula.

15.16 MODIFICATED GEARS

Designing a pair of gears is based on determining how many teeth are needed for a given transmission ratio and on operational distance D (the sum of the two gear radii). Once these kinematics have been defined, all the necessary strength tests are carried out.

If gears without cutting interference need to be cut with fewer teeth than those the tooth cutter is able to cut, which is defined by the following:

$$n_{min} \geq \frac{2K'}{\sin^2 \vartheta_0} \geq \frac{2 \dfrac{a'}{m_0}}{\sin^2 \vartheta_0},$$

(ϑ_0 and m_0 known tooth cutter parameters), then they would have to be cut ad hoc at high cost (for each gear with the same angle and pitch but with a different tooth number, a tooth cutter would have to be made).

If a sprocket is cut with a number of teeth as per the above equation then the gearing is normal and the operational inter-axle distance coincides, apart from the clearance necessary for lubrication (explained below), with the sum of the radii of the pitch circles cut gears.

However, the tooth cutter can still be used for cutting by resorting to adjusted gear cutting. In this way gears with fewer teeth can be cut without cutting interference and without substantially reducing the coating factor, while at the same time obtaining gears with robust teeth.

In the formula above, note that if the tooth cutter is used for cutting, reducing n_{min} can only be done by modifying a'. Thus the cutting addendum and dedendum values must be changed, but overall tooth height remains the same.

The displacement comes about by distancing the gear center by the amount necessary to avoid interference.

In Figure 15.37, point C, the pitch cut position, is displaced driven by the same velocity thought to belong to the cutter than to the gear. When the gear slows, despite the velocity of the cutter, there is a displacement of pitch from the gear center. On the other hand, when the gear speeds up the pitch displaces in the opposite direction.

In any case, the pitch is a straight parallel to aa (the straight line reference) at xm_0 from it in one direction or the other. In particular, xm_0 becomes positive if the cutting pitch nears the tip of the cutter, negative if it distances from the cutter tip.

As xm_0 displaces, the addendum of cutter a' changes from m_0 to $m_0(1 - x)$.

$$n \geq \frac{2(1-x)}{\sin^2 \vartheta_0},$$

$$a' = m_0(1 - x) = K'm_0.$$

By substituting $K' = (1 - x)$ in $n_{min} = 2K'/\sin^2 \vartheta_0$, then

$$n_{min} = \frac{2(1-x)}{\sin^2 \vartheta_0} = n_0(1 - x),$$

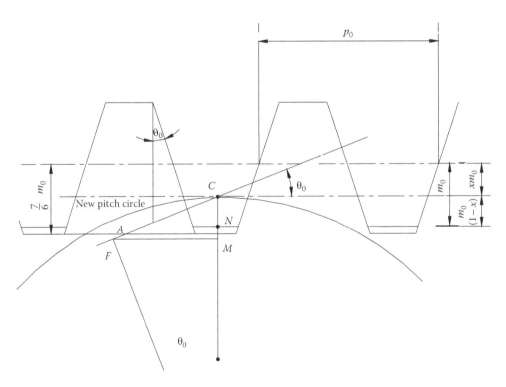

FIGURE 15.37 Position of the pitch cut in modificated gears.

with $n_0 = 2/\sin^2\vartheta_0$.

From the reasoning in paragraph 15.13 and from Figure 15.37, note that non-interference is $\overline{TC} \geq \overline{AC}$ transmuting into $\overline{MC} \geq \overline{NC}$ where $\overline{MC}$ always equals $\rho t g \vartheta_0 \sin \vartheta_0$, whereas $\overline{NC}$ equals $m_0(1-x)$ (see Figure 15.37) so:

$$\rho t g \vartheta_0 \sin \vartheta_0 \geq m_0(1-x),$$

from which

$$R_0 \frac{\sin \vartheta_0}{\cos \vartheta_0} \cos \vartheta_0 \sin \vartheta_0 \geq m_0 \left(1-x\right),$$

and finally

$$n_{\min} \geq \frac{2}{\sin^2 \vartheta_0}.$$

Extracting x:

$$x = 1 - \frac{n}{n_0},$$

with $n_0 = 2/\sin^2 \vartheta_0$ ($K' = 1$).

The most general case of making modified gears is when for the sprocket cutting and gear cutting have different displacement values of the cutter. The two gears obviously will have a different tooth thickness compared to their normal cutting pitch.

Cinematically ideal meshing (without clearance nor allowance) of the two gears is when the sum $(s_1 + s_2)$ of the tooth thickness at the operational pitch circle is equal to the circular pitch p (as we shall see, minimal clearance is allowed for lubrication so as not to overload the bearings and therefore the shafts). This leads to being able to position the two gear axles correctly.

When the gear teeth are cut thus, they come into contact during meshing, so their cutting pitches do not coincide with the working pitches. This means that radii R and R' of the working pitches do not coincide with radii R_0 and R'_0 of the cutting pitch circle.

Although it is impossible to have meshing with a sum of the tooth thickness larger than the circular pitch (there would be penetration) it *is* possible (but not correct) to create an only cinematically correct mesh (involute profiles) in which the sum of the tooth thickness is less than the circular pitch. In this case, there would be impact and subsequent tooth breakage.

Since the radius of the base circle is a characteristic of the gear manufacture, a variation of the working pitch circle leads to a variation in pressure angle ϑ, which, not coinciding with the cutting pitch circle, is different to the pressure angle ϑ_0 of the cutter.

In Figure 15.38, the radii of the basic circles of the two gears are

$$\rho = R_0 \cos \vartheta_0 \quad \rho' = R_0' \cos \vartheta_0,$$

and since ρ never changes:

$$\rho = R \cos \vartheta \quad \rho' = R' \cos \vartheta$$

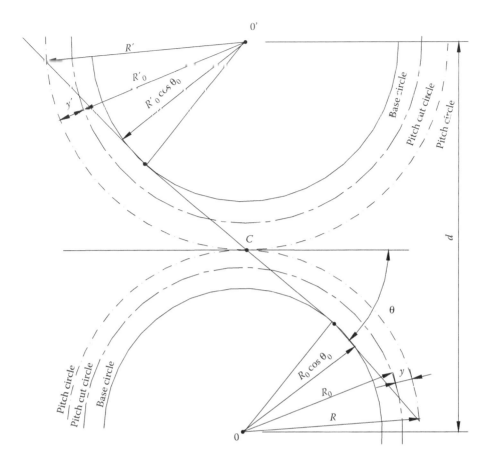

FIGURE 15.38   Gear layout for modified gears.

The pitch circle radii with the two gears enmeshed are

$$R = \frac{\rho}{\cos\vartheta} = R_0\,\frac{\cos\vartheta_0}{\cos\vartheta} \quad R' = \frac{\rho'}{\cos\vartheta} = R'_0\,\frac{\cos\vartheta_0}{\cos\vartheta}.$$

The consequent circle pitch is

$$p = \frac{2\pi R}{n} = \frac{2\pi R'}{n'} = \pi m = \pi m_0\,\frac{\cos\vartheta_0}{\cos\vartheta}.$$

Correct meshing (without clearance or penetration) of the two gears is

$$s + s' = \pi m = \pi m_0\,\frac{\cos\vartheta_0}{\cos\vartheta},$$

in which s and s' are the tooth thickness at the operational pitch circle of the two enmeshed gears. Expressing s and s' as functions of the cutting characteristics, the operational pressure angle ϑ can thus be derived.

The general formula for tooth thickness in paragraph 15.7 page 356 is:

$$s = 2r(ev\gamma - ev\vartheta).$$

In this case, it must refer to the radii R and R' of the pitch circles. The values of $ev\gamma$ and $ev\gamma'$ are also gear "construction factors" (as mentioned above), whereas ϑ is an operational characteristic.

The construction parameters calculation must obviously refer to the cutter with pitch p_0 and pressure angle ϑ_0.

The operational conditions will be

$$s = 2R\left(ev\gamma - ev\vartheta\right) = \frac{2nm}{2}\left(ev\gamma - ev\vartheta\right),$$

$$s' = 2R'\left(ev\gamma' - ev\vartheta\right) = \frac{2n'm}{2}\left(ev\gamma' - ev\vartheta\right).$$

By substituting in $s + s' = \pi m$, we obtain the equation for the operational pressure angle. The values of $ev\gamma$ and $ev\gamma'$ are obtained from the cutting condition.

For construction on the cutting pitch circle the tooth thicknesses (s_0, s_0') are

$$\begin{cases} s_0 = \dfrac{\pi m_0}{2} + 2xm_0 tg\vartheta_0 \\ s_0' = \dfrac{\pi m_0}{2} + 2x'm_0 tg\vartheta_0 \end{cases}.$$

But so too

$$\begin{cases} s_0 = 2R_0\left(ev\gamma - ev\vartheta_0\right) = nm_0\left(ev\gamma - ev\vartheta_0\right) \\ s_0' = 2R_0'\left(ev\gamma' - ev\vartheta_0\right) = n'm_0\left(ev\gamma' - ev\vartheta_0\right) \end{cases}.$$

Equalizing the two expressions we obtain $ev\gamma$ and $ev\gamma'$:

$$ev\gamma = ev\vartheta_0 + \frac{2}{n}\left(\frac{\pi}{4} + xtg\vartheta_0\right) \quad ev\gamma' = ev\vartheta_0 + \frac{2}{n'}\left(\frac{\pi}{4} + x'tg\vartheta_0\right),$$

and by substituting into the ideal meshing condition $(s + s' = p)$, we obtain

$$ev\vartheta = ev\vartheta_0 + 2tg\vartheta_0 \frac{x + x'}{n + n'}.$$

Thus, ϑ leads to the module:

$$m = m_0 \frac{\cos\vartheta_0}{\cos\vartheta},$$

and therefore the pitch circle radii:

$$R = \frac{1}{2}nm = \frac{1}{2}nm_0\frac{\cos\vartheta_0}{\cos\vartheta} \quad R' = \frac{1}{2}n'm = \frac{1}{2}n'm_0\frac{\cos\vartheta_0}{\cos\vartheta}.$$

The center distance is the sum of the pitch radii:

$$d = R + R',$$

which is the inter-axle design value to ensure good gear meshing.

One case envisaged by the norms is when $x = -x'$. This is a special case, but the necessary formulae may be derived from those above.

Note that the operational pitch circles when the two gears mesh, coincide with the cutting pitch circles or rather with those defined by the cutter. Correct meshing is $s + s' = p$ that leads to

$$s_0 + s_0' = \pi m_0.$$

The transmission ratio limit $\tau_{\text{lim}} = n/n'$ with the minimum number of teeth at the interference limit between gear and sprocket for a certain displacement is

$$\tau_{\text{lim}} = \frac{1-x}{1+x}.$$

The most widespread use of this ideal meshing $(x = -x')$ is given according to Lasche when $x = 0.5$, so

- For the small gear $a = 1.5$ m and $u = 0.7$ m
- For the large gear $a' = 0.5$ m and $u' = 1.7$ m

In this case τ_{lim} is

$$\tau_{\text{lim}} = \frac{1-0.5}{1+0.5} = \frac{1}{3}.$$

Going back to the general case $(x \neq x')$, to calculate the arc of action and other characteristics of the mesh, the addendum and dedendum of the gears must be known.

Referring to the symbols in Figure 15.39, the dedendum of the gear with center O is given by the sum of y and z. The addendum of the other gear (center O') is determined to have a radial clearance of $1/6\, m_0$ between this value and the dedendum gear of center O.

The radius difference between the operational pitch circle and the cutting pitch circle is

$$y = R - R_0 = \frac{R_0 \cos \vartheta_0}{\cos \vartheta} - R_0 = \frac{n}{2}\left(\frac{\cos \vartheta_0}{\cos \vartheta} - 1\right)m_0,$$

$$y' = R' - R_0' = \frac{R_0' \cos \vartheta_0}{\cos \vartheta} - R_0 = \frac{n'}{2}\left(\frac{\cos \vartheta_0}{\cos \vartheta} - 1\right)m_0.$$

Since

$$\eta = \frac{n}{2}\left(\frac{\cos \vartheta_0}{\cos \vartheta} - 1\right) \quad \eta' = \frac{n'}{2}\left(\frac{\cos \vartheta_0}{\cos \vartheta} - 1\right),$$

we obtain

$$y = \eta m_0 \quad y' = \eta' m_0.$$

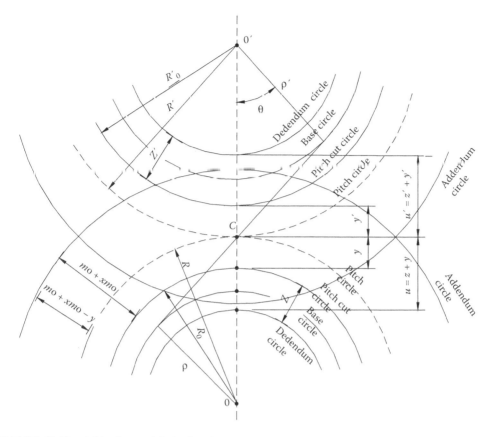

FIGURE 15.39 Addendum and dedendum in modified gears.

The distance between the cutting pitch circle and the dedendum circle is

$$z = \frac{7}{6}m_0 - xm_0 = \left(\frac{7}{6} - x\right)m_0 \quad z' = \frac{7}{6}m_0 - x'm_0 = \left(\frac{7}{6} - x'\right)m_0.$$

Consequently, the radial distances between the operational pitch circle and the dedendum circle (dedendum) are obtained from the relations:

$$u = z + y = \left(\frac{7}{6} - x\right)m_0 + \eta m_0 = \left(\frac{7}{6} - x + \eta\right)m_0,$$

$$u' = z' + y' = \left(\frac{7}{6} - x'\right)m_0 + \eta' m_0 = \left(\frac{7}{6} - x' + \eta'\right)m_0.$$

So, allowing a radial clearance of $1/6\,m_0$ between the addendum circle of the pinion and the dedendum circle of the other gear, the addenda are obtained by subtraction this value from the dedenda:

$$a = u' - \frac{1}{6}m_0 = \left(1 - x' + \eta'\right)m_0 \quad a' = u - \frac{1}{6}m_0 = \left(1 - x + \eta\right)m_0.$$

Thus the gear addenda have been calculated as long as there is no interference (i.e., a tooth tip does not touch the dedendum circle of the other gear).

The verification of this must be done referring to cutting conditions and particularly to addendum of the gear in this condition that defines fiscally the constructed gear

Referring to the tooth cutter, from the pitch circle there would be a clearance of $m_0 + xm_0$ (Figure 15.40).

Thus, for the construction conditions, the addenda are

$$\bar{a} = \left(1 + x - \eta\right)m_0 \quad \bar{a}' = \left(1 + x' - \eta'\right)m_0.$$

Ensuring no interference, we need

$$\bar{a} \leq a \quad \bar{a}' \leq a'.$$

In other words, it must be verified that the addenda thus obtained are less or at most equal to the addenda defined by operational conditions.

If the above conditions are not satisfied the teeth would have to be lowered, i.e., the external gear radii would have to be shortened (stubbing) so that teeth height would be reduced by at least:

$$\bar{a} - a = \bar{a}' - a' = \left[\left(x + x'\right) - \left(\eta + \eta'\right)\right]m_0 = \left[\left(x + x'\right) - \xi\right]m_0 = \chi m_0.$$

The radii of the addendum circles will be

$$R_e = R + a \quad R_e' = R' + a'.$$

The contact ratio is expressed by

$$\varepsilon = \frac{1}{2\pi}\left[\sqrt{\left(\frac{n' + 2K'}{\cos\vartheta}\right)^2 - n'^2} + \sqrt{\left(\frac{n + 2K}{\cos\vartheta}\right)^2 - n^2} - \left(n + n'\right)tg\vartheta\right],$$

with $K(a = m)$ and $K'(a' = K'm)$ equal to

$$K = \left(1 - x' + \eta'\right)\frac{\cos\vartheta}{\cos\vartheta_0} \quad K' = \left(1 - x + \eta\right)\frac{\cos\vartheta}{\cos\vartheta_0}.$$

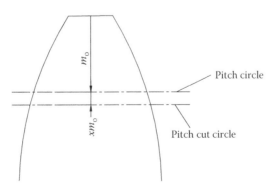

FIGURE 15.40   Position of the pitch cut circle.

Generally, when tooth profiles are produced by the cutter with a certain interference the result is the tooth profile wears away which means the loss of a contact segment equal to (Figure 15.41):

$$\overline{TA} = \rho\phi \quad \overline{T'A'} = \rho'\phi',$$

with ϕ and ϕ' obtainable from the diagram in Figure 15.31.

The contact ratio can be calculated from the formula above reported only if the corresponding addendum circle meets the action within the segment $\overline{AA'}$.

Otherwise, the K and K' values for the action arc formula would have to be those when the addendum circles pass through A and A'.

The action segment can be calculated more simply:

$$\overline{TT'} - \left(\rho\phi - \rho'\phi'\right),$$

and therefore the corresponding arc.

It should be noted that gears cannot be cut with fewer than eight teeth per $\vartheta = 15°$ and seven per $\vartheta = 20°$, because the teeth would be pointed, i.e., zero tip width.

To verify the sprocket tooth tip width where $r = nm_0/2 + (1 + x)m_0$, $s = 2r(ev\gamma - ev\vartheta_e)$ would have to be applied taking into account that radius r corresponding to pressure angle ϑ_e is defined by

$$r \cos \vartheta_e = \frac{nm_0}{2} \cos \vartheta_0,$$

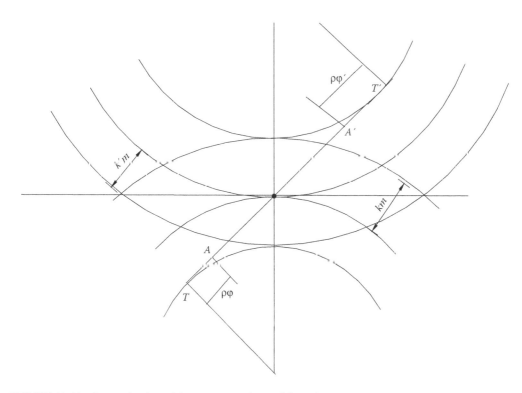

FIGURE 15.41   Determination of the contact arc in modificated gears.

and so as we have already seen:

$$ev\gamma = ev\vartheta_0 + \frac{2}{n}\left(\frac{\pi}{4} + n\,tg\,\vartheta_0\right).$$

To allow lubrication, in practice some inter-teeth clearance is needed. This is obtained by very slightly distancing the axles by Δd.

If ΔR is the radial displacement of the gear center, tooth width diminution (in the direction of the pressure line) is $2{\cdot}\Delta R{\cdot}\sin\vartheta$ for one gear and $2{\cdot}\Delta R'{\cdot}\sin\vartheta$ for the other.

The perpendicular clearance at the surface of the meshing gears is

$$\Delta s = 2\sin\vartheta\left(\Delta R + \Delta R'\right).$$

Given:

$$\Delta d = \Delta R + \Delta R',$$

then

$$\Delta s = 2\Delta d\sin\vartheta.$$

Given Δs from case to case:

$$\Delta d = \frac{\Delta s}{2\sin\vartheta}.$$

The inter-axle distance is $R + R'$ with the addition of Δd defined by Δs. If for example a 100 micron oil film (Δs) was required between the teeth the formula would provide Δd.

If d is known, Δs is fixed and $\Delta d = \Delta s / 2\sin\vartheta$, in which ϑ is approximated to ϑ^* (maybe equal to ϑ_0 or greater).

For meshed gears with zero clearance, inter-teeth distance is

$$d - \Delta d = \frac{1}{2}\frac{\cos\vartheta_0}{\cos\vartheta}\left(n + n'\right)m_0.$$

Given d, Δd, ϑ_0, n, n', and m_0, we obtain ϑ, and with

$$ev\vartheta = ev\vartheta_0 + 2tg\vartheta_0\frac{x + x'}{n + n'},$$

we obtain $(x + x')$, which conditions the displacement values calculated according to pre determined criteria. If the starting value of ϑ^* is different to deduced ϑ, the calculation must be redone by tweaking x and x' and using ϑ as the new ϑ^* until two successive iterations converge on the same value. So, with the latest ϑ all the other parameters are calculated. If the sum of x and x' are always made to be equal, iterating for ϑ can be done using inter-axle d.

There are various ways of finding x and x', so from

$$ev\vartheta = ev\vartheta_0 + 2tg\vartheta_0\frac{x + x'}{n + n'},$$

note that it combines x, x', n, n', ϑ_0, ϑ, of which usually ϑ_0, n, n' are known. The other parameters are defined by imposing two conditions to obtain the best proportions.

According to Schiebel, the two conditions are

- Exploiting contact over all the sprocket tooth surface assuming an involute profile
- Equal tooth width at the involute points closest to the base circles.

The first condition leads to defining the approach part of the pressure segment and to

$$(R + a')^2 - R'^2 + (R \sin \vartheta - \rho\varphi)^2 + 2R'(R \sin \vartheta - \rho\varphi)\sin\vartheta$$

The second condition leads to equalling the widths at the start of the involute, which in the case of non-interference becomes:

$$\rho ev\gamma = \rho' ev\gamma'.$$

Together with the previously reported $ev\vartheta$, these two equations lead to ϑ, x, x'. To make sizing simpler, Schiebel calculated x and x' for the case of $\vartheta_0 = 15°$, for 6–160 sprocket teeth, and for 8–160 large gear teeth. Other authors, depending on specific gear use, have imposed different conditions that lead to different equations combining $ev\vartheta$ to find better proportions.

15.17 GERMAN NORMS ON MODIFICATED GEARING

The concepts described above can be found in German norms governing the proportions of modificated gearing.

The German norms provide a simple method of calculating the x and x' displacement for the cutting of sprocket and gear. They assume the tool is a rack cutter and the two gears are known as gear zero and gear V:

- For gear zero, $x = 0$, $x' = 0$ (i.e., no modification)
- For gear V, $x \neq 0$, $x' \neq 0$.

Meshing different gears, the following could happen:

- Zero—zero pair (result of meshing two zero gears)
- Zero—V pair (result of meshing two V gears, one with displacement x, the other $x' = -x$)
- V pair (result of meshing 2 V gears with displacements x and $x' \neq -x$, especially when the large gear is $x' = 0$)

In the first two cases the inter-axle distance of the two gears with zero clearance is equal to the half-product of module m_0 of the cutter by the sum of the two gears' teeth:

$$d = \frac{1}{2}\left(n + n'\right)m_0.$$

In the third case (V pair), the displacement values are defined by supposing a cutter displacement at the limit of interference condition where $x = 1 - n/n_0$.

As long as $n_0 = 30$ for $\vartheta_0 = 15°$ and $n_0 = 17$ for $\vartheta_0 = 20°$, then strictly speaking, the values resulting from the x and x' displacements at the limit of interference should be for $\vartheta_0 = 15°$

$$x \geq \frac{30 - n}{30} \quad x' \geq \frac{30 - n'}{30},$$

$\vartheta_0 = 20°$

$$x \geq \frac{17 - n}{17} \quad x' \geq \frac{17 - n'}{17}.$$

Since the involute part closest to the fundamental circle needs to be eliminated (as explained above), it is achieved by having slight interference in the cutting phase at the expense of a slight reduction in contact ratio. For the slight interference, the value of the displacement (numerator) in cutting phase is reduced according to the experience of constructors:

for $\vartheta_0 = 15°$

$$x \geq \frac{25 - n}{30} \quad x' \geq \frac{25 - n'}{30},$$

for $\vartheta_0 = 20°$

$$x \geq \frac{14 - n}{17} \quad x' \geq \frac{14 - n'}{17}.$$

Reducing x compared to that defined at the limit of cutting interference determines the interference itself (cutter addendum increased) with the consequent removal of a part of the involute during gear manufacture. As we have already seen, to have meshing without clearance, the axles of the V pair should be set at a distance calculable as functions of $n + n'$, $x + x'$, ϑ_0, and m_0 using

$$ev\vartheta = ev\vartheta_0 + 2tg\vartheta_0 \frac{x + x'}{n + n'},$$

$$R = \frac{1}{2} nm_0 \frac{\cos \vartheta_0}{\cos \vartheta} \quad R' = \frac{1}{2} n'm_0 \frac{\cos \vartheta_0}{\cos \vartheta} \quad d = R + R'.$$

Defining x and x' first, the sum of $x + x'$ turns out to be proportional to $n + n'$ and so the inter-axle distance is

$$d = \left(\frac{n + n'}{2} + \xi \right) m_0,$$

where ξ is expressed by the equation seen above:

$$m_0 \xi = \left(\eta + \eta' \right) m_0 = y + y'.$$

German norms provide a chart of χ and ξ, which have been cited in the previous pages as a function of $n + n'$. Table 15.2 shows part of the chart (see Giovannozzi, Vol. II):

The chart values are only valid if both are V gears since only in the V–V pair are χ and ξ functions of $n + n'$.

This necessarily happens when both gears have fewer teeth than those (25 or 14) prescribed by the norms.

TABLE 15.2

$n + n'$	χ		ξ	
	$\vartheta = 15°$	$\vartheta = 20°$	$\vartheta = 15°$	$\vartheta = 20°$
50	0.000		0.000	
48	0.002		0.065	
46	0.008		0.125	
44	0.014		0.186	

Should one gear have more teeth it can be compensated for by a negative displacement:

$$x' = \frac{25 - n'}{30} \quad x' = \frac{14 - n'}{17},$$

but also more simply as a zero gear, a displacement of $x' = 0$.

In the first case of "more teeth," the chart applies, but in the second "zero gear" it no longer holds that χ and ξ are functions of $(n + n')$ and so the general formula (above) must be applied.

Summarizing, once the transmission ratio and cutter (m_0 and θ_0) are chosen, the norms help calculate the number of teeth n of the smaller gear for a given displacement xm_0. Knowing τ, n^1 of the larger gear can be calculated and so too x^1, then by entering with $(n + n')$ into the chart, the inter-axle distance d can be calculated by

$$d = \left(\frac{n + n'}{2} + \xi \right) m_0 \quad \text{and the stubbing } \chi m_0.$$

15.18   LOADS WHEN ENGAGING GEARS

Calculation of mesh resistance and the corresponding shaft load (friction loads are disregarded), and consequently the direction of the force transmitted by the teeth, follows the pressure line of action and therefore passes through C. So, if the teeth are involute, the direction of this force is constant and coincides with the straight line of contacts (Figure 15.42).

Peripheral forces Q being equal, overall force N acting on a tooth is maximum when the inclination of its line of action is maximum on the plane at a tangent to the pitch circles (pitch cylinders).

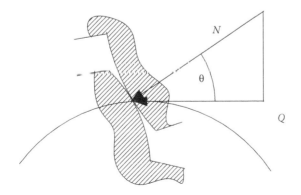

FIGURE 15.42   Load for engaging gears.

Given θ, the operational pressure angle, force N transmitted between teeth is

$$N \quad \frac{Q}{\cos \vartheta},$$

This force also has a radial component R equal to

$$R = Qtag\vartheta.$$

Forces N, Q, and R are uniformly distributed over the axial width of the tooth.

15.19 VERIFICATION CALCULATIONS FOR GEAR RESISTANCE

Calculating gear resistance is still based on simple models to which the norms also refer.

The first issue is defining the forces acting on each pair of meshing teeth. If the contact ratio is more than 1 which should be the case for the correct operation of two enmeshed gears, then more than one pair of teeth are in contact, at least for a certain enmeshing span, so allotting total forces N to the various enmeshed teeth pairs, and taking into account all the factors in that allotment, is not always a simple calculation. Apart from the static non-determination of the problem, all the deformation components (elastic, Hertzian, oil pressure, thermal) including manufacturer's errors should be taken into account, which complicates the issue without however providing a reliable evaluation. The difficulty of an exact evaluation of the load between the teeth ends up being due to spur gears (Figure 15.43). In this case when the contact ratio is between 1 and 2, from $-\varepsilon_1$ to $(\varepsilon_2 - 1)$ and from $(1 - \varepsilon_1)$ to ε_2, two pairs of teeth will be in contact. In the central part of the pressure segment, from $(\varepsilon_2 - 1)$ to $(1-\varepsilon_1)$ only one pair of teeth will be in contact. For value of ε between 2 and 3 there should be part of the action arc with three pairs of teeth in contact and part with two pairs (near the center).

It should be noted that the design scale in Figure 15.43 is $p \cos \vartheta$: segments of length 1 are the pitch base circle.

Again referring to the simplest case, note that when $\varepsilon = \varepsilon_2$ (point B), total force has most leverage at the tooth base where we assume it is fixed, but at pitch distance along the pressure line another tooth A' is enmeshed and so not all the perpendicular force $N = Q/\cos \vartheta$ is exerted on the extreme tooth, only a part. Only between A' and B' is a single tooth enmeshed, so it takes all the force of $N = Q/\cos \vartheta$. In B' (worst condition for the central part), the leverage of N is already much lower compared to contact of the two teeth at the extremity of the pressure segment. The calculation can be carried out for maximum leverage but with the load shared across two teeth or maximum load but with less leverage.

Ideally, with perfect manufacture, calculating the theoretical load allotment assumes that the sums of the elastic displacement of points (due to loads, lubrication, Hertzian stresses etc.) along the pressure line have the same value for all the enmeshed tooth pairs. This calculation, which disregards lubrication deformation, was carried out by Karas with interesting results but not easily applicable to a large number of teeth. In any case, the main principle is that the resistance calculation must take into account load variation on the teeth.

As mentioned above, the norms have simplified the situation in which it is supposed that a single tooth takes all the load N with maximum leverage (Lewis's hypothesis) or with leverage relative to contact at the extreme point of contact of a single pair (BSS norms). Once the force acting on the tooth is defined, the calculation consists in verifying the tooth's breaking at bending and wear due to local contact compression. In time, even with the same basic formula, the various norms have adopted coefficients that take into account all the factors involved in tooth durability. Any corrective measures have evolved with technological and scientific progress.

15.19.1 CALCULATING TOOTH BENDING

As touched on earlier, calculating the breaking point of a tooth due to bending according to Lewis's work at the end of the nineteenth century involved the hypothesis that a single tooth took

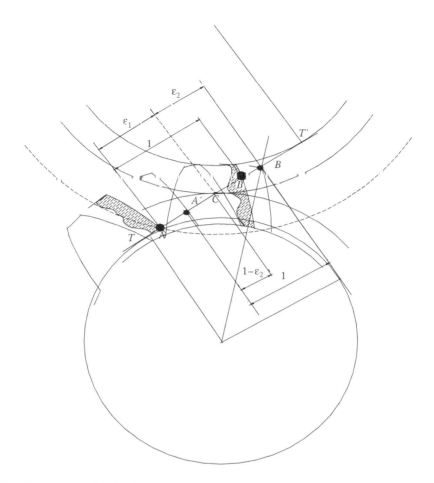

FIGURE 15.43 Contact of the teeth during the engaging.

all the force $N = \cos \vartheta$ applied to its tip. In his model Lewis schematized the tooth as a cantilever fixed at its base to which he applied Saint-Venant's theory to calculate the stress. The calculation was static and took no account (in the first versions) of any dynamic effects; in more recent applications of the same method adopted by various norms these effects are accounted for by special coefficients that reduce the admissible tension or increase nominal acting load, coefficients typical of fatigue design. It will not go unnoticed that the method has its drawbacks even in its basic premise (a tooth, which in reality is a solid stump, is considered as a Saint-Venant beam). However, the method's validity is justified by its simplicity of application and by the validity of its practical results.

When contact occurs at the tooth tip, due to the involute's properties, the angle that force N forms with the perpendicular to the tooth axis is $\beta \neq \vartheta_e$ relative to the involute extreme (Figure 15.44).

To calculate β from the figure:

$$\beta = \vartheta_e - \left(ev\gamma - ev\vartheta_e \right),$$

and since $ev\vartheta_e = tg\vartheta_e - \vartheta_e$, then

$$\beta = \vartheta_e - \left[ev\gamma \quad \left(tg\vartheta_e - \vartheta_e \right) \right] = tg\vartheta_e - ev\gamma.$$

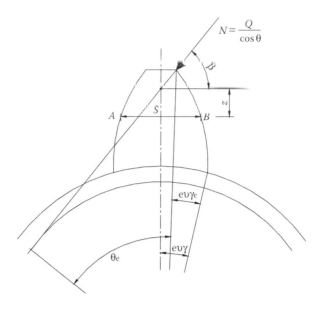

FIGURE 15.44   Extreme position of the load.

A generic tooth cross-section AB of width s with anomaly z is stressed by:

- a shearing force of $T = Q \cos \beta / \cos \vartheta$
- a compression force of $R = T tg\beta$
- A bending moment of $M = Tz$

According to Saint-Venant, the normal stress is

$$\sigma(z) = \pm \frac{M}{I} y + \frac{R}{bs}.$$

In this case, $W = I/y$ equals $1/6 \cdot bs^2$, where b is tooth face width. So, substituting:

$$\sigma(z) = \pm \frac{6Tz}{bs^2} + \frac{T tg\beta}{bs},$$

or

$$\sigma(z) = \frac{T}{bs}\left[\pm \frac{6z}{s} + tg\beta \right].$$

Neglecting tension due to the compression force and assuming $\beta = \theta$ then:

$$\sigma(z) = \frac{6Qz}{s^2 b}.$$

If the tooth had a constant cross-section, the maximum value of σ would be assumed for the fixed end of the tooth (at the base). Generally, maximum tension is in the section in which a parabola of type $z = Ks^2$ ($K = $ const, the axis originating at point M and coinciding with the tooth axis), is at a tangent to the tooth profile itself or intersects it (Figure 15.45).

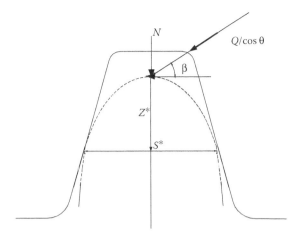

FIGURE 15.45 Model for Lewis calculation.

Having treated the tooth as a Saint-Venant solid, the parabola represents the tooth profile (solid), so in all the sections (z) there would be the same stress.

At every point on the parabola and at every z level there is the same σ. Outside of these there are values higher than σ. In practice, the most stressed section is next to the fixed end. Only in special cases is it close to the top of the tooth where its width can become so small as to compensate for the reduced distance of force Q.

If z^* and s^* are the values of z and s for the most stressed section:

$$\sigma_{max} = \frac{6Qz^*}{s^{*2}b}\frac{m}{m}.$$

Making:

$$y = \frac{1}{6}\frac{s^{*2}}{z^*m},$$

we obtain Lewis's formula:

$$Q = ybm\sigma,$$

in which y becomes the Lewis coefficient being a function of pressure angle ϑ, proportioning and tooth number.

Lewis made a table of y as a function of ϑ and n for proportioning normal and stubbing of teeth. Obtaining the modified teeth means that y must be found case by case identifying the anomaly z^* of the most stressed section and calculating the relative value of s^*.

The formula above may transform itself at the design stage such that knowing the power and rpm n, teeth module m must be calculated. In fact, the peripheral force is

$$Q = \frac{M}{R} = \frac{M \cdot 2}{nm},$$

introducing the ratio between the face width and the module:

$$\lambda = \frac{b}{m},$$

with λ assumed between 4 and 12, such that face width is neither too much limited (otherwise high pressures would be generated—contact pressures as high as 2000 bar), nor excessive, because during operation it would be more difficult to maintain contact over the whole width of the tooth as a result of poor manufacture. However, for less loaded and slower λ could be between 6 and 8.

Substituting $Q = ybm\sigma$ into $Q = 2M/nm$, we obtain

$$\frac{2M}{nm} = \sigma\lambda m^2 y,$$

from which

$$m^3 = \frac{2M}{n\sigma\lambda y}.$$

Then making

$$K = \sqrt[3]{\frac{2}{ny}},$$

finally:

$$m = K\sqrt[3]{\frac{M}{\lambda\sigma}}.$$

As to the σ values to use for the calculation, in Lewis's original work they were not given any particular attention because at that time very little was known about machine component fatigue. As knowledge grew, correction coefficients were found that took into account important factors such as surface condition, meshing velocity, tooth size, degree of lubrication, heat treatment, etc. So, the values of allowable stress must be reduced in order to determine the coefficients of the norms. Manufacturers, especially in the past, used conventional admissible tensions, based on experience, which they often failed to disclose.

The value of m obtained by applying the formula above should be compared to the next highest unified in the norms. So, the calculation is reiterated having corrected allowable stress σ due to peripheral velocity v, itself a function of the module, until for two subsequent iterations an m is found that is less than the one assumed by the norm. Convergence towards the exact module is ensured by the fact that stress σ in the formula is below the cube root and so its influence on the module is slight. This also highlights that dimensions (proportional to m) vary very little with load.

The durability of a gear mesh depends greatly on the factors above. The calculation must take into account stress fatigue and its variation over time, cycle number, and the coefficients of notch sensitivity that depends not only on the size of the tooth base but also on the surface conditions and the type of heat treatment. For a first approximation, a reduced allowable stress bending is used according to coefficients that take into account the various factors. Current norms through tables and curves, apply various coefficients analytically to the base formulae, which, according to the type of mesh (often classified by mesh velocity), take into account the factors above. In the first formulation of Lewis's method, the factors were accounted for more empirically, applying formulae that either increased stressing load (like impact effect) or reduced admissible tension due to the cyclical load (fatigue).

The formulae below take into account these two effects and can arrive at a first approximation dimensioning that must then be mediated by the norms before being put into practice.

According to ASME (American Society of Mechanical Engineers), load increase may be calculated thus:

$$\Delta Q = \frac{AV}{(V + 0.15)\sqrt{A}} \quad \text{since} \quad A = (Q + Ceb),$$

with peripheral velocity V in meters per second.

Coefficient C takes into account the type of material and proportioning, whereas e takes account of surface conditions that are a function of various applications (slow and fast gears).

To take velocity into account (in practice, cycle number), the literature uses several formulae. Here we report those mostly used for high velocities. The coefficient of the reduction α of allowable stress is given by

$$\alpha = \frac{5.6}{5.6 + \sqrt{V}}.$$

These formulae, however, have been surpassed by various norms (BSS, AGMA, DIN, UNI) but can quantitively highlight the effects of the contributory parameters then concluding that it is more opportune to refer to the norms.

15.19.2   Teeth Wear Calculation

Visual inspection of out-of-use gears nearly always shows that this is due to excessive wear. Magnifying the teeth shows a variation in profile that is no longer involute but shows irregularities mostly in the central zone. So, the bending calculation alone is insufficient for dimensioning of the teeth; local Hertzian pressure between the teeth should also be accounted for as in Reye's hypothesis as the cause of the tooth wear. Approximating the surfaces of two gears to two cylinders in contact along their generatrixes, making b the length of the smallest gear (pinion), with $Q/\cos \vartheta$ the force along pressure line, the contact stresses are

$$\sigma_{max} = 0.417\sqrt{\frac{QE\rho}{b\cos\vartheta}}.$$

In the case of gears with external teeth (Figure 15.46), the relative curvature is

$$\rho = \frac{1}{TM} + \frac{1}{MT'},$$

given T and T' the two involute centers of curvature.

If δ is the coordinate of point of contact M from the center of instantaneous rotation C, then

$$\overline{TM} = R\sin\vartheta + \delta \quad \overline{MT'} = R'\sin\vartheta - \delta,$$

and substituting into ρ, we get

$$\rho = \frac{(R + R')\sin\vartheta}{(R\sin\vartheta + \delta)(R'\sin\vartheta - \delta)}.$$

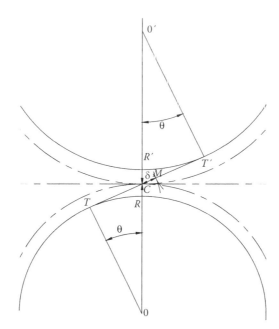

FIGURE 15.46 Scheme for wear calculation.

Analyzing the ρ curves as a function of δ shows how similar it is to specific creep or rather, at contact points T and T' ($\delta = -R \sin \vartheta$, $\delta' = R' \sin \vartheta$)$\rho$ and therefore σ_{max} become infinite. This confirms that even from a wear point of view it would be better to remove that part of the profile immediately next to base circle (see German norms).

Deriving and making the denominator zero, minimum ρ is

$$\delta = \delta_0 = \frac{1}{2}(R' - R)\sin \vartheta.$$

At the halfway point H of segment $\overline{TT'}$.

In practice, the part of most wear is right in the center, which, according to the bending calculation above, should be the least stressed (Figure 15.47). This incongruity may be explained by what follows.

When a pair of teeth are enmeshed at the extreme of their action arc (extreme part of the involute) there is always another pair enmeshed simultaneously. When contact is close to the center of

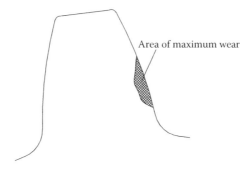

Area of maximum wear

FIGURE 15.47 Qualitative maximum wear's zone

instantaneous rotation, load is at its highest from the point of view of contact pressure (minimum enmeshed pairs at the limit is one).

The teeth are continuously lubricated so maximum pressure depends on the laws of the hydrodynamics of the allocation of lubricant pressures and not only on hertzian pressures. According to lubrication theory, to have film pressure, the relative velocity must be other than zero. At point C, velocity is zero, so in that moment there is no lubrication.

Due to the above, the calculation of σ_{max} according to Hertz is carried out for the contact point C of the pitch circles where $\delta = 0$. Substituting $\delta = 0$ in the expression of ρ then

$$\rho = \frac{(R + R')\sin\vartheta}{RR'\sin^2\vartheta} = \frac{R + R'}{RR'\sin\vartheta}.$$

Dividing the numerator and denominator by the largest gear radius R', we obtain

$$\rho = \frac{1 + \dfrac{R}{R'}}{R\sin\vartheta} = \frac{1 + \tau}{R\sin\vartheta},$$

With $\tau = R/R'$.
Substituting into Hertz's formula from above, then

$$\sigma_{max} = 0.417\sqrt{\frac{QE(1+\tau)}{b\cos\vartheta R\sin\vartheta}}.$$

Recalling that $1/2\cdot\sin 2\vartheta = \sin\vartheta\cos\vartheta$, we obtain

$$\sigma_{max} = 0.59\sqrt{\frac{QE(1+\tau)}{b\sin 2\vartheta R}}.$$

Knowing the gear materials, we know the module of elasticity, so

$$\frac{1}{E} = \frac{1}{2}\left(\frac{1}{E_1} + \frac{1}{E_2}\right).$$

We can make:

$$Q = Cbm,$$

with

$$C = \frac{\sin 2\vartheta}{0.696}\frac{\sigma_{max}^2}{E}\frac{n}{1+\tau}.$$

Buckingham's table contains the admissible values for $\sigma_{max}^2/0.696E$ and σ_{max} for a certain number of cases. They are only indicative values.

For an idea only, $\sigma_{max} = 0.4 - 0.5\ H_d$, where H_d is the Brinell hardness of the least hard contact materials.

Also in this case, force Q is proportional to the product bm according to a coefficient of geometric and durability characteristics of the material.

Gear inspection reveals two wear types:

*   Involute profile variation when lubrication is insufficient due to that outlined above
*   Pitting, when teeth surfaces deteriorate due to abundant lubrication

Pitting is due to (also according to Hertz theory) the fact that when one elastic solid is pressed against another, the point of maximum stress is at their common perpendicular at a certain depth. High loads initiate the beginning of internal damage with the formation of small cracks that propagate more or less rapidly to the surface. Due to high pressure, the oil penetrates the cracks and material is expelled.

This phenomenon therefore depends on Hertzian elastic stresses caused by oil pressure (thousands of atmospheres); this interdependence is caused by the fact that elastic deformation produces a variation in film form and the corresponding pressures due to lubrication influence these elastic deformations.

Pitting signals the end of the lifespan of the gearing.

Another phenomenon is scuffing, similar to surface wear, which happens when the lubricant film is missing or because of high temperatures or excessive loads, where one gear metal solders itself temporarily to the other and on release removes a part of the other gear.

15.20   NOTES ON THE "AMERICAN" NORMS FOR CALCULATING TOOTHING (AGMA)

The norms for gear calculations are semi-empirical and based on practical experience of the factors that can influence inter-gear wear. They are based on general concepts of the type seen above and they intervene by correcting tension values through coefficients (>1) that account for dynamic overloading, tooth size, assembly errors, and manufacturing imperfections. They also correct stress bending and stress wear.

The requirements of the United States norms (AGMA) are outlined below for cylindrical spur gears and helical gears, using the same symbols and diagrams as they do (different to the ones used thus far) so as to facilitate their application.

15.20.1   BENDING TEST

Even in this case the first test is for bending. The following formula ascribable to Lewis is for bending tension:

$$\sigma = \frac{W_t}{FmY},$$

where W_t is the tangential force, m is the metric module, F is the face width, and Y is the Lewis coefficient.

In the past, velocity coefficients were used to deal with dynamic effects, so they are similar to the ones we have already encountered:

$$K_v = \frac{3.05}{3.05 + V} \qquad K_v = \frac{6.1}{6.1 + V},$$

to insert in the following

$$\sigma = \frac{W_t}{FmY} \cdot K_v.$$

This provides a first approximation of toothing with F values three to five times the module. The next test is for wear.

15.20.2 WEAR VERIFICATION

Starting with the formulae for Hertzian contact stress and following the reasoning above brings about a formula similar to one encountered before. It uses the same symbols cited by the norms:

$$\sigma_0 = -C_p \cdot \left[\frac{W_t \left(\dfrac{1}{r_1} + \dfrac{1}{r_2} \right)}{C_v F \cos \Phi} \right]^{1/2},$$

given

$$C_p = \left(\frac{1}{\left(\dfrac{1 - \upsilon_p^2}{E_p} + \dfrac{1 - \upsilon_G^2}{E_G} \right)^{1/2}} \right)^2,$$

where C_v replaces the previously used K_v and Φ is the pressure angle, r_1 and r_2 replace the distances $\overline{TC}$ and $\overline{CT'}$, and υ and E are the Poisson module and the module of elasticity of the two gears (subscript p for the pinion, subscript G for the Gear).

15.20.3 CORRECTIVE COEFFICIENTS FOR FATIGUE

According to the AGMA norms, the previous formulae for bending and wear should be corrected by coefficients that account for different factors influencing endurance. The norms provide these coefficients in reference tables and diagrams that can also be found in the book: Shigley, J.E., Mischke, C.R. and Budynas, R.G., *Mechanical Engineering Design,* 7th ed., McGraw-Hill, 2003.

Here follow some notes on the formulae to highlight how various factors influence strength and wear.

To calculate bending, tension s (as per the norms) is:

$$\sigma = W_t K_a K_s K_m \frac{1}{K_v FmJ},$$

where
K_a is the application factor
K_s is the size factor
K_m is the load distribution factor
$J = Y/(K_f m_N)$ is the geometric factor with $m_N = P_N/(0.95Z)$, in which Z is the length of the line of action

To calculate the dynamic coefficient K_v, the norms cite the following:

$$K_v' = C_v = \left(\frac{\Lambda}{A + \sqrt{200V}} \right)^B,$$

$$A = 50 + 56(1 - B) \quad B = (12 - Q_v)^{2/3} / 4,$$

where V is peripheral velocity in meters per second, and Q_v (transmission accuracy level number) is the coefficient of gear manufacture quality (correcting for vibrations, deformations, etc.). It varies from 0–11.

A value of 11 is for a gear cut and mounted with precision operating at minimal vibration. The norms have diagrams that supply values of K_v as a function of peripheral velocity for various values of Q_v.

To Calculate Wear:

$$\sigma_c = -C_p \cdot \left[\frac{W_t C_a C_s C_m C_f}{C_v F d I} \right]^{1/2}.$$

The minus sign is for compression stress. The various subscript C symbols replace those hitherto seen for K.

$C_v = K_v$ is the dynamic factor seen before, and I is a geometric factor defined by

$$I = \frac{\cos \Phi_t \sin \Phi_t}{2m_N} + \frac{m_G}{m_G \pm 1},$$

where the plus sign is for the external gears, the minus for the internal gears. The formula is more generally valid for helical teeth where Φ_t is the frontal pressure angle linked to the pressure angle Φ:

$$\tan \Phi_t = \tan \Phi / \cos \Psi$$

where Φ is the inclination angle of the helix. For $\Psi = 0$ and $\Phi_t = \Phi$, the formula is valid for spur gears.

The transmission ratio of the gearing is $m_G = N_G/N_P$ (N_G and N_P are the teeth numbers for the gear and pinion). C_a (K_a) takes into account any overloading, its choice left to the experience of the designer given the specifics of the application. C_s (K_s) takes into account the fact that larger gears are more likely to have material flaws. C_m (K_m), the load distribution factor, takes into account the non-uniform distribution of the load over the whole tooth due to manufacturing defects, the mis-alignment of axles and teeth elastic deformation. The geometric factor (J, I) operates through the Lewis coefficient, and takes into account the fatigue stress concentration (K_f) and the face contact ratio (m_N). C_f the surface factor, accounts for the surface condition of the teeth. The AGMA norms give no values for this coefficient, suggesting only assuming it to be greater than one when surface defect factors are noted.

For the calculation of safety coefficient, the bending stress σ and contact stress σ_c values are then compared with the allowable stress limits (σ_{all}, $\sigma_{c\,all}$), which are a function of bending fatigue resistance (S_t) and contact fatigue resistance (S_c) of the material. The S_t and S_c values are tabulated and are functions of various parameters like hardness and heat treatment, tooling. Finally, the expressions for tension limits are

$$\sigma_{\text{all}} = S_{\text{t}} K_{\text{L}} / K_{\text{T}} K_{\text{R}} \quad \sigma_{\text{c all}} = S_{\text{c}} C_{\text{L}} C_{\text{H}} / C_{\text{T}} C_{\text{R}},$$

where K_{L}, C_{L} is the life factor, C_{H} is hardness factor, K_{T}, C_{T} is the temperature factor, K_{R}, C_{R} is the reliability factor.

The life factor accounts for gear durability and equals 1 for cycle numbers over 10^7. The AGMA norms supply values for such coefficients for cycles other than 10^7.

The reliability factor is assumed as 0.99 for cycles over 10^7. There are diagrams that supply values for other cycles.

The hardness factor is only used for the gear and not the pinion and equals (only valid for hardness ratio values between pinion and gear of greater than 1.70):

$$C_{\text{H}} = 1 + A(m_{\text{G}} - 1) \quad \text{with} \quad A = 8.98(10^{-3})(H_{\text{BP}} / H_{\text{BG}}) - 8.29(10^{-3}),$$

where H_{BP} and H_{BG} are the Brinell hardness values of the pinion and gear, and $m_{\text{G}} = N_{\text{G}}/N_{\text{P}}$ is the transmission ratio of the gearing (N_{G} and N_{P} number of gear and pinion teeth).

The temperature factor has rising values over 250°C (at the discretion of the designer).

The ratio between bending fatigue resistance σ_{all} and bending tension σ defines the safety coefficient of bending S_{F}. Analogously, the ratio between contact fatigue resistance $\sigma_{\text{c all}}$ and contact tension $\sigma_{\text{c all}}$ defines the safety coefficient of wear S_{H}.

Italian gear makers mostly refer to the ISO 6336 norms that provide details of coefficients and methods for the various tests in a similar form to those already seen.

15.21   DESIGN PHASES FOR A GEARING

As we have seen in previous paragraphs, designing a gear pair must start from their operating conditions, i.e., the required torque, peripheral velocity, and oil temperature. Next come material choice, manufacturing quality, heat treatment type, and oil choice. Last, there is tooth geometry, and so using the new symbols:

- Calculating the minimum number of pinion teeth:

$$n_{\text{min}} = \frac{2(1-x)}{\sin^2 \vartheta_0} \quad \text{(for non-modificated gear } x = 0\text{)}$$

- Calculating the number of gear teeth knowing the required transmission ratio $\tau = n/n'$
- Calculating the module through an iterative calculation using:

$$m = K \sqrt[3]{\frac{M}{\lambda \sigma}},$$

and finding m_0;

- $R'_0 = n' m_0 / 2$;
- Calculating the pitch radii of the pinion $R_0 = n m_0 / 2$ and $R'_0 = n' m_0 / 2$
- Calculating center distance $d = R_0 + R'_0 + \xi m_0$ (for non-modificated gear $\xi = 0$)

- Calculating the initial cylinder from which the gear will be cut (finding χ only for modificated gear)
- For pinions, the bonding verification according to the norms using K-type coefficients (perhaps referring to AGMA)
- For pinions, the wear verification according to the norms using C-type coefficients (perhaps referring to AGMA)
- Determining the contact ratio ε
- Constructing specific creep diagrams for the enmeshed gear pair

16 Press and Shrink Fits

16.1 INTRODUCTION

A very commonly used method for stably linking two components one within the other is to size them such that they interfere with each other when cold. For example, in a shaft-hub assembly, the hub aperture is slightly smaller than the shaft size.

The join is made by heating the component with a hole (hub) that dilates, allowing it to be mounted onto the shaft. The heat removed, the assembly cools and the hub contracts creating a forced join between shaft and hub.

To carry out this join the condition $D < D_a$ (Figure 16.1) must be satisfied. If the two components are steel, the diameter difference is in thousands of millimeters to create a stable join.

The effectiveness of this join relies exclusively on the difference between the two diameters. At the design stage, the pressure at play must be evaluated to discover at what point the join might no longer hold, or if the elastic limits of the components will be exceeded.

Forced junctions are often used in machine-making, both for joining shafts and hubs more securely than is obtained with keys whenever rapidly changing high torque is involved, as well as for joining components of the same part (the two halves of a flywheel, the toothing and cogwheels of a large gearing, a railway wheel and its hub). Pipe riming is a forced junction for pipes that must resist high or very high pressures.

Apart from in special cases, these junctions are fixed by design, the two components never requiring dismantling.

Below, reference is made to methods of making the junctions, and for complete designs refer to Giovannozzi, vol. I.

16.2 FIXING FLYWHEEL COMPONENTS

When the two parts of a flywheel that must operate at an angular velocity of ω are fixed, the formulae and reasoning behind the design are easily applicable to other practical cases.

The objectives in designing a flywheel are to identify any interference between components; to ensure that they do not separate in the course of operation; and to size the components (junction plates and slot flywheel area) so they remain integral even in dynamic conditions.

Consider the case of two pairs of plates (plate hereafter) that connect the two parts of a flywheel crown (Figure 16.2) with plate length l_p and distance l_v between the supports at the plate perimeter in the slot.

If $l_p < l_v$, insertion of the plate in the slot could only happen after it was heated.

Heating the plate lengthens it, making it possible to position it in its seating. During cooling, plate length diminishes bringing the two facing flywheel components into contact.

Before manufacture, interference must be calculated:

$$i = l_v - l_p > 0,$$

depending entirely on component geometry.

When the flywheel is at operating velocity, the two facing components will tend to separate, and consequently the plate, which at rest locked the two components together, is subject to a tensile stress of force T (Figure 16.3).

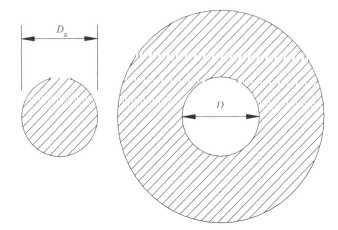

FIGURE 16.1   Shaft–hub assembly.

In boundary conditions, the plate could lengthen until the two facing components separated. If that were to happen, the forced junction would cease to be and the flywheel would no longer be stable. If material behavior were perfectly elastic, the sum of length variations would be constant and equal to interference. Interference is always equal to the sum of length variations and in operation remains unaltered. Interference would vary, however, if one of the materials yielded plastically.

It is easy to intuit that if the plate and slot were the same size, no force would be generated between the two facing flywheel components. However, interference as defined above and built in at assembly means that these two facing components are subject to a force F_0 which, in the absence of other forces, is equal to tensile stress T on the plate:

$$F_0 = T.$$

For equilibrium, F_0 is also the compression force exerted in that rim section between the supports of the plate perimeter.

As long as $F_0 > 0$, the join remains stable. Were this not the case, there would be no forces to prevent the components from separating and the whole would be unstable.

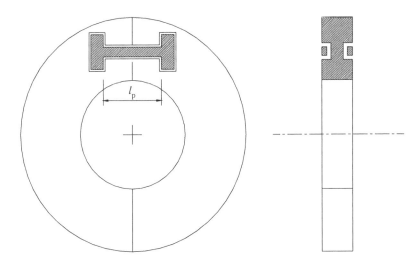

FIGURE 16.2   Connecting of two pairs of a flywheel.

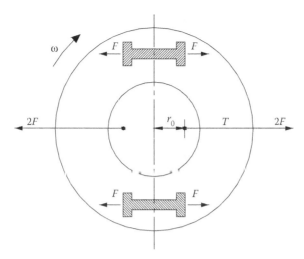

FIGURE 16.3   Acting loads.

Suppose that flywheel of mass m_v is subject to a rotational velocity of ω. On each of the half rims of the flywheel there would be a centrifugal force of

$$2F = \frac{m_v}{2}\omega^2 r_0 .$$

Supposing the rim mass were very large compared to the hub and spokes and this force were applied to the center of gravity of the rim of average radius R, or at a distance of r_0 from the center, then $r_0 = 2R/\pi$.

If there were two connection plates, force F would be exerted on each, equal to half of that exerted on the half rim.

After hot assembly, the cooling process produces perpendicular/normal tensions within the joined components and surface pressures in the facing components in contact.

During flywheel rotation due to external force F, the initial elastic reaction F_0 of the rim varies up to F'_0, which at its upper limit could be 0 (facing components separate). So, force T to which the flywheel plate is subject during rotation changes from F_0 to:

$$T = F'_0 + F \quad \text{where} \quad F'_0 < F_0$$

To size the components, F'_0 and T must be calculated. To keep the system stable, $F'_0 > 0$ is necessary.

$$F'_0 > 0 \quad \sigma = \frac{T}{S} < \frac{\sigma_{Sn}}{m},$$

with plate straight section area S, material yield limit σ_{Sn}, and security coefficient (chosen wisely) m.

When calculating F'_0 and T it should be remembered that even for elastic deformation, interference remains unchanged.

S is the area of the reacting section and E is the elasticity modulus of the rim plate. Initially, the lengths of plate and slot are l_p and l_v, the difference between them is the interference.

With the flywheel immobile, force F_0 will elongate the plate by

$$\lambda = \frac{F_0 l}{ES},$$

whereas the corresponding rim section will shorten by

$$\lambda' = \frac{F_0 l}{E'S'},$$

where in both equations l is identical but generally E and S are different.

However, between the distance of the plate supports and the length of the isolated plate there is overall interference of the two parts defined by their geometries at the time of manufacture and equal to

$$i = \lambda + \lambda' = \frac{F_0 l}{ES} + \frac{F_0 l}{E'S'}.$$

Now, supposing the flywheel is set in motion, the plate would be subject to a tensile stress of $F'_0 + F$ (F'_0 contact force, F external force) lengthening it by

$$\lambda_1 = \frac{F'_0 + F}{ES} l,$$

whereas the rim section would be subject to deformation:

$$\lambda'_1 = \frac{F'_0}{E'S'} l.$$

Overall interference of the two components is

$$\lambda_1 + \lambda'_1 = \frac{F'_0 + F}{ES} l + \frac{F'_0}{E'S'} l = i.$$

Taking into account that in the absence of permanent deformation, the interference of the conjoined parts remains the same (interference is defined by geometry) then

$$\frac{F_0 l}{ES} + \frac{F_0 l}{E'S'} = \frac{F'_0 + F}{ES} l + \frac{F'_0}{E'S'} l.$$

From this equilibrium F'_0 (a new force exchanged by the two facing components and resulting from the external force F):

$$F'_0 = F_0 - \frac{F}{1 + \dfrac{ES}{E'S'}}.$$

The overall tensile stress $T = F'_0 + F$ exerted on the plate is

$$T = F_0 + \frac{F}{1 + \dfrac{E'S'}{ES}}.$$

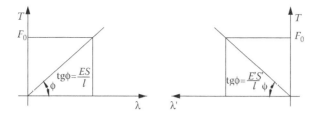

FIGURE 16.4 Forces versus displacements.

This equation highlights that the plate is not subject to the whole force F, but only a fraction:

$$\frac{F}{1 + \dfrac{E'S'}{ES}}.$$

If the plate was infinitely rigid ($ES = \infty$)—totally undeformable even adding force F—the rim would continue to exert an unaltered force F_0. The result would again be $T = F_0 + F$.

If the rim was infinitely rigid ($E'S' = \infty$), only the plate would deform, and that pull would be F_0. Apart from analytically, F'_0 and T can be calculated graphically.

Considering the relation above $\lambda = F_0 l / ES$, then

$$F_0 = \frac{ES}{l}\lambda.$$

This relation is of the $F = Kx$ type and shows that the plate behaves like a spring with rigidity ES/l (see Chapter 13, Springs). Plotting T as a function of λ, would produce a straight line of angular coefficient φ passing through the coordinate λ and F_0; that straight line represents the plate characteristics.

The same can be done for the rim, but the λ^{l} values are the reverse of the λ values because, while the plate elongates, the corresponding rim section shortens (Figure 16.4).

If the system is immobile, the plate and rim have $T = F_0$ in common. Uniting the two diagrams produces Figure 16.5.

The two straight lines are the characteristics of the plate and rim. When the flywheel rotates, the centrifugal external force F diminishes the force exerted on the facing surfaces of the rim by F_0 and F'_0, but increases force T on the plate by F_0 and F'_0.

Referring to Figure 16.6, the plate's load point during rotation is at $(\lambda_1, F'_0 + F)$, whereas for the rim the load point is at (λ'_1, F'_0).

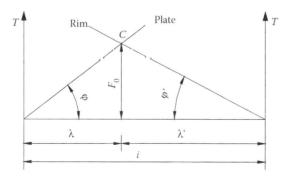

FIGURE 16.5 Loads on the rim and on the plate in rest condition.

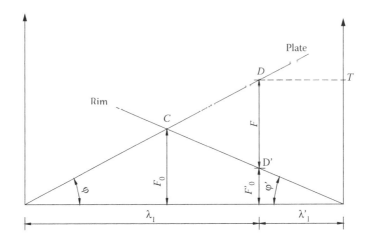

FIGURE 16.6 Loads on the rim and on the plate in working condition.

To determine operating conditions, force F (known) is plotted as far as it can be read between the two straight lines φ and φ'. The ordinate for the rim characteristic is F'_0, whereas the ordinate for the plate characteristic is $T = (F'_0 + F)$. The abscissa represent the deformations λ and λ'.

From this diagram it is easy to arrive at the previously derived formulae. It suffices to look at the various triangles to see that

$$tg\varphi = \frac{ES}{l} \quad tg\varphi' = \frac{E'S'}{l}.$$

The simplicity of this plate/rim calculation for lengthening due to a force, should not detract from the results which, by substituting (ES/l), $(E'S'/l)$ for rigidity, are the generic rigidities:

$$C = \frac{F_0}{\lambda_0} \quad C' = \frac{F_0}{\lambda'_0},$$

of the two conjoined components.

So, the previously derived formulae can be written thus

$$F'_0 = F_0 - \frac{F}{1 + \dfrac{C}{C'}} = \frac{i}{\dfrac{1}{C} + \dfrac{1}{C'}} - \frac{F}{1 + \dfrac{C}{C'}}, \tag{16.1}$$

$$T = F_0 + \frac{F}{1 + \dfrac{C}{C'}} = \frac{i}{\dfrac{1}{C} + \dfrac{1}{C'}} + \frac{F}{1 + \dfrac{C'}{C}}. \tag{16.2}$$

If, as often happens, it is required that the conjoined component's own tension of rigidity C (here specifically the plate) should vary very little when subject to external force F, the C'/C ratio should be high. This means that what serves as the plate should not be rigid compared to the other component.

In some practical cases (connecting-rod big end join to shaft), by playing with the rigidity values of the components, load variations compared to initial pre-loading can be limited, which noticeably reduces the danger of fatigue breaking. Good junctions must ensure durability so that any pre-loading must be the minimum that is indispensible to correct function. High pre-loading does ensure joint durability, but at the cost of limiting the possible absorption of load variations.

As mentioned above, as the external force F grows, the residual fitting force F'_0 between the two components diminishes to zero.

To determine the boundary force F_{max} of F for which F'_0 would zero itself, it suffices if $F'_0 = 0$ (16.1):

$$F_{max} = F_0 \left(1 + \frac{C}{C'} \right).$$

This relation provides the initial fitting force F_0 such that the boundary force F_{max} can be applied without the two components separating.

Furthermore, with a forced fitting with initial interference i, it is possible to determine the effects as i varies Δi. In practice this can happen either because the most compressed part heats up compared to the strained part (e.g., steam pipe flange bolts, $\Delta i > 0$), or because of plastic lengthening of the strained part (e.g., excessive external overloading, or heat sliding, $\Delta i < 0$). If C and C' remain unchanged in these cases as usually happens, the only effect of Δi variations is reduced to the initial tightening:

$$\Delta F_0 = \frac{\Delta i}{\dfrac{1}{C} + \dfrac{1}{C'}},$$

so

$$F_0 = \frac{i}{\dfrac{1}{C} + \dfrac{1}{C'}}.$$

When an external force F is applied, with C and C' unchanged, the same force variations $C\Delta\lambda$ and $C'\Delta\lambda$ when interference was i, are found for the two components.

So, the new forces of the two components of rigidity C' and C are

$$F'_0 = (F_0 + \Delta F_0) - C'\Delta\lambda = \frac{i + \Delta i}{\dfrac{1}{C} + \dfrac{1}{C'}} - \frac{F}{1 + \dfrac{C}{C'}},$$

$$T = (F_0 + \Delta F_0) - C\Delta\lambda = \frac{i + \Delta i}{\dfrac{1}{C} + \dfrac{1}{C'}} - \frac{F}{1 + \dfrac{C'}{C}}.$$

Thus, the force values are the same as before except they have both increased by the same value ΔF_0.

The graph in Figure 16.6 shows what happens when, after an initial tightening F_0 below the plastic limits, the system is externally overloaded by F such that one component is permanently deformed (Figure 16.7). In this case, having exceeded the elastic limit, the plastic deformation produced is λ_p.

In the T diagram, λ represents force-deformation curve for the components (flywheel, plate). In the case of plastic deformation λ_p of the plate, the overall force exerted corresponding to that deformation can be found in the diagram by reading force F between the two characteristic lines. The plate undergoes permanent deformation λ_p equal to the horizontal distance between P identified by F and the lengthening of the straight line due to the plate's condition.

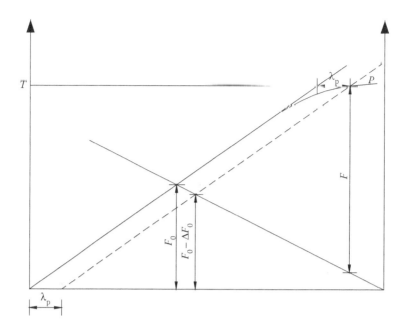

FIGURE 16.7   Forces in plastic deformation case.

Given that rigidity C remains constant, and once the permanent lengthening of the plate's straight line has been produced, it is replaced by its parallel from point P, which is at distance λ_p from the plate's straight line with just one consequence—the loss of ΔF_0 from the initial tightening.

The interference variation is $\Delta_i = -\lambda_p$, so the new interference is

$$i' = i - \lambda_p.$$

Thus, the system could be thought of as a forced fitting with different interferences and with a statically exchanged force of less than F_0.

Forces F'_0 and T are still expressed by the relations for $\Delta_i > 0$ in which $\Delta_i = -\lambda_p$ are substituted:

$$F'_0 = \frac{i - \lambda_p}{\dfrac{1}{C} + \dfrac{1}{C'}} - \frac{F}{1 + \dfrac{C}{C'}},$$

$$T = \frac{i - \lambda_p}{\dfrac{1}{C} + \dfrac{1}{C'}} + \frac{F}{1 + \dfrac{C'}{C}}.$$

Obviously, the operational range of the forced fitting is limited, so more attention should be given to its application. Overloading flywheels could occur when control systems malfunction and when opposing torque declines.

16.3   FORCED HUB-SHAFT SHRINKING

Consider the frequent case of having to force (by heat or pressure) a hub onto a shaft (Figure 16.8). d is the contact diameter between the shaft and hub, and b is the hub length. There should be an interference i of

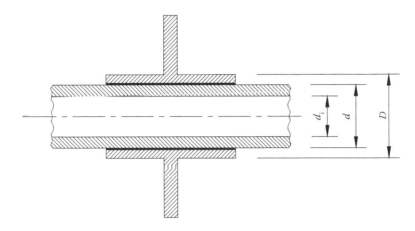

FIGURE 16.8 Hub-shaft shrinking.

$$i = d_a - d_m,$$

where d_a is the shaft's external diameter and d_m is the hub's external diameter. Interference between the shaft and hub during assembly generates a certain pressure p. The moment due to tangential friction forces (coefficient f) is given by the product of tangential force and the arm $d/2$, according to

$$M_t = (f \cdot p \cdot \pi \cdot d \cdot b) \cdot \frac{d}{2}.$$

Between this moment and the one to transmit, a safety coefficient n should be considered that varies between 1 and 2. The equation is

$$nM_t = fp\pi \frac{d^2}{2} b.$$

So, normal pressure p to transmit uniform moment M_t between shaft and hub is

$$p = \frac{2nM_t}{f\pi d^2 b}$$

Designing the junction consists in evaluating:

• The interference i required to generate the pressure p needed to transmit moment M
• The stress generated by the interference to verify component (shaft, hub) durability

To determine the interference required, any shaft or hub displacements due to absolute radial tension p will be analyzed.

Stress generated by external loads in thick rotating discs is expressed in formulae by Giovanozzi, Vol. II, p. 198. In the case of an immobile disc ($\omega = 0$), $C = D = 0$, and radial and tangential tensions are expressed by

$$\sigma_r = A - \frac{B}{\rho^2} \quad \sigma_t = A + \frac{B}{\rho^2}.$$

With integration constants A and B that depend on surrounding conditions and the generic variable $\rho = r/R$ from the ratio of generic radius r and the external radius R of the disc.

Referring to Figure 16.8, in which the shaft is hollow, system geometry is defined by the parameters α and β from the ratio between the internal radius r_i and the external radius r_e $(r_i < r_e)$

$$\alpha = \frac{r_{i(hub)}}{r_{e(hub)}} = \frac{d}{D},$$

$$\beta = \frac{r_{i(shaft)}}{r_{e(shaft)}} = \frac{d_i}{d}.$$

with $d_i < d < D$.

The variation in the hub's internal diameter due to the radial pressure generated when the two components (shaft and hub) are mounted with a certain interference (Figure 16.9) can be determined as follows.

The σ_r of the hub's external radius is zero, so

$$\text{for } \rho = \frac{r}{r_e} = 1 \quad \text{then } \sigma_r = 0.$$

But, the σ_r of the hub's internal radius r_i is equal to the pressure between the shaft and hub with a changed sign due to the convention that when the tension is a compression type, then the sign is negative:

$$\text{for } \rho = \frac{r}{r_e} = \alpha \quad \text{then } \sigma_r = -p.$$

As regards deformations:

$$\varepsilon_r = \frac{d\xi}{dr} \quad \varepsilon_t = \frac{2\pi(r + \xi) - 2\pi r}{2\pi r} = \frac{\xi}{r}.$$

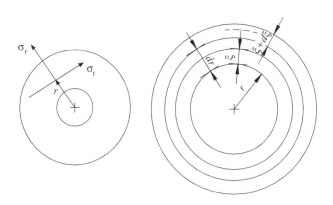

FIGURE 16.9 Nomenclature for stress and strain.

The boundary point thought to belong to the hub would move by

$$\xi_m = \varepsilon_{tm} \cdot \frac{d}{2}.$$

The same point thought to belong to the shaft would move by

$$\xi_a = \varepsilon_{ta} \cdot \frac{d}{2}.$$

For symmetry, since the axial tension is zero, the main radial and tangential tensions are

$$\varepsilon_r = \frac{1}{E}\left(\sigma_r - \frac{\sigma_t}{m}\right) \quad \varepsilon_t = \frac{1}{E}\left(\sigma_t - \frac{\sigma_r}{m}\right).$$

So, to calculate tension the A and B constants need to be found to make σ_r the hub's internal and external radii.

The external radius would be

$$0 = A - \frac{B}{1} \Rightarrow A = B.$$

The internal radius would be

$$-p = A - \frac{A}{\alpha^2} = A\left(1 - \frac{1}{\alpha^2}\right) = A\left(\frac{\alpha^2 - 1}{\alpha^2}\right).$$

So

$$A = B = p\frac{\alpha^2}{1 - \alpha^2}.$$

By substituting radial and tangential tension, the general relations are

$$\sigma_r = p\frac{\alpha^2}{1 - \alpha^2}\left(1 - \frac{1}{\rho^2}\right) \quad \sigma_t = p\frac{\alpha^2}{1 - \alpha^2}\left(1 + \frac{1}{\rho^2}\right).$$

At the shaft-hub contact point, for radius r_i for which $\rho = r_i / r_e = \alpha$:

$$\sigma_r = p\frac{\alpha^2}{1 - \alpha^2}\left(1 - \frac{1}{\alpha^2}\right) = -p,$$

$$\sigma_t = p\frac{\alpha^2}{1 - \alpha^2}\left(1 + \frac{1}{\alpha^2}\right) = p\frac{1 + \alpha^2}{1 - \alpha^2}.$$

Once σ_r and σ_t are known, and making $1/m = \nu$, the hub's unit tangential deformation is

$$\varepsilon_{tm} = \frac{1}{E}\left(p\frac{1+\alpha^2}{1-\alpha^2} + \nu p\right) = \frac{p}{E}\left(\frac{1+\alpha^2}{1-\alpha^2} + \nu\right).$$

So, the movement $\xi_m = \varepsilon_{tm}\cdot d/2$ would be

$$\xi_m = \frac{pd}{E2}\left(\frac{1+\alpha^2}{1-\alpha^2} + \nu\right).$$

Knowing the hub radius variation, the shaft radius variation can be found analogously starting from the boundary conditions of σ_r:

$$\text{for } \rho = \beta \quad \text{then } \sigma_r = 0,$$

$$\text{for } \rho = 1 \quad \text{then } \sigma_r = -p.$$

Substituting the first condition in $\sigma_r = A - B/\rho^2$, we get

$$0 = A - \frac{B}{\beta^2}.$$

Substituting the second condition and also the relation $A = B/\beta^2$ we get

$$-p = \frac{B}{\beta^2} - B = B\left(\frac{1}{\beta^2} - 1\right) = B\left(\frac{1-\beta^2}{\beta^2}\right).$$

Finally, we get

$$A = -\frac{p}{1-\beta^2} \quad B = -p\frac{\beta^2}{1-\beta^2}.$$

Substituting σ_r and σ_t in the two expressions we get the general relations:

$$\sigma_t = -\frac{p}{1-\beta^2}\left(1 + \frac{\beta^2}{\rho^2}\right),$$

$$\sigma_r = -\frac{p}{1-\beta^2} - \frac{-p\beta^2}{1-\beta^2}\frac{1}{\rho^2} = -\frac{p}{1-\beta^2}\left(1 - \frac{\beta^2}{\rho^2}\right).$$

For the external radius where $\rho = 1$, they become

$$\sigma_r = -\frac{p}{1-\beta^2}(1-\beta^2) = -p,$$

$$\sigma_t = -\frac{p}{1-\beta^2}(1+\beta^2).$$

Substituting $\varepsilon_t = (\sigma_t - \nu\sigma_r)/E$, we get

$$\varepsilon_{ta} = \frac{1}{E'}\left(-p\frac{1+\beta^2}{1-\beta^2}+\nu p\right) = -\frac{p}{E'}\left(\frac{1+\beta^2}{1-\beta^2}-\nu\right).$$

The movement $\xi_a = \xi_{ta}\cdot d/2$ becomes

$$\xi_a = -\frac{p}{E'}\frac{d}{2}\left(\frac{1+\beta^2}{1-\beta^2}-\nu\right).$$

Overall, the shaft and hub interference relative to the diameter is

$$i = \left(\left|\xi_a\right|+\left|\xi_m\right|\right)\cdot 2.$$

Substituting ξ_a and ξ_m, we get

$$i = \frac{d}{2}\left[\frac{p}{E'}\left(\frac{1+\beta^2}{1-\beta^2}-\nu\right)+\frac{p}{E}\left(\frac{1+\alpha^2}{1-\alpha^2}+\nu\right)\right]\cdot 2.$$

This is the most general expression. If, as often happens, the hub and shaft have the same elastic modulus $E' = E$, the i expression becomes

$$i = \frac{dp}{E}\left(\frac{1+\beta^2}{1-\beta^2}+\frac{1+\alpha^2}{1-\alpha^2}\right).$$

In the case of a solid shaft where $d_i = 0$ and $\beta = 0$, we get

$$i = \frac{2dp}{E}\left(\frac{1}{1-\alpha^2}\right).$$

It should be emphasized that p is exclusively a function of interference, apart from geometry and material.

The expression helps determine p or i given that it must be one of the two.

Using the σ_r and σ_t expressions:

$$\text{hub's internal radius}\quad\begin{cases}\sigma_r = p\\[2mm]\sigma_t = p\dfrac{1+\alpha^2}{1-\alpha^2}\end{cases},$$

$$\text{shaft's internal radius}\quad\begin{cases}\sigma_r = 0\\[2mm]\sigma_t = -\dfrac{2p}{1-\beta^2}\end{cases},$$

$$\text{shaft's external radius} \begin{cases} \sigma_r = -p \\ \sigma_t = p\dfrac{1+\beta^2}{1-\beta^2} \end{cases}$$

Note that σ_t is hub tensile stress and shaft compression.

The ideal tensions on the hub's and shaft's internal radius adopting the hypothesis of breakage at maximum expansion are

$$\text{for the hub} \quad \sigma_i = p\left[\frac{1+\alpha^2}{1-\alpha^2}+\nu\right],$$

$$\text{for the shaft} \quad \sigma_i' = -\frac{2p}{1-\beta^2}.$$

Adopting the theory of maximum tangential tensions analogously:

$$\text{for the hub} \quad \sigma_i = +\frac{2p}{1-\alpha^2},$$

$$\text{for the shaft} \quad \sigma_i' = -\frac{2p}{1-\beta^2}.$$

As a function of σ_i or σ_i', p can be obtained from these relations by subsequent substitution of i to determine the relationship between interference and maximum ideal tension in the hub or shaft.

The calculations above refer to elastic behavior of the material. High interference levels can cause permanent plastic deformation that undermines the hypotheses on which the calculations were based. At these high levels, the material's characteristics change in the most stressed zones (e.g., the hub's internal radius), changing almost continuously from one radius to another.

The rational calculation of such a connection/junction must take into account these conditions. So, material behavior hypotheses must be made in passing from one spoke to another. Knowing the variation in plastic modulus of the spoke, a calculation could be made that would account for the variation, which in turn is a function of the applied stress. Using an iterative calculation, tension values are close to their real counterparts. More simply but less accurately, the material can be subdivided into strata (circular rims) to which an elastic modulus is assigned, imposing equilibrium and congruence conditions to each strata and defining the calculation according to the boundary conditions.

To know how much interference there is, simply refer to the case when both hub and shaft are steel ($E = E' = 200,000 \text{ N/mm}^2$) and the shaft is solid.

From previous equations and adopting the maximum expansion theory:

$$\frac{i}{d} = \frac{\sigma_i}{E}\frac{2}{(1+\nu)+\alpha^2(1-\nu)} = -\frac{\sigma_i'}{E}\frac{1}{1-\alpha^2}.$$

Analogously, adopting the theory of maximum tangential tension:

$$\frac{i}{d} = \frac{\sigma_i}{E} = -\frac{\sigma_i'}{E} \frac{1}{1-\alpha^2},$$

and considering that generally hub thickness is $s \cong 0.3d$ and correspondingly $\alpha = 1/(1 + 2 \cdot 0.3)$ $= 0.625$, the two relations become:

$$\frac{i}{d} = 1.27 \frac{\sigma_i}{E} = -1.64 \frac{\sigma_i'}{E} \quad \frac{i}{d} = \frac{\sigma_i}{E} = -1.64 \frac{\sigma_i'}{E}.$$

The formulae show that σ_i' are lower in absolute value than σ_i. This means that because of the forced fitting, the hub is more stressed than the shaft.

Suppose that due to the forced fitting, the hub reaches an admissible tension of 200 N/mm², the two breakage hypotheses would provide interference percentages of the order:

$$\frac{i}{d} = 1.27 \cdot \frac{200}{200,000} = 0.127\% \quad \frac{i}{d} = \frac{200}{200,000} = 0.1\%.$$

So, the interference used in the forced fitting of steel hubs onto steel shafts on average is between 1 and 1.3 per thousand. In any case, from time to time it is a good idea to bear in mind the junction type as well as the material type of hub and shaft. Ideally, both components would be able to reach tensions that would guarantee equal lifespan. To do that, one must be able to take into account all three tensions exerted on the shaft (twisting, bending, and tensile stress).

Shafts experience constant bending and twisting, which are at their highest at the outer shaft surface; if the shaft is hollow, the forces due to forced fitting might be lower. Remember that in an ideal design every point should achieve the same safety conditions and every component should have the same lifespan.

16.4 THE RELATIONSHIP BETWEEN INTERFERENCE AND THERMAL VARIATION

From physics we know that the final length l of a free shaft of initial length l_0, which undergoes thermal variation ΔT, is

$$l = l_0(1 + \alpha \Delta T),$$

where α is the linear expansion coefficient of the material.

The formula can provide the variation in length:

$$l - l_0 = l_0 \alpha \Delta T,$$

or:

$$\frac{\Delta l}{l_0} = \alpha \Delta T.$$

The linear expansion coefficient of carbon steel, which is usually used for making the plates and rims for forced fittings, is $\alpha = 1.2 \times 10^{-5}$ 1/°C.

To carry out an assembly with 0.13% interference, a temperature increase would suffice of the shortest component of

$$\Delta T = \frac{1.3 \cdot 10^{-3}}{1.2 \cdot 10^{-5}} = 110°C$$

In practice, to facilitate assembly, one component must be slightly larger but not such that the steel's structural characteristics are challenged; temperatures usually do not exceed 300°C.

Should higher thermal shifts be required, the other component is cooled in either liquid air or nitrogen. With liquid air the temperature drops to around −160°C.

17 Pressure Tubes

17.1 INTRODUCTION

Pipes usually have a circular cross-section and they are joined up by special junctions or rings to be used to transport fluids (liquids, vapors, gas) and more unusually pulvirulent or granular material.

According to UNI standards, pipes are conventionally defined by their nominal diameter, which is the effective internal diameter in millimeters and acts as a reference for all the connectors (flanges, valves, etc.).

Pipe strength is calculated according to a fictional nominal pressure that takes into account not only effective pressure but also the pipe's operating conditions.

In the table on page 334 of Giovanozzi, vol. II (a useful reference for pipe, junction and flange designs) is a collection of nominal pressures in kg/cm^2 for junctions. The values in brackets are ceiling values to be avoided and refer only to pipe junctions and not to other components (flanges, valves, taps, etc.).

There are three other effective operating values that correspond to the nominal pressures equal to 100%, 80%, and 64% that refer to the following uses:

- Non-chemically dangerous fluids at temperatures below 120°C
- Non-dangerous fluids at temperatures between 120°C and 300°C and dangerous fluids below 120°C
- Non-dangerous fluids at 300°C–400°C and very dangerous fluids at lower temperatures

For these three conditions, and for each nominal pressure there is a fixed test pressure P_0, as set out in the table in Giovanozzi, Vol. II, p. 334.

For operating pressures lower than $10 N/cm^2$, P_0 must be higher than $10 N/cm^2$ over the operating pressure. For vacuum pipe components, the test pressure must be at least $15 N/cm^2$.

In extreme operating conditions, in poor pipe conditions and in any other negative situations, lower pressures are prescribed, whereas for special steel pipes that are stronger than normal carbon steel ones, higher operating pressures are allowed in the second and third conditions defined above.

The following paragraphs set out pipe and flange calculations that give an idea of what the reference rules report and which can help in sizing up components and mechanical parts.

17.2 SIZING AND VERIFYING PIPE STRENGTH

Given the conveyed fluid capacity Q in m^3/s, and the internal diameter $d_i = 2r_i$ of the pipe, then

$$Q = \frac{\pi d_i^2}{4} V,$$

in which V is the average velocity of the fluid.

The value of V varies from case to case from 0.5–1 m/s in piston water pump intake pipes, to 100 m/s for turbine condensing conduits. Since load loss and the stress due to pump thrusting are proportional to density, the general concept is, the denser the fluid the lower the velocity.

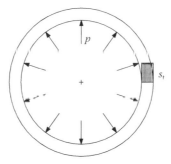

FIGURE 17.1   Qualitative tangential stress distribution in narrow pipes.

With r_i the internal pipe diameter and $r_e = r_i + s$ the external pipe diameter given a thickness of s, then there are two cases of strength calculation:

- Narrow pipes where $r_e / r_i < 1.1$
- Wide pipes where $r_e / r_i > 1.1$

As in the case of forced fittings, β is the ratio:

$$\beta = \frac{r_i}{r_e}.$$

For narrow pipes, the tangential tension σ_t is considered constant over the whole tube thickness (Figure 17.1). So, the Marriotte formula can now be applied:

$$\sigma_t = \frac{p d_i}{2s},$$

where d_i is the internal pipe diameter, s is thickness, and p is the internal fluid pressure.

For wide pipes, the distribution of σ_t is not constant and as in the case of forced fitting it behaves like the type in Figure 17.2.

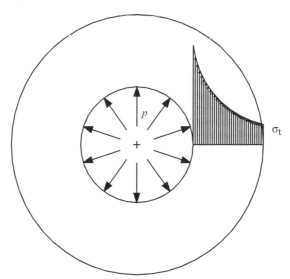

FIGURE 17.2   Qualitative tangential stress distribution in wide pipes.

By introducing the parameter $\rho = r/r_e$, it is now possible to apply the same formulae for forced fittings:

$$\sigma_r = A - \frac{B}{\rho^2} \quad \sigma_t = A + \frac{B}{\rho^2}.$$

The most dangerous tensions have an internal diameter of $(\rho = \beta = d_i/d_e + 2s)$:

$$\sigma_t = \sigma_i = \frac{1+\beta^2}{1-\beta^2}\, p.$$

Note how these formulae are exactly the same as those obtained using the rotating disc theory for constant thickness (Giovanozzi, vol. II). When the r_e/r_i ratio does not have high values $(1.1 < r_e/r_i < 1.4)$ a simpler formula can be used. The radial movement ξ $(\xi/r = \varepsilon_t)$ can be hypothesized as constant for all radii.

Since $\sigma_t = E\varepsilon_t$, and substituting $\xi = \varepsilon_t \cdot r$, then

$$\xi = \frac{\sigma_t \cdot r}{E} \quad \text{and so } \sigma_t = \frac{\xi \cdot E}{r} \text{ or } \sigma_t \cdot r = \xi \cdot E = \text{const.}$$

This last expression is the equilateral hyperbole equation that represents stress behavior as radius varies. The total load $2P$ per unit length perpendicular to a diameter plane is

$$P = \int_{r_i}^{r_e} \sigma_t dr = \int_{r_i}^{r_e} \frac{\xi E}{r} dr = \xi E \cdot \ln \frac{r_e}{r_i} = \sigma_t \cdot r \cdot \ln \frac{r_e}{r_i},$$

and $P = pr_i$ since the pressure is uniform:

$$\sigma_t = \frac{pr_i}{r \ln \dfrac{r_e}{r_i}}.$$

As regards the overall tension on the pipe, apart from σ_t and σ_r, σ_z must be taken into account, acting along the pipe axis. Because of the pipe's head pressure that produces an overall tensile stress of $p\pi d_i^2/4$, a tensile stress is generated parallel to the tube's axis:

$$\sigma_z = p \frac{\frac{\pi}{4} d_i^2}{\frac{\pi}{4}\left[(d_i + 2s)^2 - d_i^2\right]} = p \frac{\beta^2}{1-\beta^2}.$$

The maximum ideal tension along the internal rim, adopting the breakage hypothesis of maximum deformation, is

$$\sigma_i = \sigma_t - \frac{\sigma_r + \sigma_z}{m} = \frac{p}{1-\beta^2}\left[(1+\beta^2) + \frac{1}{m}(1-2\beta^2)\right].$$

Resolving this relation for β^2 and making $1/\beta = 1 + 2s/d_i$, we obtain s/d_i, and so s as a function of σ_i/p and m (Poisson modulus):

$$s = \frac{d_i}{2}\left[\sqrt{\frac{\dfrac{\sigma_i}{p}+\dfrac{m-2}{m}}{\dfrac{\sigma_i}{p}-\dfrac{m+1}{m}}}-1\right].$$

When longitudinal tension σ_z is not applied, from analogous calculations reported above:

$$s = \frac{d_i}{2}\left[\sqrt{\frac{\dfrac{\sigma_i}{p}+\dfrac{m-1}{m}}{\dfrac{\sigma_i}{p}-\dfrac{m+1}{m}}}-1\right].$$

17.3   STRENGTH VERIFICATION FOR PRESSURIZED VESSELS

The formula above can also be used to design pressurized vessels, which in most cases have spherical or elliptical ends of the same material welded along their circumference.

Sizing the spherical cap, especially when tank length is not long compared to cap radius R ($f/2R < 0.1$), a first approximation can be carried out using the formulae for flat circular plates along a perimeter with a load equal to pressure p uniformly distributed. In these conditions, the maximum bending tension at the plate center of thickness s is

$$\sigma_{max} = \pm\frac{3(3+\nu)}{8}\,p\,\frac{R^2}{s^2},$$

and the maximum ideal tension adopting the hypothesis of maximum expansion is equal to $(1-\nu)\,\sigma_{max}$.

The maximum axial deformation at the cap center is

$$f = \frac{3(5+\nu)(1-\nu)}{16}\,p\cdot\frac{R^4}{Es^3}.$$

For ν equal to 0.3, it proves to be about four times that of a plate of equal size fixed along the outer edge (veta is Poisson modulus, E is elastic modulus of the material).

In cases where both pipe and tank are used in high temperature environments, any possible tensions that could be created through temperature variations must be taken into account. So, to avoid any critical conditions one must predict the possible thermal deformation that must be absorbed by modifying the whole system, particularly to avoid any static indetermination. This calculation can be applied to reactors, or better, to tanks used for chemical reactions in which high pressures and temperatures develop.

The calculation is carried out using the formulae for wide pipes in which the axial tension must always be accounted for due to the surface pressures from the end caps that may be welded or bolted. Usually, at least one end is flanged to allow disassembly for servicing of the active part of the reactor or for loading the catalyzers.

Due to the high pressures generated by chemical reactions, the walls and caps are always thick and made of stainless steel with high deformability prior to breaking (AISI). To calculate the walls, the formulae for wide pipes (above) are used.

To calculate the end caps, a first approximation can refer to the plate theory in various constraints. For an idea of maximum thickness, the formula for fixed circular caps can be used. If the constraints are close to the fixed part, the maximum stress at the fixed part is

$$\sigma_{max} = \pm \frac{3}{4} p \cdot \frac{R^2}{s^2},$$

about 1.5 times less than the maximum relative to the fixed condition.

For reactors, evaluating the tension due to the thermal gradient between inner and outer surfaces of the cylinder is important because of the significant stress added to the internal pressure. Thermal tension can be calculated using the formulae for "constant thickness discs" in Giovanozzi, vol. II, p. 227. They are part of another design course that should be studied. In these cases there could be "creep" phenomena that over time can create critical situations.

In highly stressed thick tanks, internal tensions can be so high as to create surface plasticization that, over time, could reach the external surface. So, highly deformable steels (AISI) are used to avoid any significant plasticization breakage. These simple formulae can only provide a first maximum sizing. Simply calculated geometry can be a basis for applying the FEM codes that for axially symmetric cases are user friendly.

17.4   SOME OIL PIPE CONSIDERATIONS

Fuel pipes are subject to checks defined by the API standards to guarantee reliability, given the consequences of any loss. Generally, these pipes are thin, so for a nominal diameter of $D_N = 500$ mm, thickness would be 6.5 mm. So, here the Marriotte formula can be applied.

In the case of clearly defined applications for refinery piping, operating pressures are relatively low (2–3 atm). For example, for a pipe of diameter 500 mm, operating pressure p of 2 atm, carbon steel type C20, yield point of 250 N/mm^2, and break point of 400 N/mm^2 (these applications use high plastic deformation steels to favor "leak before break"), the yielding of the pipe longitudinally (plastic deformation followed by yielding) may take place when a pipe degenerates to a thickness of 0.25 mm.

As regards the formation of holes over time (years), a simple calculation can estimate that pipe yield may occur for thickness s that reduces proportionally with the diameter d_1 of the hole according to the approximate formula (valid for hole sizes which are not too small):

$$s = \frac{3pd_f}{8\tau_{max}},$$

where τ_{max} is the maximum tangential tension permissible, about $0.55\ \sigma_s$ (σ_s yield point of material, about 250 N/mm^2).

In this particular case ($p = 2$ atm), you could have a hole of diameter d_f 2 mm when the surrounding thickness degrades to $1/8 \times 10^{-2}$ mm (a little more than a thousandth of a millimeter). For a hole of 20 mm the surrounding thickness needs to have degraded to $1/8 \times 10^{-1}$ mm (ten times more than the previous one). This means reductions that highlight how pipes yield to holes as their thickness thins to almost a hole.

The calculation above, with all its limitations due to its simplistic form (hole like a semi-rim-fixed plate) highlights how pressure tests (run at 1.5 times the operating pressure p) can pinpoint holes that are practically already there. Information about pipe reliability is minimal, as a meagre thickness (for a hole of 2 mm, about 2×10^{-3} mm) to qualify a pipe as usable when instead it is practically unusable.

The points above highlight that even low pressure pipes that develop small holes in the metal pipe walls can avoid fluid loss because the protective coating (e.g., tar) is sufficiently strong. However, the coatings themselves are subject to a number of weaknesses depending on the operating conditions. Apart from physical "laceration," a coating could age through long exposure to the sun or through a high number of extreme thermal cycles, or there could be manufacturing defects such as air bubbles.

The operating conditions of the pipe considered (low pressure, temperatures more or less constant, heat insulated) might lead one to think that any yielding of the coating could be due to a manufacturing defect such as an air bubble that would first swell then burst, allowing oxidants to enter (e.g., water) and lead to the formation of holes.

Furthermore, it should be underlined that this type of pipe not only operates at low pressure but is also subject to practically constant pressure and temperature, so it is not subject to stress fatigue. This means that any rupture could only happen if stress were higher than yield load of the material.

The pipes most at risk for security (inflammability) are those that convey fluids with low flash points, such as crude oil. The less dangerous ones, however, are those that convey steam, water, and highly inflammable fuels like fuel oils. A strict inspection program forces the company to pay special attention to the tarred and insulated pipes that convey the most dangerous fluids (lower flash point) and develop a maintenance program that checks the entire pipework, even those less at risk of flash points.

For tarred and insulated crude pipelines, the inspection begins with a close look at the coating over the entire pipeline with subsequent insulation wherever the pipe shows signs of oxidation, followed by a thermographic investigation to evaluate for any incipient losses; should any anomalies arise, the investigation must bare the area by removing more coating.

Because of the difficulty of identifying points subject to regular external corrosion, an investigation should first identify those tracts at greatest risk of corrosion (sharp direction changes, diameter variations, contact with water, etc.), by continual inspection at $360°$.

For bare pipelines, that are easy to inspect from outside and do not convey low flashpoint fluids, maintenance must be programmed using non-destructive checks (NDC) that are suitable to guarantee reliability by monitoring any cracks. In all cases, as mentioned before, all the critical points of the pipeline must be identified (areas where corrosion could begin), and having subdivided the area into pipe lengths from weld to weld, a risk assessment and an evaluation of remaining lifespan must be carried out.

17.4 INSTABILITY IN EXTERNALLY PRESSED PIPES

In the calculations of pipes (especially thin pipes) pressed externally, or those subject to internal vacuum, it is not sufficient to consider only the tension compression $\sigma = pd_i/2s$ that could give rise to instability. For an idea of what physically happens, imagine a straw. If you blow into a straw, sealing the other end, nothing happens; but if you suck, it starts to deform and flatten.

The elastic deformation of a pipe, which can happen under vacuum, comes in different forms depending on pipe shape and material (Figure 17.3). On the left (Figure 17.3), the deformation behaves as if it were divided into four parts (highlighted in Figure 17.4), the inflexion points being dots in the figure.

To find the critical pressure, an approximation could be made from the Eulerian critical load on beams. It is known that for a shaft the critical load is:

$$P_{crit} = \frac{\pi^2 E J_{min}}{l_0^2}.$$

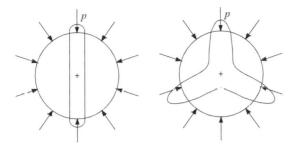

FIGURE 17.3   Qualitative pipe shapes in instability conditions.

In the case in point (Figure 17.4), because of the particular form of deformation (two lobes):

$$l_0 = 0.5 \frac{\pi d}{2}.$$

The inertial moment J_{min}, per unit length is

$$J_{min} = \frac{1}{12} \cdot 1 \cdot s^3.$$

By substituting in the P_{crit} formula:

$$P_{crit} = \frac{\pi^2 E \dfrac{1}{12} \cdot 1 \cdot s^3}{0.25 \dfrac{\pi^2 d^2}{2}} = \frac{4}{3} E \left(\frac{s}{d} \right)^2 s.$$

To determine critical σ_{crit} divide by s:

$$\sigma_{crit} = \frac{P_{crit}}{s} = 1.33 E \left(\frac{s}{d} \right)^2$$

Instability problem emerged for values of the slenderness ratio $\lambda = l_0 / \rho_{zz} > 50$.

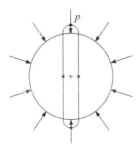

FIGURE 17.4   First deformation mode in instability conditions.

In the case in point $l_0 = \pi d/4$ (two lobes), and ρ_{zz} is

$$1 \cdot s \cdot \mu_{zz}^2 = \frac{1}{12} 1 \cdot s^3 \quad \rho_{zz}^2 \quad \frac{1}{12} s^2 \to \rho_{zz} = 0.289 \cdot s$$

Substituting in the λ expression:

$$\lambda = \frac{\pi d}{2 \cdot 0.289 \cdot s} = 1.36 \frac{d}{s} > 50,$$

for which instability occurs when $d/s > 37$.

The critical internal pressure p_{crit} for this case is

$$p_{crit} = \frac{1.33}{4} E \left(\frac{s}{r} \right)^3.$$

In reality, if there are multiple flanged pipes with their supports, rigidity is increased, so the danger of instability lessens. In this case, the deformation on the right in Figure 17.3 is more probable.

For more than two lobes, by the instability theory of the pipes, the critical pressure is:

$$p_{crit} = 0.11(n^2 - 1)E \left(\frac{s}{r} \right)^3 \qquad\qquad n = 3,4,5\dots \text{ and } 2r = d.$$

Here, it is deformed into three lobes (Figure 17.5) where length l_0 is $l_0 = pd/6$.

$$P_{crit} = 3E \left(\frac{s}{d} \right)^2 s \text{ and } p_{crit} = 0.88E \left(\frac{s}{r} \right)^3,$$

thus obtaining for the P_{crit} a coefficient that changes from 1.33 to 3, and the corresponding d/s decreases to 18.

The formulae above apply also to vacuum vessels, whose calculations must take into account the vessel bottom that can deform into a shape with several lobes. The calculation of $d/s > 18$ could be considered a reference value before using more accurate calculations and models.

Figure 17.6 shows the case of a steel tank vacuum deformation that has several lobes. Calculations show that these were caused by an effect of the load bearing structure revealing a slightly higher vacuum value than if there had been no boundary effects.

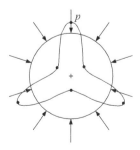

FIGURE 17.5   Second deformation mode in instability conditions.

FIGURE 17.6   Case of a steel tank vacuum deformation.

As can be easily seen, only need very low vacuum levels are needed compared with the values for positive internal pressures to cause a collapse. So, for a steel vessel (σ_{sn} = 400 N/mm²) of 200 mm diameter and 4 mm thick, collapse would be caused by 1/3 of the force required for rupture.

Figure 17.7 shows the first deformation mode obtained by finite element method (FEM) for the vessel in the previous figure. Under vacuum conditions (external pressure), the vessel needs to be appropriately ribbed internally (see sub-oceanic structures).

17.5   CALCULATIONS FOR FLANGE BOLTS

The logic behind flange bolt calculations is the same as that used for flange joints. The junction is a forced fitting with a gasket interposed between the two opposing flange surfaces. It is not easy to calculate the forces acting on the bolts—or rather, the calculation is uncertain. Other uncertainty factors have induced the regulations to refer to simpler methods for evaluating load.

To determine the external forces, we suppose that the fluid pressure is that of a pipe with a larger diameter, so we refer to a calculation that is 2/3B larger (Figure 17.8):

$$D' = D_{N} + \frac{2B}{3},$$

and so to a force on each of the n bolts of

$$\frac{F}{n} = \frac{1}{n}\frac{\pi D'^{2}}{400}P_{N}.$$

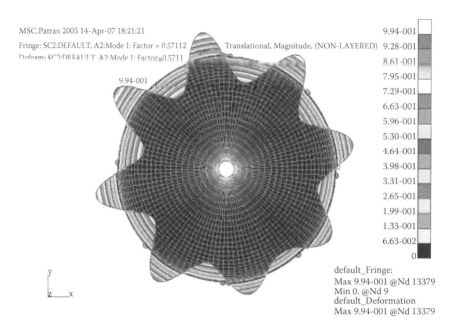

MSC.Patran 2005 14-Apr-07 18:21:21

Fringe: SC2:DEFAULT, A2:Mode 1: Factor = 0.57112   Translational, Magnitude, (NON-LAYERED)

Deform: SC2:DEFAULT, A2:Mode 1: Factor=0.5711

9.94-001

| 9.94-001 |
| 9.28-001 |
| 8.61-001 |
| 7.95-001 |
| 7.29-001 |
| 6.63-001 |
| 5.96-001 |
| 5.30-001 |
| 4.64-001 |
| 3.98-001 |
| 3.31-001 |
| 2.65-001 |
| 1.99-001 |
| 1.33-001 |
| 6.63-002 |
| 0 |

default_Fringe:
Max 9.94-001 @Nd 13379
Min 0. @Nd 9
default_Deformation
Max 9.94-001 @Nd 13379

FIGURE 17.7   Results by FEM application.

According to standard UNI 2215 (formula reference parameters), the maximum acceptable load on each bolt is

$$P = \frac{\pi}{4}(d_n - c)^2 \frac{R}{K},$$

with diameter d_n of the core, coefficient c depending on the bolt core diameter (see standard), and acceptable load R/K depending on steel type (see standard). The simplistic calculation of the static type does not account for important factors such as high temperatures, thrust pressures due to pumps or valves in the circuit, etc. For pressures over 100 atm or temperatures higher than 400°C the simplistic calculation above is not applicable, so the join must be sized with more rigorous criteria.

17.6   FIXED FLANGE CALCULATIONS

The standards also refer to flanges with a very simple approximate calculation.

The model used is that of a beam of height h fixed along the entire length of its circumference of diameter d_1 (Figure 17.9).

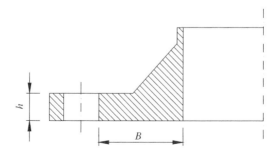

FIGURE 17.8   Section of a fixed flange.

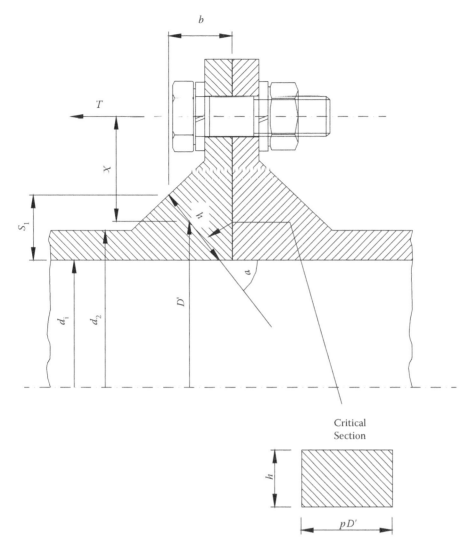

FIGURE 17.9 Fixed flange.

Total load derives from the action of the bolts, which must overcome force F that tends to separate the flange, but which for safety is raised by a 1.15 coefficient. It bends the entire beam and the maximum bending moment is

$$M = 1.15F \cdot x = \frac{\pi D' h^2}{6} \sigma_f,$$

with

$$F = \pi/400 \cdot D'^2 PN.$$

$\sigma_f = \beta \sigma$ in which:

σ_f is the safety bending stress of the material ($\sigma_f = 37.5$ N/mm^2 for Cast Iron, $\sigma_f = 100$ N/mm^2 for rolled steel)

β is a proportionality coefficient greater than 1 reported (1,5 for Cast Iron, 1 for rolled steel)

σ is the safety tensile traction stress of the material (25 N/mm² for cast iron, 100 N/mm² for rolled steel)

Since $h/b = \cos\alpha = s_1/\sin\alpha$ are the geometric values linked to the following relation, we deduce:

$$b = \cos\alpha\sqrt{\frac{3}{2}1.15\frac{D'P_{\mathrm{N}}\cdot x}{100\sigma_{\mathrm{f}}}} \quad s_1 = btg\alpha.$$

The standards for the most common applications use $\alpha = 37°$, so

$$b = 0.106\sqrt{\varphi\frac{d_iP_{\mathrm{N}}\cdot x}{\sigma_{\mathrm{f}}}} \quad s_1 = 0.079\sqrt{\varphi\frac{d_iP_{\mathrm{N}}\cdot x}{\sigma_{\mathrm{f}}}} = 0.75b,$$

with $\varphi = D'/d_i$ and from practical data $1 + P_{\mathrm{N}}/A$, where $A = 150$ for cast iron, $A = 500$ for molten steel, and $A = 700$ for forged and sheet steel. The calculation above is very approximate, the most obvious evidence being to treat the flange (perhaps approximated to a plate) as a beam. However, the simple calculation does have justifiable load uncertainties because of the gasket, which is the fourth system component (pipe, flange, bolts, gasket).

In fairly significant joints such as in piping for dangerous fluids, the geometric parameters that derive from the simple calculation must be treated like first approximation data and must be substituted by more specific data if a closer analysis of the system is required.

17.7 FREE FLANGE CALCULATION

The so-called free flanges are those that can be disassembled (Figure 17.10). To size them, a maximum calculation like the one for cup springs (Chapter 13) is used.

Considering the two forces $F/2$ acting at the center of gravity of the semicircumference (Figure 17.8), they form a torque with moment:

$$M = \frac{F}{2}\left(\frac{D_2}{\pi} - \frac{D_1}{\pi}\right).$$

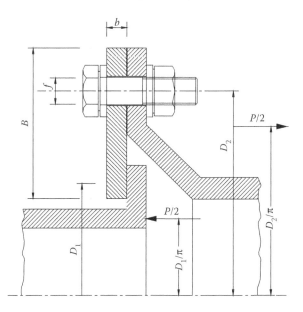

FIGURE 17.10 Free flange.

Referring to the symbols in Figure 17.8, the maximum bending is

$$\sigma_f = \frac{M}{W} = \frac{\dfrac{F}{2}\left(\dfrac{D_2}{\pi} - \dfrac{D_1}{\pi}\right)}{\dfrac{1}{6}2(B - f)b^2}.^*$$

As regards P, which tends to separate the flange halves, this is calculated referring to the pipe's external diameter D_E, or $P = p/400 \cdot D_E^2 PN$. Based on practical data, the standards assume that only a fraction (0.59) of the load P acts on the circumference.

By substituting $0.59P$ instead of F, the flange thickness can be found. The fixed flange values of σ_f can be used.

The component calculations examined thus far have been static. For dangerous fluid pipes where any fluid loss could injure people, sizing the components (bolts, flanges, welds) must take into account the complexity of the system of which it is a part, and must comply with the various norms. In other words, all the other elements in the circuit must be taken into account (pumps, valves, etc.) that could produce dynamic loads: switching on of a pump, or the more or less quick closure of valve(s).

The codes currently on the market for a given geometry provide for the constraints of a system and its various components and for dynamic analyses that determine the system's own frequencies, including possible system thrusts.

17.8 FLANGE BOLTS

The flange bolt calculation follows the logic of flange joints. This means a forced fitting joint between which there is often a soft gasket. Calculating the exact forces exerted on the bolts is not easy, or at least might be uncertain. Other factors of uncertainty oblige the norms to refer to simple methods for evaluating loads.

To calculate the external forces, the fluid pressure is assumed to be that for a higher diameter pipe, so the calculation diameter is increased by $2/3B$ (Figure 17.11):

$$D' = D_N + \frac{2B}{3},$$

so the force on each bolt is

$$\frac{F}{n} = \frac{1}{n}\frac{\pi D'^2}{400}P_N.$$

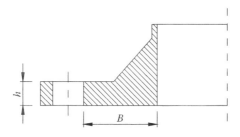

FIGURE 17.11 Symbols in a fixed flange.

According to UNI 2215 (refer to standards for parameter values), the maximum acceptable load on each bolt is

$$P = \frac{\pi}{4}(d_n - c)^2 \frac{R}{K}.$$

- Bolt diameter d_n
- Coefficient dependent on bolt diameter c (see norm)
- Safety coefficient K (see norm)
- Acceptable load R/K depending on steel type (see norm)

The simplistic calculation (static), does not take into account significant factors that could present in some applications, such as high temperatures, thrust pressures from pumps and valves, fluid hammer, etc.

For high pressures over 100 atm or temperatures above 400°C, the simplistic calculation above is not applicable so the joint must be sized with more rigorous criteria as indicated later.

17.9   NOTES ON HOT METAL CREEP

Before drawing this chapter to a close, we wish to note a phenomenon that applies to all pressurized containers (the latest standards consider all pipes as pressurized containers): the "creep" of hot materials. This phenomenon concerns materials over a certain temperature (as we will see later) and concerns many components in the chemical and petrochemical industries.

Creep occurs in materials subject to loads over certain temperatures when they show progressive plastic deformation over time. Hot creep is basically an accumulation of deformations of a material subject for long periods to loads and temperatures over a certain creep threshold.

In general, creep can take place at any temperature, but only if operating temperatures (T) exceed 0.4 times the material's melting point (Tf) can significant alterations occur ($T > 0.4\ Tf$). For steel, creep is insignificant below 300°C, whereas for aluminum its effects are already visible at 100°C.

When there is creep, deformation can no longer only be defined as a function of applied load ($\varepsilon = f(\sigma)$), but as the variability of it with time t and operating temperature T of the material ($\varepsilon = f(\sigma, t, T)$). Creep propagation velocity or deformation velocity $\dot{\varepsilon}$, generally varies with load as well as with the operating temperature of the material.

Numerous mechanical components operate at high temperatures, and under load their constituent metals are subject to permanent plastic deformation that can lead to breakage. Creep analysis is of particular practical importance on account of the huge number of applications of metallic materials at high temperatures, e.g., steam generators, chemical and petrochemical plants, gas turbines, and high pressure piping.

So, it is obvious that the design criteria for mechanical structures or components operating at high temperatures differ considerably from those used at room temperature. Creep analysis is carried out by experimentally determining the deformation ε curve as a function of time t from creep test in which load and temperature testing remain constant until the breakage (Figure 17.12).

The creep curve varies considerably with test loads and temperatures. Figure 17.12 shows two curves at different temperatures. At low test temperatures, the creep curves are simpler and linear compared to those at higher temperatures ($T > 0.4Tf$) with the same loads. This shows a significant difference in behavior at low and high temperatures. The low temperature creep curve has a single phase, whereas for the high temperatures it shows three distinct phases, referred to as primary, secondary, and tertiary creep.

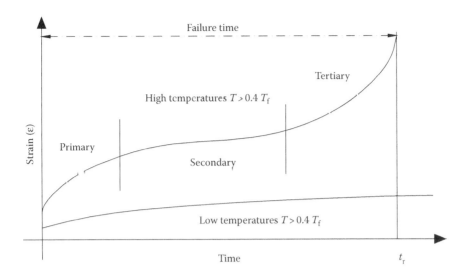

FIGURE 17.12 Characteristic curve in creep analysis.

From a qualitative point of view, the overall trend of the creep curves at high temperatures over various times is closely related to work hardening as a result of previous plastic deformation, leading to a decline in deformation as well as to softening, which in turn replaces the prior work hardening.

In terms of creep trend, for a fixed temperature, "creep limit" is defined as when the constant unit load causing permanent plastic deformation or creep ceases, after infinite time or a very long defined time or (in practice) grows with a constant velocity tending to zero, and for less time than that defined. The "load limit for creep over time" is defined, by contrast, as that load that leads to material breakage due to advancing plastic deformation within a pre-fixed time.

Plotting "load limit for creep over time" versus time, curves analogous to those of Wöhler are produced.

To analyze hot creep, extended time trials (1000 h) or short ones (10 h) can be carried out.

From the results of creep tests for various materials, creep limit graphs are reported for creep limits defined by the organization or laboratory performing the analyses.

17.9.1 CREEP TESTS

As described above, creep tests are generally run keeping the temperature 0 and mono-axial load or force σ constant. Figure 17.13 shows the graphs of deformation ε and that of deformation velocity $\dot{\varepsilon} = d\varepsilon / dt$ both as functions of test time t.

These graphs highlight three subsequent instant deformation ε_0 phases in which overall deformation is the sum of one elastic ε_0 and one plastic component ε_{0p}. At the start ($t = 0$), as a result of a large reduction in the resistance modulus E of the material due to the high temperature, the load applied to the test sample produces an immediately considerable elastic deformation ε_0. In this case, the deformation velocity ($\dot{\varepsilon} = d\varepsilon / dt$) is theoretically infinite due to the significant increase in deformation during an infinitesimal time.

Primary creep (I) reveals a decreasing deformation velocity so, over time, the material deforms less quickly up to a constant value defining the next phase. At low temperatures, this trend is continuous and indefinite up to breakage. At high temperatures, the continuous decrease in creep velocity leads to a condition that is independent of the time and deformation velocity, constituting the second creep phase.

Secondary creep (II) reveals a minimal and constant deformation velocity ($\dot{\varepsilon}$) in which there is generally considerable deformation. At the end of II, deformation velocity tends to rise again.

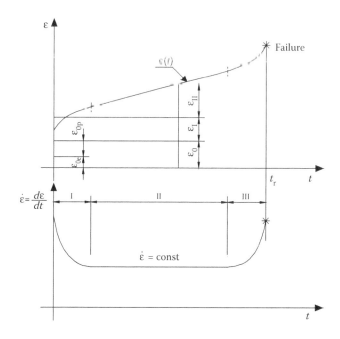

FIGURE 17.13   Curves $\varepsilon(t)$ and $d\varepsilon/dt$ versus t.

Subsequent to the primary and secondary phases there is a significant increase in deformation with a consequent rise in relative velocity $\dot{\varepsilon}$. This behavior is typical of tertiary creep (III) when it proceeds rapidly until the material is no longer able to deal with deformation and breaks completely.

This phase is connected microscopically and macroscopically to a reduction in the resistant cross-section, as well as to material softening. At the end of this phase and close to breakage, there are various instability phenomena that produce many transverse neck-ins with consequent breakage.

In many materials with significant tertiary creep it is often difficult to apply the most co mmon constituent factors to analyze creep because of the non-homogeneity of the load σ and deformation ε. These constituent relations presuppose a constant tension on the test sample (σ = cost).

In these cases, the high level of approximation that constituent factors employ must be taken into account.

For hot creep, deformation is significantly influenced by the load σ, apart from temperature (Figure 17.14).

At low temperatures, the influence of load on deformation is almost insignificant compared to that at high temperatures. At high temperatures, even a small increase in load from σ_3 to σ_2, leads to a creep curve change towards much greater deformation with a consequent significant reduction in lifespan and break point.

As load σ decreases, there is a decrease in secondary creep velocity $\dot{\varepsilon}$ to a steady state and an increase in secondary creep amplitude at the cost of the tertiary and primary phases, even though to a much lesser extent.

The experimental curves can extrapolated by fitting the experimental data with analytical curves. Generally, total deformation (ε_{tot}) to which the test sample is subject from instant to instant is a function of the type:

$$\varepsilon_{tot} = \varepsilon_0 + \varepsilon(t),$$

where ε_0 is the initial elastic deformation to which the test sample is subject from moment to moment and $\varepsilon(t)$ is creep deformation at a generic moment.

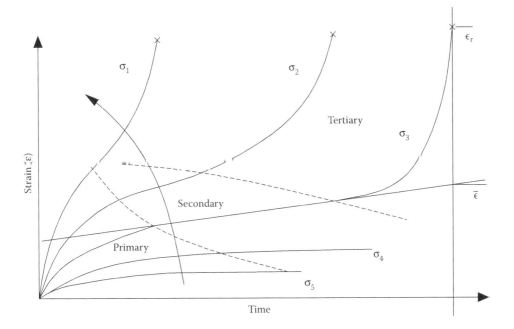

FIGURE 17.14 Creep curves as stress parameter.

Plotting the elastic deformation ε_0 curve can be done via the creep curve (Figure 17.15.a) considering the deformation at the instant of load application, or via the load deformation curve (σ, ε), which depends on test temperature T (Figure 17.15b).

The dependence of ε_0 on load σ and operating temperature is expressed as

$$\varepsilon_0 = f_1(\sigma,T).$$

Creep deformation is a function of time t as well as load σ and temperature T, so

$$\varepsilon_c = f_1(\sigma,T,t).$$

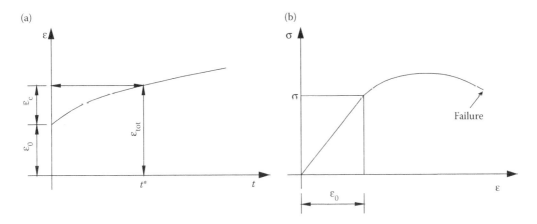

FIGURE 17.15 Curve ε versus t (a), curve σ versus ε (b)

At low temperatures ($T < 0.4T_f$), many crystalline materials show a common and linear increase in creep deformation ε_c with the logarithm of time t. So, in these conditions the deformation trend is a function of time of type:

$$\varepsilon(t) = \varepsilon_{tot} - \varepsilon_0 = \alpha_1 \ln(\alpha_2 \, t + 1),$$

where α_1 and α_2 are two constants depending on the material, load, and operating temperature.

For temperatures lower than $0.3T_f$, these constants are not subject to significant changes as functions of $\dot{\varepsilon}$ and T, whereas for temperatures between $0.3T_f$ and $0.4T_f$, variations are more marked.

Differentiating the previous equation, deformation velocity is

$$\dot{\varepsilon} = \frac{\alpha_1 \cdot \alpha_2}{\alpha_2 \cdot t + 1}.$$

From an analytical point of view, this expression coincides with the experimental results of a continual decrease in deformation velocity at low temperatures. For temperatures of $T > 0.4 \, Tf$, the previous expressions are no longer suitable for describing deformation ε over time. These were in fact the result of analyzing material trends at low temperatures, where deformation mechanisms are completely different.

This is the reason why the creep trend of materials is generally described by different analytical relations as a function of operating temperature and, in some cases, also as a function of deformation phase—primary, secondary, and tertiary.

Among the analytical expressions applied to hot creep, the most famous and widely used is McVetty's:

$$\dot{\varepsilon} = \dot{\varepsilon}_0 + c \cdot e^{-\alpha t},$$

where $\dot{\varepsilon}$ represents the velocity variation of unit deformation ε with time t, and $\dot{\varepsilon}_0$ and c are two constants.

By integrating:

$$\varepsilon = \varepsilon_0 + \dot{\varepsilon}_0 t - \frac{c}{\alpha} \cdot e^{-\alpha t}.$$

For $t \to \infty$, $\varepsilon = \varepsilon_0 + \dot{\varepsilon}_0 \, t$, there is a linear deformation e increase with time according to the asymptote of curve $\varepsilon(t)$, which is identical to the curve itself after a certain point. Various authors present other formulae connected to certain representations of the experimental results.

Determining lower creep limits based on long and more extended tests are obviously more reliable and precise. However, compared to what happens when determining the fatigue limit of a certain material, from experience, there is a more or less constant relation between the long-term results and the short-term. So, it is possible to approximate creep limit well on only short-term tests using variously fixed rules by various norms, laboratories, and organizations.

17.9.2 Microstructural Aspects of the Deformation Phases

The three creep phases are linked on a microscopic scale by transformations that, although they start with stable and dynamically balanced microstructure deformations, subsequently lead to degradation and total breakdown.

Material deformation is the consequence of complex mechanisms at the microstructural level, for example, lattice defect movements. These originate from external variables such as temperature and load, as well as from the material's own crystalline lattice characteristics.

Secondary creep is significant given that it reveals deformation velocity $\dot{\varepsilon}_s$, known also as creep resistance. As explained above, $\dot{\varepsilon}_s$ is only a function of external and internal (θ, σ) conditions, as well as being independent of time, given that it remains constant throughout the whole of the secondary phase.

High temperature deformation mechanisms are

- High temperature plasticity with dislocation glide or dislocation glide and dislocation climb
- Deformation by vacuum diffusion both at the edge of crystals and within the lattice

There are two types of dislocation mechanisms, glide and climb, giving rise to dislocation creep.

During dislocation glide the material is subject to much lesser loads than those that cause crystalline lattice collapse.

The consequent motion of the dislocation produces permanent deformation that might be impeded by crystal edges or other dislocations.

Dislocation glide generally occurs at temperatures below $0.5T_f$ and with loads higher than those for yield.

Glide velocity is proportional to force per unit length, which acts along the line of dislocation according to the dislocation mobility coefficient, a function of temperature.

It follows that the tertiary phase deformation velocity depends on applied load.

Dislocation climb is that movement that occurs at higher temperatures than those for dislocation glide and by vacuum diffusion towards the dislocation nucleus edge where there is most stress. As a consequence, atoms move towards less stressed zones and the dislocation nucleus shifts in the crystalline lattice.

Essential to this mechanism, atomic diffusion can take place either within the crystalline lattice or along the dislocation nucleus, thus constituting two different sub-mechanisms. These two atomic diffusion mechanisms are known as "high temperature climb" and "low temperature climb."

The first happens at high temperatures $(T > 0.6Tf)$ and low stress levels; experimentally, deformation is controlled by diffusion in the crystalline lattice.

17.9.3 Physical Interpretation of Creep

According to the physical interpretation of the Bailey–Orowan model, creep is the result of competition between two opposing mechanisms, recovery, and work-hardening. Recovery is the phenomenon produced when a material subject to stress becomes more ductile, reacquiring its deformation capacity.

Usually, the conditions necessary to activate recovery are high temperatures, this suitable and variable activation energy being Q, typical of thermally activated phenomena.

Work-hardening is the mechanism that makes the material ever harder and difficult to deform.

Work-hardening might take place when, during a tensile stress test, the test sample requires more than its load σ to increase the deformation between yield and breakage. After load σ is applied, elastic deformation occurs, but applying a subsequent load makes the material harder and less deformable. Recovery does not happen at low temperatures and creep velocity decreases. However, at high temperatures, recovery might occur (if thermally activated), thus generating secondary creep during which work-hardening acts simultaneously.

So, during secondary creep, recovery occurs at the same time as work-hardening, the two opposite mechanisms of deformation. With increasing operating temperatures, there is an increasing risk of activating recovery, which tends to overwhelm work-hardening.

By contrast with the first two phases of creep, it is not possible to analyze the tertiary phase with Bailey–Orowan model since it is the result of micro- and macroinstability. In particular, microstructural defects such as cavities, crystal-edge defects, and different breakage types prevent definition of a single model.

Moreover, during tertiary creep the test sample is subject to continual transverse necking-in with its consequent increase in load, and this must be taken into account. Because creep velocity $\dot{\varepsilon}$ depends on load, there is a significant increase in deformation ε as well as $\dot{\varepsilon}$ close to the structural defects in the material. This leads to a considerable increase in the number and size of microstructural defects, which in turn leads to a decrease in the transverse cross-section of the test sample.

Deformation tends to increase just where there is a smaller cross-section because local stress is greater than in any other part of the test sample. So, a continuous degradation mechanism is triggered, leading to ever-greater deformation up to the final collapse of the test sample.

17.10   BRIEF NOTES ON TESTING CREEP

As seen above, when operating temperatures near 0.4–0.5 melting point, creep can occur even at values less than yield tension and variations in Young's modulus.

If the temperature is not high, initial deformation ε_0 remains constant ($\varepsilon = \varepsilon(t) = $ cost. $d\varepsilon/dt = 0$) and the deformation disappears as soon as the load is removed. Increasing the temperature and keeping the load constant means that at a certain temperature, further deformation than ε_0 will start, which will vary with time. During primary creep, growth velocity after an initial phase of increase slows to a subsequent steady state ($d\varepsilon/dt = $ cost) and finally in the tertiary phase velocity grows again until collapse.

So, the main parameters influencing creep and in particular deformation are load, temperature, and time. Creep is usually plotted such that for a given load, unit deformation is a function of time against temperature (T in Kelvin), or for a given temperature unit, deformation is a function of time against load (applied tension).

Studying and characterizing materials due to creep, taking into account the four parameters, would take forever (practically impossible). The designer needs data that allow the testing of structural danger when a material must work or has worked under creep.

To link deformation under different operating conditions (deformation imposes the main constraints in practice) to other parameters and obtain laboratory data with relatively fast tests, we refer to the Larson–Miller relation $PLM = T(C + \log t_r)$ that defines a parameter as a function of the material's characteristics ($C = 20$ for ferrite steels), temperature T [K] and breakage time in hours. It summarizes the physical observation of which loads are critical for which high or low temperatures and for which times.

So, characterizing the material could be carried out by subjecting it to a certain load at high temperatures for brief periods. Via the PLM, it is possible to create a $P = f$ master curve for any given material, and having decided on a behavior model, to use the master curve for calculations of durability and lifespan percentage spent. (For apparatus under pressure, inspection organizations provide guidelines for creep calculations.)

The tension values reported correspond to the creep limit at a given temperature between 25°C and 35°C (according to different recommended guidelines) where the applied load produces a deformation of 0.001%/hr or at the end of the test there is a permanent deformation of 0.2% (the smallest of the two). So, just as in fatigue graphs, the lifespan spent can be related to creep limit corresponding to 1×10^6 hr operation. This can provide damage per hours of operation in different conditions and trace the new curve (acceptable stress-time) for the subsequent hours of operation.

The recommended guidelines of a number of inspection organizations (refer to them for further information) also take into consideration the calculation of the lifespan fraction spent due to fatigue and creep damage. By assigning life-cycle percentage due to fatigue to the x-axis, and lifespan percentage spent in hours to the y-axis, a linear curve can be plotted for each material, identifying breaking point.

18 Welded Joints

18.1 INTRODUCTION

Welding is a technique for permanently joining metal components by using a molten material that can be either a layer of the components (base material) or a filler metal. The weld must guarantee that the mechanical characteristics of the join are at least equal to those of the base materials.

Of all the fixed joints, welding is economical, simple and the least bulky. Compared with riveting, there is no weakening of the structure by drilling. Its disadvantage is the dependence on manpower and the uncertainty of the weld's behavior under stress. Furthermore, the use of significant heat could be conducive to cracking in the base material as well as in the weld itself.

When the base material is melted the weld is called *autogenous*, if not it is *heterogeneous*.

Autogenous welding is either by *melting* or by *pressure*.

In welding by *melting*, the join happens by melting base material together with filler metal into a melt bath that then solidifies into a weld bead. The two are distinguished by their heat source:

- Arc welding: heat is produced by a voltage arc between the base material and a metal electrode that gradually melts, providing the filler metal. It is quite commonly used in metal structural work and pressurized containers. Norms UNI 1307/1 and /2 specify welding procedures and join types.
- Arc welding can use:
  - Jacketed electrodes
  - Infusible electrodes in an inert gas (TIG)
  - Submerged arc
  - Continuous flow in inert gas (MIG)
  - Continuous flow in an active gas (MAG)
  - Continuous flow without gas
  - Electrogas
  - Plasma
- Gas: heat is produced by the combustion of gas (acetylene, hydrogen). The weld material is wire or rods.
- Thermite: heat is produced by the chemical reaction between aluminum and iron oxide (components of thermite) and the filler metal is molten iron, the reaction by-product.
- Electron beam: heat is produced by a focused beam of electrons that penetrates the base material. Carried out in a vacuum, it can handle thicknesses greater than 100 mm.
- Laser: heat is produced by a focused laser (monochromatic light) beam and can handle thicknesses up to 15 mm.

In *pressure* welding, the base materials are heated locally to a plastic state and joined by mechanical pressure. The heat source determines the type:

- Heating element: heat is produced by the Joule effect due to ohmic resistance that opposes the flow of electric current across the surfaces in contact.
- Spark: heat is produced by voltaic arcs formed by potential difference between the two surfaces to weld.

- Fire: once the base materials have been brought to plasticity they are joined by hammering or mechanical pressure.

Brazing and *braze welding* are *heterogeneous* welding techniques. Brazing is when the join is made by melting a filler metal between the very hot surfaces to be joined. It is used when the base material's temperature must be limited; when the base materials are difficult to weld, or are totally different making welding impossible; or when aesthetics of the join are a priority or indispensible.

In braze welding, the method is very similar to gas welding except that the base materials are not close to melting and a filler metal is used with a melting point lower than the base materials. The welded join is where the base materials are joined; the minor surfaces delimiting the base materials are called edges, rims, or heads, and the major surfaces are called faces.

The weld can be subdivided into three parts (Figure 18.1):

- Base metal, when there is no variation in material
- Transition, when the base metal and filler metal have melted together and then begun to re-crystallize
- Filling, when the filler metal is chemically and physically similar to the base steel but has significant extensibility

Depending on external surface, welds are distinguished as planes, convexes, and concaves (Figure 18.2). Concaves are the most efficient because they avoid the weld notch on the outer edge, which improves join fatigue.

The norm relating to welding procedures is UNI 1307/1 "Terminology for Welding Metals—Welding Procedures 1986," and the norm relating to types of weld is UNI 1307/2 "Terminology for Welding Metals—Weld Types 1987" (Figure 18.3).

The qualification for this technology and its procedure is the certification ASME PQR/WPS, and for a manual welder, ASME WPQ.

18.2   WELDING DEFECTS

There are two types of welding defects:

- Lack of homogeneity (chemical or structural) between the welded zones and/or a thermally altered zone and base material, which can be identified by mechanical or metallographic tests.

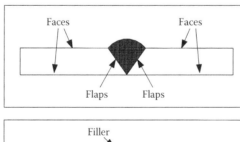
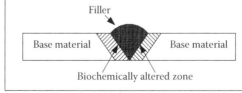

FIGURE 18.1   Parts of a weld.

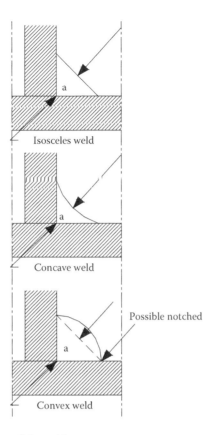

FIGURE 18.2 External surfaces of the welds.

- Geometric discontinuity, identifiable by non-destructive checks. According to type and origin, these are distinguished as: (see Figure 18.4)
  Lack of penetration and melting: edge discontinuities caused by no melting of both parts or of one part.
  Clinker pollution: this derives from a loss of control of the melt bath, being particularly frequent on resumption and in first passes. When the first pass has a curved profile,

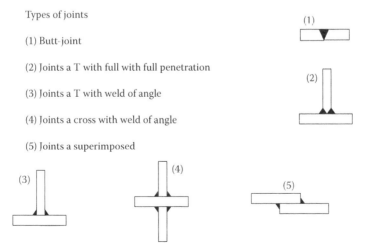

FIGURE 18.3 Common types of welded joints.

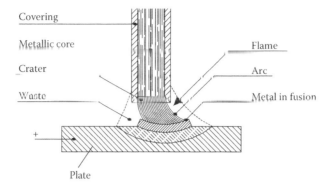

Covering

Metallic core

Crater

Waste

Flame

Arc

Metal in fusion

+

Plate

FIGURE 18.4 Welded zone.

the next pass only melts the uppermost layer of the first without getting to the corners and edges of the curvature, meaning a lack of local penetration. If all the clinker were removed, there would be an elongated semi-weld because of the lack of melting in that zone. Limited penetration and melting are due to excessively fast passes and incomplete caulker filling.

Semi-weld: semi-weld is similar to a lack of melting with the interposition of oxide.

Blowholes: blowholes are spherical gas-filled cavities with 1mm diameters also called *pores,* which can be isolated or in groups. If they are elongated they are called *woodworm.* They are frequent in first passes using basic rods and may be caused by a too-long arc or by long electrodes. They can also be caused by voltage drops or rusty edges.

Magnetic repulsion: this phenomenon occurs independently of the electrode type and is when the electric arc deviates instead of running with the weld bead. The welder loses control of the melt bath causing semi-welds or clinker. It can be diminished but not completely eliminated if the arc is doubled.

Hot cracks: generally in the melt zone, these are high-temperature breaks during join solidification. They happen in steels when C and impurity (mainly S and P) content is high in the base materials.

Cold cracks: generally in the melt zone and thermally altered zones, these happen when the join temperature falls below 100°C. One cause is humidity within the electrode jacket, which can hydrogenate the weld causing fragility in the part adjacent to the weld bead.

18.3   WELD TESTS

To evaluate weld strength, welds are subject to mechanical tests of tensile stress, folding endurance, bending, and crushing.

To verify weld status, non-destructive tests are carried out: x-rays, acoustic radiation, and liquid penetration. The results are measured against the acceptable limits imposed by the norms (UNI, ASME, IIW). In all important applications and in accord with the norms, welds are x-rayed for quality certification. When the installation is tested, reference is made to these x-rays.

In some cases, when a defect is detected, the weld must be re-done. The following are two examples of re-welding.

- Lamina cracks: the area is made mirror clean, a penetrating red liquid is sprayed on the area and left for 15 min, the area is rinsed with water, a penetrating white liquid is sprayed and left for 5 min. Any cracks will be highlighted. At each end of the crack a hole is drilled which prevents further propagation of the crack. Along the whole length of the crack a grinder is used to make a V-shaped indentation that can then be welded.

- Tubing cracks: the weld is removed and the area is cleaned. If little has been removed the area is re-welded. If a lot has been removed, the area is re-created by repeated over-beading.

18.4   WELD STRENGTH

Weld strength depends on different factors. Subjecting a weld (with filler metal) to tensile stress ought to cause a break in the weld at a stress load that is slightly less than if the workpiece metal were continuous, that is, un-welded. The break occurs not along the weld bead but along its edge (Figure 18.5). The loss of weld strength is due to the workpiece metal heating up and cooling non-homogeneously. Metal hardening does not greatly influence the tensile stress breaking point, but it does lower (sometimes substantially, according to the steel) the fatigue limit.

From this point of view, gas welds are better than electrical welds because they cool more slowly. In welds subject to fatigue, bead plasticity is more important than great tensile stress resistance. Welds cause internal tensions because thermal deformation is impeded by constraints of the material components.

During cooling, the weld bead cannot contract and therefore remains under tensile stress along its whole length. It is better to let the thermal dilations take their course whenever possible and then, if needed, straighten any deformed parts. It is even possible to impose preventative deformations or particular expedients on the parts needing welding.

The final characteristics of the weld can be significantly improved by subsequent thermal treatment (normalization) to reduce internal tension: *hot hammering*, which refines the grain size, and *surface finishings* (*grinding*), which reduce any discontinuities of form with their consequent intaglios.

18.5   STATIC CALCULATION

For the variety of stresses to which the weld bead is subject, it is impossible in practice to calculate exactly the tension within it, nor can experience of its shape reveal the non-homogeneities and discontinuities within a filler metal whose characteristics are imperfectly understood, as are those of the zone in proximity to the weld bead. Furthermore, the amount of thermal tension due to cooling is uncertain, and this has a great influence on fatigue resistance.

However, the FEM codes are able to make close approximations to real deformation and tensile stress, as well as to factor in form all of which can help the designer. Following the norms, the calculation again refers to the simple formulae that derive from Saint Venant's theory which are easy to use. The tension calculated refers to an ideal resistant cross-section with sides a and l such that they are less than $\lambda < 1$ of the acceptable tensile stress of the workpiece metal. a is the critical cross-section (in head welds it is the minimum thickness between the sheets/plates; in corner welds it is the height of the right-angled triangle inscribed in the bead), with total length L, and effective length $l = L - 2a$,

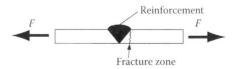

FIGURE 18.5   Fracture zone.

The tension due to tensile stress P or shearing T are

$$\sigma = \sum \frac{P}{(a*l)} \qquad \tau = \sum \frac{T}{(a*l)},$$

The tension due to a bending moment M_f or to a twisting moment M_t are

$$\sigma = \frac{M_f}{W_f} \qquad \tau = \frac{M_t}{W_t},$$

with the resistance modulus W referring to the resistant cross section.

The overall normal tensile stress σ and the total tangential τ make up the conventional ideal tensile stress according to

$$\sigma_i = \sqrt{(\sigma^2 + \tau^2)}.$$

18.6 STATIC CALCULATION ACCORDING TO NORM CNR-UNI 10011

As an example, this paragraph refers to the norm CNR 10011, but it is clear that there are different norms that indicate the guidelines, such as DIN or AWS where the welding symbols are also specified.

As regards the static and fatigue resistance of a weld the current norm is UNI EN 1993: Eurocode 3 (Designing Steel Structures), which has replaced CNR-UNI 10011 (1988) (Steel Constructions— Instructions for Calculations, Building, Testing and Maintenance) since 2005. However, reference will be made to CNR-UNI 10011 because it is the norm most commonly used by Italian designers.

The norm CNR-UNI 10011,

- Refers to the no alloyed steel
- Refers to steel Fe360, Fe430, Fe 510 UNI 7070 (actually as S235,S275,S355, UNI-EN 10025)
- Requires the qualification of welding procedure
- Requires the compliance of electrodes to the classes of the UNI 5132
- Defines the following classes of welding:
  - Class I and Class II for butt joints or at full penetration
  - Single class for joints with strings corner

	R_m [MP$_a$]	$R_{\alpha m}$[1] [MP$_a$]	$\sigma_{\alpha mm}$[2] [MP$_a$]	
			$t \leq 40$	$t > 40$
Fe360	360	235	160	140
Fe430	430	275	190	170
Fe510	510	355	240	210

18.7 STATIC CALCULATION—HEAD AND COMPLETE PENETRATION WELD

Figure 18.6 shows the following tensions are at play in the weld bead:

- $\sigma_{\perp}$ tension—tensile stress and/or compression perpendicular to the longitudinal cross-section of the bead
- $\tau_{\parallel}$—tangential tension in the longitudinal cross-section of the bead

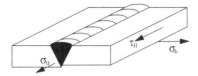

FIGURE 18.6   Stresses at play in the weld bead.

- $\sigma_{\parallel}$ tension—tensile stress and/or compression parallel to the bead axis
- L: bead length
- a: least thickness connected or thickness of completely penetrated material
- H: total width of workpiece metal + filler metal

The resistant cross section of the bead is

1. For stresses from normal traction forces to the bead axis and for stresses from shearing force $A_{res} = La$
2. For stresses from tensile forces parallel to the bead axis, $A_{res} = Ha$

$$\sqrt{\sigma_{\perp}^2 + \sigma_{\parallel}^2 - \sigma_{\perp}\sigma_{\parallel} + 3\tau_{\parallel}^2} \leq \begin{cases} \sigma_{adm} \\ 0.85\sigma_{adm} \end{cases}.$$

The static verification of the weld bead is subject to the following limitations: the upper limit is valid if the weld is class I; the lower limit is valid if the weld is class II.

18.8   STATIC CALCULATION—WELDS WITH CORNER BEADS

Figure 18.7 shows the following tensions are at play in the weld bead:

- $\sigma_{\perp}$ tension—tensile stress and/or compression perpendicular to the bead axis.
- $\tau_{\parallel}$ tangential tension depending on bead axis.
- $\sigma_{\parallel}$ tension—tensile stress and/or compression in the transverse cross-section of the bead: not considered
- L: bead length
- a: throat height (of the triangle circumscribed by the transverse section of the bead
- p: bead foot

The resistant cross-section, both for tensions from tensile stress perpendicular to the bead axis and tensions from shearing along the axis, is the "throat section": area = La.

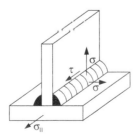

FIGURE 18.7   Stresses for a T joint.

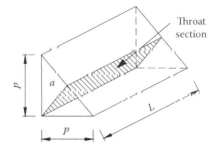

FIGURE 18.8   Throat section.

If the bead is symmetrical $a = p/\sqrt{2}$ (Figure 18.8).

The "throat section" should be turned over on one side of the bead so as to identify the tension components $\sigma_\perp$, $\tau_\parallel$, $\tau_\perp$ (Figure 18.9).

Note that depending on which way it is turned over, $\sigma_\perp$ becomes $\tau_\perp$ and vice versa; $\tau_\parallel$ remains unchanged.

The static verification provides for the following limitations:

- If $\sigma_\perp$, $\tau_\parallel$ and $\tau_\perp$ are present:

$$\sqrt{\sigma_\perp^2 + \tau_\perp^2 + \sigma_\parallel^2} \leq \begin{cases} 0.85\sigma_{adm} \\ 0.70\sigma_{adm} \end{cases}$$

$$|\sigma_\perp| + |\tau_\perp| \leq \begin{cases} \sigma_{adm} \\ 0.85\sigma_{adm} \end{cases}.$$

- If $\sigma_\perp$ and $\tau_\perp$ are present:

$$|\sigma_\perp|, |\tau_\perp|, |\tau_\parallel| \leq \begin{cases} 0.85\sigma_{adm} \\ 0.70\sigma_{adm} \end{cases}.$$

- If only $\sigma_\perp$ or $\tau_\parallel$ or $\tau_\perp$ are present:

$$|\sigma_\perp| \quad and \quad |\tau_\perp| \leq \begin{cases} 0.85\sigma_{adm} \\ 0.70\sigma_{adm} \end{cases}.$$

The upper limit is valid for Fe360; the lower limit is valid for Fe430 and Fe510.

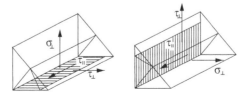

FIGURE 18.9   Stress components.

18.9   LOADS ON THE BEAD

The forces (and any moments) on the bead are calculated by balancing the external loads. Once calculated, the corresponding tension components $\sigma_\perp$, $\tau_\parallel$ e $\tau_\perp$ can be calculated, which are used in the static verification of the most stressed weld. Obviously, the same "throat section" orientation should be used throughout. The figures from 18.10 to 18.15 show different examples of stresses on the bead.

Examples:

$$2F = P \rightarrow P/2$$

Reversing the section on the support:

$$\tau_\parallel = \frac{F}{La} = \frac{P}{2La}$$

• Reversing the section on the plate:

$$\tau_\parallel = \frac{F}{La} = \frac{P}{2La}$$

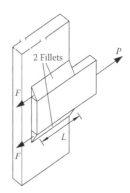

FIGURE 18.10   Joint loaded by axial load.

$$2F_1 = P \rightarrow F_1 = P/2$$

$$F_2 L = Pb \rightarrow F_2 = Pb/L$$

• Reversing the section on the support:

$$\tau_\parallel = \frac{F_1}{ha} = \frac{P}{2ha}, \quad \tau_\parallel = \frac{F_2}{ha} = \frac{Pb}{haL}$$

• Reversing the section on the plate:

$$\tau_\parallel = \frac{F_1}{ha} = \frac{P}{2ha}, \quad \tau_\parallel = \frac{F_2}{ha} = \frac{Pb}{haL}$$

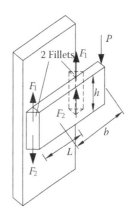

FIGURE 18.11   Joint loaded by transversal load.

$$Fh = M \rightarrow F = M_e/h$$

• Reverse th section on the support:

$$\tau_\parallel = \frac{F}{La} = \frac{M_e}{Lah},$$

• Reverse th section on the plate:

$$\tau_\parallel = \frac{F}{La} = \frac{M_e}{Lah},$$

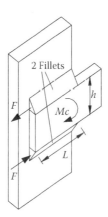

FIGURE 18.12   Joint loaded by bending moment.

$$\uparrow \quad F = P$$

$$\circlearrowright \quad M = Pb$$

- Reverse the selection on the support:

$$\tau_\parallel = \frac{F}{ha} = \frac{P}{ha}, \quad \tau_\perp = \frac{M}{ah^2/6} = \frac{6Pb}{ah^2}$$

- Reverse the selection on the plate:

$$\tau_\parallel = \frac{F}{ha} = \frac{P}{ha}, \quad \sigma_\perp = \frac{M}{ah^2/6} = \frac{6Pb}{ah^2}$$

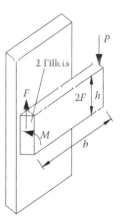

FIGURE 18.13 Joint loaded by transversal load and moment.

$$\swarrow \quad 2F_1 = P \rightarrow F_1 = P/2$$

$$\circlearrowright \quad F_2 h = Pb \rightarrow F_2 = Pb/h$$

- Reversing the section on the support:

$$\tau_\perp = \frac{F_1}{La} = \frac{P}{2La}, \quad \tau_\parallel = \frac{F_2}{La} = \frac{Pb}{Lah}$$

- Reversing the section on the plate:

$$\sigma_\perp = \frac{F_1}{La} = \frac{P}{2La}, \quad \tau_\parallel = \frac{F_2}{La} = \frac{Pb}{Lah}$$

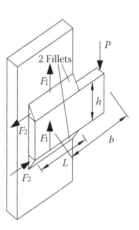

FIGURE 18.14 Joint loaded by transversal load.

$$\circlearrowright \quad FL = M_e \rightarrow F = M_e/L$$

- Reverse the section on the support:

$$\tau_\parallel = \frac{F}{ha} = \frac{M_e}{haL},$$

- Reverse the section on the plate:

$$\tau_\parallel = \frac{F}{ha} = \frac{M_e}{haL},$$

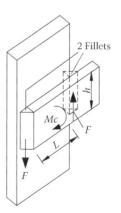

FIGURE 18.15 Joint loaded by bending moment.

Apart from their use in the welds reported above, they are also very common in fluid tubing. Generally, these are inter-tube welds that must conform to the various norms (UNI, API, and ASME). Of particular importance are the welding of gas pipes and the tubing for dangerous fluids, especially in joining tubes of different diameters (critical), or that change direction or cross-section. Generally, for the static calculation, nominal tubing tensions are used (see chapter "Tubing") assuming the precautionary safety coefficients (over 4) depending on fluid pressure, which takes into account the decline in strength characteristics due to salting.

18.10 FATIGUE CALCULATION

To give some idea of how weld fatigue develops, Giovannozzi, vol. I, reports the old German norms relating to calculations for welding bridges. They refer to tension σ_s of the workpiece metal, which is compared to the maximum tension having multiplied coefficient γ, a function of the ratio between the minimum tension σ_{min} and the maximum tension σ_{max}.

More precisely, σ_{max} is multiplied by coefficient γ, a function of the load ratio R ($R = \sigma_{min}/\sigma_{max} \leq 1$). The law of variation of γ is $\gamma = 1 - xR$, where $x = 0.2 - 0.5$ depending on steel and bridge type.

The justification for such a norm derives from the fact that, using the Weyrauch–Kommerell diagram, for materials (usually quality steels) that are simple (with horizontal piece that begins at the ordinate axis) as in Figure 18.16.

The safety coefficient is the ratio between the minimum and maximum tensions for the same minimum tension, or referring to yield tension as the ratio between the point B ordinate (σ_s) and the point B' ordinate ($\sigma_{B'}$), given that

$$(\sigma_{B'}) = \sigma_{max} - \sigma_{min} tg\beta = \sigma_{max}\left(1 - \frac{\sigma_{min}}{\sigma_{max}}\tan\beta\right) = \sigma_{max}(1 - R\tan\beta) = \sigma_{max}\gamma \leq \sigma_s.$$

The same reasoning is applied to tensions from shearing, and therefore when there are normal and tangential tensions:

- The values of γ_n and γ_t, respectively for σ and τ
- The terms $\sigma = \gamma_n\sigma_{max}$ and $\tau = \gamma_n\tau_{max}$

- The ideal stress $\sigma_i = \sigma/2 + 1/2\sqrt{(\sigma^2 + 4\tau^2)}$

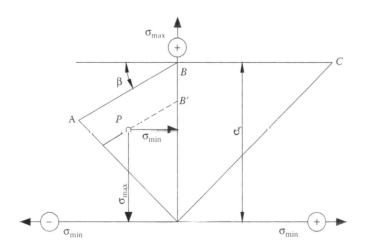

FIGURE 18.16 Weyrauch–Kommerell's diagram.

Thus calculated, this ideal tension is compared to the yield tension of the workpiece metal, having divided the latter by the coefficient α, a function of the weld type, and load ratio R.

The norm reports a whole series of values among which:

- $α = 1$ for welding heads subject to tensile stress
- $α < 1$ in other cases
- $α > 1$ only a special case

18.11   FATIGUE CALCULATION ACCORDING TO NORM CNR-UNI 10011

SN curves on a log-log graph are formed of sections with intervals governed by the equation:

$$\Delta\sigma^m \times N = \text{constant},$$

in which m has different values depending on stress type and cycle number.

Unlike that reported above, the fatigue resistance of welds does not vary with the ratio of stress R and/or average tension. This is due to the high value of residual tensile stress in the real welded structure that modifies the local stress ratio applied to the material in the zone where the crack was triggered.

18.11.1   Fatigue Resistance—σN Curves with Δσ Stress (Figure 18.17)

- Every weld is characterized by $\Delta\sigma_A$, depending on the acceptable stress of constant amplitude for $N = 2 \times 10^6$ cycles
- Welds characterized by $\Delta\sigma_A$ correspond to a group of curves σN with intervals governed by $\Delta\sigma^m \times N = \text{constant}$

These curves are defined by:
- A group of parallel straight lines with $m = 3$ in the field of $10^4 \leq N \leq N_D$ cycles;
- $N_D = 5 \times 10^6$ cycles with $\Delta\sigma_A > 56$ MPa;

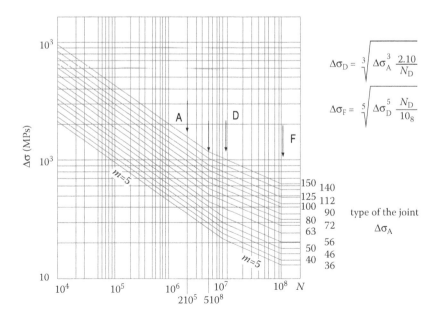

FIGURE 18.17   σN curves with Δσ stress.

- $N_D = 10^7$ cycles with $\Delta\sigma_A < 56$ MPa;
- A group of parallel straight lines with $m = 5$ in the field of $N_D \leq N \leq 10^8$ cycles;
- a group of parallel straight lines from $N_F = 10^8$ cycles.

18.11.2 FATIGUE RESISTANCE—σN CURVES WITH $\Delta\tau$ STRESS (FIGURE 18.18)

- Welds with $\Delta\tau$ stress have a single curve
- SN (the only category 80 weld) governed by $\Delta\tau_m \times N = $ constant and in particular:
  - a straight line with $m = 5$ in the field of $10^4 \leq N \leq 10^8$ cycles
  - a horizontal straight line starting at $N_F = 10^8$ cycles

18.11.3 FATIGUE RESISTANCE—INFLUENCE OF THICKNESS ON ACCEPTABLE Δ

Fatigue resistance $(\Delta\sigma_A)$ is for the weld category thickness $t \leq 25$ mm; for $t > 25$ mm they can be corrected with

$$\Delta\sigma_{A,t} = \Delta\sigma_A \sqrt[4]{\frac{25}{t}},$$

$\Delta\sigma_A$: Δ of acceptable tension (weld category)

- t: thickness in millimeters of the part under most stress $\Delta\sigma_{A,t}$: Δ of corrected acceptable tension
- Fatigue resistance—details

In definitive for the applications:

No verification of fatigue is necessary if:

- All $\Delta\sigma$ are less than 26 MPa or of $\Delta\sigma_D$
- All $\Delta\tau$ are less than 35 MPa
- The total number of cycles is less than $N = 10^4$ cycles

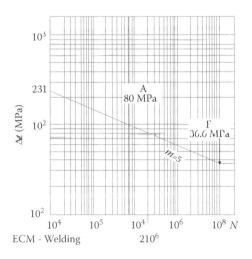

FIGURE 18.18 σN curves with $\Delta\tau$ stress.

$$\text{Miner's rule:} \qquad D = \sum \frac{N_i}{N_i^*} = \sum \frac{\alpha_i N \text{ tot}}{N_i^*}$$

$$\text{If } \Delta\sigma < \Delta\sigma_D e \qquad N_i^* = \frac{\Delta\sigma_A^3 \cdot 2 \cdot 10^6}{\Delta\sigma_i^3}$$

$$\text{If } \Delta\sigma_i < \Delta\sigma_D e \; \Delta\sigma_A > 56 \text{ MPa} \qquad N_i^* = \frac{\Delta\sigma_A^5 \cdot 5 \cdot 10^6}{\Delta\sigma_i^7}$$

$$\text{If } \Delta\sigma_i < \Delta\sigma_D e \; \Delta\sigma_A \leq 56 \text{ MPa} \qquad N_i^* = \frac{\Delta\sigma_D^5 \cdot 10^7}{\Delta\sigma_i^5}$$

$$\text{In the case of } \Delta\tau: \qquad N_i^* = \frac{\Delta\tau_A^5 \cdot 2 \cdot 10^6}{\Delta\tau_i^5}$$

$$CS = \frac{1}{D} \text{ (in terms of life)}$$

FIGURE 18.19   Miner's rule.

For Δ constant stress, the fatigue limit is $\Delta\sigma_D$ ($\Delta\tau_D$).

For variable stress amplitude, Miner's rule (Figure 18.19) is used; every Δ below $\Delta\sigma_F$ (or $\Delta\tau_F$) may be disregarded.

The Δ equivalent method (Figure 18.20) can also be used: to calculate $\Delta\sigma_{eq}$, $m = 3$ is assumed throughout the field ($m = 5$ per $\Delta\tau_{eq}$).

18.12   SPOT WELDING

Spot welding is very commonly used in the automobile industry. It is particularly suitable for welding sheet metal and when there are low stresses and no seal is required. It is not used for pressurized containers or similar because it is a discontinuous weld, for low pressures, using roller welding is used.

Of the two automated types of welding, spot and roller, the current industry preference (especially where thickness is minimal) is for spot welding, perhaps also because it is more economical.

Spot welding is a particularly simple method that works by passing an electric current between two cylindrical copper electrodes (hemispherical or cone), compressing the sheet metal to be

Δ equivalent method

- Used to calculate the $\Delta\sigma_{eq}$ (or $\Delta\tau_{eq}$) equivalent to the spectrum of the applied Δ stress;
- $\Delta\sigma_{eq}$ $\Delta\tau_{eq}$) is the Δ of the stress at amplitude constant, that applied for the number of cycles N equal the total number of the spectrum cycles, produces the same fatifue damage:

$$\Delta\sigma_{eq} = \sqrt[3]{\sum \frac{\Delta\sigma_i^3 N_i}{N}}, \qquad \Delta\tau_{eq} = \sqrt[5]{\sum \frac{\Delta\tau_i^5 N_i}{N}}$$

- The fatigue verification is made by confrontation between the $\Delta\sigma_{eq}$ (or $\Delta\tau_{eq}$) and the $\Delta\sigma_{res}$ (or $\Delta\tau_{res}$), according to the value taken by the corresponding SN line for the total number N cycles of the spectrum
$$\Delta\sigma_{eq} \leq \Delta\sigma_{res}, \qquad \Delta\tau_{eq} \leq \Delta\tau_{res}$$

FIGURE 18.20 Δ equivalent method.

welded. High resistance in the contact zone causes fusion, which on cooling corresponds to the weld spot. In the case of roller welding the electrodes are discs whose thickness corresponds to the thickness of the weld bead. Spot welding highlights how weld quality is a function of the type of material, its thickness, the contact pressure, the timing of current flow (100th sec), and its intensity. Clearly, these parameters are adjusted to produce the best outcome.

Without going into details of materials, timing, etc., for which ample detail is available based on years of experience through the American norms RWMA and AWS, it is worth focusing on some general concepts that are also common to other types of weld.

Welded plate thickness is a function of steel type. Often mild steels are used. Generally, milder is the steel the higer is the thickness that can be welded but better is not to exceed the 4 mm. In the auto industry, spot welding is especially used for assembling body parts where the thickness is in terms of less one millimeter. The type of material also governs other parameters such as time and current intensity. Generally, the higher the current the shorter the time.

As in all other types of welding, the heat developed affects the resistance, so the best results are a combination of the aforementioned parameters, once the type of steel and its thickness have been established. Stainless steel welding also requires either titanium or niobium, which act as stabilizers against the formation of carbide ligands. Spot welding quality is also affected by the type of electrode.

In making a weld, geometric parameters come into play, such as sheet thickness, the electrode tip diameter, and the distance between spots. For particulars, refer to the norms and other specialized publications.

18.12.1 Static Calculation

As mentioned above, spot welds are used where stress levels are not high, and their static calculation can be carried out according to the previous concepts that refer to spot diameter and sheet thickness. For tougher materials, the static calculation requires a different spot diameter. If the spot is subject a shear force T, the average stress, if S is the surface area of the spot is $\tau = T/S$. If the spot is subject to a traction force T (plates face to face) the average stress is $\sigma_c = T/S$. Generally, the norms refer to empirical formulae rather than to static tests to determine spot size. As example the AWS uses the following formula:

$$d = 4\sqrt{s},$$

where s is the least thickness of the two sheets to be welded. Figure 18.21 shows the spot diameters d defined by the various norms as a function of the thickness s of the plates.

As regards resistance to tensile strength and shearing, the norms cite the percentage reduction in resistance due to a decline in the steel's characteristics depending on quality. This varies from 20% for mild steels up to 40% for high carbon steels, which are less suited to this type of welding. As a general rule, an increase in local tension should also be taken into account due to the shape of the welding point, which has little effect on the static resistance but which may influence fatigue resistance.

Tests on spot weld breaks show that they happen directly in the tip (the tip cross-section breaks down), or the sheet steel around the spot breaks down (the tip remains whole), due to thermal alterations that modify the steel's resistance characteristics.

Spot welding, as we have said, is used when stress is generally low, and in practice one single line of spots is used, or more unusually, two staggered lines to avoid any weakening in the sheet metal that might lead to an anomalous break (the material breaks, not the spot). In practice, the use of thin sheets and the various welds above help minimize the effects of bending or other more complex forms of stress.

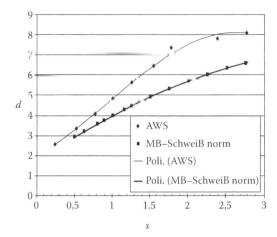

FIGURE 18.21   Confront *d-s* curve for different norms.

18.12.2   FATIGUE CALCULATION

The major auto manufacturers refer to summarised yheir guide lines for the calculation of spot weld fatigue. Here we report general indications by B. Atzori et al.,[*] about the fatigue calculation. In testing for spot weld fatigue, it has been hypothesized that there is a unified dispersion band (Figure 18.22) independent of particular geometry or steel type (this hypothesis is also valid for continuous bead welds).

The results of fatigue tests for constant amplitudes of stress are shown on the double logarithm diagram of Wöhler. This plots relative ratio stress amplitude σ_a/σ_A (where σ_A represents the stress amplitude corresponding to a reference life N_A of 2×10^6 cycles) *(ordinate)* against cycle number *(abscissa)*. Figure 18.22 evaluates the experimental data in terms of weld survival probability. The equation of the 3 lines in the log-log diagram is: $N = N_A(\sigma_a/\sigma_A)^{-k}$ being that N_A and σ_A are the values of the refferring *A* point

Fatigue resistance is the same for mild steels as for low alloy steels with high yield. Even welds between different steel types do not worsen fatigue resistance. Weld fatigue resistance depends exclusively on specific geometry. The parameter with most effect on fatigue resistance is the ratio of inter-spot distance to spot diameter. As this ratio decreases, fatigue resistance rises (given the same conditions). The i/d parameter has a saturation value of ~3.5 below which fatigue resistance does not rise. For $d = 6$, and a standard sheet metal thickness for auto bodies of 1 mm, i would be 21 mm.

For sheet metal up to a thickness of 2 mm, for infinite life designs, or a cycle number over 2,000,000, the following are valid:

	Stress Amplitude (σ_A) [N/mm²] Corresponding to the Fatigue Limit for $R = 0$	
Probability of Survival	$i/d > 3.5$	$i/d \leq 3.5$
50% (for experimentation)	$90/(i/d)$	25
97.72% (for design)	$60/(i/d)$	16

By experimental welded spot data, calculations, such as the one in Figure 18.23, can be performed for different values of survival probability (50%, 97.7%), for constant stress amplitude and

[*] Blarasin, A., Atzori, B., Filippa, P., and Gastaldi, G., Design fatigue data for joining steel sheets in the automotive industry, International Congress and Exposition Detroit, Michigan, Feb. 26 to Mar. 2, 1990. SAE Technical Paper 900742.

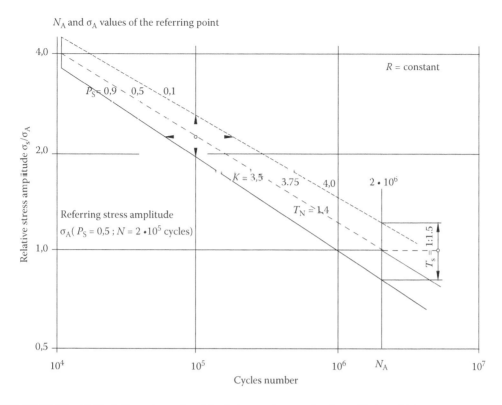

FIGURE 18.22 Unified dispersion strip of the fatigue endurance for the welded steel joint.

stress ratio ($R = 0$). Therefore it is possible to link the amplitude of stress fatigue corresponding to certain durations as a function of i/d.

Experimental data conclude that from the total number of spots, it is not the number of lines that matters but the distance between the first and last line; nor does the disposition of spots affect fatigue resistance. The trend is that fatigue resistance increases as distance increases between the spot lines.

18.13 RIVETED JOINT

Norms CNR-UNI 10011 give indications about the calculation of these joints. Here are reported, as example, only basic indications for some particular case.

Nailing is a fixed link that is cheap and convenient but has decreased in importance with the development of welding. Typical uses of nailing are in high-pressure tanks for fluids, and steel construction to join plates and metallic elements. Each nail has a head and a shank that has a slightly truncated cone (Figure 18.24). The materials commonly used are non-alloyed steel, aluminum and brass. The pre-heated nail is introduced into the hole drilled on the edges of parts to be joined and is then heat-hammered to form the other head. During the cooling, the nail contracts and then the two heads make a considerable compression on the plates to be connected.

In the case of $d \leq 10$ mm nails (rivets) the riveting is made at cold. The holes are obtained by drilling and subsequent boring, operations of fundamental importance because the nails are susceptible to fatigue in the zone of the holes. In riveting that is not subject to dynamic effects, only friction opposes the sliding of parts.

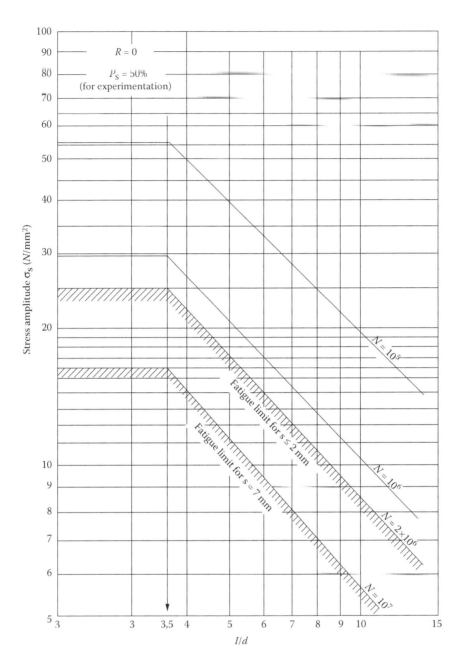

FIGURE 18.23 σ_s-l/d curves for $P_s = 50\%$.

The traction stress on the nail for the cooling effect and on the plates depends on the mode of working of the nailing at hot. If the plates were infinitely stiff and the nail had no permanent plastic deformation the result would be:

$$\sigma = E \cdot \alpha \cdot \Delta T$$

being that:

    E is the modulus of the nail material
    α is the linear expansion coefficient

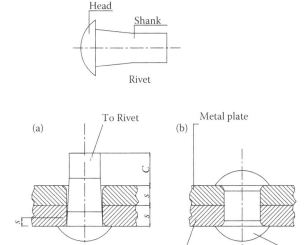

FIGURE 18.24   Rivets. (a) before the rivet, (b) after the rivet.

ΔT is the difference between the temperature at the time of hammering and the ambient temperature

In effect the elasticity of metal and especially the plastic hot deformation lower the theoretical value of σ by an amount difficult to assess because this depends on other experimental factors such as the mode of the hammering and the velocity of the cooling. Having determined the value of σ and the coefficient of friction f between the plate pressed by the nail, the maximum force N transmitted by one nail is:

$$N = f\,\frac{\pi d^2}{4}\,\sigma, \tag{18.1}$$

where d = diameter of the nail.

Rivets placed in operation are usually subject to vibration and fatigue stress that encourage the sliding. In this way there will be shear actions on the shank of nails that increases up to take over completely the friction effect. In this case between the stem and the hole can produce dangerous impact stresses. In the case of nails performed in the construction of boilers and tanks to ensure the seal, at the outer edges of the sheets do not finish in a right angle but at a certain angle ($\approx18°$ on the plane of the plates). These inclined edges are caulkled trying not to create hazardous carvings. Resin or varnish is often sandwiched between the surfaces of two plates.

In structural terms it is distinguished double nailing overlap joint and simple nailing overlap joint. Both can have only one row of nails, two rows facing or staggered, or three staggered rows. In the first (double nailing overlap joint), as each plate is subjected to an action of friction on each of the two sides, the force transmitted by effect of the nail is twice that provided by (18.1). The second (simple nailing overlap joint) acts as a tensile load and a couple for the equilibrium equation, $N \times s/2$, with the load on the contact plane of two plates. Assuming equal thickness s of the two plates, there is a total torque $N \times s$. When the riveted joint is under load, among the heads of nails and plates, in addition to the normal pressure due to cooling of the nail, there will be a tangential action T (without the riveting two-lug). To balance the rotation of the nail, each head must grow a moment $s \times T$ which is exercised between the head of the nail and the plate. The nail is then submitted to traction, bending, and shear (Figure 18.25).

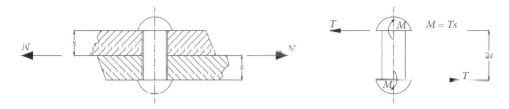

FIGURE 18.25   Loads on the nail.

In the case of an overlap joint with two rows of nails (see Figure 18.26), the moment $M = N \times s$ (equivalent when $M' = F_e c$), the external force F_e in the axis direction of the nail causes a decrease in the clamping force.

To consider the effect of F_e you can do the same considerations made about the forced junctions (see Chapter 16). If

$$\Delta l = l \cdot (1 + \alpha \Delta T) = \lambda,$$

the force F_0 caused by the nail cooling is

$$F_0 = \frac{EA\Delta l}{l},$$

this force F_0 causes a shortening λ' in the thickness of the two sheets as:

$$\lambda' = \frac{F_0 l}{E'A'},$$

where

$$A' = bp - n\frac{\pi d^2}{4} \quad (p \text{ pitch of the reverts}),$$

E' modulus of elasticity of the plats

When the external force F_e acts, the clamping force on the plate, decreases, going from F_0 to F', it is possible to calculate the limit value of F_e for which there is no detachment of plates:

$$F' = F_0 - F' = F_0 - \frac{F_e}{1 + \dfrac{EA}{E'A'}} = 0.$$

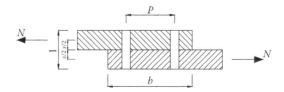

FIGURE 18.26   Joint with two rows of nails.

For the above, the nails are designed to resist a parting by friction action at shear loads N. Therefore for shear stresses:

$$N = n \cdot \frac{\pi d^2}{4} \tau,$$

where

- N is the maximum strain transmissible
- n is the number of nails
- d is the diameter of the nails
- τ is the average voltage of excision (being the nail in round $\tau_{max} = 3/2\tau$)

According to this equation we assume that all nails absorb an equal amount of load. In reality, due to the elastic deformation of the plates, the first row of nails is more overloaded than the next. To reflect this fact, lower admissible stresses are adopted with regard to the number of rows of nails. Just because all the nails do not work the same way we should not over-do the number of rows. If the two plates are connected to the same section, the distribution of loads is symmetrical about the centerline of the joints (a in Figure 18.27), otherwise the greater load is absorbed by the rows of nails placed on the side of the thicker sheet (b in Figure 18.27).

The nails must be calculated so that if a failure does occur it will only rip out the nails, not tear the sheet. For this reason the distance from the outer edge of the nail plate must be twice the diameter of the nail. With reference to the sheet, the stress τ is: $\tau_{str} = F/(s\,2d)$ (Figure 18.28).

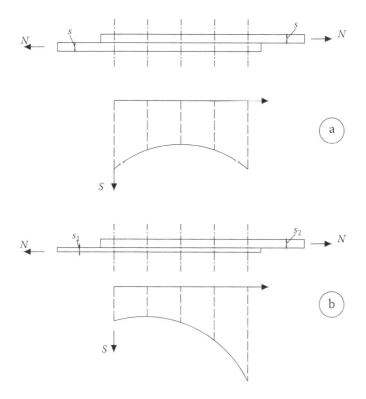

FIGURE 18.27 Joint with different rows of nails.

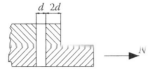

FIGURE 18.28 Tear verification.

It should also verify that the p_0 pressure of the stem against the walls of the hole is less than the yield stress of steel.

Referring to Figure 18.29 and 18.30, the distributed pressure p is

$$p(\varphi) = p_0 \cos^2 \varphi,$$

Being φ the angular coordinate of a generic point of the hole, it is possible write the following equation:

$$2 \int_0^{\pi/2} p(\varphi) \cdot \sin \varphi \cdot r \cdot d\varphi = 2 \int_0^{\pi/2} p_0 \cdot \cos^2 \varphi \cdot \sin \varphi \cdot r \cdot d\varphi = \frac{F}{n \cdot s}.$$

Where s is the plate thickness and n the nails number for each row,
The average value p_m of the pressure is

$$p_m = \frac{N}{nsd}.$$

The value of p_0 obtained must be compared with the Brinnell hardness of the steel reduced reduced by a of coefficient 2.5 ($p_0 \le H_d/2.5$)

In the case of dynamic loading therefore to determine the resistance to fatigue (need Smith or Haig diagram of the steel) it is important assess the geometry of the connection components. For a nail with a semi-spherical head, the notch effect is high (safety factor equal to 3) while it is smaller in the case of countersunk head (safety factor equal to 1.3).

In the case of bending moment, it is necessary to find the normal load on each nail. As an example, we can consider the case of a joint of one beam into a second that creates a bending pure moment M_i (Figure 18.31) Assuming that all nails work under the same friction conditions, it is easy to allocate the loads on the row of nails as the neutral axis is also the centroidal forces axis. Supposing sliding instead, as a first approximation it is assumed that the axis nn around which the second beam rotates coincides with the lower edge of it. In this case the stress in the nail (or the row of nails) is proportional to the distance y from the axis nn. According to equilibrium equation is easy to evaluate the load for each row and therefore on each nail. The overload on the nail can be determinate by the above considerations.

FIGURE 18.29 Pressure distribution.

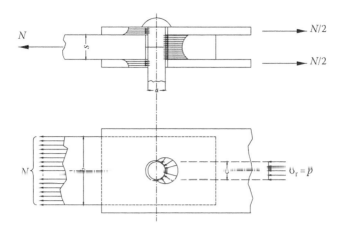

FIGURE 18.30 Pressure on the shank.

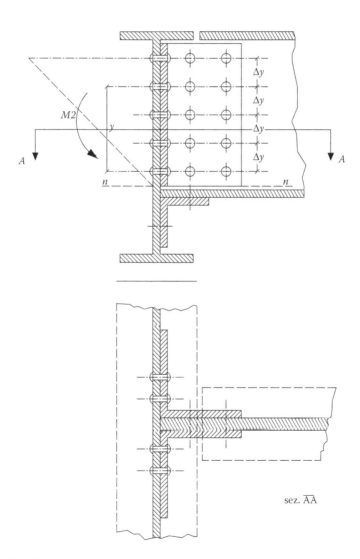

FIGURE 18.31 Bending moment.

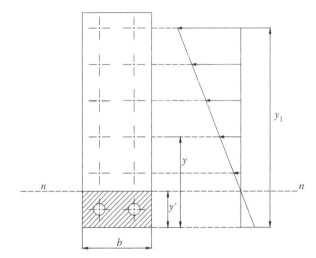

FIGURE 18.32   Traction and compression zones.

The calculation can be improved. Consider F as the result of the traction tension and C the result of the compressive tensions that are equal and opposite so as to balance the internal moment M_i. F is distributed on all rows of the junction, with C on a strip area (see Figure 18.32) of the lower edge of beam, producing a high pressure. C could be allocated as a linear pressure over an area $b\,y'$ (height y' unknown), while F is distributed only on the nails out of the compression zone. The y' unknown value can be determined by iterative method imposing the equality of static moments.

19 Couplings

19.1 PRELIMINARY NOTIONS AND TYPES OF JOINTS

This chapter deals with the main types of the many joints on the market. As in Giovannozzi, Vol. I, those examined here are the most interesting from a didactic-educational point of view because of the calculations involved

Only drawings of the simplest calculations and joints will be included, for the sake of brevity. Many drawings of joint constructions still on the market can be found in the above volume.

Joints are mechanical components which join shafts, generally coaxially (placed one after the other), which must rotate at the same angular velocity. The connection is permanent, and cannot be modified while the machine is working, as can happen with clutches, which, as will be seen in the next chapter, permit the removal of the connection or its setting up even while the shafts rotate (by engaging/disengaging).

There are two main categories of joints:

- Fixed joints (rigid, semi-elastic and elastic)
- Mobile joints

Whatever the type, joints are mechanical systems of various components and in their simplest form, are made of two rigid bodies each connected to one end of two shafts (drive and driven), and sandwiched between them is a part with variable characteristics (depending on the type).

In particular, rigid fixed joints join two shafts or sections of shafts so that neither can move independently. Shafts and their supports must be perfectly coaxial. They are used to join long transmission shafts or engines to slave machines, as long as there is no risk of axial shifting or vibration transmission by either shaft. This category includes sleeve couplings, cylindrical shell joints (bolted ring), Sellers coupling, disk couplings, ring-disc couplings.

Elastic and semi-elastic fixed joints connect two shafts by means of elastically deformable components. This type of coupling can join shafts that are not perfectly coaxial, or when slight radial, axial, or peripheral movements must be allowed between them. These joints limit any after-effects resulting from violent shock or abrupt or periodical variation of torque, by reducing or absorbing any vibrations. Also in this category are all the pin joints, Pomini joints, and Zodel joints.

Mobile joints are designed to allow the connected shafts longitudinal and transverse movement as well as quite sharp angular deviations without compromising their elastic deformation. They are used to join shafts belonging to different machines whose relative positions can vary, subject even to quite wide variations during operation (expansion geared joints, Oldham coupling, Cardano coupling).

In addition to these joints, safety joints are widely used. However, their aim is to prevent any twisting that exceeds the fixed limit and could endanger any other important mechanical components. They are used to control machines that function intermittently (compensators, shearing machines, hammers, etc.) using flywheel kinetic energy. Safety joints may be of the pin or clutch type.

Joint size (like clutch size) is set by simple empirical formulae according to the shaft diameters involved, formulae that are simply the approximate representation of manufacturer's data supplied with the component. It may seem obvious that sizing a joint (or clutch) to design depends on the designer and his capacity to identify the relationship between the moment of drive, total forces and their amounts, and the designer's ability to match his calculations with the manufacturer's loads so that his design can handle any eventuality.

Manufacturers generally choose joints that can handle well over the average torque, by using coefficients appropriate to the type of machine and its use.

Examples of calculations are given in the following pages, where M is the drive moment constant, and d is shaft diameter.

Generally, joints are chosen according to use, and so the data needed to insert into the tables are supplied by the manufacturers: drive power and shaft revolutions. With this data, the joint design must take into account the forces generated at the shaft flanges and on the component between them. The following chapters examine various types of joints, mainly to highlight how to outline the design and estimate its performance.

The sequence is always the same even though not be mentioned again. Knowing the input data, the stress applied to the various components is determined. By actually analyzing the joint, simple calculations are made and simple algorithms applied to them to determine stress and any deformations (in this course they are the results of the Saint-Venant theory).

To help apply the procedure, it is useful to outline briefly even those joints that are not widely used anymore, having been replaced by new joints that are better designed and made of better materials.

The aim of analyzing the joints (referred to later in this chapter) is to give an idea of the many "mechanical systems" on the market that can permanently join coaxial (or almost coaxial) shafts.

Joining non-coaxial shafts can be achieved using other components (gears, chains, belts) depending on power and shaft orientation. Among the joints that allow slight shaft displacements are chain or belt joints whose mechanical component is a double chain or a double toothed belt that interlocks half on one shaft, half on the other.

For these types of joint no calculations are made, as they refer to the inter-flange component, which in some cases (see Springs) takes whole chapters.

19.2   RIGID JOINTS

19.2.1   BOX COUPLING/SLEEVE JOINT

This is made of a one-piece perforated cast iron sleeve inserted at the ends of the two shafts to be joined and attached to them by spanners/wrenches (Figure 19.1). As protection from the protruding of the keys, sheet metal is fixed to the sleeve by self-tapping countersunk screws.

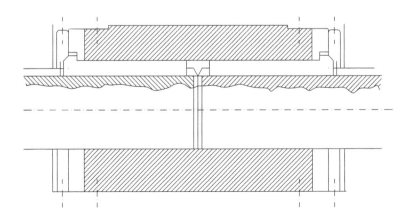

FIGURE 19.1   Box coupling.

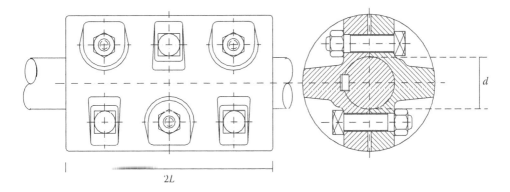

FIGURE 19.2 Cylindrical joint with bolts and spline.

The use of such joints is very limited because of the off-centeredness due to the [CN] and the play between the sleeve and shafts. Off-centeredness and play cause vibrations so the use of these joints is limited to shafts that rotate slowly and with moderate power (1–3 kW).

19.2.2 CYLINDRICAL JOINTS

These are made of two cast-iron cylindrical shells, internally reamed to the exact diameter of the shafts to be joined. There is a slight play between the two semi-shell facing surfaces, which are bolted together at the ends of the shafts so they have sufficient adhesion to transmit the torque for shaft diameters up to 50 mm. For greater diameters, bolts and splines are used (Figures 19.2 and 19.3). For protection, a cylindrical sheet metal casing is often used (Pomini, Lohmann, and Stolterfoht construction).

The two semi-shells can either be bolted together or joined by rings. In the latter case, the two semi-shells are joined by force fitting two steel rings to the outer surface of the shells, which is slightly conical (1:30–1:40). When subjected to shock or axial thrust, these joints spontaneously slacken and therefore are less safe than bolted shell joints.

In the case of the bolted shell joint (Figure 19.2), calculating its strength is especially about sizing the bolts. With M proportional to the friction coefficient f, the stress P necessary to transmit the moment M entirely by friction depends on the law by which the pressure along the contact arc

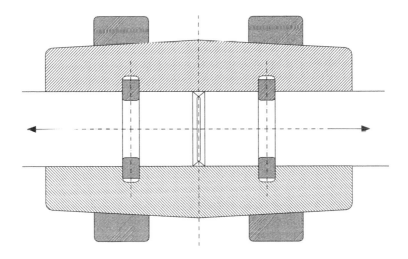

FIGURE 19.3 Cylindrical joint with rings.

between shells and shafts is shared. Assuming a uniform distribution of pressure ($p = \cos t$), the maximum transmittable moment by friction is

$$M = \frac{1}{2} fp\pi dL \frac{d}{2} - \frac{1}{2} fp\pi r^2 l,$$

$$M = \frac{1}{2} fp\pi dL \frac{d}{2} = \frac{1}{2} fp\pi r^2 L.$$

The pull stress that fastens the two semi-shells at the end of the first shaft, which counts for half the total pull, is given by

$$\frac{P}{2} = prL,$$

and, consequently,

$$\frac{P}{M} = \frac{2prL}{\pi fpr^2 L} \Rightarrow P = \frac{2M}{\pi rf} = \frac{4M}{\pi df}.$$

If n indicates bolt number and d_n the core diameter, the stress generated in each bolt when fastening the two semi-shells will be

$$\sigma = \frac{P}{A} = \frac{16M}{\pi^2 dfnd_n^2}.$$

As already mentioned, this calculation is valid only if the stress distribution is constant. In practice, constant stress is not easy to obtain because it would be necessary to have perfect shells and great tightening of the bolts. To compensate such uncertainty, a low friction coefficient is used, in particular a product value $\pi f = 1$.

Another joint, considered a by-product of box coupling, is the Sellers joint (Figure 19.4). This uses an externally conical bush wedge within an external sleeve with a conical internal surface. At the extremity of the shafts, two bushes with opposing conical shapes dovetail. They are cut along a meridian to have transverse elasticity.

The axial force fitting is through tightening the three bolts in the three holes that partly perforate the sleeve, and partly perforate the bushes. Transmitting the moment is produced by the friction between the two surfaces in contact. However, to avoid any rotation, bush splines opposite the meridian can be used to increase bush elasticity.

The bolt tightness, assuming that transmitting the moment is done by friction, can be calculated in the following way.

The transmitted moment M, with constant pressure p, is

$$M = \pi \frac{L}{\cos \alpha} pfR_m^2,$$

f being the friction coefficient, p the pressure between the surfaces and R_m the mean radius of the cone.

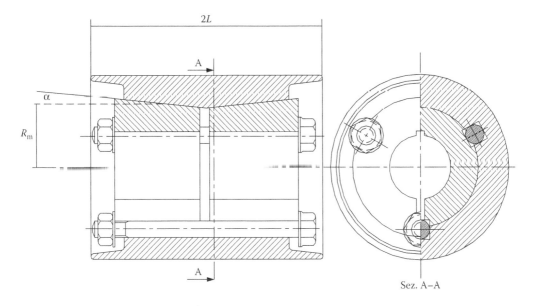

FIGURE 19.4 Sellers joint.

Total bolt tightness T, assuming $tg\varphi = f$ is expressed by

$$T = 2\pi R_m pL \frac{\sin(\alpha + \varphi)}{\cos\alpha\cos\varphi},$$

therefore,

$$T = M/R_m \left(\frac{\sin(\alpha + \varphi)}{\sin\varphi} \right).$$

Assuming three bolts are used, if σ_{amm} is the acceptable stress, the diameter of the core is

$$d_n = \sqrt{\frac{4T}{10\sigma_{amm}}} \cong \sqrt{\frac{5M}{\sigma_{amm}R_m}},$$

having $3\pi = 10, f = 0.3$, and considering the values of α used in practice (a few degrees).

This type of joint allows good centering even with shafts of different diameters and has no protruding parts, but it is bulky in the radial sense and expensive because of its complexity.

Predictably, these joints can be used when the shaft rotation speed is relatively low (not exceeding 600 rpm).

19.2.3 DISC AND FLANGE COUPLINGS

Disc couplings are used to transmit high moments even in difficult conditions when collisions and overloads are present. They are made of two cast iron discs, with a hub coupled to the middle of the shaft (Figure 19.5). The discs are joined through a ring of bolts that fasten the two surfaces in

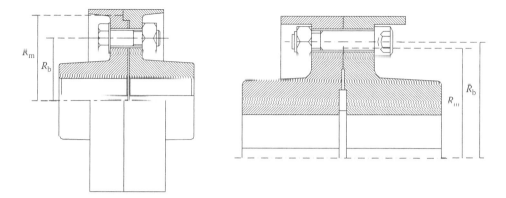

FIGURE 19.5 Disc couplings.

contact, transmitting the moment through tangential friction forces between the surfaces, perpendicular to the joint axis. The moment can also be transmitted through the bolts which, in this case, must be calibrated to do the job.

Each semi-shell is connected to its respective shaft either by a cotter bolt or spline, sometimes integrated with axial stop devices. When the moment has to be transmitted by friction, the two discs must be compressed together by tightening the bolts to produce the necessary tangential friction. To ensure this friction has the maximum arm of force (leverage) possible, a ring is coupled peripherally to each disc, leaving some play between the bolt stem and the disc holes. When the bolts have to transmit the moment, their stems must be calibrated to require a slight force of entry into the holes (coupling K6 h6).

Centering these types of joints can be done in different ways. Figure 19.5 shows one type with a ring-shaped ridge on one of the discs, and a corresponding axial notch deeper than the ridge on the other disc. Centering can also be done on the disc periphery. The advantage is that any detailed work is concentrated only in that area. Another way of centering is to place a ring in slots made in both parts of the joint.

In the previous examples, to release the connection and therefore split the joint, one of the parts must be moved axially to separate the ridge and ring from their slot.

To overcome this drawback, joints with an external centering ring in two parts (Figure 19.6a) are used where the thickness of the semi-rings is different if the peripheral or central areas are considered. Moreover, in each semi-ring a threaded hole is made into which an extractor can be screwed to dismantle the joint.

When the two joint discs are forged into one piece with the shafts, the joint is called a flange (Figure 19.6b). Thus, the linkage between hub and shaft, the weak point of connection by discs, is eliminated. These joints are used for large diameters ($d = 25 - 700$ mm) and work with strong impacts.

Even with flange joints, centering is ensured by a central ridge, while coupling is ensured by the usual male–female parts.

In proportion to transmitted moment, flange joints are smaller radially than disc joints, lacking a hub. The maximum size of a flange joint can be obtained using the following empirical formulae:

$$D = (2 - 1.8)d \quad s = (0.3 - 0.25)d \quad \hbar = (2 - 12)\,\text{mm}.$$

$$d_b = 10\,\text{mm}\sqrt{\frac{d}{10\,\text{mm}}} \quad \text{(and anyway always} \leq 0.2\text{)}.$$

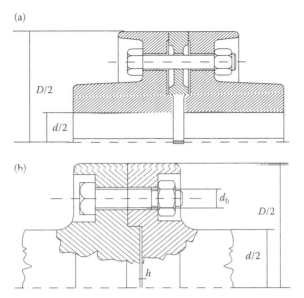

FIGURE 19.6 Flange couplings.

If moment transmission was due to friction, the calculation would be similar to that used for the friction clutch.

The strength calculation of a disc joint (Figure 19.5) can be made in two different ways depending on the moment transmission method. When moment transmission, M_t, is due to friction, the mean radius, R_m, of the contact ring of the two discs, P the total tightness of all the bolts, f the friction coefficient between the two discs, then the value of the total pull to transmit the moment is equal to

$$P = \frac{M_t}{fR_m},$$

with n the number of bolts, and d_n their core diameter, the normal strain on each bolt is

$$\sigma = \frac{M_t}{fR_m n \frac{\pi}{4} d_n^2}.$$

When the moment transmission, M_t, is the result of the bolts shearing, R_b being the radius of the circumference of the bolt ring, total shearing force is

$$T = \frac{M_t}{R_b},$$

and knowing that in a flat circular section the maximum tangential strain is equal to 4/3 of the mean tangential strain, then

$$\tau = \frac{16}{3} \frac{M_t}{R_b n \pi d_b^2},$$

d_b being the calibrated diameter of the bolt.

If $\tau = 0.7\sigma$, $d_n/d_b = 0.9$, $R_m/R_b \sim 1$ and $f = 0.2 - 0.3$, the ratio of the maximum transmittable moments of the bolts, respectively, through tensile stress and by shearing is lower than 1/2, or

$$\frac{M_t}{M_t^*} = f \frac{d_n^2}{d_b^2} \frac{\sigma}{\tau} \le \frac{1}{2}.$$

So, when the bolts shear, the moment transmission is about double what would be expected with the bolts working under tensile stress. However, in different circumstances, it is better to choose friction joints, since shear joints are more expensive because of the greater accuracy required in their manufacture (hole reaming/boring for the bolts, their location, bolt calibration). For tensile stress bolts, Pomini's formula is

$$d_b = 8 \text{ mm}\sqrt{\frac{d}{10 \text{ mm}}}.$$

The calculations made so far are only an outline. Clearly, different orientations, which could increase the proportions of the various elements, should be taken into account, as well as the importance of the product and real uncertainties.

When shafts are connected with these joints, moments of bending can be generated by centrifugal forces or errors in the axial plane that strain the linked components. Obviously, the greater the distance from the support bearings, the greater the phenomenon.

In order to attract the attention of the designer to considering that these types of effect may strain the shaft linkages unexpectedly, let us consider a flange joint more closely.

For hypotheses that will be established, the following calculation refers to a bolt linkage where the coupling tolerances are such to only allow tensile stress on the bolts. So, one might suppose that in a moment of bending, the coupling faces of the flanges would remain flat. Thus, there would be a compressed area and another that would tend to want to separate if were it not for the bolts experiencing tensile stress.

Calculation of the bolt stress and contact pressure on the compressed area of the flange, according to the previous hypotheses, is made by determining the neutral axes for rigid beams. In this case, if the flange and bolt materials are the same, as usually happens, the elastic modulo are the same. Therefore, the neutral axis is the equation of the static moments of the compressed area and the stressed area.

With reference to Figure 19.7, if M'_b, M''_b, M_f are the static moments, respectively, of stressed bolts, non-working bolts, and part of the compressed flange, and y is the distance of the neutral axis from the flange center, the equation of the static moment is

$$M'_b - M''_b = n \frac{\pi d_b^2}{4} y = M_f - 2M''_b.$$

Remembering that area A of a circular segment is (symbols as per Figure 19.7)

$$A = \frac{1}{2}\left[\frac{D}{2}(s - l) + hl\right],$$

and its barycentric distance from the flange center is

$$y_0 = \frac{l^3}{12A},$$

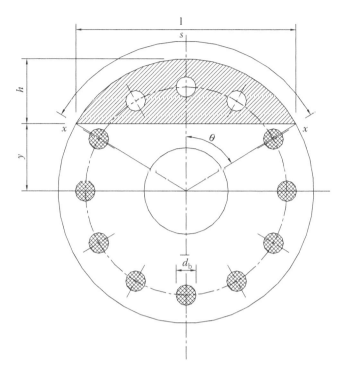

FIGURE 19.7 Disc couplings by bolts.

the previous expression becomes

$$M_f - 2M_b'' = \frac{l^3}{12} - Ay - 2M_b'' = n\frac{\pi d_b^2}{4}y.$$

As a function of angle θ:

$$s = D\theta \quad l = D\sin\theta \quad h = \frac{D}{2}(1 - \cos\theta) \quad y = \frac{D}{2}\cos\theta.$$

Therefore,

$$\frac{2}{3}\frac{\sin^3\theta}{\cos\theta} - (\theta - \sin\theta \cdot \cos\theta) - \frac{16M_b''}{D^3\cos\theta} = n\pi\left(\frac{d_b}{D}\right)^2,$$

which by trial and error produces θ and therefore the position of the neutral axis.

When bending strain is frequent, a more rational design can be accomplished by setting a forced linkage to generate compression stress between the flanges by tightening the bolts to the immobile shaft. In operation, this stress will ensure the flanges maintain contact over their whole surfaces. To calculate this, refer to forced linkages (Chapter 16).

19.3 FIXED SEMI ELASTIC AND ELASTIC JOINTS

These joints are used to avoid collisions and any abrupt wrenching. A simple example could clarify the need for elastic joints in some mechanical applications. Elastic joints are able to absorb quantities of energy to avoid collision and consequent overload through interposed elastic components.

With reference to Figure 19.8, consider a flywheel of polar moment I, linked to an engine by a shaft of elastic constant K. Suppose one of the extremities, S, rotates at angular velocity sum of constant ω_m (mean angular velocity) and sinusoidally variable with time $\omega = \omega_m + \omega(t)$.

Rotation is the sum of that which is constant ω_m, and that which is sinusoidal $\overline{\vartheta} = \vartheta \sin \omega t$. Because of sinusoidal oscillation, the flywheel oscillates equally $(\overline{\vartheta} - \vartheta' \sin \omega t)$ with amplitude ϑ' that can be calculated as follows.

By definition the elastic constant K of the shaft driving the flywheel is

$$K = \left| \frac{M}{\Delta \vartheta} \right|.$$

Elastic moment M is

$$M = -K \Delta \vartheta = -K(\overline{\vartheta}' - \overline{\vartheta}).$$

Torque is positive, being:

$$M_{\mathrm{I}} = I \frac{d^2 \overline{\vartheta}'}{dt^2}.$$

Balancing the two, we get

$$-K(\overline{\vartheta}' - \overline{\vartheta}) = I \frac{d^2 \overline{\vartheta}'}{dt^2}.$$

By substituting its value in the derivative and the ϑ and ϑ' expressions, we have

$$-K \sin \omega t (\vartheta' - \vartheta) = -I \omega^2 \vartheta' \sin \omega t,$$

which results in

$$\vartheta'(I\omega^2 - K) = -K\vartheta \Rightarrow \vartheta' = \frac{\vartheta K}{K - I\omega^2} = \frac{\vartheta}{1 - \frac{I}{K}\omega^2},$$

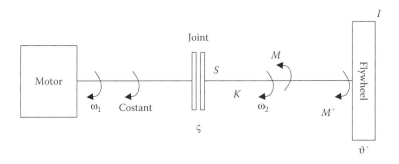

FIGURE 19.8 Scheme for working joint.

Finally, by introducing the flywheel oscillation $\omega'^2 = K/I$, we can write

$$\vartheta' = \frac{\vartheta}{1 - \dfrac{\omega^2}{\omega'^2}}.$$

By plotting ϑ'/ϑ as a function of $(\omega/\omega')^2$, we get Figure 19.9.

From this, we deduce that, in order to reduce flywheel oscillation amplitude ϑ', $(\omega/\omega')^2$ must be greater than 1. That means that ω', or K, must be kept low so drive-shaft rigidity must be low, meaning that the joint must be deformable. This is possible by interposing an elastic joint.

The graph also shows that $\omega' > \omega$ must be avoided or, even worse, that $\omega = \omega'$, since in the former case the flywheel oscillation would be greater than S, and in the latter the oscillation would tend to infinity.

Moreover, with elastic or semi-elastic joints, slight errors in the axial plane or angulation are tolerable without overloading the shafts or supports. Furthermore, eventual shocks can also be absorbed. The energy absorbed for a maximum deformation is

$$A = \frac{1}{2} M\vartheta,$$

where M is the torque of maximum deformation and ϑ is the relative rotation of the two halves of the joint due to the deformation of the joint's elastic components. In a collision, with no elastic component, ideal elastic behaviour would be

$$M_{\max} = M\left(1 + \sqrt{1 + \frac{2\vartheta}{\vartheta_{st}}}\right),$$

where M is the applied moment, ϑ the recovery angle, and ϑ_{st} the drive-shaft deformation under moment M. Therefore, even for little play, the load values would be double those applied with consequent transmission overload.

Mass production of this type of joint is significant. Specialized manufacturers use rubber or synthetic materials of different shapes as joint or damping components (see the Rotex joint in Figure 19.10, for example) where the damping component is an elastic ring with involute profiled

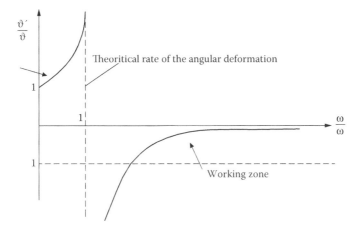

FIGURE 19.9 Angular deformation as function of the angular velocity.

FIGURE 19.10   Rotex joint.

gear teeth). The special mechanical and hysteresis characteristics of these materials permit significant vibration absorption and prevent resonance during normal operation, also because there is no linear correspondence in these components between transmitted moment and relative joint rotation. Should non-linear behavior be insufficient to obtain increasing rigidity, the deformable components could be suitably shaped (Figure 19.11).

It is not possible to make a rigorous analytical size calculation for this type of joint because the shape and contact conditions of the deformable components change during operation. Usually, joint characteristics are experimentally detected on prototypes.

19.4   PEG JOINTS

These linkage components are made of two semi-joints each with a flange hub. From each flange, pegs project axially at regular distances (Figure 19.12). The flanges also have a series of holes and the intermediate rubber sleeves carried by the pegs of one semi-joint located in the holes opposite the other semi-joint alternately to obtain a perfectly balanced system. Hinges/pins and the rubber sleeves allow slight axial play.

In the moment of overload, the elastic components have to absorb a certain energy in the form of potential elastic energy and return it partly or fully in the next instant, creating a sort of barrier to abrupt acceleration and overload, thus regulating the motion of that protected component.

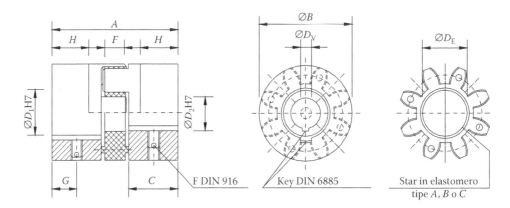

FIGURE 19.11   Section of Rotex joint.

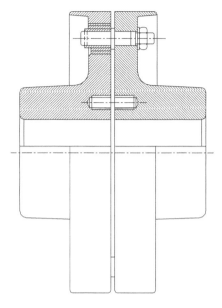

FIGURE 19.12 Peg joint.

The rubber-coated joint pegs are tested for bending since they undergo a peripheral stress Q spread over the whole length of the peg equal to

$$\frac{Q}{n} = \frac{2M}{nD_m},$$

where D_m is the peg diameter, n the number of pegs, and M the transmitted moment.

Figure 19.13 shows a Norton and a Rupex joint. Both use peg sleeves with convex external surfaces that allow the absorption of angular alignment errors.

In the joint shown in Figure 19.14 (by Elco), the elastic sleeves have a special shape that produces non-linear behavior over its whole operational range due to the progressively increasing cross-sectional area under compression, as shown.

N-Eupex elastic joints, manufactured by Flender (Figure 19.14), are used in the mechanical industry, especially where perfect power transmission is required even for slight non-axial shaft alignment. These joints mount elastic pads with high damping capacity that have a high critical

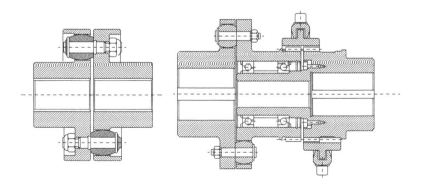

FIGURE 19.13 Norton and Rupex joints.

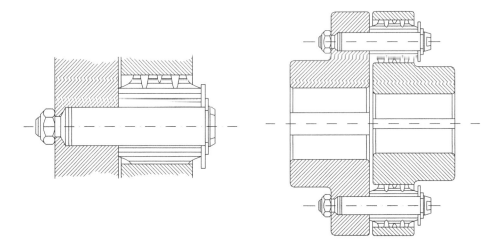

FIGURE 19.14 Elco joint.

velocity threshold to limit increased resonance and protect coupled engine components from system shocks. They are small, light, and have low mass inertia moment, possessing torsional flexibility that allows them to absorb the critical torsional vibrations of the system. Moreover, they are extremely safe even with unexpected overloads, their safety is limited only by the admissible loads of the cast iron components. Table 19.1 shows an example supplied by the manufacturers, which allows a joint to be easily chosen.

19.5 SPRING JOINTS

In these joints, the elastic component of the joint interposed between the two shafts is made of various types of springs. There are different spring shapes, their choice based on how much power is transmitted by the shafts. Different joints are taken into consideration in the following pages and, consistent with the aim of this collection of notes, those analyzed here by calculating their elastic component are those already included in Giovannozzi, Vol. I, which are still widely produced and used.

The calculation utilizes system analysis and extrapolates from the definitions of constraints and loads. This procedure can also be used for the many other similar joints on the market.

19.5.1 BiBBY JOINT

The Bibby joint is a classic sprung joint. The two mortised discs on the shafts (drive and driven) have two radial slots inside, which a continuous S-shaped steel strip winds. When the moment is applied, the belt settles on the walls of the slots that have variable width in the axial direction (Figure 19.15).

The shape of the teeth produced by the slots varies rigidity as the applied moment increases. Under load, due to the non-axiality of the opposing teeth, elastic torsion is also produced in the belt that further varies rigidity.

For load $M_t = 0$, the resulting condition in Figure 19.16 and Figure 19.17 show that the opposing teeth do not touch and the thin strip is rectilinear. The moment applied, there is a continuous settling-in of the strip on the teeth of the two semi-joints with a consequent decrease in the free flexure length and a continuous increase in system rigidity.

The overall deformation of the spring is that of a first phase when the strip settles to the curvature of the supporting tooth and a second phase when the free length of the strip (non-settled) bends.

In practice, if the tooth profile is an arc of a circle ($R = \cos t$), in the first phase the bending moment is constant and inversely proportional to the arc's radius. With reference to Figure 19.18, $l = x' + x$ indicates the free length that varies as the thin strip settles.w

TABLE 19.1
Table for N-EUPEX (Flender) Joints

N–EUPEX–DS
Flexible couplings
Type EDS and EDS
for flange mounting to pulleys
or flywheels

EDS in two parts

DDS in three parts

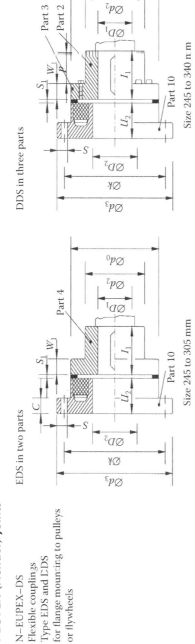

Size 245 to 305 mm

Size 245 to 340 n m

N-EUPEX Couplings		Nominal Values	Torque	Speed	Bore D_1		d_3	k	s		# off	c	Flange Fitting Dimentions								Mass Moments of Inertia J		Weight Total
Type	Size	P_N / n	T_{KN} N_m	n_{max} 1/min rpm	Part 4 da mm	Part 2 a mm	h_8	k	a	alt	off	c	d_2	D_2 B_7	l_1	d_2	v_2	v_1	a_1	F	Part 4 or 2/3 kgm²	Part 10 kgm²	kg
EDS	245	0.21	2000	2750	38	90	314	282	18	M16	8	20	245	115	90	150	55	55	2...6	–	0.082	0.16	26.1
EDS	272	0.29	2800	2450	48	100	344	312	18	M16	8	20	272	130	100	165	60	18	3...8	–	0.132	0.24	32.6
EDS	305	0.41	3900	2250	55	110	380	348	18	M16	9	22	305	145	110	180	65	20	3...8	–	0.208	0.4	42.8
DDS	245	0.21	2000	2750	24	85	314	282	18	M16	8	20	245	115	90	138	55		2...6	6	0.115	0.16	29.2
DDS	272	0.29	2800	2450	32	95	344	312	18	M16	8	22	272	130	100	155	60		3...8	6	0.2	0.24	38.2
DDS	305	0.41	3900	2150	55	105	380	348	18	M16	9	22	305	145	110	172	65		3...8	7	0.3	0.4	47.6
DDS	340	0.58	5500	1950	48 / 100	100 / 120	430	390	22	M20	9	25	340	170	125	165 / 200	70		3...8	7	0.49 / 0.53	0.7	65.3

[1] Upon failure of the flexible elements the coupling parts rotate freely to each other. There is non contact metal to metal contact.

[2] P_N – nominal power rating in kW; n – speed in r.p.m.

[3] Mass moments of inertia and weights refer to couplings with medium-sized bores.

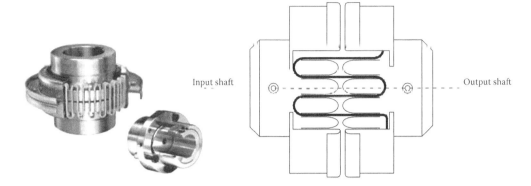

FIGURE 19.15   Bibby joint.

Because of the symmetry at point C of the line taken by the elastic strip:

$$\frac{d^2y}{dx^2} = \frac{M}{EJ} = 0 \rightarrow M = 0,$$

therefore, point C behaves like a hinge.

The two halves of the thin strip transmit a concentrated peripheral force P only at C. The value of P is defined by the condition that produces moment M corresponding to tooth curvature at points B and B' where contact ends. That is

$$\frac{1}{R} = \frac{M}{EJ} = \frac{Px}{EJ} \quad Px = \frac{EJ}{R}.$$

If $R = \text{cost}$, the whole second member is constant and the free bending length $2x$ is inversely proportional to the force P transmitted. As P increases, the thin strip tends to settle more on the tooth, and the deformation calculation can refer to the three shift components (Figure 19.19): the first component (f_1) is due to settling; the second (f_2) due to the tangency variation of the last contact point on the tooth; the third (f_3) due to the free bending of the tooth.

As Figure 19.19 shows, drawing the tangent in B and projecting A and B on the symmetry axis, the three components, f_1, f_2, and f_3, are obtained:

$$f_1 = y \quad f_2 = xtg\gamma = x\frac{dy}{dx} \quad f_3 = \frac{Px^3}{3EJ} = \frac{x^2}{3R}.$$

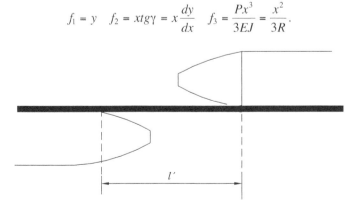

FIGURE 19.16   Initial position of the opposing teeth.

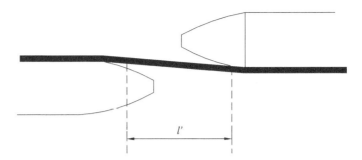

FIGURE 19.17 Position of the opposing teeth when a moment is applied.

If the profile were circular, we could write

$$y = R^2 - (R^2 - x'^2)^{1/2}.$$

However, since the curvature profile of the tooth is weak, we can replace it with an osculating parabola to make the calculation easier:

$$y = a * x'^2.$$

Deriving twice we obtain constant $a*$ which equals $1/2R$ so

$$f_1 = y = \frac{(l-x)^2}{2R} \quad f_2 = x\frac{dy}{dx} = x\frac{l-x}{R}.$$

To sum up, the total deflection $f_t = f_1 + f_2 + f_3$ is

$$f_t = \frac{(l-x)^2}{2R} + x\frac{(l-x)}{R} + \frac{x^2}{3R},$$

and the deflection for the deformation of the two parts is

$$2f = \frac{3l^2 - x^2}{3R}.$$

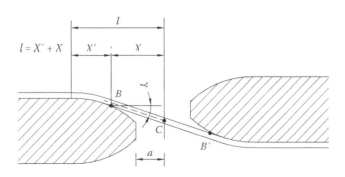

FIGURE 19.18 Position of the strip in working conditions.

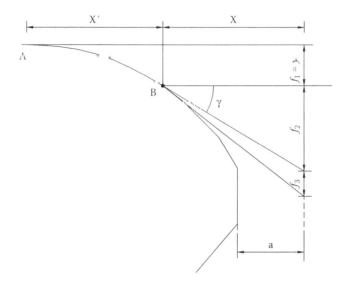

FIGURE 19.19 Scheme for the calculation.

If $f_0 = l^2/2R$ is the deflection that the thin plate would have if the tooth lengthened to the center line and the whole thin plate settled on the tooth, $P_0 = EJ/lR$ is that load which, when exerted at the center line, makes the curvature of the tooth and thin plate coincide at point A (i.e., that load which starts to settle the spring on the tooth), and $\varphi = f/f_0$, $\psi = P/P_0$, $\alpha = a/l$ are dimensionless parameters, then the spring deformation phases determine the characteristics of the following joint.

In the first phase ($0 \leq P \leq P_0$), applying moment $M_t \neq 0$, until the thin plate mimics the tooth curvature at A, the deflection will be equal to that of a beam fixed at one end with free length l:

$$ f = \frac{Pl^3}{3EJ} \qquad \varphi = \frac{Pl^3 2R}{3EJl^2}. $$

The deflection is linear and in particular when $P = P_0$, we obviously have

$$ \varphi = \frac{\dfrac{l^2}{3R}}{\dfrac{l^2}{2R}} = \frac{2}{3} \qquad \psi = 1. $$

As already mentioned, in this phase the trend of φ is linear.

In the second phase ($P > P_0$), the deflection expression is the one previously found, that is, $f = (3l^2 - x^2)/\delta R$ and φ are

$$ \varphi = \frac{\dfrac{3l^2 - x^2}{\delta R}}{\dfrac{l^2}{2R}} = \frac{l^2 - \dfrac{x^2}{3}}{l^2} = 1 - \frac{x^2}{3l^2}. $$

Recalling that

$$\frac{1}{R} = \frac{Px}{EJ} = \frac{P_0 l}{EJ},$$

we have

$$\frac{x}{l} = \frac{P_0}{P} = \frac{1}{\psi} \quad \varphi = 1 - \frac{1}{3\psi^2} \quad \psi = \frac{1}{\sqrt{3}} \frac{1}{\sqrt{1-\varphi}}.$$

In the second deformation phase, the joint characteristics are no longer linear.

In the third phase, the spring is settled and $x = a$. The expression of the deflection with the three contributions

$$f = \frac{(l-x)^2}{2R} + \frac{x(l-x)}{R} + \frac{Px^3}{3EJ},$$

becomes

$$f = \frac{l^2 - a^2}{2R} + \frac{Pa^3}{3EJ},$$

and φ equals

$$\varphi = \frac{\dfrac{l^2 - a^2}{2R} + \dfrac{Pa^3}{3EJ}}{\dfrac{l^2}{2R}} = \frac{2R}{l^2}\left(\frac{l^2 - a^2}{2R} + \frac{Pa^3}{3EJ}\right) = 1 - \left(\frac{a}{l}\right)^2 + \frac{2}{3}\frac{RPa^3}{EJl^2}.$$

By substituting $P_0 = EJ/lR$, we have

$$\varphi = (1 - \alpha^2) + \frac{2}{3}\alpha^3\psi \quad \psi = [\varphi - (1 - \alpha^2)]\frac{3}{2\alpha^3}.$$

So, the characteristic is linear again with a different inclination producing the same trend shown in Figure 19.20.

As regards the validity limits of ψ of the second phase, note that the lowest is $\psi = 1$. As regards the highest, note that in the third phase, the thin plate behaves like a beam fixed in B and with a free length equal to a. Thus

$$\frac{1}{R} = \frac{M}{EJ} = \frac{Pa}{EJ} \rightarrow P = \frac{EJ}{aR},$$

and recalling that $P_0 = EJ/lR$ is valid, we deduce that

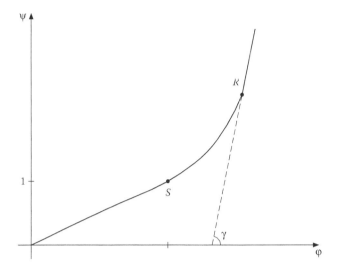

FIGURE 19.20   Characteristic of the Bibby joint.

$$\psi = \frac{P}{P_0} = \frac{\dfrac{EJ}{aR}}{\dfrac{EJ}{lR}} = \frac{l}{a} = \frac{1}{\alpha}.$$

Therefore, the relation found for the second phase is valid in the field $1 \leq \psi \leq 1/\alpha$. At the extreme $\psi = 1/\alpha$, the parameter φ equals

$$\varphi = 1 - \frac{\alpha^3}{3}.$$

In points S and R, transition from linear to non-linear and from non-linear to linear, the segments of curve have the same tangency. If we consider the curve relating to the second segment:

$$tg\gamma = \left(\frac{d\psi}{d\varphi}\right)_R = \frac{d\left(\dfrac{1}{\sqrt{3}}\dfrac{1}{\sqrt{1-\varphi}}\right)}{d\varphi} = \frac{3}{2\alpha^3}.$$

Instead, considering the curve relating to the third phase:

$$tg\gamma = \left(\frac{d\psi}{d\varphi}\right)_R = \frac{d\left(\left[\varphi - (1-\alpha^2)\right]\dfrac{3}{2\alpha^3}\right)}{d\varphi} = \frac{3}{2\alpha^3},$$

meaning that the values coincide.

Analogously, we can demonstrate that the tangencies are equal even at point S.

The characteristics of the Bibby joint are defined as:

- In the field $0 \leq \psi \leq 1$ from $\psi = (3/2)\varphi$
- In the field $1 \leq \psi \leq 1/\alpha$ from $\psi = (1/\sqrt{3})\,(1/\sqrt{1-\varphi})$
- In the field $1/\alpha \leq \psi \leq \psi_{max}$ from $\psi = [\varphi - (1-\alpha^2)]3/2\alpha^3$

As shown, the calculation does not consider the torsional deformations undergone by the belt that increase joint rigidity. However, it approximates the real behavior of such joints at the design phase and defines the frequency field typical of the system which, as already seen, shows constant values in the first and third phase, but variable values in the second deformation phase.

For variable characteristic joints, there are no harmonic oscillations and therefore they are said not to resonate since rigidity varies from point to point.

If I_1 and I_2 indicate the moment of inertia of the masses upstream and downstream of the joint, and K_1, K_2, and K_3 indicate rigidity in the three functioning phases, respectively, the torsional vibration frequency will be

- In phase one $\omega = K_1((I_1 + I_2)/(I_1 I_2))$ with $K_1 = 3/2$
- In phase two the motion is not harmonic, K_2 and the frequency would vary with deformation
- In phase three $\omega = K_2((I_1 + I_2)/(I_1 I_2))$ with $K_2 = 3/(2\alpha^3)$

19.5.2 Voith Maurer Joint

In Voith Maurer joints, the deformable component is like the one shown in Figure 19.21a. Because of symmetry, the calculation is that for C springs with the center fixed. On close inspection, note that the spring arms are attached to two flanges integral with the drive and driven shafts.

In Figure 19.21b, the deflection is made up of the deformation due to the straight arm flexure due to force Q/n and the torsional of C due to torsional moment Qr/n, hypothesising a perfectly circular-shaped spring and constant Qr/n.

The first part of the deflection is given by

$$f_1 = \frac{Qr^3}{3nEJ},$$

considering the straight part fixed in B.

$$f_1 = \frac{Qr^3}{3nEJ}.$$

To calculate C rotation, and referring to torsional springs:

$$\Delta\varphi = \lambda_1 \frac{\pi Qr^2}{2nGJ_p},$$

Consequently, the total deflection is

$$f_1 = \frac{2Qr^3}{3nEJ} + \lambda_1 \frac{\pi Qr^3}{nGJ_p},$$

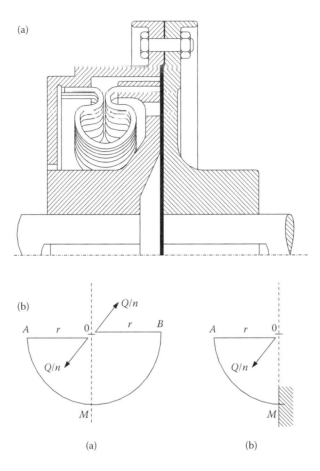

FIGURE 19.21 Voith Maurer joint.

where λ^1 is the coefficient, function $d/2r$, shown in the diagram on page 473, which takes into account the curvature of the component. In this case, the elastic component corresponds to one spring coil mentioned on page 473 of length $\pi r/4$.

With these characteristically linear joints, their nominal operating conditions should not be close to any possible resonance. To achieve this, the joint must be chosen such that system frequency, calculated with formulae above, must be outside the frequency range of possible external exciter harmonics.

As for durability, the straight section undergoes a maximum bending moment equal to Qr/n, and the curved part undergoes a constant torsional moment also equal to Qr/n. Since $Q/n = P$, for a circular cross-section, the tension and bending forces are

$$\sigma \approx 10P\frac{r}{d^3} \quad \tau = 0.5\sigma.$$

To calculate system frequency, the reference point is rigidity that can be inferred from the deflection expression. It is also obvious that the masses to be considered are those of the whole kinematic chain.

19.5.3 FORST JOINT

In the Forst joint, whose operating principle is the same as the above, the flexible component still has a circular cross-section (Figure 19.22).

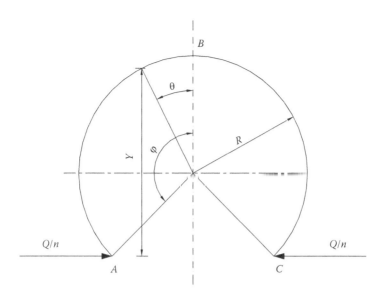

FIGURE 19.22 Spring of the Forst joint.

In simple terms, the system is a series of C springs of center of gravity radius R whose ends are fixed to the flanges of the drive shaft and driven shaft, but whose C springs are set on a plane parallel to that of the flanges. The characteristics are obtained by again calculating deflection due to a certain moment.

Calculating the joint's semi-ring spring follows the scheme in Figure 19.22.

In Figure 19.22, Q is the peripheral force equal to the ratio of the transmittable moment divided by the distance of the spring fixtures from the shaft center, n being the number of springs.

In a first approximation, the spring cross-section can be assumed constant irrespective of the curvature effect. As an effect of system symmetry, the median section does not rotate and can be considered fixed. Applying the theorem of the ellipse of elasticity, the hypothesized deflection due to component bending is given by

$$f = \frac{2Q}{nE}\int \frac{y^2}{J}\,ds,$$

and, since $y = R(\cos\vartheta - \cos\varphi)$, with reference to Figure 19.22, then

$$f = \frac{Qr^3}{nEJ}(\varphi + 2\varphi\cos^2\varphi - 1.5\sin 2\varphi).$$

A more sophisticated calculation may be necessary when the spring shows significant cross section variability and when the inertia radius of the median section is large compared to the mean radius of the spring.

Having defined the cross-section variability law, the previous integral must be calculated to take into account the first effect after substituting the moment of inertia which, if the spring cross-section is rectangular of width b and setting a linear height variation with θ so that it passes through h for $\theta = 0$ at h^1 for $\theta = \varphi$, then

$$J = \frac{1}{12}b[(h^1 + (h - h^1)\theta/\varphi]^3.$$

As regards loads, the maximum moment is in the fixed section, so

$$M_f = \frac{Q}{n} R(1 - \cos\varphi).$$

If the component is very curved (ratio between spring radius R and section radius of inertia ρ lower than 15), as often happens to minimize volume and all the resulting forces of inertia, the following formula must be applied to calculate load:

$$\sigma = \frac{M_f y}{Se(R_n - y)},$$

- y the generic distance from the neutral axis
- R_n the neutral radius (distance from the neutral axis to the center of the solid)
- e the distance of the center of gravity axis from the neutral axis, equal to $(R - R_n)$
- S the area of straight cross-section of the solid

For the circular spring cross-section with radius d, the neutral radius

$$R_n = \frac{d^2}{4(2R - \sqrt{4R^2 - d^2})},$$

and the edge loads are calculated by making $y = d/2 + e$ and $y = d/2 - e$.

For the rectangular spring section of height h, the neutral radius is given by

$$R_n = \frac{h}{\ln(R_e / R_i)},$$

with $R_e = R + h/2$ and $R_i = R - h/2$ and the edge loads are calculated by making $y = h/2 + e$ and $y = h/2 - e$. Tensile stress must be added to the loads calculated above for the fixed section. If the ratio R/ρ is higher than 15, $\sigma = M_f / W$ is applied to calculate bending tension. Apart from the elastic joints analyzed so far, there are others on the market that use torsion springs numbering 3, or less frequently 6. The spring ends are set into flange seats on both shafts so that during transmission the spring's coaxiality is ensured.

In other joints, the elastic component consists of 3 (6) "packets" of steel leaves laid at 120° (60°) fixed at the base of one of the shaft hubs, interlocking with suitably shaped spokes. On the other side the "packets" are seated and fixed to the inside of the bell fixed to the other shaft.

Both in the first and second case, the moment "passes" *via* the elastic components. The forces applied to each spring are always equal to the ratio between the transmittable moment and the distance of the fixture furthest from the shaft center.

Having defined the applied force, the load and characteristics can be calculated as per the Spring chapter 13 for torsion springs (the first case) and for bending springs (the second).

In joints with thin layer springs, the centrifugal force effect tending to increase tensile stress and consequently joint rigidity may not be negligible. The outline calculations for thin layer springs are sufficient to evaluate the above-mentioned effect quantitively.

19.6   MOBILE JOINTS

Mobile joints are used when the relative positions of the driven and drive shafts are not fixed, i.e., when the shafts are incidental or parallel but not axial. When in each instant the transmission ratio between the two shafts is constant they are defined as constant-velocity joints.

In practice, in less significant cases when constant velocity is not required, the joints used could be the more economical ones with constant mean transmission ratio with limited variations depending on joint geometry.

In the following pages there will be examples of these joints, but for those who are really interested we recommend specialist publications. The volume G. Bongiovanni, G. Roccati "Giunti Articolati" [Articulated Joints] Levrotto & Bella Torino [5] is highly recommended.

19.6.1 CARDAN (UNIVERSAL) JOINT

Universal joints (Figure 19.23) are used when motion is transmitted between two shafts with parallel or incidental axis (forming angle α). They are used in vehicle transmission systems and in certain machine tool controls.

As will be shown later, simple universal joints are non constant-velocity—the drive-shaft angular velocity is not always equal to that of the driven shaft, and the two angular velocities coincide only in specific positions.

Referring to Figure 19.24, if the joint is rotating, the fork ends (B_1, B_2) of drive shaft $a-a$, will draw a perpendicular circle whose diameter is $2R$ and segment B_1B_2 is a trace of the circle in the design plane.

The driven shaft fork ends $(B'_1B'_2)$ will draw a circle (for equal forks) that will still be equal to $2R$ in the plane perpendicular to the axis of the driven shaft $b-b$ containing the ends of the fork themselves. Projecting the second circle perpendicularly to the engine axis $a-a$ will create an ellipse.

In defining driven shaft velocity as a function of prop shaft velocity, it has to be noted that while one end of the driven fork rotates perpendicularly to the axis (driven) along the equation circumference:

$$\frac{x^2}{R^2} + \frac{y^2}{R^2} = 1,$$

the same end projected instant by instant perpendicularly to the engine axis, draws the ellipse of the R and $R \cos \alpha$ semi-axis.

Counting the rotation angles φ starting from the one perpendicular to the plane containing the two axes (driven and drive) passing through the cross center (Y-axis), point B_2 (coinciding with the beginning for the circle and ellipse) for a rotation of 90° will be at the end of the circle perpendicular

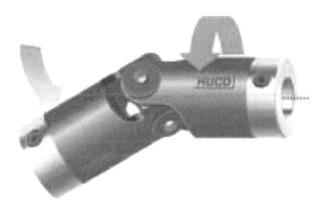

FIGURE 19.23 Cardan joint.

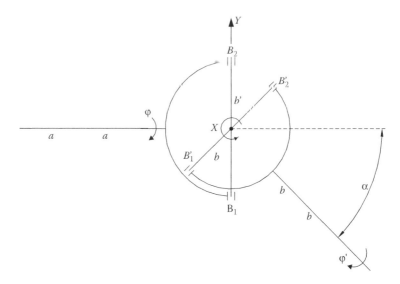

FIGURE 19.24 Scheme of the Cardan joint.

to the engine axis and at the end of the ellipse perpendicular to the driven axis. Therefore, it will describe different spaces (circumference and ellipse arcs) sharing the same ordinates but different abscissa. On the Y-axis, the motion will be harmonic, the distance made by the B_2 projection on the X-axis (perpendicular to Y-axis) in a complete lap ($\varphi = 2\pi$) of the drive shaft will be $2R$, instead the distance on the same X-axis for the projection of the same point (B_2) as a point of the other cylinder (axis b–b) in the same time will be $2R \cos \alpha$. At a rotation of the B_2 point of the driven shaft on the perpendicular plane to the engine axis, φ' will correspond (on the same plane) to a rotation φ of the driven axis. For the points of the two curves (circumference and ellipse tangents on the Y-axis), generally:

$$tg\varphi = x/y \quad tg\varphi' = x'/y'.$$

Because of above, $y = y'$, so, $tg\varphi = tg\varphi' \, x/x'$.

By substituting the ratio x/x' with what is found from the two curve equation, we obtain

$$tg\varphi' = tg\varphi \cos \alpha.$$

The ratio between the tangents of the two angles is equal to $\cos \alpha$. Generally, in practice, the angle of incidence α is $<20°$.

Table 19.2 shows the assumed tangent values of φ and φ' and $\Delta\varphi = \varphi - \varphi'$, and in the case of considerable angles with $\alpha = 20°$.

It is also possible to draw an instant by instant diagram of $\Delta\varphi$ as a function of φ. The result is the periodic trend in Figure 19.25.

As you see, the drive shaft and driven shaft have the same rotation only after period $\pi/2$. The maximum phase shift between φ and φ' is for $\varphi = \pi/4$ and equals about $20°$.

The same observations can be made for transmission ratio:

$$\tau = \frac{\omega'}{\omega} = \frac{d\varphi'/dt}{d\varphi/dt}$$

TABLE 19.2
Table of $\Delta\varphi$ as a Function of φ

φ	$tg\varphi$	$tg\varphi'$	φ'	$\Delta\varphi'$
0	0	0	0	0
$\pi/4$	1	0.94	$\cong 43°$	$\cong 2°$
$\pi/2$	∞	∞	$\pi/2$	0

Consider the ratio found earlier: $tg\varphi' = \cos\alpha\, tg\varphi$, and derive both members. We obtain

$$\frac{d}{dt}(tg\varphi') = \frac{d}{dt}(\cos\alpha\, tg\varphi).$$

Developing the derivatives and recalling that

$$\frac{d}{dt} tg\varphi = \frac{d\varphi}{dt}\frac{1}{\cos^2\varphi},$$

then

$$\frac{d\varphi'}{dt}\frac{1}{\cos^2\varphi} = \frac{\cos\alpha}{\cos^2\varphi}\frac{d\varphi}{dt},$$

that is

$$\frac{\omega'}{\omega} = \cos\alpha\frac{\cos^2\varphi'}{\cos^2\varphi}.$$

Recalling that

$$\frac{1}{\cos^2\varphi^1} = \frac{\sin^2\varphi' + \cos^2\varphi'}{\cos^2\varphi'} = 1 + tg^2\varphi' = 1 + \cos^2\alpha tg^2\varphi,$$

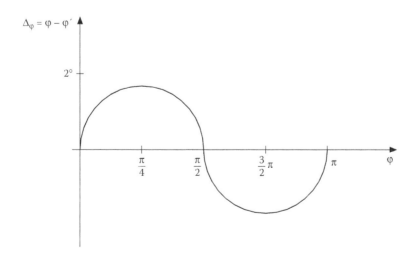

FIGURE 19.25 $\Delta\varphi$ as a function of φ for the Cardan joint.

and that $\tau = \omega' / \omega$, we have

$$\frac{\cos^2 \varphi}{\cos^2 \varphi'} = \cos^2 \varphi (1 + \cos^2 \alpha \, tg^2 \, \varphi)$$

$$= \cos^2 \varphi \left(1 + \cos^2 \alpha \, \frac{\sin^2 \varphi}{\cos^2 \varphi} \right)$$

$$= \cos^2 \varphi + \cos^2 \alpha \sin^2 \varphi.$$

Adding and subtracting $\sin^2 \alpha \sin^2 \varphi$, we obtain

$$\cos^2 \varphi + \cos^2 \alpha \sin^2 \varphi + \sin^2 \alpha \sin^2 \varphi - \sin^2 \alpha \sin^2 \varphi$$

$$= \cos^2 \varphi + \sin^2 \varphi (\cos^2 \alpha + \sin^2 \alpha) - \sin^2 \alpha \sin^2 \varphi$$

$$= \cos^2 \varphi + \sin^2 \varphi \sin^2 \alpha \sin^2 \varphi$$

$$= 1 - \sin^2 \alpha \sin^2 \varphi,$$

and, finally,

$$\tau = \frac{\omega'}{\omega} = \cos \alpha \, \frac{\cos^2 \varphi'}{\cos^2 \varphi} = \cos \alpha \, \frac{1}{1 - \sin^2 \alpha \sin^2 \varphi} = \frac{\cos \alpha}{1 - \sin^2 \alpha \sin^2 \varphi},$$

and, consequently,

$$\text{for } \varphi = 0 \quad \tau = \cos \alpha \cong 0.94,$$

$$\text{for } \varphi = \frac{\pi}{2} \quad \tau = \frac{1}{\cos \alpha} \cong 1.06.$$

Plotting τ as a function of φ, we obtain the graph in Figure 19.26.

If we suppose drive-shaft angular velocity ω is constant, driven shaft angular acceleration is

$$\frac{d\omega'}{dt} = \omega^2 \, \frac{\sin^2 \alpha \cdot \cos \alpha \cdot \sin 2\varphi}{(1 - \sin^2 \alpha \cdot \sin^2 \varphi)^2}.$$

Such acceleration cancels out for φ multiples of $\pi 2$, and assumes equal but opposite values for supplementary φ values. Every quarter of a revolution (φ from 0 to 90°, φ from 90° to 190°, etc.), it has a maximum or a minimum, of equal absolute value, corresponding to a certain value of $\overline{\varphi}$ of φ defined by the ratio

$$\sin^2 \overline{\varphi} - \sin^2 \overline{\varphi} tg 2\overline{\varphi} = \frac{1}{\sin^2 \alpha},$$

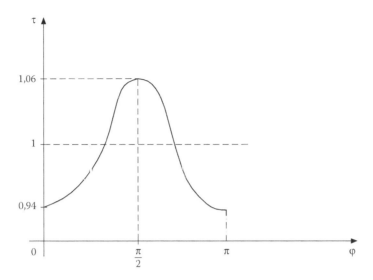

FIGURE 19.26 τ as a function of φ for the Cardan joint.

that can be obtained by cancelling out the derivative of $d\omega'/dt$ in relation to φ.

For small α values, such as those usually used in universal joints, $\varphi = 45°$ approximately and therefore:

$$\left(\frac{d\omega'}{dt}\right)_{max} \cong \frac{\sin^2 \alpha \cos \alpha}{\left(1 - \frac{\sin^2 \alpha}{2}\right)^2}\omega^2 \cong \sin^2 \alpha \cdot \cos \alpha \cdot \omega^2 \cong \alpha^2 \omega^2.$$

Acceleration is therefore related to the square of the angle of incidence. This is one of the reasons why α must be kept as low as possible.

In practice, because of deformability, maximum acceleration is much lower than in theory. In any case, the simple universal joint is limited to α angles not greater than 10°–15°.

If a wider angle is necessary, a double universal joint is used so that the mean angle of incidence doubles. Figure 19.27 shows the diagrams of the incident and parallel axes used in these cases.

The double universal joint is a constant velocity joint; this solution provides a double phase shift that cancels out the initial phase shift. To verify this, in general the universal joint ratio is $tg\varphi_C = \cos \alpha tg\varphi_M$.

If motion is supplied by a, shaft a'' will be driven in relation to a, therefore

$$tg\varphi'' = \cos \alpha tg\varphi.$$

Analogously, shaft a' will be driven in relation to a'' and therefore for a phase shift of 90°:

$$tg(\varphi' + 90°) = \cos \alpha tg(\varphi'' + 90),$$

and since $tg(\varphi + 90) = \cot \varphi$, we obtain

$$\cot g\varphi' = \cos \alpha \cot g\varphi'' \quad tg\varphi'' = \cos \alpha tg\varphi'.$$

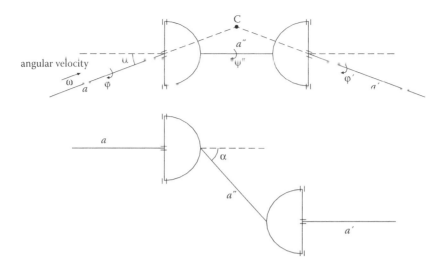

FIGURE 19.27 Double Cardan joint.

Consequently, $tg\varphi = tg\varphi'$, so the joint is a constant velocity one.

The same applies to parallel axes. With $tg\varphi'' = \cos \alpha tg\varphi$ for a phase shift of 90°:

$$tg(\varphi' + 90) = \cos(-\alpha)tg(\varphi'' + 90),$$

or

$$\cot g\varphi' = \cos \alpha \cot g\varphi'',$$

so even in this case we have $tg\varphi = tg\varphi'$.

Finally, a double universal joint has two advantages: it can assume $\alpha = 20°–30°$, and it is a constant velocity joint.

In the case of skew shafts, constant velocity transmission can be achieved if the α angles formed between the extreme and intermediate axes are equal through suitable layout.

In a double universal joint, the intermediate shaft is accelerated or slowed, so it is better if it is light to avoid that inertia damage. In automobile deployment, the prop shaft is often in two parts, which can be distanced or drawn together by a telescopic device with a splined linkage so that angle α can be made equal at the entrance and exit.

Many joints based on the universal joint principle are on the market and in industrial use. In the auto industry, the Rzeppa joint is widely used where the drive and driven components are linked by a crown of balls (held in a cage) with the drive shaft race on one side and the driven shaft race on the other. Perfect constant velocity is thus obtained.

19.6.2 OLDHAM JOINT

The Oldham joint is infrequently used because it only transmits low power. However, it is used when the shaft axes to be linked must be parallel but not coincident. The downside is it produces considerable inertia and in the most traditional versions the friction generated by parts in contact produce low performance.

It is made of three discs, one on the drive shaft, one on the driven, and one is intermediate (Figure 19.28).

FIGURE 19.28 Oldham joint.

The end discs have a central diametric spline. The intermediate one has two ribs in the opposite faces at 90° to each other (Figure 19.29).

It can be demonstrated that the angular velocity of the intermediate disc center is double that of the drive shaft. In Figure 19.30, for one revolution y, points A move to D and points B move to C. The intersection of straight lines $\overline{CC}$ and $\overline{DD}$ is none other than the cross center describing a circumference of diameter e.

If the disc rotates $\pi/2$ instead of y, the cross center moves from O_1 to O_2 and the velocity of the intermediate disc center is double that of the drive shaft velocity (when the drive shaft revolves $\pi/2$, the disc center revolves π).

Note the intermediate disc motion is the sum of two motions: one is rotary, the other is rectilinear along the line linking the centers of the two shafts. The intermediate disc rotates at an angular velocity equal to that common to the two shafts. However, its center rotates at double the velocity.

If P is the intermediate component weight, ω the angular velocity of the drive and driven shafts, and e the eccentricity of their axes, the intermediate component undergoes a centrifugal force:

$$F_c = \frac{P}{g}\frac{e}{2}(2\omega)^2 = \frac{2P}{g}e\omega^2,$$

which it discharges to the shaft supports, causing supplementary friction vibrations. This joint can therefore only be used at low revolutions. Checking the vibration mode should ensure that resonance is avoided close to nominal operation.

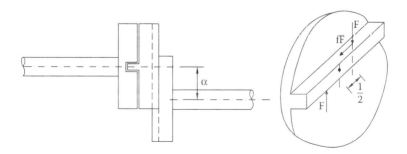

FIGURE 19.29 Scheme of the Oldham joint.

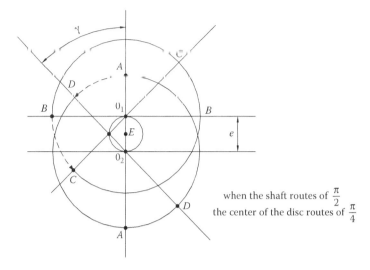

FIGURE 19.30 Scheme to evaluate the velocity of the intermediate disc center.

To determine the performance of an Oldham joint refer to Figure 19.29. The forces indicated are the resulting pressure on the ribs of the intermediate disc assumed exerted at a distance $l/2$ from the disc center. Performance is expressed by

$$\eta = \frac{P_u}{P_u + P_p},$$

where P_u is the output power and P_p the power lost.

Referring to Figure 19.29, P_u equals:

$$P_u = C\omega = 2F\frac{l}{2}\omega,$$

(in the expression, factor 2 refers to the fact that on the ribs of the other side of the disc the same forces occur).

Each force F produces a tangential component fF, due to creep between the ribs and splines of the end disc, and consequently the power lost is

$$P_p = 4fFv.$$

Considering that the distance of one rotation is equal to $s = 4e$, linear velocity v can be expressed as a function of ω, so

$$v = \frac{s}{t} = \frac{2\omega e}{\pi},$$

therefore,

$$P_p = \frac{8fF\omega e}{\pi}.$$

By substituting P_u and P_p in the performance expression and dividing by $Fl\omega$, we finally obtain

$$\eta = \frac{Fl\omega}{Fl\omega + \dfrac{8fF\omega e}{\pi}} = \frac{1}{1 + \dfrac{8fe}{l\pi}} \approx 1 - \frac{8fe}{l\pi}.$$

The centrifugal force and performance expressions above help choose the size and type of joint. They show that by modifying geometrical parameters one can reduce the damaging effects of eccentricity, and also how to limit the power lost by acting on l and eccentricity e. By apposite choice of Oldham joint parameters, the rotation velocity range normal for these joints can be extended.

The most recent Oldham joints using roller cages for alternative motion achieve better results.

Analogously, the Schmit joint has a flange fixed to the drive shaft linked to an intermediate ring by three connecting rods. The other side of the ring is linked to the driven shaft by three more connecting rods. This kinematic mechanism is two articulated quadrilaterals, one integral with the entry flange and one integral with the exit flange, with equal instant angular velocities, and therefore the joint has an absolutely constant velocity. The position of the intermediate ring center is always fixed and consequently there are no forces of inertia to produce those unwanted transmission vibrations.

Other joints used in motion transmission between non-axial shafts, where one is inside the other (railways), are "dancing ring" joints and "lyre" joints whose details can be found in specialist books (G. Bongiovanni and G. Roccati).

20 Clutches

20.1 INTRODUCTION

Compared to joints, clutches create temporary controllable linkages with intermediate stages up to total disconnection. The connection or disconnection mechanism must be simple and rapid. Some clutches, like those for security systems, cannot transmit torque over a pre-fixed value, or surplus clutches that disconnect when one of the two shafts begins to slow. According to requirements, clutches can connect or disconnect at rest or in motion relative to the parts to connect.

Clutches can first be classified into those with teeth and those by friction. The former can only engage when the shafts are almost at rest, whereas they can be disengaged in motion. The latter can be engaged or disengaged in motion without excessive or uncontrollable loads.

A clutch with teeth was encountered in Chapter 12 (Hirth teeth). Of all the clutches on the market, in this chapter we will look at friction clutches because they are more interesting didactically and from an evolutionary perspective. Please refer to Giovanozzi, Vol. I, for more details on tooth clutches.

Nearly all clutches can be classified according to:

- Number of fiction surfaces
- Friction surface shape
- Method of exerting pressure
- Method of engagement (radial or axial)
- Fiction material

Table 20.1 shows a simplified clutch chart.

Automatic clutches require no intervention since their components develop the forces needed, such as centrifugal or friction forces. They are classified thus:

- Surplus clutches allow power to be transmitted from a specific side of the shaft and not vice versa.
- Safety or slip clutches cease to transmit power once a pre-determined maximum is reached.
- Centrifugal clutches depend on the centrifugal force of rotation to provide inter-friction disc pressure.

20.2 FRICTION CLUTCHES

Friction clutches permit gear changing while the vehicle is moving. Since the transmittable moment depends on clutch characteristics (contact pressure, surface number, friction material, etc.), it is well definable and so the drive-train components are designed accordingly to avoid overloading that could lead to clutch slip with consequent overheating and degradation of the friction material. To guarantee this, clutches are usually sized with higher than necessary limits, just as the drive-train components are.

Apart from transmitting torque from the crankshaft to the driveshaft (clutch fully disengaged), the friction clutch must also be able to disconnect the engine from the transmission when gear has to be changed or when the vehicle must be stopped without switching off the engine (clutch fully engaged). In more recent applications, the "clutch" also absorbs torque peaks during fast gear changing. It must also work as a torsional damper limiting torsional amplitude if there are

TABLE 20.1

Surface	Engagement Method	Clutch Type
Flat	Axial	- Disc
		- Stromag
Conical	Axial	- Simple cone
	Radial	- Double cone
		- Dohman-Leblanc
Cylindrical	Radial	- Suspension
		- Expansion
		- Logarithmic

any resonances between the excitation frequency and the kinematic chain frequencies of the transmission.

Clutches connect components with initially different velocities, so in the time necessary for the two velocities to match, work is lost to friction with a corresponding power loss and consequent heating that is all the greater the longer the time.

Analyzing the work lost allows us to see how to vary the moment to minimize the work lost to friction and therefore to clutch heating.

Figure 20.1 shows ω as the crankshaft velocity, the driveshaft velocity ω', M^* the moment transmitted by the engine, M the moment transmitted by the clutch to the driveshaft, and M_R the resistant torque.

In the following considerations, constant ω will be hypothesized over the entire clutch movement.

At the beginning of engagement, driveshaft velocity ω' is zero and only after time T will it have the same velocity as the crankshaft ($\omega = \omega'$). When totally engaged, $\omega = \omega'$ and for equilibrium $M^* = M$. So reference will be made to M.

In time T, total engine work supplied is

$$L_m = \int_0^T M\omega dt = \omega \int_0^T M dt.$$

Of engine work L_m, a part serves to overcome moment M_R acting on the driveshaft, a part is used to accelerate the driveshaft and a part is lost to friction.

Clutches allow the gradual transmission of power to the driveshaft: 0 to M with continuity. As long as M is lower than M_R, the shaft remains at rest so all the crankshaft work is lost to friction. When the pressure between the discs reaches a value for which M is greater than M_R, the driveshaft begins rotating from 0 to the velocity of the crankshaft. During this operation, there is slippage and any power lost is due to moment M_R because of the relative velocity. At the end of this phase, the two shafts reach the same velocity and all power is transmitted to the driveshaft.

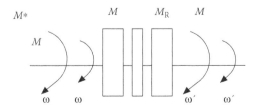

FIGURE 20.1 Scheme in working conditions for a clutch.

So, t_0 is the time when $M < M_R$, and the work lost to friction is

$$L_{p_1} = \int_0^{t_0} M\omega\, dt = \omega \int_0^{t_0} M\, dt,$$

supposing ω constant during engagement.

After time $t = t_0$, engine moment exceeds M_R, so the driveshaft begins rotating at a variable velocity until it reaches equilibrium with the moments applied. Over this operation the work lost is

$$L_{p_2} = \int_{t_0}^{T} M(\omega - \omega')\, dt,$$

which may also be written as

$$L_{p_2} = \int_{t_0}^{T} (M - M_R)(\omega - \omega')\, dt + \int_{t_0}^{T} M_R(\omega - \omega')\, dt. \qquad (*)$$

The dynamic equilibrium is

$$M - M_R = J\frac{d\omega'}{dt},$$

with J the polar moment of inertia of the driveshaft and so the first term of $(*)$ is

$$\int_{t_0}^{T} (M - M_R)(\omega - \omega')\, dt = \int_{t_0}^{T} J(\omega - \omega')\frac{d\omega'}{dt}\, dt$$

$$= \int_0^{\omega} J(\omega - \omega')\, d\omega' = J\left[\omega\omega' - \frac{\omega'^2}{2}\right]_0^{\omega}$$

$$= J\left[\omega^2 - \frac{\omega^2}{2}\right] = \frac{1}{2}J\omega^2,$$

having taken into account that when $t = t_0$ then $\omega' = 0$ and when $t = T$ then $\omega' = \omega$.

Finally, by substituting in L_{p_2}, then

$$L_{p_2} = \int_{t_0}^{T} M_R(\omega - \omega')\, dt + \frac{1}{2}J\omega^2.$$

Total work lost is the sum of L_{p_1} and L_{p_2}:

$$L_p = L_{p_1} + L_{p_2} = \int_0^{t_0} M\omega dt + \int_{t_0}^{T} M_R(\omega - \omega')dt + \frac{1}{2}J\omega'^2. \quad (**)$$

Replacing this sum with ΔL:

$$L_p = \Delta L + \frac{1}{2}J\omega^2,$$

which shows that the work lost is greater than the kinetic energy that the driveshaft has at the end of engagement.

For $M_R = 0$, the work L_p lost to fiction is lower:

$$L_p = \frac{1}{2}J\omega^2.$$

Now consider engine work. We saw earlier that

$$L_m = \int_0^T M\omega dt.$$

Applying the motion value and hypothesizing $M_R = 0$:

$$M = J\frac{d\omega'}{dt},$$

and substituting into the integral:

$$L_m = \int_0^T J\omega\frac{d\omega'}{dt}dt.$$

Changing the integration limits:

$$L_m = \int_0^\omega J\omega d\omega = J\omega\left[\omega'\right]_0^\omega = J\omega^2.$$

Calculating the minimum energy dissipated during engagement:

$$L_m = L_R + L_p + \Delta E_c.$$

Given that $\Delta E_c = 1/2 \cdot J\omega^2$.

When $M_R = 0$, obviously $L_R = 0$, and also $L_p = 1/2 \cdot J\omega^2$, and $L_m = J\omega^2$ consequently:

$$J\omega^2 = \frac{1}{2}J\omega^2 + \frac{1}{2}J\omega^2,$$

which means that the energy balance is satisfied. At the fringe when $M_R = 0$, this is the minimum energy dissipated.

When $M_R \neq 0$ and the two integrals (**) are not zero:

$$\Delta L = \int_0^{t_0} M\omega \, dt + \int_{t_0}^{T} M_R(\omega - \omega')dt.$$

The first integral is the work during $t = t_0$, which is as small as the rapidity with which M grows to M_R. If from the beginning $M > M_R$, this integral would become zero with $t_0 = 0$.

The second integral may be written $M_R (\omega - \omega')_m (T - t_0)$ in the interval $t_0 - T$ when mean relative velocity is $(\omega - \omega')_m$. Therefore, it becomes smaller the higher $d\omega/dt$ becomes.

For the motion quantity principle:

$$M - M_R = J \frac{d\omega'}{dt},$$

which highlights that $d\omega'/dt$ is proportional to $M - M_R$ and that the second integral becomes smaller the higher M becomes.

Finally, the work lost to friction through slippage can be reduced by applying high moment M and minimum engagement times. This contrasts with the need to apply moments gradually to avoid impact and high over-loading.

To reduce dynamic loading, the clutch needs to be designed for a maximum moment M_{max} transmittable by the clutch not much higher than moment M_R from the crankshaft, and it must be done such that engagement is soft and progressive.

The curve in Figure 20.2 shows the qualitative trend of distance λ between the mobile fiction surfaces and the position of incipient contact with fixed friction surfaces relative to the axial shift s of the sliding collar.

Point A defines the play OA between the active fiction surfaces. Due to elastic deformation, beyond distance B the pressure is such as to create elastic deformation between the parts in contact, and so distance λ is negative. Its minimum is C, known as the "dead spot." The final value of λ is reached at point D where the lever system that exerts the axial force has been relaxed.

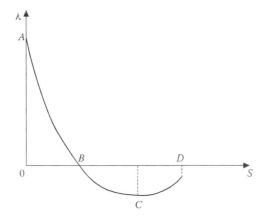

FIGURE 20.2 Qualitative trend of the distance λ between the mobile friction surfaces and the axial shift s.

20.3   FLAT PLATE CLUTCHES

The active surfaces in flat plate clutches are symmetrical discs and nearly always have a circular crown. Their operating principle is simple: a force perpendicular to the disc plane generates by way of the friction material a tangential force leading to the moment of torque. This perpendicular force can be generated through springs, levers, and electromagnetic forces.

The disc was pressed against the flywheel *via* coil springs in the very first applications (see Giovanozzi, Vol. I, p. 764); nowadays, however, it is done by diaphragms, or rather springs similar to Belleville, but with notches (about 20) in the direction of the cone generatrix as far as the least diameter circle, to reduce rigidity compared to a classical Belleville. Each notch on the side of the greatest diameter circle finishes with a circular zone for centering and at the same time, having a suitable radius, reduces the effect of notching in each plate generated by those notches. The behavior of each plate under the applied load is of a bending trapezoid-shaped spring fixed at its ring end.

Diaphragm springs have the advantage of being more balanced, less bulky, able to continue working even after a long lifespan, and they have reduced pedal pressure.

Figures 20.3 and 20.4 shows the friction disc of a dry clutch. It is made of a material with a high friction coefficient (similar to that used in brake linings). The disc is pressed against the flywheel by springs (usually plate springs). Operated by a mechanical system, depressing a pedal causes the disc to separate from the flywheel, leaving the two shafts (crank and drive) independent.

The friction material is subject to wear, particularly when the pedal is only slightly depressed. As a result, the force of the springs is weaker and the disc slips, heating up and deteriorating. Disengaging the clutch at the same time as fast acceleration also speeds up wear.

There are clutches that are submerged in a liquid (usually oil) to improve the lifespan and efficiency of the friction disc. They are used when, especially in the auto industry, limited dimensions, and proportionally high power are required. The easy substitution of their friction material has also contributed to their increasing popularity.

The most commonly used friction material is Ferodo, but other materials are used on the active disc surfaces from wood to polymers. These materials can work dry or in oil.

In the first case, for similar active areas, the transmittable moment is clearly greater, but in the second case heat is more easily dissipated.

The active surfaces are symmetrical and flat (discs) usually with a circular crown. How axial forces are generated, friction material type, number of friction surfaces, and lifespan, all influence the probabilities of the hypotheses assumed for their calculations.

If $R_i = xR_e$ is the internal friction disc radius compared to the external radius and is the pressure $p = p(r)$ on a generic radius, then the calculation can follow these two hypotheses:

FIGURE 20.3   Two parts of a friction disc.

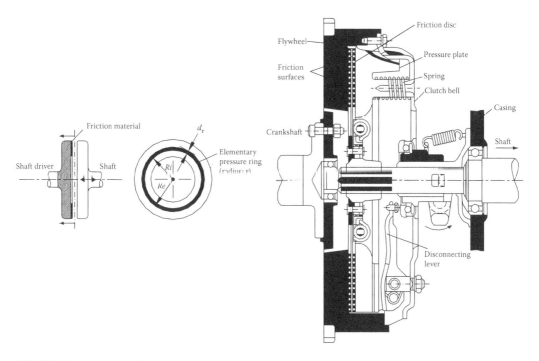

FIGURE 20.4 Friction disc system.

- Constant engagement pressure
- Volume of removed friction material because of wear proportional to the power lost to friction (Reye's hypothesis).

In the first case (constant engagement pressure), $p = \cos t = p_0$. Total compression on the circular crown is

$$P = \int_{R_i}^{R_e} pdA = \int_{R_i}^{R_e} p_0 dA.$$

For element x of the circular crown, its area is

$$dA = 2\pi r\, dr,$$

so the force P expression is

$$P = p_0 2\pi \int_{R_i}^{R_e} rdr = p_0 2\pi \frac{R_e^2 - R_i^2}{2} = \pi p_0 (R_e^2 - R_i^2).$$

If dT is the tangential force ($dT = fdP$), the transmittable moment between two contact surfaces is

$$M = \int_{R_i}^{R_e} rfdP = \int_{R_i}^{R_e} rfpdA = fp_0 \int_{R_i}^{R_e} r2\pi rdr = fp_0 2\pi \int_{R_i}^{R_e} r^2 dr = fp_0 2\pi \frac{R_e^3 - R_i^3}{3}.$$

Radius R_{eq} to which force fP is applied and which is capable of providing the same moment proves to be different to the mean radius. So

$$R_{eq} = \frac{M}{fP} = \frac{fp_0 2\pi \frac{R_e^3 - R_i^3}{3}}{f\pi p_0 (R_e^2 - R_i^2)} = \frac{2}{3} \frac{R_e^3 - R_i^3}{R_e^2 - R_i^2},$$

and therefore can be expressed by

$$R_{eq} = \frac{2}{3} R_e \left(1 + \frac{x^2}{1 + x}\right),$$

whereas the mean radius is

$$R_m = \frac{1}{2}(R_i + R_e) = \frac{1}{2} R_e (1 + x).$$

As these two expressions show, R_m tends to R_{eq} as x tends to 1, or the crown becomes ever narrower ($R_i = xR_e$). The x values used in practice vary from 0.5 to 0.7.

If the number of fiction surfaces is n, the transmittable moment is

$$M_{tot} = nM = nfPR_{eq}.$$

The number of discs $n_d = n/2$ required to transmit moment M_{tot} is

$$n_d = \frac{M_{tot}}{2 fPR_{eq}}.$$

The second case is based on Reye's hypothesis according to which the material removed through wear is proportional to the power lost to friction:

$$dL = K\,dV.$$

Since

$$dV = dA\delta \text{ and } dL = f \cdot dP \cdot \upsilon \cdot dt = f \cdot p \cdot dA \cdot \omega \cdot r \cdot dt,$$

then

$$f \cdot p \cdot dA \cdot \omega \cdot r \cdot dt = K \cdot dA \cdot \delta,$$

from which

$$p = \frac{K \frac{\delta}{dt}}{f\omega r}.$$

With the further hypothesis that $\delta / dt = \cos t$ (δ / dt represents the thickness of material removed in unit time, unit consumption), then

$$p = \frac{K'}{r} \Rightarrow pr = \cos = p_0 r_0.$$

The axial force, therefore, is

$$P = \int_{R_i}^{R_e} p dA = \int_{R_i}^{R_e} \frac{p_0 r_0}{r} 2\pi r dr = 2\pi p_0 r_0 (R_e - R_i).$$

Analogously, the transmittable moment of two discs in contact is

$$M = \int_{R_i}^{R_e} fpr dA = \int_{R_i}^{R_e} r f \frac{p_0 r_0}{r} 2\pi \cdot r \cdot dr = 2\pi f p_0 r_0 \frac{R_e^2 - R_i^2}{2}.$$

Referring to R_{eq}, and with $M = f P R_{eq}$, we obtain

$$R_{eq} = \frac{M}{fP} = \frac{\pi f p_0 r_0 (R_e^2 - R_i^2)}{f 2\pi p_0 r_0 (R_e - R_i)} = \frac{R_e + R_i}{2}.$$

So, in Reye's hypothesis, the equivalent radius is equal to the mean radius

$$M_{tot} = nM = 2n_d f P R_{eq},$$

and the number of discs required to transmit the torque is

$$n_d = \frac{M_{tot}}{2 f P R_{eq}}.$$

To avoid clutch overheating and therefore a decline in the friction coefficient, it is necessary to limit the work lost or the work lost per unit active surface. Roughly speaking, it ought to be between 200 and 350 Nm/cm^2, depending on friction material. So, it must be

$$L_s = 1 / nS \int_0^{T_0} M(\omega - \omega')dt \le 200 - 350 \text{ Nm/cm}^2.$$

As regards the friction coefficient to use in the calculation, it should be recalled that these vary with temperature. However, provided they do not exceed 200°C, their variation is limited, and therefore practically constant values can be used for linings and sintered metals, being 0.2–0.45 for dry clutches and 0.05–0.2 for wet ones.

Operational demands also influence the choice of materials. For ferodes and sintered metals, p values are between 0.1 and 0.2 N/mm^2, with higher values for wet clutches.

Staying with auto clutches, the friction disc is pressed against the flywheel by coil springs in the very first applications (Giovanozzi, Vol. I, p. 764) and today by diaphragm or rather Belleville-like springs (see Chapter 13).

Diaphragm springs have the advantage of being more balanced, less bulky, able to continue working even after a long lifespan, and they have reduced pedal pressure.

Another important component of the clutch system is the drive disc that contains friction gaskets, cylindrical coil springs (anti-grab), and a damper made from friction discs pressed by disc springs to the hub. With a certain angular play, the hub marries the splines.

When designing a clutch, the force of engagement must be soft, the friction coefficient must not vary much with temperature, contact surface wear must be limited, and after a certain lifespan it must still work.

So, sizing a clutch with upgraded torque coefficients between 1.2 and 1.3 consists of:

- Determining the maximum axial force
- Sizing the diaphragm springs with geometric parameters so that between a new and used disc the engine's maximum torque can be transmitted with safety margins of 10%–15%
- Fixing the pedal load P so that for cables the value is 0.7 P and for hydraulics is 0.9 P

On average, pedal loads are between 250 and 300 N with pedal travel (including any deformation in the kinematic chain) between 150 and 200 mm.

The "anti-grab" torsional springs also need sizing with the formulae reported in Chapter 13; the axial forces exerted on them are easy to calculate given the spacing at which they are mounted (about half the external diameter of the friction gasket).

Sizing the various components well should take into account the data reported above and should adhere to the limits cited for the work specifically lost to friction. A good clutch is the result of choices, also based on experience. The many variables of the clutch system must be combined such that they meet different needs and the limits cited above.

For greater detail on clutches and friction materials refer to the ample documentation provided by manufacturers.

20.4   CONE AND DOUBLE CONE CLUTCHES

Clutches with conical friction surfaces can transmit moment through axial forces that are less than those for frontal clutches.

Conical clutches are used in automobiles and in particular in synchronizers.

In conical clutches, a bell is mounted on the driveshaft into which the cone fits through a system of levers and springs. The cone with its friction surface is spline-mounted onto the crankshaft. The cone angles vary from 10° to 15°.

To guarantee constant pressure, the parts in contact must be suitably rigid. The bell should be designed so that under load there are no deformations that could bring about any altered performance.

Calculation of the transmittable moment M as a function of the perpendicular force N is carried out according to the following model (Figure 20.5).

If the clutch is engaged while the system is at rest, the friction components are in opposite directions compared to the relative velocity of the two surfaces and therefore in a direction coinciding with the cone generatrix. So

$$F = N \sin \alpha + fN \cos \alpha = N(\sin \alpha + f \cos \alpha).$$

If disengagement occurs with the vehicle at rest then the friction component's sign changes:

$$F = N \sin \alpha - fN \cos \alpha = N(\sin \alpha - f \cos \alpha).$$

So

$$F = N(\sin \alpha \pm f \cos \alpha).$$

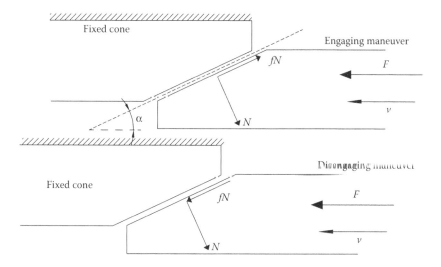

FIGURE 20.5   Model for a double cone clutches.

If engagement occurs with the vehicle in motion, the velocity resultant practically coincides with the tangential direction (the axial velocity of the clutch is relatively small compared to the peripheral velocity) and the axial component of the friction force is almost non-existent. So

$$F = N \sin \alpha.$$

If disengagement occurs in motion, in the moment when the axial force is missing, the components perpendicular to the friction surfaces are also missing and disengagement occurs automatically. The maneuver can also occur with continuity. If the axial force is generated by a spring, F is a function of the vector that gradually changes from engagement in motion to disengagement. In this case, if the resistant moment were close to the transmittable one, slippage would occur, leading to disengagement.

For engagement at rest it may be concluded that

$$N = \frac{F}{\sin \alpha + f \cos \alpha},$$

whereas engagement in motion is

$$N = \frac{F}{\sin \alpha}.$$

If maximum axial force F is spring generated it always remains the same depending entirely on maximum spring deformation. So, the maximum transmittable peripheral force is theoretically always equal to

$$Q = f \frac{F}{\sin \alpha}.$$

Approaching slippage, the cone generated friction components are missing, and so for equilibrium, N changes from $N = F / (\sin \alpha + f \cos \alpha)$ to $N = F / \sin \alpha$.

For $tg\,\alpha < f$, the two parts are blocked; unblocking them requires an axial force opposite to the force exerted by the spring:

$$F' = N'(f\cos\alpha - \sin\alpha),$$

N' being the perpendicular reaction.

In practice, for real deformation of the various parts to occur, the force is different to N, as experience with cones confirms.

Using a calculation similar to that for flat disc clutches, it is possible to verify that the transmittable moment for each case examined ($\rho = \cos t$, $P\,r = \cos t$) is equal to that of a clutch with the same internal and external radii divided by $\sin\alpha$. This shows how, in the case of conical friction surfaces, it is possible to transmit the same moments but with less pressure compared to flat surface clutches.

To avoid clutch shudder, this type of clutch should work with relatively low pressures (p from 10 to 50 N/cm²) and friction materials like linings or sintered metals.

Conical surfaces make the clutch more "grabby" so they are no longer in use for auto applications, at least in transmissions, where the preference is for flat discs.

They are used in synchronizers where increasing synchronization torque is possible by increasing the contact surfaces (multiple cones), as it is not possible to increase the synchronizer's diameter.

The permissible wear for friction surfaces in these applications for the purposes of compactness does not exceed 0.15 mm, which corresponds to a minimum axial distance of about 1.5 mm. Synchronizers are nearly always wet and their friction coefficients vary from 0.08 to 0.12 with pressures varying from 3 (steel-bronze) to 6 (steel-steel). For mean velocities of 6 m/s the specific heat energy for the two friction materials should not exceed 0.1 N/mm² and 0.55 N/mm².

Lastly, for minimum bell thickness, the following formula is useful:

$$s = R_\mathrm{m}\,\frac{\sqrt{10p}}{\sqrt{\sigma_\mathrm{amm}}},$$

which derives from the simplistic hypothesis of rigid disc and load applied to the mean radius of the conical crown, providing values capable of ensuring the disc is sufficiently rigid.

20.5   RADIAL BLOCK CLUTCHES

Today, these clutches are not in widespread use. As in all clutches, the three essential components are those united to the shafts and the mobile clutch discs. In radial block clutches the mobile discs are blocks that under pressure press against the external bell of the driveshaft by means of levers and springs.

Because they are little used, the diagram of a clutch used in the past will serve as our example whose spring size ensures the necessary load to transmit the moment.

Figure 20.6 shows Dolmen–Leblanc clutch. The calculation for this requires finding the force needed by the transmission for a given torque, and the size of the blocks, bell, and springs that ensure that engagement transmits the moment.

The peripheral force at the mean radius R_m of the friction surfaces is

$$Q = \frac{M}{R_\mathrm{m}}.$$

The overall perpendicular force N on the friction surfaces is linked to Q in

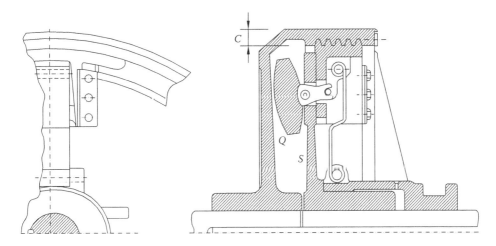

FIGURE 20.6 Dolmen–Leblanc clutch.

$$Q = \frac{fN}{m},$$

where m is a safety coefficient that assumes values from 1.5 to 2.5.

With i the number of blocks, and iP the radial force exerted by the springs to generate force N, then

$$iP = N \sin \alpha.$$

Since $N = Qm/f$ can be deduced from the expression above

$$iP = \frac{Qm}{f} \sin \alpha = \frac{mQ}{f^*},$$

with $f^* = f/\sin \alpha$, the so-called virtual friction coefficient.

Since $N = Qm/f$, the perpendicular force on each block is equal to that number divided by the number of blocks, so

$$\frac{N}{i} = \frac{m}{i} \frac{Q}{f}.$$

The area on which this force is exerted is $2na' \cdot a''$, with n the number of block teeth, a'' the width of the active tooth sides (in the cross-section with a meridian surface), and a' the peripheral length of the teeth.

The specific pressure on the tooth sides is

$$P = \frac{N}{2ina'a''},$$

which should not exceed 200 N/cm^2.

The springs that guarantee the pressure needed are S or C shaped. The calculation is applicable to both types. Due to symmetry, the center line is not subject to any rotation under force and so it can be considered fixed for both spring types in this calculation model.

The load on each spring can be found by knowing that the moment is transmitted via the tangential force Q applied at the mean radius R_m of the teeth embedded in the bell. If α is the external tooth surface inclined to the vertical, the total force the springs i must exert is

$$iP = mM / (fR_m),$$

with the safety coefficient m assumed around 1.5–2.0 and f the friction coefficient. With an α of $20°$ and if $f = 0.1$, the value of $f \sin \alpha = 0.3$.

Having calculated the load P on each spring, their possible deformation characteristics can be calculated such that they continue to exert force P between the active surfaces of the blocks and bell.

Referring to Figure 20.6, for a spring undergoing perpendicular force P and a maximum bending moment Pa, if A is the straight cross-section area and J the moment of inertia of the spring cross-section, then there is a tension of

$$\sigma = \frac{P}{A} \pm \frac{Pa}{J} y,$$

given that y is the distance of the neutral axis of the most loaded fiber.

These springs may have circular or rectangular cross-sections and so the previous formula needs adapting.

As regards deformation, if the spring is not too thin ($l/2\rho \leq 20$), it is calculated by taking into account that total deformation that is made up of a component due to the bending of a length a, and another due to the rotation of length l multiplied by a:

$$y_0 = 2\left(\frac{Pa^3}{3EJ} + \frac{Pa^2l}{2EJ} \right).$$

If the spring is too thin ($l/2\rho \geq 40$), the vertical deformations produce a variation of moment that, this time, and contrasting with the previous calculation, is not constant over the whole length l. In this case, the moment must be expressed as a function of the x coordinate and integrated in the elastic line equation.

Referring to Figure 20.7, the moment is

$$M(x) = P(a + y),$$

and so

$$EJ \frac{d^2y}{dx^2} = -P(a + y).$$

Integrating the equation leads to a particular integral $y = -a$, and with $k^2 = P/EJ$, to the homogenous integral $y = C_1 e^{kx} + C_2 e^{-kx}$.

Finally, we can also express:

$$y = A \sin kx + B \cos kx - a,$$

given that $K^2 = P/EJ$.

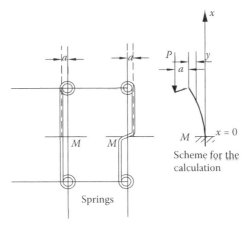

FIGURE 20.7 Springs in a Dolmen–Leblanc clutch.

The constants A and B are determined by the boundary conditions since $y' = 0$ for $x = 0$ and $y = 0$ for $x = l/2$. By having substituted in the expression for elastic deformation $\beta = Kl/2$, then

$$A = 0 \quad B = \frac{a}{\cos\beta} \quad y = a\left(\frac{\cos Kx}{\cos\beta} - 1\right),$$

and at point B as an absolute value:

$$\left(\frac{dy}{dx}\right)_{x=l/2} = \left[\frac{Ka\sin Kx}{\cos\beta}\right]_{x=l/2} = \frac{Ka\sin\dfrac{Kl}{2}}{\cos\beta} = Katg\,\beta$$

Since β is small, generating $tg\beta$ in series, it is:

$$tg\beta = \beta + \frac{\beta^3}{3} + \ldots$$

Just considering the first term of the series, y_0 is found first. For the second term:

$$y_0 = 2\frac{Pa^2}{EJ}\left(\frac{a}{3} + \frac{l}{2} + \frac{Pl^3}{48EJ}\right).$$

Therefore, the vector is that of the leaning beam with an increase in length l to $l + 4f_0$.
The bending moment that is now variable is at maximum at M, equal to

$$M_{max} = -EJ\left(\frac{d^2y}{dx^2}\right)_{x=0} = EJ\frac{K^2a}{\cos\beta} = \frac{Pa}{\cos\beta}.$$

Generating the series $\cos\beta$ is:

$$\cos\beta = 1 - \frac{\beta^2}{2} + \ldots \rightarrow \frac{1}{\cos\beta} = 1 + \frac{\beta^2}{2} + \ldots$$

and focusing on the second term:

$$M_{max} = Pa\left(1 + \frac{Pl^2}{8EJ}\right).$$

So, summarizing further generation of the calculation reveals a third term compared to those found with the first approximation. So:

$$(P^2 a^2 l^3)/(24 E^2 J^2),$$

so the exact vector can be expressed by

$$y_0 = 2\left[\left(\frac{Pa^3}{3EJ} + \frac{Pa^2 l}{2EJ}\right) + \frac{P^2 a^2 l^3}{24 E^2 J^2}\right].$$

With a less precise calculation for slim values between 20 and 40, the third term could be taken into account for the length $l/2$ subject to a constant moment equal to Pa, which produces a shift in direction y of $Pal^2/8EJ$, corresponding to a shift in direction z of $(P^2 a^2 l^3)/(32 E^2 J^2)$. According to this hypothesis, y can be written

$$y_0 = 2\left[\left(\frac{Pa^3}{3EJ} + \frac{Pa^2 l}{2EJ} + \frac{P^2 a^2 l^3}{32 E^2 J^2}\right)\right].$$

As regards load, for springs larger than 40, the moment is not constant, and the cross section corresponding to the maximum can be obtained deriving the curvature expression and making it equal 0.

So, the maximum moment is

$$M_{max} = Pa\left(1 + \frac{Pl^2}{8EJ}\right),$$

with the consequent loads from the previous formula, just by substituting Pa with M_{max}.

Loads are produced in the bell due to centrifugal effects that are also subject to loads from the radial effects of the blocks.

If R_m is the mean radius, and s is the mean thickness of the bell with axial length L and density ρ, with the velocity v at radius R_m, then the mean pressure acting on the bell per unit axial length can be expressed as

$$p = \rho s v^2 / R_m + \frac{iP}{2\pi R_m L},$$

to which, from calculations previously seen (Thin pipes, Chapter 17) corresponds a tension of

$$\sigma = \rho v^2 + \frac{iP}{2\pi sL}.$$

Pressure p for transmitting the moment is around 200 N/cm². This can be calculated by dividing force N (known) by the total contact surface, which, if S is the surface of each tooth and n is the number of teeth, is $2nS$.

So

$$p = N / 2nS \leq 200 \text{ N/cm}^2.$$

20.6   SURPLUS CLUTCHES

In some applications, mechanical systems must be used that only allow motion in one rotational direction, disengaging any shafts in the opposite direction. Surplus clutches are able to do this. Uni-directional torque continues to be transmitted by way of friction and by combining other system components. This clutch-type is made up of active components and surfaces that they act on combined in suitable ways. So, both the driveshaft bell and the flywheel, together with the free components, must have the right design to work well.

Since the connection can become rigid, the active surfaces of the driveshaft, the crankshaft, and the free components must always be in contact.

20.6.1   CYLINDRICAL ROLLERS

When the active components are cylindrical rollers, the calculation is carried out as described in Figure 20.8. Roller equilibrium imposes that the resultant of all the forces and moment is zero. Taking into account the perpendicular and friction forces, that equilibrium means that the forces F arising at the contact points are equal and opposite and point in the same direction as the conjoined points of contact.

If the angle is ψ between the tangents and races at the two contact points with the roller as long as it stays immobile, then $\psi/2$ must be less than the friction angle ($\psi/2 < \varphi$).

Expressing P (perpendicular force) and T (friction component) as a function of F then

$$P = F\cos\frac{\psi}{2} \quad T = F\sin\frac{\psi}{2} = Ptg\frac{\psi}{2}.$$

The force that produces the moment in the blocked conditions is clearly T. So, given the number of rollers n, the total torque transmitted to the blocked clutch is

$$M = nTR = nPtg\frac{\psi}{2}R.$$

Taking into account that the greater Hertzian loads are produced where the relative curve is maximum—A—where $\rho = 1/r$, then

$$\sigma = 0.418\sqrt{\frac{PE}{rl}} \quad P = 5.75\frac{\sigma^2}{E}rl.$$

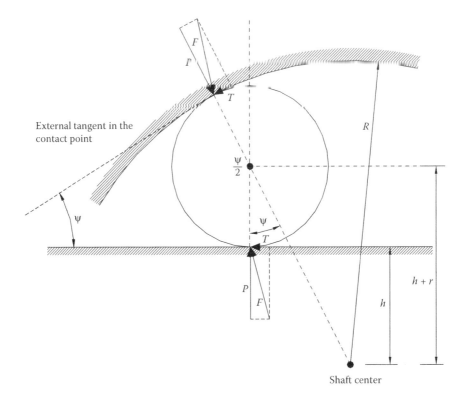

FIGURE 20.8   Scheme of a cylindrical roller.

By substitution in the moment relation:

$$M = 5.75\frac{\sigma^2}{E}nRltg\frac{\psi}{2}.$$

Using the notation in Figure 20.8:

$$\cos\psi - \frac{h+r}{R-r}.$$

Elastic deformation (expansion of the outer race, local flattening, deformation of the inner race) produces a different angle ψ' equal to

$$\cos\psi' = \frac{\left(h-\Delta h\right)+\left(r-\Delta r\right)}{\left(R+\Delta R\right)-\left(r-\Delta r\right)}.$$

Finding this relation must be solved by trial and error because P, used to calculate deformation, depends in turn on ψ'. It is advisable, however, to keep $\psi \leq 6°$ and $\psi \leq 9°$.

20.6.2   NON-CYLINDRICAL ROLLERS, BLOCKED

When the rollers that create the peaking are not cylindrical (Figure 20.9), contact between the various parts is guaranteed by a circular elastic ring consisting of a coil spring fixed to the two lateral grooves of each free component. Both races, external and internal, are circular.

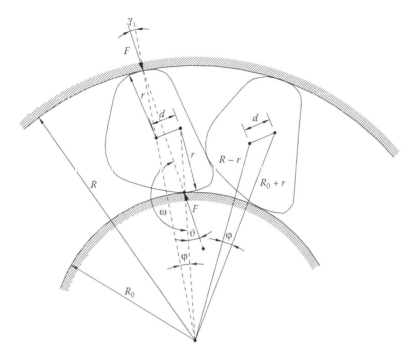

FIGURE 20.9 Scheme of non-cylindrical rollers.

The external race is the driver, and the different centers of curvature of the outer surfaces of the mobile components cause the block when the components rotate in the same direction as the transmitted torque.

With the same reasoning, note that the races transmit a direct force F according to the conjoined points of contact with the roller. Since the contact surfaces are different, angles ϑ and γ of the conjoined points of contact with the perpendiculars are different, their difference being $\varphi = \vartheta - \gamma$ (Fs being inclined at γ and ϑ compared to the perpendiculars at the points of contact).

Applying Carnot's theorem to a triangle with sides $R_0 + r$, $R - r$, and d, we obtain

$$\cos\varphi = \frac{(R_0 + r)^2 + (R - r)^2 - d^2}{2(R_0 + r)(R - r)}.$$

From Nepero's formula for γ:

$$tg\frac{\vartheta - \gamma}{2} = ctg\left(\frac{\varphi}{2} + \gamma\right) = \frac{R - R_0}{R + R_0} ctg\frac{\varphi}{2}.$$

So, from $\vartheta - \gamma = \varphi$, we get ϑ.

The greatest Hertzian loads are when there is contact with the inner race where the relative curvature is greatest. If n is the number of free components, and asserting $T/P = tg\vartheta$, the torque for an engaged clutch is

$$M = nTR_0 = nPtg\vartheta R_0.$$

Using a calculation analogous to the preceding one, the relation becomes

$$M = 5.75 \frac{\sigma^2}{E} nrR_0 l \frac{R_0}{R_0 + r} tg\vartheta.$$

In which σ is the Hertzian stress
The angle ϑ should not exceed $3°–4°$.

20.7 SAFETY CLUTCHES

Safety clutches are capable of limiting the transmitted torque, hence their other name "torque limiters." For example, Figure 20.10 shows a safety cluch with ball bearings that, if well calibrated, can
guarantee optimum transmission of maximum torque. This transmission is due to the mutual operation between two sets of bearings in discs on the input and output shafts. The disc on the output
shaft can slide axially, pushing against the bearings by effect of a Belleville spring (Chapter 13),
with its external diameter engaged with a seating in the disc itself; the internal diameter is fixed
with play to the shaft so that it can be pre-loaded by using a metal ring. When the torque exceeds
the pre-set limit, the bearings overload the spring, which yields, allowing the disc to slide axially, in
turn tripping a microswitch to stop the engine and set off an alarm.

By choosing appropriate spring characteristics, the residual load can re-arm the system once the
temporary overload terminates. Again, spring choice can lead to a hooking system with manual
re-arming.

The torque limiters on the market can be mounted on shafts up to a maximum of 10,000 rpm
since they are compact and axial symmetric.

The calculation for the active mechanical components of the limiter focuses on a torque that is
1.5 times the maximum unhooking torque (M_s) defined in turn according to the downstream characteristics of the transmission.

The tangential force $F_t = M_s / R$, with R the mean radius of the circumference of the bearing races.

If f is the friction coefficient and i the number of bearings, the perpendicular force on each bearing
is $F_u = F_t / (if)$. In the case of steel-on-steel contact with little lubrication, f may be assumed to be 0.15.

The resulting Hertzian loads can be calculated once bearing and race dimensions are known. In
practice, the mating ratio C (the ratio between the race radius d' and the bearing radius d), varies
from 0.92 to 0.95. For steel-on-steel, the maximum tension is with variable c almost linear between
1114 and 975 $(N/mm^2)^{2/3}$.

$$\sigma_0 = c \cdot \sqrt[3]{\frac{F_n}{d^2}},$$

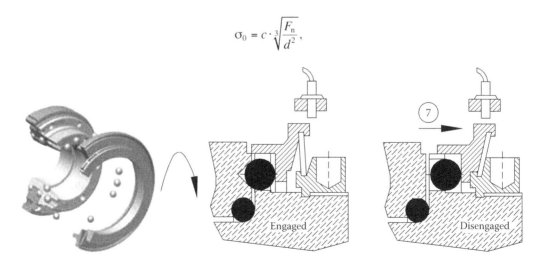

FIGURE 20.10 Safety clutches.

At the design stage, the preceding formula can calculate the maximum transmittable moment once the maximum admissible flattening tension is assumed (for surface treated steels $\sigma_{amm} = 0.4\text{--}0.5\ H_d$).

The number of bearings is a function of their size and the unhooking torque. The ratio between the mean race circumference radius R and the bearing radius d, is variable between 5 and 8.

As previously stated, the crucial element in the system is the disc spring. Recalling Belleville springs (Chapter 13), these must be sized to give rise to the qualitative characteristic in Figure 20.11 with $h/b = 2.5$ and pre-loaded so as to work when $f/b = 2\text{--}3$.

Acting on the geometric characteristics of the spring it is possible to limit the intervention time (tenths of a second) to ensure operating conditions with minimal sliding and least overheating.

The spring's durability must be calculated so that for the maximum operational value the tensions do not exceed the admissible tensions.

The mobile disc calculation uses the same model as the case of the mobile flange. Supposing the load on one side is distributed along the circumference in contact with the spring of radius R_c, and on the other along the mean contact circumference R, then the bending moment is

$$M_f = iF_n(R_e - R) / \pi.$$

If s is the mean thickness of the mobile disc, the durability modulus $W = (R_c-R)\ s^2/3$. There is a corresponding stress of

$$\sigma = 3iF_n / s^2.$$

20.8   CENTRIFUGAL CLUTCHES

Centrifugal clutches (Figure 20.12) provide gradual motion without grabbing and are an automatic means of coupling. They can be designed so that transmission commences once a pre-set torque is reached. Thus, they are used in a variety of mechanical systems, including boat engines, compressors, pumps, and scooters. As well as automatic activation, they can also de-activate automatically.

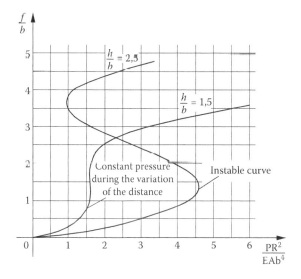

FIGURE 20.11   Qualitative characteristic of the springs used in safety clutches.

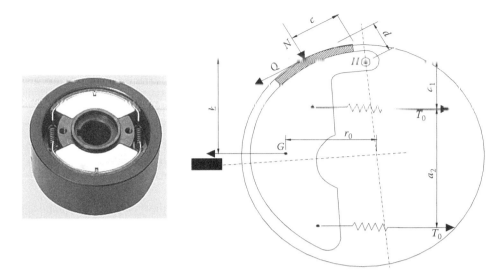

FIGURE 20.12 Centrifugal clutches.

For certain applications engagement can be further retarded, making it even softer by using springs that counteract the centrifugal force. Moreover, the centrifugal forces can be amplified or reduced by lever systems.

These clutches are used in electromechanical construction and in particular in coupling synchronous motors. So at start-up, they avoid dangerous power surges and corresponding increases in temperature.

Below are some examples to illustrate their functioning principles and provide design criteria for engineers on how to carry out or improve a design; it is also suggested that users consult manufacturers' documentation.

A recently built clutch by Hillard Corporation incorporates all the previously mentioned details (Roccati [5]). There are the three essential components: the external bell connected to the driveshaft; three equal masses hinged together as levers and also hinged to a suitably shaped disc integral to the crankshaft; friction surfaces positioned on the outermost part of the block.

As the shaft gradually accelerates, centrifugal force pushes the friction surfaces onto the external bell that, at a certain point having overcome the contrasting force of the springs and after an initial phase of slippage, begins to rotate.

The return of the masses to the still position is ensured by the springs. A clutch produced by Siemens in the early 1950s was conceived along these criteria. For the relative calculations refer to Figure 20.12 and the symbols there reported.

If m is the mass, r_0 the distance between their centers of gravity G from the shaft axis in that position when the friction surfaces come into contact with the crown, with d the arm of the tangential force Q transmitted, and ω the angular velocity, then for the equilibrium of moments:

$$Nc \pm Qd = mr_0\omega^2 b - T_0(a_1 + a_2).$$

If f is the friction coefficient, the maximum transmittable tangential force by a block is $N = Q_{max}/f$, and in the fringe condition of sliding we have

$$Q_{max}\left(\frac{c}{f} \pm d\right) = mr_0\omega^2 \cdot b - T_0(a_1 + a_2)$$

then

$$Q_{max} = \frac{mr_0\omega^2 \cdot b - T_0(a_1 + a_2)}{\dfrac{C}{f} \pm d}.$$

To make the transmittable moment depend on revolutions per minute, the following must not happen: $c/f - d \leq 0$, because $c = fd$, Q becomes infinite and the clutch would work as a stop to friction.

The angular velocity ω_0 at first contact is defined by annulling the numerator of Q_{max}, or by

$$T_0(a_1 + a_2) = mr_0\omega_0^2 b.$$

For values higher than ω_0, it being practically impossible to vary spring movement, the effect is constant, and so the peripheral force grows with the centrifugal force.

The calculation mentioned previously relies on knowing the application points of forces N and Q that, if the contact surfaces are small, could be approximated to the center of gravity.

If the length of the friction surfaces is long and the mass is not appropriately shaped, the constant pressure that defines the center of gravity is not acceptable, so establishing that center of gravity requires a pressure distribution hypothesis and means the use of calculations similar to those used for the resultants of drum braking (Chapter 21).

Using this type of clutch in systems with limited error margins for engaging torque, it is better to use a clutch with greater mass and therefore more active surfaces to facilitate the constant pressure hypothesis, even though this somewhat complicates its construction. There are also other clutches on the market with quite different shapes but which still adhere to the same principles. The reader's intelligence could also create such adaptations. The Piaggio clutch (Figure 20.13) is also a centrifugal clutch and it is easy to identify the three main components.

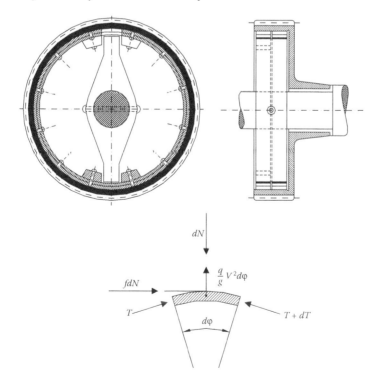

FIGURE 20.13 Piaggio clutch.

The forces that transmit moment to the driveshaft are generated by a cut-out band whose external surface is coated with friction material and which, during its active phase, creates pressure on the driveshaft bell. Integral with the crankshaft there is an arm that rests on projections of the band, which in turn correspond with the band cut-outs. In reality and for symmetry and load equilibrium, there are two bands side by side. Figure 20.12 shows that when the crankshaft rotates, one half of the band works in compression and the band opposite is in traction.

Hypothesizing that the moment transmitted by the driveshaft is entrusted to centrifugal force (excluding the deformation produced by the arm), the equilibrium equation for a band section is

$$
\begin{cases}
dN = \dfrac{q}{g} v^2 d\varphi + 2T \sin \dfrac{d\varphi}{2} + dT \sin \dfrac{d\varphi}{2}, \\[2mm]
fdN = dT \cos \dfrac{d\varphi}{2}
\end{cases}
$$

q being the band weight per unit length, g gravitational acceleration, v the band's peripheral velocity, f the friction coefficient between band and drum, R the drum's internal radius, and T the band tension. Aligning the trigonometrical terms ($\sin d\varphi/2 = d\varphi/2$, $\cos d\varphi/2 = 1$), then

$$
\begin{cases}
dN = \dfrac{q}{g} v^2 d\varphi + T d\varphi + dT \dfrac{d\varphi}{2}, \\[2mm]
fdN = dT
\end{cases}
$$

and neglecting $dTd\varphi/2$ because it is infinitesimal, the vertical and horizontal equilibrium equations are

$$
\begin{cases}
\dfrac{q}{g} v^2 d\varphi \pm T d\varphi = dN \\[2mm]
dT = fdN
\end{cases}.
$$

In the first equation the $+$ and $-$ signs refer to band compression and traction. So, when the band is in traction and as long as it does not detach from the crown or as long as $dN > 0$ is always true, then

$$
\frac{q}{g} v^2 d\varphi - T d\varphi > 0,
$$

or rather T should always be less than $(q/g)v^2$.

So, if the band is tight, even the maximum tension of its most-loaded extreme must satisfy the relation:

$$
T_1 < \frac{q}{g} v^2.
$$

From the equilibrium equation above, highlighting dN and substituting $dN = dT/f$ from the previous equation:

$$
fd\varphi = \frac{dT}{\dfrac{q}{g} v^2 \pm T}.
$$

From the moment that $(q/g)v^2 = \cos t$, then

$$d\left(T \pm \frac{q}{g}v^2\right) = dT,$$

so by substituting

$$\int d\varphi = \frac{d\left(T \pm \frac{q}{g}v^2\right)}{\frac{q}{g}v^2 \pm T},$$

and then

$$f\varphi = \pm \log\left(T \pm \frac{q}{g}v^2\right) + C.$$

At the fringe $T = 0$ for $\varphi = 0$, the integration constant is obtained:

$$C = \pm \log\frac{q}{g}v^2,$$

and by substitution:

$$f\varphi = \pm \log\left(T \pm \frac{q}{g}v^2\right) \pm \log\frac{q}{g}v^2. \quad (*)$$

For the band in compression, the + sign is valid so

$$f\varphi = \log\left(T + \frac{q}{g}v^2\right) - \log\frac{q}{g}v^2 \qquad e^{f\varphi} = \frac{T + \frac{q}{g}v^2}{\frac{q}{g}v^2}.$$

If ϑ is the angle embraced by half the band ($\vartheta \cong \pi$), the corresponding tension T_1 is

$$T_1 = e^{f\varphi}\frac{q}{g}v^2 - \frac{q}{g}v^2,$$

and also

$$T_1 = \frac{q}{g}v^2(e^{f\varphi} - 1).$$

For tight bands, the − sign of (*) is valid, so

$$f\varphi = \log \frac{\frac{q}{g}v^2}{T - \frac{q}{g}v^2},$$

If T_1^* is the corresponding traction tension, then

$$e^{f\vartheta}T_1^* - e^{f\vartheta}\frac{q}{g}v^2 = \frac{q}{g}v^2,$$

for which finally:

$$e^{f\vartheta}T_1^* = \frac{q}{g}v^2(e^{f\vartheta} - 1) \quad T_1^* = \frac{q}{g}v^2\left(1 - \frac{1}{e^{f\vartheta}}\right).$$

Summarizing:

$$T_1 = \frac{q}{g}v^2(e^{f\vartheta} - 1) \quad \text{for the compressed part,}$$

$$T_1^* = \frac{q}{g}v^2\left(1 - \frac{1}{e^{f\vartheta}}\right) \quad \text{for the tight part,}$$

and so the maximum transmittable torque is

$$M = (T_1 + T_1^*)R = \frac{q}{g}v^2\left(e^{f\vartheta} - \frac{1}{e^{f\vartheta}}\right)R$$

FIGURE 20.14 Desch Centrex clutch.

As an example of the latest generation of centrifugal automatic clutches, Desch Centrex (Figure 20.14) makes dry safety clutches with automatic engagement and disengagement depending on velocity. Seven models are mass produced to handle up to 4500 Nm.

As a drive clutch, the Centrex clutch has definite advantages if the engine has to accelerate to high inertial moments or load moments. At rest and in drive the engine and operating machine are completely separate until the engagement velocity is reached.

Torque transmission only happens when the pre-set engagement velocity is exceeded. During operation, the centrifugal clutch guarantees that the machine components are not subject to damage by overloading, separating the engine and operating machine when the engine blocks.

They are used mostly in the following sectors: street sweepers, boat engines, ventilators, fans, compressors, pump transmissions, concrete mixers, drum brakes, vibrators, centrifugal pumps, etc.

21 Brakes

21.1 INTRODUCTORY CONCEPTS

There are three categories of brakes: disc brakes, drum/shoe brakes, and ribbon brakes. Currently, the first are widespread throughout the automobile industry and are replacing drum brakes. The second are used in the railway industry and lifts. The third are used wherever rotation velocities are low so that high pressures are not required for braking.

From the dynamic equilibrium equation:

$$\frac{d\omega}{dt} = \frac{M_m - M_r - M_f}{I}, \tag{21.1}$$

in which:

- M_m is the driving torque
- M_r is the resistant torque
- M_f is the braking torque
- I is the total inertial moment of the car

(everything reduced to the same axle)

Once the angular velocity with time t reduction $(\omega_0 - \omega)$ is imposed, the braking moment is calculated as follows:

$$\int_0^t M_f dt = I(\omega_0 - \omega) + \int_0^t \left(M_m - M_r\right) dt.$$

For constant M_m, M_r, and M_f:

$$\omega_0 - \omega = \frac{t}{I}(M_f + M_r - M_m),$$

of which

$$M_f = I\frac{\omega_0 - \omega}{t} + M_m - M_r,$$

and in this case, motion is slowing down if

$$M_f > M_m - M_r.$$

In design practice, this relation is used:

$$M_f = n(M_m - M_r).$$

529

With n the coefficient of the loaded braking moment, depending on brake and brake system type.

If a precise calculation is required of a braking moment within a specified time, it is necessary to evaluate inertial torque by accounting for the overall moving mass. So, all the masses are reduced to the same shaft by imposing equivalence of kinetic energy.

For a car with mass m_r travelling at velocity v and with moment characterized by a rotational inertia I_i with respect to the (rotational) axes that rotate at angular velocity ω, the equivalent total mass m' is calculated as follows:

$$m'v^2 = m_r v^2 + I_r \omega^2 \Rightarrow m' = m_r + I_r \frac{\omega^2}{v^2}.$$

If τ is the transmission ratio between the axle of applied mass and the rotating axle of radius R and $v = \tau \omega R$, then

$$m' = m_r + \frac{I_r}{\tau^2 R^2}.$$

Reducing the entire rotating mass with $M' = \Sigma m'$, and naming/labelling driving power as P, resistant force as Q, and braking force as F, reduced to the same point, the dynamic equilibrium equation can be written as

$$M' \frac{dv}{dt} = P - Q - F. \tag{21.2}$$

Equations (21.1) and (21.2) are formally identical. However, the second one is preferred for global translation motion, the other for other types of motion.

After this brief general introduction to brake issues, more detail will be given on the three brake types: disc brakes, drum brakes, and ribbon brakes.

21.2 DISC BRAKES

In disc brakes (Figure 21.1), braking is achieved on the flat surfaces of a disc by the axial pressure of one or more brake pads (Figure 21.2) made of a high friction coefficient material. Apart from generating friction, they must have high thermal conductivity to prevent excessive heating in the contact zone. They are usually shaped in the form of a circular rim sector. Once, the main brake pad component was asbestos fiber in a binding resin but with the outlawing of asbestos it was replaced by a composite of materials each fulfilling a specific role (noise reduction, friction, thermal conductivity).

The brake pads are mounted in a calliper that may be fixed or mobile. The disc is squeezed by two equal forces on either side so braking does not produce axial forces on it.

Depending on the friction surface area needed for strong braking, we can distinguish between complete or partial disc brakes. Those with circular friction rims (complete) are used for heavy slow vehicles and in aeronautics. For most cars, brakes are nearly always partial with the active surfaces limited to circular sectors.

The main reasons why the discs brakes are replacing drum brakes:

- Less fading (when a rise in temperature reduces braking efficiency; it happens because of the evaporation of some brake pad components which then form a film that reduces friction)
- They work better in any environmental condition and they are self-cleaning

FIGURE 21.1 Disc brake.

- Less suspended mass and therefore fewer drawbacks for inertia
- Less clearance variation with the temperature variation change of brake components (disc brake play is influenced by disc thickness so temperature is less of a factor given that it is proportional to initial dimensions and consequently mostly radial)
- Brake pad wear is uniform

With fixed callipers and the disc fixed axially, there are two identical opposite facing pistons whereas with floating callipers on a fixed disc or fixed callipers and a floating disc there is just one piston that squeezes both disc sides, either because the calliper moves or the disc moves (Figures 21.3 and 21.4). For floating callipers, one brake pad is connected to a piston operated by hydraulic fluid under pressure, whereas the other is connected to the calliper. Activating the brakes *via* the hydraulic fluid, the piston pushes the brake pad toward the disc. By reaction, the calliper is pushed in the opposite direction, pushing the other pad against the opposite side of the disc.

For automobiles, the most common solution is floating callipers, which are less bulky and more economical. Their efficiency depends on how easily they slide in their groove. Floating callipers

FIGURE 21.2 Brake pads.

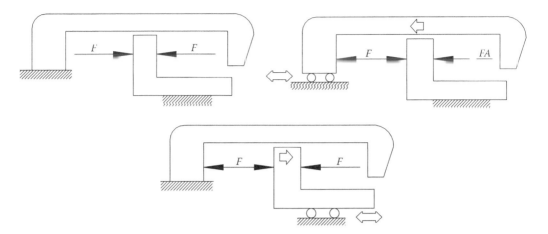

FIGURE 21.3 Different solutions for disc brakes.

can be slowed by rust, encrustations, etc., which means they must be lubricated and appropriately protected so that cooling is not compromised.

The fluid pressure operating the brakes comes from the force of the driver's foot on the pedal. When that force ceases, the pressure stops and a spring sends the piston back to its initial position in the wheel cylinder. Braking for heavy and very heavy vehicles (and recently also for automobiles) is power aided. This means that the driver's foot pressure is added to atmospheric pressure that acts on a piston kept previously depressed by engine vacuum. In handbrakes, the squeezing of the brake pads is carried out mechanically by a cam.

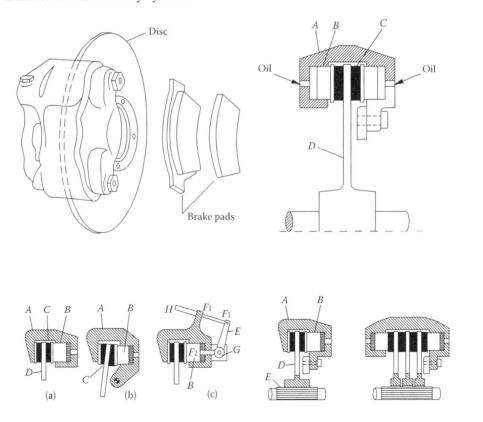

FIGURE 21.4 Common solutions for disc brakes.

FIGURE 21.5 Ventilated discs.

Disc brake size is less of a problem compared to the corresponding size of a brake drum, and so the tangential friction forces with less distance must be greater. This also means greater normal forces and consequently the need for of power assisting. The heat generated during braking heats up the disc. Heating up is partially dissipated by ventilation and conduction; however, it is partially stored. This requires that the disc has appropriate mass so the heat stored does not reach excessive levels. In some cases, "ventilated discs" made up of two identical facing discs connected by radial or inclined ribs are used (see Figure 21.5). This design solution provides channels into which air enters from the inside to be pushed out of the disc periphery by the wheel's rotation. The heat exchange surface area is thus much bigger with consequent lower rise in temperature. Other ways to help with cooling are use of holes and channels. The holes are made axially in the pad contact zone and they work by lightening the disc. The channels consist of helical grooves that help conduct friction residues away from the disc.

Figure 21.6 shows a twin disc by Valeo used in large industrial vehicles (10–13 tons). This fixed calliper hydraulic disc is capable of generating a braking torque of 17,000 Nm per axle. Disc temperature under normal braking is around 300°C, but under heavy braking can reach 600°C–700°C.

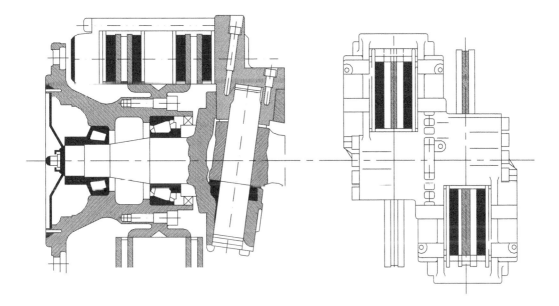

FIGURE 21.6 Twin disc by Valeo.

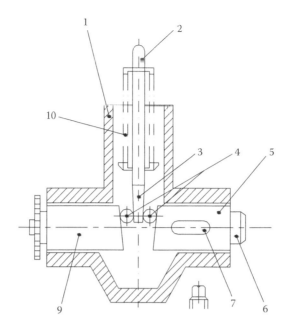

FIGURE 21.7 Compressed air braking system.

The operating pressure is about 1470 N/cm^2. This brake weighs about 15 kg less than an equivalent drum brake and has half the number of components. Their common use on even industrial vehicles attests to their reliability.

Since about 1950, heavy industrial vehicles use compressed air disc brakes for which the same general considerations are valid (fixed and floating calliper). Apart from the advantages mentioned earlier—lighter and fewer components—they also have the advantage over drum brakes of much faster brake pad renewal and facilitated disc face grinding, which becomes necessary when the disc shows grooving and vitrifaction.

According to American literature, the limit threshold for the ratio between average braking power and pad area is equal to 950 Nm/(cm^{2}s), given that this average power is defined as that required for stopping (equal to the initial live forces) for the whole duration of the braking (rapport between initial braking velocity and the supposed constant deceleration). The average power is also equal to the product of the braking force divided by average velocity (half the initial velocity). This does not indicate efficient braking in all conditions, but may serve as a good approximation.

Vehicles with compressed air braking usually use spring activation, as shown in Figure 21.7. Air pressure *via* a piston keeps a spring compressed (10), and the conic shaft (2) acting through rollers (4) on two pistons (5 and 9) activates the brake when that pressure drops. This type of brake is used for railway vehicles and by some heavy road vehicles. They are used because they default to total stop, being a passive brake—meaning that at rest and during braking they are activated by a spring. So, not braking means compressed air is pressured up, in turn activating mechanisms that release the brake.

21.3 PERFORMANCE AND DIMENSION ANALYSES

Designing the active components of a disc brake makes use of the same calculation used for designing the surfaces of a clutch friction plate in the following phases:

- Determining the braking moment
- Defining geometry
- Verifying resistance to thermo-mechanical loads

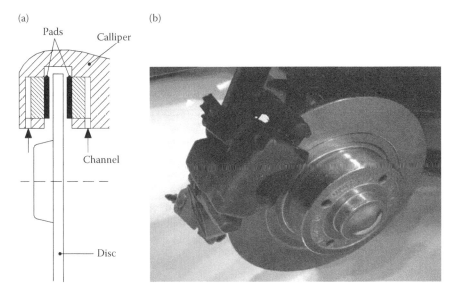

FIGURE 21.8   (a) Scheme of a disc brake. (b) The brake system.

In the case of facing friction surfaces of the type depicted in Figure 21.8, it is easy to understand that moment is transmitted by squeezing the two disc surfaces with force F.

As we have already seen in Chapter 19 on clutches, the following are the reference hypotheses:

- Constant pressure on both the active surfaces ($p = const$).
- Variable pressure according to Reye's hypothesis (volume of removed material is proportional to fiction $pr = const$).

Which hypothesis to adopt depends on the active surfaces and on the degree of machining used in making them.

Generally, it is assumed that pressure is constant over small active surfaces and that there is great flexibility in those components on which this pressure is exerted. For large surfaces (circular rims) as in "complete" brakes and after some usage, the second hypothesis is more appropriate, in which pressure is inversely proportional to the radius.

For sizing up a disc brake, the calculation nearly always assumes constant pressure ($p = const$) over the whole active surface.

Given the braking moment (external) at the wheel axle, the geometric parameters can be calculated for the wheel at rest. As we saw for transmitted moment in clutches, also the braking moment of a disc brake can be achieved by acting on different contact surfaces. If n is the number of contact surfaces between the discs and pads, the total axial force is

$$P = n\frac{\alpha}{2}p_0(R_e^2 - R_i^2),$$

in which p_0 is contact pressure, α is the angle that defines the surface sector of contact, R_e is the external radius, and R_i is the internal radius of the contact surface.

If f is the friction coefficient, the total braking moment exerted by the pads on the discs is

$$M_t = nM = nfPR_m.$$

$R_m = 1\backslash2(R_e + R_i)$ (equivalent radius) equals about 0.7–0.8 R_e, which corresponds to about one-third of the vehicle's wheel radius, R_r. As we can see, for an equal friction coefficient and wheel radius, high braking moments can be obtained by acting on disc number and contact surface area. In heavy vehicle and aeronautical applications, there could be a number of discs per wheel and the angle α could be equal to 2π (aeronautics).

For cars, there is only one disc per wheel ($n = 2$), so angle α varies between $40°$ and $60°$.

As regards the specific pressure p_0 value to apply, it can reach a maximum of 100 N/cm^2 (remembering that the maximum value for clutches was 10 N/cm^2). In any case, this maximum value is, above all, a function of the product pv, which represents the power lost to friction during braking, which also varies with the pads' friction material. Nevertheless, it should always be kept below 300 N/cm^2 m/s.

A further verification must be made of the specific work lost as the ratio between the work lost to friction during braking and the total active surface. If n is the number of pairs of facing surfaces each of area S, the total active surface area is nS, and given that the relative angular velocity between disc and pad equals the angular velocity of the disc, then the specific work lost is

$$L_s = \int_{ti}^{tf} M\omega_i/nS \cdot dt,$$

for the time necessary that initial velocity becomes 0. To avoid overheating, the specific work lost should not exceed 140 Nm/cm^2.

The friction coefficient f values depend on the pad material and therefore they are provided by the manufacturer. These values generally decrease linearly with temperature (see Figure 21.9) and relative velocity.

During braking, the temperature rises but relative velocity decreases, which implies using an average behavior of the material (e.g., Ferodo agglomerate), and assuming that for inexcessive velocities (pad/wheel velocity about 21 m/s) and for temperatures over $300°C$, the friction coefficient drops by 50% almost linearly from 0.5 (room temperature) to a nearly constant 0.2. The values above mentioned are only approximate values, so designing the braking system needs the manufacturers' values, depending on the friction material.

The values cited above impose the contact area dimensions, so for a fixed value of α, the values of R_e and R_i can be calculated.

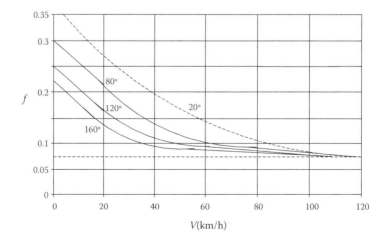

FIGURE 21.9   Friction coefficient f versus velocity (parameter temperature).

In disc brakes, after a first approximate sizing with the values supplied, the verification of resistance is carried out taking into account centrifugal load and thermal stress.

Because of high operating pressure and the high relative velocities between disc and pad, high temperature rise occurs. The work developed during braking is, in part, externally dissipated, but some of it increases internal disc energy.

Discs are usually made of high wear resistant Ni-Cr steels that turn into increasingly wear-resistant special thermal treatments (case hardening). Special cast irons treated with Ni or Ni-Cr may even be used.

In the latest applications, particularly in sports cars, the discs are made of carbon fiber with a substantial improvement in reliability indices (ratio between the variation of braking torque and friction torque).

For vehicles, it is necessary to evaluate the power dissipated that takes into account velocity and road camber.

One consequence of power dissipated is an increase in disc temperature that makes it expand in all directions. Temperature rise ΔT can even reach 800°C with a consequent increase in thickness s of the disc given by

$$\Delta s = \alpha s \Delta T,$$

where α is the thermal expansion coefficient of the material ($\alpha = 1.2 \times 10^{-5}$ 1/°C). For metals, this coefficient varies with temperature as in Figure 21.9.

Expansion with rising temperature varies the clearance between disc and calliper, which in turn varies pedal travel and may involve pad contact on the disc. (A similar problem happens with the brake-shoes of drum brakes.) For this reason, they need an automatic system for taking up clearance to maintain braking performance even after long braking.

For more details on these automatic systems refer to specialized publications. In any case, it is necessary to allow some clearance between disc and pad as a function of the disc material and its thickness. For 21 mm steel discs and a thermal range to 700°C, the required play Δs is about 0.17 mm, which coincides with those (0.15–0.3) found elsewhere.

Evaluating disc resistance means checking how the disc resists centrifugal force and high thermal stress. Compared to drum brakes, the pressure of the callipers on the pads and so on the discs is much higher with consequent higher temperatures.

The heat is, however, dissipated due to the system's structure (not boxed as drum brakes) helped too by holes and grooves on the two disc surfaces. This also helps eliminate any high frequency vibration or whistling during braking (vibrations due to the contact of imperfect pad and disc planes).

Calculating the average increase in temperature during braking takes into account that total vehicle energy (rotating and translating masses) is transformed into heat energy. After the initial transitory phase during which the temperature of mass M of the disc rises by ΔT, the heat must be dissipated *via* the disc surfaces.

So, the total mass M of the discs has a temperature increase of

$$\Delta T = E/(cM),$$

given c, the specific heat of the disc material (for steel or cast iron assume $c = 500$ J/kg °C), and the amount of heat dissipated *via* the disc surfaces S (temperature T_2) to the outside (temperature T_1) is

$$Q = h_c S(T_2 - T_1),$$

where h_c is the thermal transmission coefficient (in air about 2–3W/m² °C).

Looking at the formula above, to keep temperature T_1 low during braking, the exposed surfaces should be ample with added ventilation. In reality, this is always a compromise, since large discs are heavy, which would eliminate one of the advantages that disc brakes have over drums. Reducing T_1 is often down to special wheel-rim profiles.

As regards thermal loads, we refer to transitory local conditions, which means what happens along the thickness of the disc during one revolution.

As we know, any hot areas in a structure with temperatures over the average are subject to compression as the colder areas are subject to traction (since hot areas expansion is obstructed by cold areas). In a structure that is uniformly heated with no temperature variation over the whole structure, there is no thermal stress.

During braking, assume that the temperature of the surface in contact with the callipers is t_1, and that the temperature of the internal disc surface t_2 is equal to that away from the calliper zone. Thus, the temperature gradient is defined in terms of the heat quantity Q over unit time per unit surface. According to Fourier:

$$Q = \frac{\lambda}{s}(t_1 - t_2),$$

with Q expressed in cal/(m²h). The material's conductivity coefficient is λ in cal/(m h °C), s is the disc thickness, t_1 is the calliper zone temperature and t_2 is the non-calliper zone temperature.

Hypothesizing an average temperature rise in the zones facing the pads of $(t_1 + t_2)/2$, and that at the moment of braking the callipers prevent the linear expansion of the disc, the maximum thermal stress σ due to either compression or traction corresponding to a given heat flow Q may be expressed thus

$$\sigma = \frac{1}{2}Qs\frac{E\alpha}{\lambda}\frac{1}{1-\nu},$$

where α [1/°C] is the material's linear expansion coefficient (Figure 21.10), and $\nu = 1/m$ is the Poisson coefficient. The maximum thermal stress due to compression or traction is proportional to disc thickness by a factor of $(E\alpha/\lambda)/(1-\nu)$, which is specific for every material (see Table 21.1).

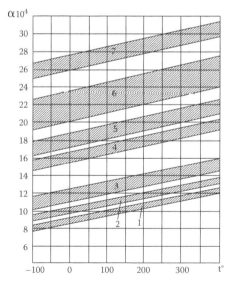

1-titanium alloy, 2-martensitic steels,
3-perlitic steel and irons, 4-austenitic steels,
5-copper basic alloy, 6-aluminum alloy,
7-magnesium alloy

FIGURE 21.10 Linear expansion coefficient α versus t.

TABLE 21.1
Thermal Strength Characteristics of Materials

Materials	$E \times 10^{-4}$ [N/mm²]	$\alpha \times 10^5$ [°C]	λ [cal/(m × h × °C)]	ν	$E\alpha/[\lambda(1-\nu)]$	σ_b [N/mm²]	$\sigma \lambda(1-\nu)/E\alpha$
Grey cast irons	8	11	35	0.15	8	30	10
Carbon steels	21	11	40	0.3	8.3	60	7.2
Alloy steels	21	12	35	0.3	10.4	120	11.5
Austenite steels	21	16	15	0.3	32	70	2.2
Aluminum alloys $\begin{cases} \text{castable} \\ \hline \text{wrought} \end{cases}$	7.2	22	150	0.33	1.6	20 / 50	12.5 / 31
Magnesium alloys $\begin{cases} \text{castable} \\ \hline \text{wrought} \end{cases}$	4.2	28	70	0.33	2.5	15 / 23	6 / 10
Bronzes	11	18	70	0.33	4.3	60	14
Titanium alloys	12	8.5	7	0.3	21	120	8.7

This reasoning aims to show how the various parameters can influence thermal variation stress. During braking, the disc is also compressed by a normal force that creates axial compressions that are added to thermal ones, which would tend to increase disc thickness (if it were free) by $\Delta s = \xi s\, \Delta T$. The sum of the two contributions could not, however, exceed the pad squeezing force, which during braking cannot exceed 50 N/mm² without destroying the pad. Finally, the disc loads created during braking from axial compression say little about the overall tension state to which the main contributor is centrifugal force that for automobiles can reach 210 N/mm².

This effect is nearly always neglected, since as pad pressure rises, rotational velocity decreases, and consequently its effect on tension.

Naturally, materials with low linear expansion coefficients are much less susceptible to the thermal stress due to the limitations of form. For the linear expansion coefficients of various metals, see Figure 21.7 where they are reported as a function of temperature.

For all materials, the $(1/1 - \nu)$ factor is about 1.5 except for gray cast iron, which is 1.18 (hence often used for discs). For a rough comparison of the thermal stress of two materials it may be useful to use $E\alpha/\lambda$.

The comparisons in Table 21.1 clearly show how light alloys are preferable in terms of thermal stress because of their low $(E\alpha/\lambda)/(1 - \nu)$ values, and the titanium and stainless steel alloys are the worst.

The resistance of a material to thermal stress σ can be expressed as the ratio β between resistance to tensile stress s_b of the material and the factor $(E\alpha/\lambda)/(1 - \nu)$, whose values are reported in the last column of Table 21.1:

Because of their high thermal stress resistance (high values of β), the special aluminum alloys are the best. The titanium and stainless steel alloys are the worst.

The issue of which material to choose for building a disc system resides in the fact that the values above are only valid for temperatures up to 400°C for which the characteristics of resistance, elasticity, linear expansion, and heat conduction are subject to only relatively small changes.

At higher temperatures these characteristics do change (disc brakes reach temperatures of 400°C–600°C) so special materials are used, such as steels alloyed to Ni, W, Mo, Ta, and nickel alloys, titanium alloys, etc.

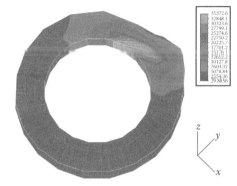

SVIEV 4.06 File:brembo 99/04/17 15.18   LC 1/   1 Vu = 7 Lo = 45 La = 45 R = 0

FIGURE 21.11   Qualitative thermal stresses on the face of the disc.

Figures 21.11 and 21.12 show the qualitative behavior of thermal stress for average temperature rises of 400 C.

In designing vehicle's brakes in order to avoid wheel spinning, it is generally necessary to take into account that the braking torque (applied by pads or by brake shoes for drums) is equal to the opposing moment provided by longitudinal road-tires contact (adhesion) forces in incipient slipping conditions or that, neglecting inertia, the following holds true:

$$T_a R = T_d R_m,$$

with force T_a applied to the external wheel radius (including the tire), and T_d the force applied to radius R_m, to which is applied the resultant of braking actions. This implies that the perpendicular forces, T_a and T_d, should vary according to the ratio:

$$\frac{F_f}{F_a} = \frac{f_a R}{f R_m},$$

where f_a is the road-tires adhesion coefficient and f is the pads-disc friction coefficient.

The R_m/R ratio depends on the free space around the disc or drum brake. In practice, this ratio is about 1/3.

Calculating F_a for each wheel during braking uses the equilibrium equation for dynamic conditions. Referring to Figure 21.13, which reproduces ideal conditions in a straight line, note that the ratio between F_{a1} on the front wheel and F_{a2} on the back is a function of geometric size. In particular, using the symbols shown in Figure 21.13, that ratio is: $(F_{a1}/F_{a2}) = (b + fh)/(a - fh)$ and highlights how different braking forces are even, in ideal conditions.

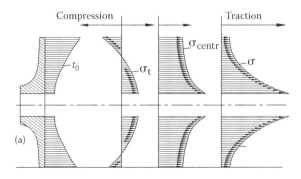

FIGURE 21.12   Qualitative components of the thermal stresses.

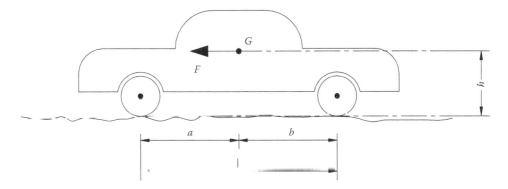

FIGURE 21.13 Scheme for the calculation of the acting forces.

It is relatively simple in practice to obtain the different braking forces operating on the piston actuators of pad pressure for disc brakes (or on the brake shoes for drum brakes) such that the ratio between the front piston surfaces and those at the back are equal to the force ratio $k = (b + fh)/(a - fh)$.

In this way, it is not necessary to have different active surfaces, so the brake is sized for a moment on each wheel i equal to mM/i, given that m, a coefficient greater than 1, which takes into account the average and not the maximum moment (the more exact the calculation, the closer k is to 1).

In practice, the first approximation is made for $m = 3$ and assumes an average moment value for M to bring the automobile to rest.

To avoid locking the back wheels, for many years braking correctors have been used that vary the braking torque at the wheels as a function of vertical load, reducing it when the vertical load decreases.

Figure 21.14 shows the operational diagram of a common brake corrector. The front and rear brakepipes use opposite corrector cylinders to piston P. The lever L acts on the piston rod and at the other end of the lever a spring M is connected to the suspension S. The load acting on the spring is proportional to suspension travel and with a suitable coefficient C to vertical load Z_R of the same suspension. A counter-spring m reacts to the action PM of spring M doing without constant H,

$$P_M + P_m = C \cdot Z_R + H.$$

The two corrector cylinder chambers are linked by a by-pass that is seen in different operating positions in Figure 21.14 (a) shows the by-pass totally open and (b) closed.

With A, a, D, d, respectively, the areas of the straight cross-section and the diameters of the piston and its rod with the by-pass closed, then the equilibrium between the axial forces is

$$C \cdot Z_R + H + (A - a) \cdot K \cdot p_F = p_R \cdot A.$$

Thus, extracting p_R:

$$p_R = K' \cdot Z_R + K'' \cdot p_F + K''',$$

Where

$$K' = \frac{C}{A}; \quad K'' = \frac{(A - a) \cdot K}{A}; \quad K''' = \frac{H}{A}.$$

Figure 21.15 shows the p_R pressure curves as a function of p_F for five different vehicle load types (B—driver only; F—fully loaded). The curves are only valid for a closed by-pass when the piston obstructs one of the balance cylinder vents.

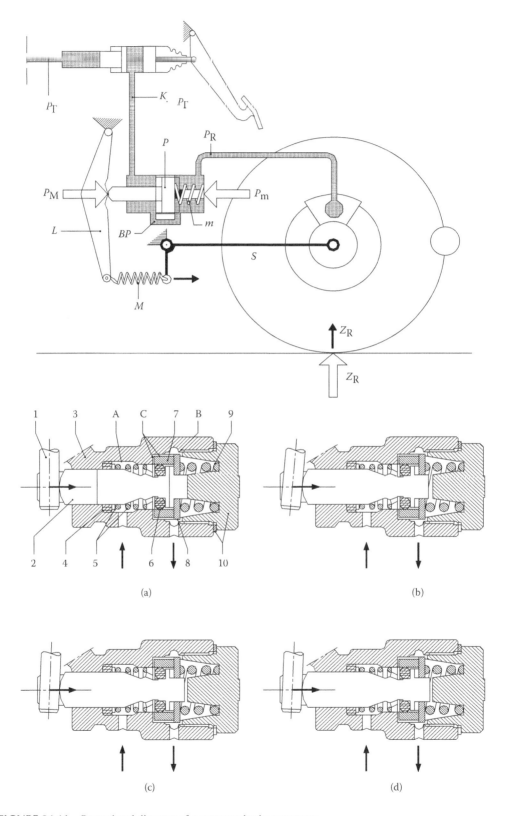

FIGURE 21.14   Operational diagram of a common brake corrector.

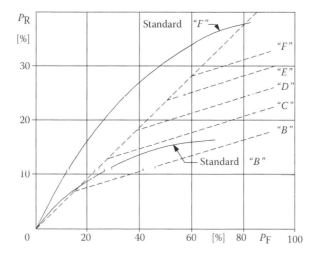

FIGURE 21.15 p_R pressure curves as a function of p_F (see figure 21.14).

Pressure p_R with an open by-pass is

$$p_R = p_F \cdot K.$$

With non-linear behavior, approximating the theoretical curves with a fragment/radian is too rough, especially for a heavily laden (F) vehicle.

The friction coefficient between the disc and pads lowers as the temperature rises, contact pressure rises, and vehicle velocity rises. It lowers almost linearly with temperature, so that with the most commonly used pad material it goes from 0.5 at 50°C to 0.2 at 400°C. There is less slope over 400°C (0.1 from 400°C to 600°C, then remaining constant). During braking, while the inter disc-pad velocity drops, the friction coefficient ought to rise, but braking produces heat so as the temperature rises the benefits of a drop in velocity are annulled.

Analogously, the traction coefficient f_a, which depends on various contact factors (tire, road surface), also varies with velocity according to the type of law shown in Figure 21.16 for dry whole roads (dashed curve) and wet roads with treaded (curve A) and untreaded (curve B) tires. The behaviors shown in the figure are qualitative even though they average out quite well for a traction coefficient that also depends on other factors like road surface. In other words, the dashed curve can shift by rotating around A depending on the road surface quality. We intuit therefore how the behaviors reported in the figure change with time during braking.

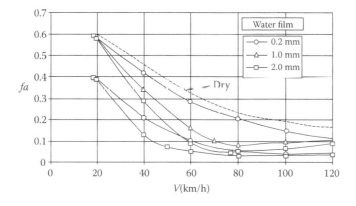

FIGURE 21.16 f_a coefficient versus velocity (parameter water film).

For maximum braking efficiency, the braking force exerted on the discs should be instantaneously proportional to the ratio between the traction f_a and friction f coefficients.

Because of the points above, maximum braking efficiency cannot be achieved autonomously so electronic control systems (ABS) intervene to vary the F_f/F_a ratio (corresponding to instantaneous traction coefficient) exerted almost continuously (15 Hz intervention frequency) on pressure p_0 between the active surfaces. Simply adopting larger piston surface areas for the front brakes without ABS, M_a/M_p would be useful for only one single value of f.

It is also necessary to take into account that the vehicle's center of gravity is not constant but depends on load amount and distribution, and that above all that locking the back brakes makes the vehicle uncontrollable, whereas the front brakes can be under the driver's control.

Car manufacturers suggest the maximum braking force for disc brakes is 40 kN requiring a pedal force of 500 N, which is not insignificant for the driver. Furthermore, there are heavy industrial vehicles that may not be stopped by such a force. So, there are mechanisms that can reduce the force required on the pedals —servobrakes—exploiting all the traction available during braking.

Figure 21.17 describes the main mechanisms of a pneumatic servobrake.

The device is powered by the engine where in the inlet manifold downstream of valve f (closed during braking, accelerator released) a high vacuum is created with respect to atmospheric pressure.

The unit itself is made up of a pneumatic cylinder whose piston of area A is coaxial to a hydraulic cylinder whose piston of area a is rigidly fixed to the pneumatic cylinder with a stem. The Valve V in the hydraulic piston allows fluid to flow towards the brake and to transmit to it the pressure created by the brake pump operated by the driver through the brake pedal.

At rest, spring M keeps two pistons at maximum travel due also to pre-loaded valve V and to the fact that at rest the brake pump outlet to the fluid reservoir S is open. When the brake pedal is depressed, pressure P_g is generated by the brake pump in the suppressor and pneumatic cylinders. For safety, this pressure would reach the brakes anyway should the servo fail. Furthermore, pressure P_g shifts the cylinder overcoming opposing spring m equalising with atmospheric pressure one of the two faces of the pneumatic piston. The other piston face is subject to inlet manifold vacuum p_m exerting force $F = (p_{atm} - p_m) \cdot A$ that equalizes with a pressure rise by the piston in the suppressor cylinder, which transmits a pressure to the brakes of

$$p_f = p_g + \frac{(p_{atm} - p_m) \cdot A - F_M}{a}.$$

Force F_M exerted by spring M is generally small so the servo-assisted force is much greater than p_g of the direct action.

At rest, $p_g = 0$, so spring m of the cylinder locks off atmospheric pressure and opening up to inlet vacuum causing the pneumatic piston to be subject on both faces to pressure p_m. Thus, the direct action of the driver as well as the servo-assisted one that follows is annulled (a characteristic of all servo systems).

Figure 21.18 shows a cross section of the type of servobrake described above in which the brake pedal operates the cylinder directly and the pump also acts as a suppressor: $C1$ and $C2$ are the pneumatic actuating chambers; E is the command and regulating module; M is the vacuum feed; (1) is the vacuum outlet; (2) is the protective casing; (3) and (6) are the regulation valve springs; (7) and (8) are atmospheric vents to $C1$ via the regulation module (E); (9) is the regulation module piston; (10) is the support pneumatic piston of membrane (11) to seal the piston hermetically to the lid (13); (12) is the reaction disc and (14) is the push rod; (15) is the piston (10) return spring; (16) is the hydraulic pump section; (17) is a three brake tube feed; (18) and (19) are compensation and pump feed holes; (21) is the hydraulic piston (24) reaction spring with a sealing ring (23) pressed by a floating ring holder (22); (21) floating ring; (25) front gasket with stem and guide bush (26), to centering the push rod (14).

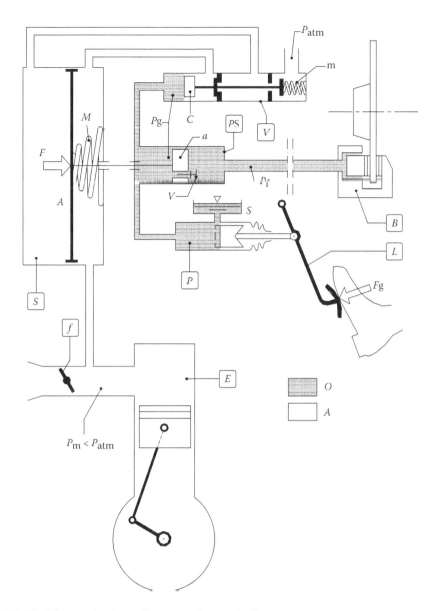

FIGURE 21.17 Main mechanisms of a pneumatic servo-brake.

When designing the whole braking system for a vehicle with disc or drum brakes, verifying the skid limits for certain values of f is insufficient because many other conditions need to be taken into consideration such as load, traction, and friction between active surfaces.

Today, even by using commercial calculation codes, it is possible to simulate a variety of braking conditions by varying the parameters, as we have seen above. So, at the drawing board it is possible to evaluate braking system efficiency for every conceivable condition.

21.4 DRUM BRAKES

In addition to disc brakes, vehicles can use drum brakes. Drum or internal block brakes are made of two internally expandable brake shoes (Figure 21.19) (normally in cast iron with a friction pad attached to them) pressed from within against a cylindrical brake drum.

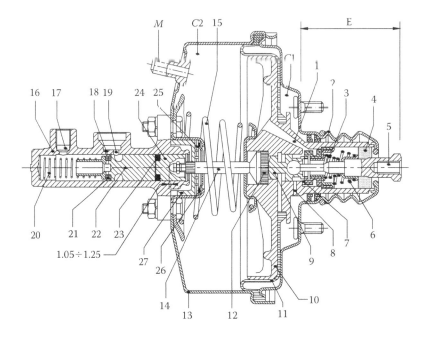

FIGURE 21.18   Cross-section of a type of servo-brake.

The drum must be able to resist the pressure from the brake shoes apart from their own centrifugal force, and should have sufficient mass to absorb and dissipate braking heat without reaching unreasonably high temperatures.

For better cooling, the external surface of the drum is usually finned (Figure 21.19). The friction pad is mounted on each shoe at an angle having an amplitude between 60° and 100°.

Depending on the direction of drum rotation, the friction forces developed between the shoes and drum may be concordant or discordant with those produced by the clamping force F with respect to the shoe pivot, which could increase or decrease shoe clamping against the drum and therefore braking torque. In both cases it is a question of leading and trailing shoe ("*auf- und ablaufende Backe*" in German, "*machoire comprimée et tendue*" in French, and "*ganascia avvolgente e ganascia svolgente*" in Italian). In other words, the shoe is leading if the friction force resultant applied by the shoe helps the external clamping force. Trailing shoe is the opposite case.

In the schematic Figure 21.20, where for simplicity we suppose that the tangential force fN (f = friction coefficient, N = resultant of perpendicular pressures) passes through the half-way point of the contact arc, bisected by the diameter perpendicular to the axis of symmetry, the aforementioned F force applied to the shoe would produce a braking torque of − leading shoe, + trailing shoe):

FIGURE 21.19   Drum brakes.

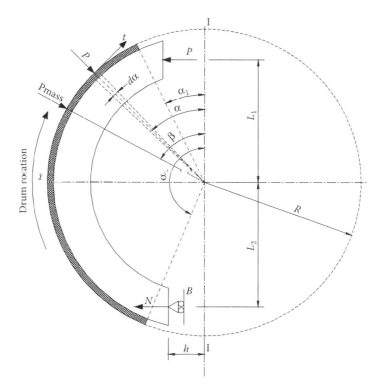

FIGURE 21.20 Schematic figure for drum brakes.

$$fN = fF \frac{1}{a} \frac{1}{1 \pm f \dfrac{b}{a}}.$$

Supposing $a \cong b$, $f = 0.3$, F being equal, the braking moments in both directions of drum rotation would be in this ratio:

$$\frac{1+f}{1-f} = \frac{1.3}{0.7} = 1.86.$$

Referring to the various constructional configurations collected in Figure 21.21, the two shoes of a drum brake move about two fixed adjacent pivots (Simplex brake), which can be reduced to one or even substituted by two supports at the shoe curve apex against a fixed surface. In this case, one shoe is leading and the other trailing, and the braking torque is the same for both forward and reverse directions of motion.

In Duplex brakes, the shoe pivots are diametrically opposed, and the shoes are both leading or trailing depending on the direction of vehicle travel. In reverse, braking torque is much less.

In a Servo brake, the two shoes are connected by a device that can slide in only one direction.

For the direction of rotation depicted in Figure 21.21, the tangential force at the free end of the shoe on the left (leading) loads that of the right-hand shoe, which is thus also leading. So, they are both leading shoes in series. In the opposite direction, no force is transmitted between the two shoes so it works like a Simplex brake.

In a Duo-Servo brake, the shoes react against each other through an intermediate mobile device and they are loaded at their other end by two identical pistons. Drum rotation produces a greater load at the pivot end of the shoe compared to the piston end so it is repelled until it stops. Thus, in both rotational directions the two shoes work in series and braking torque is the same.

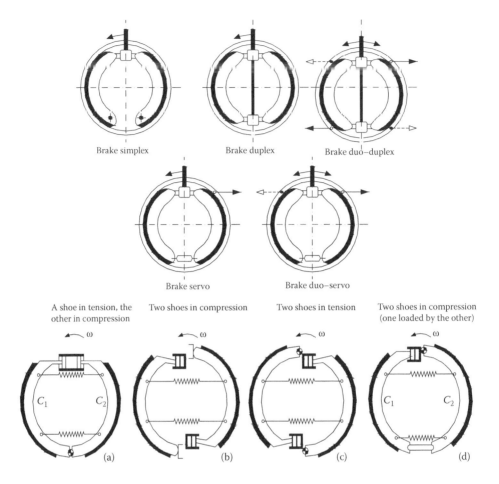

Brake simplex   Brake duplex   Brake duo–duplex

Brake servo   Brake duo–servo

A shoe in tension, the   Two shoes in compression   Two shoes in tension   Two shoes in compression
other in compression   (one loaded by the other)

C_1 C_2 (a)   (b)   (c) C_1 C_2 (d)

FIGURE 21.21   Various constructional configurations for drum brakes.

As already referred to in the diagrams, shoes can have fixed pivots or floating ones which react by contact with a fixed support or with each other held in place by a force which pushes them against the drum. As the friction surface wears away, the resulting clearance is taken up by cams placed either half-way along the contact arc or on supports or cylinders.

As stated above, blocks (shoes) can be outside the drum in which case loads are applied through a series of levers. They are used for braking in lifts and cranes and with the appropriate leverage systems in railways. More generally, they are used in those cases when there are no particular space restrictions. The load at the end of the control lever is usually applied by an appropriate electromagnet. There are interesting applications of these brakes in cable cars. Figure 21.22 shows the entire braking system of a cable car built by Agudio.

Braking occurs by pressure between pads (6) and the main cable (1). The pads attached to the fixed (2) and mobile (3) shoes are brought into contact with the cable through Belleville springs (4) that find their counterpart in an external cylinder "short tube" (4) and in an internal gudgeon (5).

When the pressure drops within the cylinder (15), the spring extension causes the shoes to clamp.

21.4.1   CALCULATIONS AND SIZING

For drum brake calculations, the following is a traditional method taken from Giovannozzi, vol. I. Later, the same calculation will be considered with reference to current methods. The two procedures will serve to help the calculation for a block brake and make comparisons.

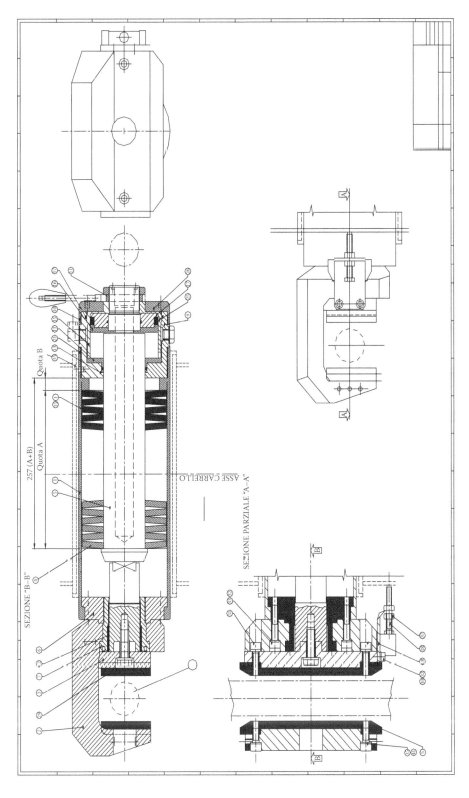

FIGURE 21.22   Braking system of a cable car built by Agudio.

The block brake calculation requires determining force F that is applied to the block and evaluating if the maximum pressure p_0 on the drum is compatible with the friction material characteristics and the maximum permissible temperature. It also requires sizing the block (internal shoe) and its gudgeon above all in terms of deformation.

Following Giovannozzi, vol. I, the mutual actions between block and drum are calculated first. This may be done assuming $p = p_0 = const$, which is well suited to limited contact surfaces (encirclement angle not higher than 60°), or Reye's hypothesis with a distribution of $p = p_0 \sin \theta$ (p_0 being the maximum pressure) valid for larger encirclement angles where θ is angle between the lever pivot connecting with the drum center and a generic radial direction.

Referring to Figure 21.23, with Q the force applied at the end of the lever l from the pivot, if b is the block width, then in the hypothesis $p_0 = const$, the resultant force of the elementary normal actions is

$$N = 2p_0 Rb \sin\beta.$$

Consequently, $T = fN$ and supposing f is constant, the braking moment is

$$M_f = 2fp_0 R^2 b\beta.$$

For angles that are so small as to retain $\sin\beta \approx \beta$, we see that $M_f \approx fNR$, and N and fN pass through the half-way point of the contact area.

The value of p_0 is determined by an equilibrium equation for the hinged point of the shoe by

$$Ql = \int_{\vartheta_0}^{\vartheta_1} p_0 br \sin(\pi - \vartheta) R d\vartheta \mp \int_{\vartheta_0}^{\vartheta_1} fp_0 [R - r\cos(\pi - \vartheta)] bR d\vartheta$$

$$= \int_{\vartheta_0}^{\vartheta_1} p_0 br \sin\vartheta R d\vartheta \mp \int_{\vartheta_0}^{\vartheta_1} fp_0 b[R + r\cos\vartheta] R d\vartheta$$

$$= p_0 brR[\cos\vartheta]_{\vartheta_1}^{\vartheta_0} \mp fp_0 bR^2(\vartheta_1 - \vartheta_0) + fp_0 bRr[\sin\vartheta]_{\vartheta_0}^{\vartheta_1}$$

$$= p_0 brR[\cos\vartheta_0 - \cos\vartheta_1] \mp fp_0 bR^2(\vartheta_1 - \vartheta_0) \mp fp_0 bRr[\sin\vartheta_1 - \sin\vartheta_0],$$

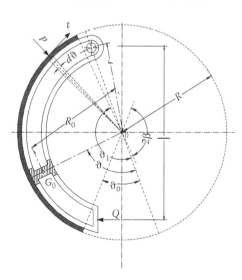

FIGURE 21.23 Scheme for calculations.

from which we obtain

$$p_0 = \frac{Ql}{R^2 b} \frac{1}{\frac{r}{R}(\cos\vartheta_0 - \cos\vartheta_1) \mp f\left[(\vartheta_1 - \vartheta_0) + \frac{r}{R}(\sin\vartheta_1 - \sin\vartheta_0)\right]},$$

in which the minus sign qualifies it as leading.

For the block resistance calculation, the bending moments M_r and M_t due to radial and tangential forces must be calculated. If R_0 is the distance of the block's straight cross-section's center of gravity from the drum, then

$$M_r = \int_{\vartheta_0}^{\vartheta_G} p_0 b R_0 R \sin(\vartheta_G - \vartheta) d\vartheta,$$

where ϑ_G is the angle that identifies the block's generic cross-section. Solving the integral, we get

$$M_r = [p_0 b R R_0 \cos(\vartheta_G - \vartheta)]_{\vartheta_0}^{\vartheta_G} = -p_0 b R R_0 \left[\cos(\vartheta_G - \vartheta_0) - 1\right],$$

or

$$M_r = p_0 b R R_0 [1 - \cos(\vartheta_G - \vartheta_0)].$$

Since $R - R_0 \cos(\vartheta_G - \vartheta_0)$ is the arm of the tangential force fP_0 with respect to A, the bending moment M_t is given by:

$$M_t = f p_0 b \int_{\vartheta_0}^{\vartheta_G} \left[R - R_0 \cos\left(\vartheta_G - \vartheta\right)\right] R d\vartheta$$

$$= f p_0 b \left\{R^2 \left(\vartheta_G - \vartheta_0\right) - \left[R R_0 \sin\left(\vartheta_G - \vartheta\right)\right]_{\vartheta_0}^{\vartheta_G}\right\}$$

$$= f p_0 b \left[R^2 \left(\vartheta_G - \vartheta_0\right) - R R_0 \sin\left(\vartheta_G - \vartheta_0\right)\right].$$

Or

$$M_t = f p_0 b R^2 \left[(\vartheta_G - \vartheta_0) - \frac{R_0}{R}\sin(\vartheta_G - \vartheta_0)\right].$$

If M_Q is the bending moment due to force Q, in the straight cross-section, the overall bending moment is

$$M = M_Q - M_r \pm M_t,$$

in which the minus sign qualifies it as leading.

The formulae above and sign specifications are also valid for external block brakes.

As we have already said, the constant pressure hypothesis over the whole contact arc is not realistic when the encirclement angle is large (80°–90°). In these cases, Reye's hypothesis is acceptable only if the block movements are small and the elastic deformations of the block due to load are negligible, then the pressure along the contact arc is $p = p_0 \sin \vartheta$, where ϑ is the angle between the lever pivot connecting with the drum center and a generic radial direction, and p_0 the maximum pressure. So, for block braking, the contact zone corresponding to ϑ values around $\vartheta = \pi/2$, is particularly efficient and it can be seen that it is useless to extend the contact zone to values of ϑ close to π and 0.

It is therefore not worth increasing the expansion of the block because there would be no benefit to the increased wear given that the outer sections are not as efficient as those closer to the approach direction.

Referring to Figure 21.23, maximum pressure p_0 can be obtained with an equilibrium equation for the pivot, the resultant of the pressures N and the braking moment M_f.

Analogously, the same procedure can be used to size the block having found out the bending moment M_r due to pressure and moment M_t due to tangential forces. Knowing all the forces exerted on the block also determines the reaction of the gudgeon so as to size it.

To facilitate things, here are Giovanozzi's formulae from volume I:

$$Ql = \int_{\vartheta_0}^{\vartheta_1} pbRd\vartheta \left[r \sin \vartheta \mp f(R + r \cos \vartheta) \right]$$

$$= \frac{1}{4} p_0 bRr \left\{ \begin{array}{l} 2(\vartheta_1 - \vartheta_0) - \left(\sin 2\vartheta_1 - \sin 2\vartheta_0 \right) \\ \mp f \left[\dfrac{4R}{r} \left(\cos \vartheta_0 - \cos \vartheta_1 \right) + \left(\cos 2\vartheta_0 - \cos 2\vartheta_1 \right) \right] \end{array} \right\},$$

$$M_r = \int_{\vartheta_0}^{\vartheta} pbRd\vartheta' \cdot R_0 \sin\left(\vartheta - \vartheta' \right)$$

$$= \frac{1}{4} p_0 bRR_0 \left\{ \sin \vartheta \left(\cos 2\vartheta_0 - \cos 2\vartheta \right) - \cos \vartheta \left[2\left(\vartheta - \vartheta_0 \right) - \left(\sin 2\vartheta - \sin 2\vartheta_0 \right) \right] \right\},$$

$$M_t = \int_{\vartheta_0}^{\vartheta} fpbRd\vartheta' \left[R - R_0 \cos(\vartheta - \vartheta') \right]$$

$$= \frac{1}{4} fp_0 bR^2 \left\{ \begin{array}{l} 4\left(\cos \vartheta_0 - \cos \vartheta \right) - \dfrac{R_0}{R} \left[\cos \vartheta \left(\cos 2\vartheta_0 - \cos 2\vartheta \right) \right. \\ \left. +2 \sin \vartheta (\vartheta - \vartheta_0) - \sin \vartheta (\sin 2\vartheta - \sin 2\vartheta_0) \right] \end{array} \right\},$$

$$M = M_a - M_r \pm M_t$$

$$M_f = \int_{\vartheta_0}^{\vartheta_1} fpbRd\vartheta \cdot R = fp_0 R^2 b \left(\cos \vartheta_0 - \cos \vartheta_1 \right),$$

Using a different approach, elastic deformation of the pad can be taken into account. Deformations on the pad are relatively large compared to the block's and drum's elastic deformation, which are almost perfectly rigid.

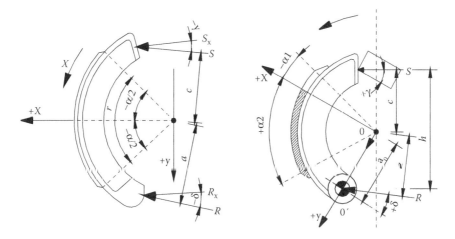

FIGURE 21.24 Scheme for calculations in a different approach.

Referring to Figure 21.24, given two right-angle axes x and y, and supposing the shoes and drum are rigid, then due to elastic deformation of pad thickness s and external radius r, the axis of the rigid component (externally cylindrical) against which the pad presses will experience a shift of s_x, s_y compared to the drum axis.

In a cross-section anomaly α with respect to axis x, the components of this shift produce a pad thickness variation in a radial direction of

$$\Delta s = s_x \cos\alpha + s_y \sin\alpha.$$

The percentage variation of thickness $\Delta s / s$ multiplied by the elastic modulus E of the pad equalises the corresponding radial pressure whose product for width b of the pad axially is the perpendicular force per unit length peripherally along the curved friction surface of radius r.

So, given that

$$P_x = \frac{Ebr}{s} s_x \quad P_y = \frac{Ebr}{s} s_y,$$

the elementary perpendicular force on arc $rd\alpha$ of the curved friction surface is

$$dN = E\frac{\Delta s}{s} br\, d\alpha = (P_x \cos\alpha + P_y \sin\alpha)\, d\alpha,$$

whereas the tangential one is fdN (f is the friction coefficient between pad and drum).

Knowing α_1, α_2 are the values of α at the edge of the pad, the braking force F (the ratio between braking moment and r) is

$$F = f\int dN = f[P_x(\sin\alpha_2 - \sin\alpha_1) + P_y(\cos\alpha_2 - \cos\alpha_1)].$$

The maximum value of $dN/d\alpha$ is obtained for β of angle α defined by the relation:

$$\tan\beta = \frac{P_y}{P_x} = \frac{s_y}{s_x},$$

which is the angle that identifies the direction of the shift resulting from s_x, s_y (approach direction).

Again referring to Figure 21.14, given the force S exerted on the shoe for braking, the force R exerted on the other end of the shoe (pivot reaction in the case of a hinged shoe), γ and δ the respective inclinations on axis x, then the equilibrium equations (according to the axes and rotation) for the shoe are

$$x)\ S\cos\gamma - P_x m - P_y q \pm fP_x q \mp fP_y n + R\cos\delta = 0,$$

$$y)\ S\sin\gamma - P_x q - P_y n \pm fP_x m \pm fP_y q + R\sin\delta = 0.$$

$$Sc \pm rF - Ra = 0,$$

where more briefly

$$m = \int_{\alpha_1}^{\alpha_2} \cos^2\alpha\, d\alpha = \frac{1}{4}[(\sin 2\alpha_2 - \sin 2\alpha_1) + 2(\alpha_2 - \alpha_1)],$$

$$n = \int_{\alpha_1}^{\alpha_2} \sin^2\alpha\, d\alpha = \frac{1}{4}[(\sin 2\alpha_2 - \sin 2\alpha_1) + 2(\alpha_1 - \alpha_2)],$$

$$q = \int_{\alpha_1}^{\alpha_2} \sin\alpha\cos\alpha\, d\alpha = \frac{1}{2}(\sin^2\alpha_2 - \sin^2\alpha_1),$$

minus and plus signs for leading and trailing shoes.

For hinged shoes, axis x is the approach direction, which is the perpendicular to the straight line connecting the drum center with the pivot. In the case of floating shoes, axis x is the bisector of the $2\bar{\alpha}$ arc encircled by the pad. In both cases, the angles are counted starting from axis x: plus for a drum rotation with shoes leading, minus for shoes trailing (clockwise and anti-clockwise in Figure 21.14).

For hinged shoes, if the shift is rigid $s_y = 0$, $P_y = 0$, and so

$$P_x = \frac{F}{f(\sin\alpha_2 - \sin\alpha_1)},$$

for which, given that

$$A = \frac{m \pm fq}{f(\sin\alpha_2 - \sin\alpha_1)} \quad B = \frac{q \mp fm}{f(\sin\alpha_2 - \sin\alpha_1)},$$

the equilibrium equations for the shoe (depending on axes and rotation) become

$$x)\ S\cos\gamma - AF + R\cos\delta = 0,$$

$$y)\ S\sin\gamma - BF + R\sin\delta = 0.$$

$$Sc \pm rF - Ra = 0. \tag{21.3}$$

Solving the third equation (21.3) in R and substituting it into the first two [x), y)], the following are obtained for transmission ratio τ_i for hinged shoes:

$$\tau_i = \frac{F}{S} = \frac{h}{Aa_0 \mp r} = \frac{a_0 \sin \gamma + c \cdot \tan \delta}{Ba_0 \mp r \cdot \tan \delta},$$

with $h = a_0 \cos \gamma + c$ distance of S with respect to the pivot. Dividing the two first equations member by member having substituted R through the third, we find

$$\tan \delta = \frac{Bh - \sin \gamma (Aa_0 \mp r)}{Ac \pm r \cos \gamma}$$

Finally, to find the maximum pressure p_0, which is obviously produced in the approach direction, it is sufficient that $p(\alpha) = p_0 \cos \alpha$, and so

$$F = fbrp_0(\sin \alpha_2 - \sin \alpha_1) = \tau_i S$$

The formulae above are simpler for pads that are symmetrical with the x-axis ($\alpha_2 = -\alpha_1 = \overline{\alpha}$). There is no need to report them here as they are easily deducible.

For floating shoes, with the axes set out as in Figure 21.24, ($\alpha_2 = -\alpha_1 = -\overline{\alpha}$), then

$$m = \frac{2\overline{\alpha} + \sin 2\overline{\alpha}}{2} \quad n = \frac{2\overline{\alpha} - \sin 2\overline{\alpha}}{2} \quad q = 0.$$

Substituting R, extracted from the third (21.3), into the two equilibrium equations for unhinged shoes, there are two relations as S, F, P_y, and by eliminating P_y, we obtain the transmission ratio for floating shoes:

$$\tau_f = \frac{F}{S} = \frac{1 \mp fL}{a \dfrac{2a + \sin 2\alpha}{4 \sin \overline{\alpha}} \left(f + \dfrac{1}{f} \right) + r(f \sin \delta - \cos \delta)},$$

given that

$$l = a \cos \gamma + c \cos \delta \quad L = u \sin \gamma + c \sin \delta.$$

Then substituting S, extracted from the third (21.3) into the first two equations [x), y)], there are two relations as F, R, P_y, and by eliminating R, we obtain the ratio $P_y / P_x = s_y / s_x = \tan \beta$, where β is the angle that defines the approach direction:

$$\tan \beta = \frac{(2\overline{\alpha} + \sin 2\overline{\alpha})(fl + L) - 4fr \sin \alpha \sin(\gamma - \delta)}{(2\overline{\alpha} - \sin 2\overline{\alpha})(l \mp fL)}.$$

Maximum pressure p_0 is on the x-axis and varies with α according to $p_0 \cos(\alpha - \beta)$. The braking force equals:

$$F = fp_0 r \int_{-\overline{\alpha}}^{\overline{\alpha}} \cos(\alpha - \beta) \, d\alpha = fp_0 br[\sin(\overline{\alpha} + \beta) + \sin(\overline{\alpha} - \beta)],$$

from which, given $F = \tau_t S$, we obtain p_0 as a function of S.

It may happen that for the floating shoe that from a certain α the inter pad and drum pressure is annulled and so the previous calculation does not apply without adaptation. So

$$dN/d\alpha = P_x \cos\alpha + P_y \sin\alpha.$$

Therefore $dN = 0$ per $tg\alpha = -P_y/P_x = -ctg\beta$, for $\alpha = 90° - \beta$.

The part of the pad extending beyond this limit is inefficient (according to the hypothesis). So, it is best to start by calculating angle β to see how efficient the arc is between α_1 and α_2; if it is negative it reduces the pad's angular width. The relations above are more than able to deal with this eventuality.

The same law can be deduced by assuming the shoe and drum are perfectly rigid whereas the pad is elastic and works with a pressure proportional to local radial pressing.

Shoe rotation around pivot O' (diagram on the right, Figure 21.24) is dynamically equivalent to the same rotation around center O of the drum (totally wear free and zero elastic pressing of the pad) with a shift equal to the above rotation for distance $\overline{OO'}$ and perpendicular to it, on which pressure variation alone depends.

This shift could also be defined by its components along two right-angle axes.

In the case of floating shoes, the approach direction is not defined a priori, in that shifts are possible along two right-angle axes, x that bisects contact arc α_0 between pad and drum, and y.

For a given approach direction defined by angle β, and for a maximum pressure p_0 and friction coefficient f between drum and pad, the trend of the pressure forces p and friction forces fp friction is defined, and therefore also the overall reaction R between the drum and brake shoes as a function p_0 and β is defined.

Making the ratio between component x and y of force $\vec{S} + \vec{R}$ equal (Figure 21.24) to $tg\gamma$, and its moment with respect to P null, there are two equations from which we obtain $tg\beta$ and p_0, and therefore the other necessary measurements. For brevity, only the results are reported here.

Given

$$A = \frac{1}{r}(a\cos\gamma + c\cos\delta) \quad B = \frac{1}{r}(a\sin\gamma + c\sin\delta),$$

$$\left.\begin{array}{c} f(\alpha_0) \\ f_1(\alpha_0) \end{array}\right\} = \frac{\alpha_0 \pm \sin\alpha_0}{4\sin\dfrac{\alpha_0}{2}} \quad \left.\begin{array}{c} f(f) \\ f_1(f) \end{array}\right\} = \frac{1}{f} \pm f,$$

(+ sign for f, − sign for f_1).

$$x = \pm A - fB \quad y = \pm f\sin\delta - \cos\delta \pm \frac{a}{r}f(\alpha)f(f),$$

(+ sign for leading shoe, − sign for trailing shoe).

So

$$F = S\frac{x}{y},$$

$$\tan\beta = \frac{f(\alpha_0)(B \pm fA) \pm f\sin\delta\sin(\delta - \gamma)}{f_1(\alpha_0)(A \mp fB)}.$$

where F, the peripheral braking force, is the ratio between the braking moment and R, and the plus signs are for leading shoes and the minus signs are for trailing shoes.

Comparing drum and disc brakes, the former are able to produce braking torques which are 2–4 times those of discs for the same overall dimensions. Disc brakes are less efficient (lower ratio of maximum friction force to maximum control force) than drums, which however have a lower regularity index.

Disc brakes wear more uniformly than drum brakes because overall their contact surfaces are smaller and so over time they work more reliably. Disc brakes pad replacement is considerably faster than that for shoes brakes.

As regards the construction and components of the two systems, the disc brake system is far simpler with half as many parts, which reduces its weight significantly (even to almost half).

Disc brakes have lower thermal capacity than drums, so for equal braking force they generate more heat. However, disc brakes can dissipate more heat to the atmosphere which means they are able to stay within acceptable behavior limits without significant deformation even at their wear limit. Disc brakes are definitely more suited to repeated braking cycles, and overall they are more commonly the brake of choice.

21.5 RIBBON BRAKES

In ribbon brakes (Figure 21.25), the drum is partly encircled by a metallic tape or ribbon, usually made of steel line with ferodo; the thickness is about 1/210th of the drum diameter, and wide enough to deal with different loads during braking.

Ribbon brakes can be subdivided into three groups: simple ribbon, differential, and double ribbon (the ribbon is doubled over half-way along its length almost doubling the surface in contact with the drum).

For ribbon brake calculations and referring to Figure 21.26, T_1, T_0 are the tensions in the two free sections of the ribbon.

Starting with

$$f d\varphi = \frac{dT}{\frac{q}{g} v^2 \pm T},$$

already used for centrifugal clutches (Chapter 19), for $v = 0$, it becomes

$$f d\varphi = \frac{dT}{T},$$

integrating and reverting to real numbers:

$$T = C e^{f\varphi}.$$

C is obtained by imposing $T = T_0$ for $\varphi = 0$, from which $C = T_0$ is obtained.

If ϑ is the encirclement angle (in radians) of the ribbon on the pulley, then

$$\frac{T_1}{T_0} = e^{f\vartheta}.$$

Since the tensions of the two free sections of the ribbon are T_1 and T_0, the braking moment is

$$M_f = (T_1 - T_0)R.$$

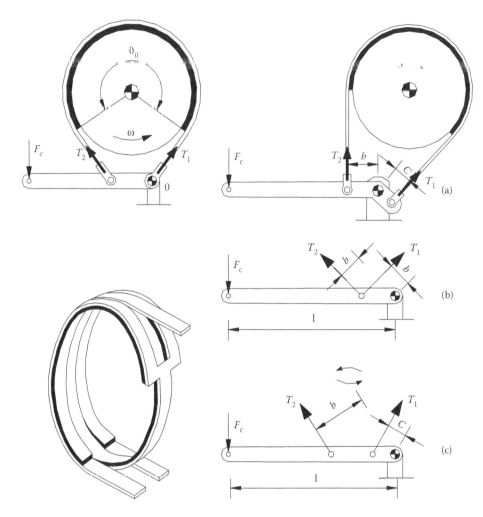

FIGURE 21.25 Ribbon brakes. Scheme for calculations in a different approach.

With $F = M_f / R$ the braking force at radius R, the equation above becomes

$$F = T_1 - T_0.$$

Since $T_1 / T_0 = e^{f\vartheta}$, then T_1 and T_0 can be expressed as a function of F:

$$\frac{T_1}{T_0} = e^{f\vartheta} \qquad \frac{F + T_0}{T_0} = e^{f\vartheta}.$$

So

$$T_0 = F \frac{1}{e^{f\vartheta} - 1}. \tag{21.4}$$

Then substituting $T_0 = T_1 / e^{f\vartheta}$, we obtain

$$T_1 - F \frac{e^{f\vartheta}}{e^{f\vartheta} - 1}.$$

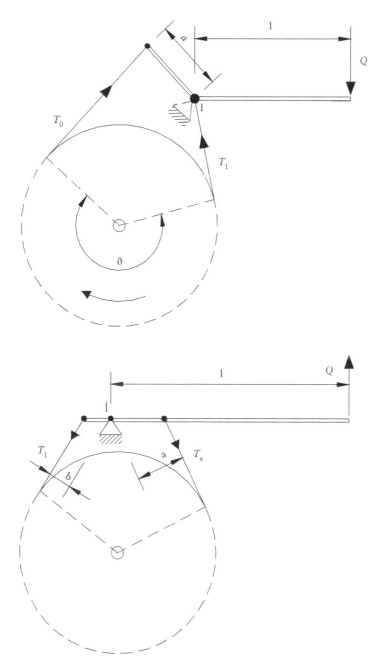

FIGURE 21.26 Scheme for calculations of ribbon brakes.

The values of $e^{f\vartheta}$, within the usual limits of f and ϑ, can be looked up in semi-logarithmic curves (Giovanozzi, Vol. 1, p. 878), where for a given f, the variation of $e^{f\vartheta}$ as a function of ϑ is represented by a straight line.

The two tensions, T_1 and T_0, are exerted on the shaft-drum with force P equal to their vector resultant of intensity expressed by

$$P = \sqrt{T_1^2 + T_0^2 - 2T_1T_0 \cos \vartheta}.$$

Ribbon brakes are usually simple with the less strained arm fixed on the lever (usually perpendicular to the lever axis). The more strained arm may be fixed to the lever pivot (as in Figure 21.15) or to another fixed point on the chassis. The value of force Q needed at the lever for a certain braking force equals the sum of moments around pivot 1:

$$Ql = T_0 a.$$

Substituting $T_0 = F/(e^{f\vartheta} - 1)$ we get:

$$Q = \frac{F}{e^{f\vartheta} - 1} \frac{a}{l},$$

and so

$$F = \frac{l}{a}(e^{f\vartheta} - 1)Q.$$

Inverting the pulley rotation, the tensions T_1 and T_0 in the two free arms of the ribbon are exchanged so the required force Q grows by $e^{f\vartheta}$. Normally, simple ribbon brakes are only used for one direction of pulley rotation.

In electric winches, the brake is permanently fixed by a weight Q (or by a spring), so that when the motor operates its effect is counterbalanced by an electromagnet at the end of the lever.

The braking torque limit is given by the load capacity of elastic joint that connects the pulley brake to the shaft.

The lever end travel s must be such that it nullifies initial play $\lambda(1-3)$ between ribbon and pulley so at angle λ the ribbon shifts from radius $R + \lambda$ to radius R. The leading edge shift of the ribbon on the lever is therefore:

$$(R + \lambda)\vartheta - R\vartheta = \lambda\vartheta,$$

and so lever end travel is

$$s = \frac{l}{a}\lambda\vartheta.$$

It should be noted that s is independent of pulley radius.

21.6   DIFFERENTIAL RIBBON BRAKES

Figure 21.27 shows construction diagrams for differential ribbon brakes. Here, tension T_1 produces a moment opposite to that of T_0.

Equilibrium of the lever around its pivot is

$$T_0 \cdot a - T_1 \cdot b = Ql.$$

Substituting T_0 and T_1 we obtain

$$Q = \frac{-F\dfrac{e^{f\vartheta}}{e^{f\vartheta} - 1}b + F\dfrac{1}{e^{f\vartheta} - 1}a}{l} = \frac{F}{e^{f\vartheta} - 1}\frac{a - be^{f\vartheta}}{l}.$$

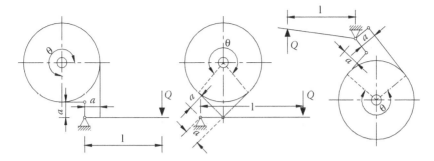

FIGURE 21.27 Differential ribbon brakes.

This relation shows that the system can stop ($a/b = e^{f\vartheta}$) or brake ($a/b > e^{f\vartheta}$).

In practice applications, the value $a/b \gg e^{f\vartheta}$ is realized with a hand control or by system control able to regulate the intensity and versus of the force Q.

Taking into account clearance λ, the lever rotation a required for braking is

$$(a - b)\alpha = \lambda\vartheta,$$

and lever end travel s is

$$s = \alpha l = \frac{l}{a - b}\lambda\vartheta.$$

Compared to the equation obtained for a simple ribbon belt whose denominator is only a, this one's is $(a - b)$, and so the travel increases by the ratio $a/(a - b)$.

Reversing the emotion the rotation force Q changes into Q',

$$Q' = \frac{F}{e^{f\vartheta} - 1}\frac{ae^{f\vartheta} - b}{l},$$

with a ratio of

$$\frac{Q'}{Q} = \frac{ae^{f\vartheta} - b}{a - be^{f\vartheta}} = e^{f\vartheta}\frac{1 - \dfrac{b}{a}e^{-f\vartheta}}{1 - \dfrac{b}{a}e^{f\vartheta}},$$

which increases by $e^{f\vartheta}$, given that $a > be^{f\vartheta}$.

So, the differential brake is even less effective than the simple one for bi-directional braking (Special Cases, Giovanozzi, Vol. 1, pg. 881).

The brake design requires sizing the ribbon, which is usually made of quality steel $R \cong 600$ N/mm². The traction test of the ribbon, on which the ferodo is fixed with countersunk nails, is carried out for maximum tension T_1, taking into account the reduction in effective traction area due to the nail holes or end attachment screws, which can also cause fatigue breaking.

If b_n and S_n are the width and thickness of the ribbon, d is the nail/screw diameter, and n is the number of rows of nails/screws (usually 2), the effective traction area is

$$A = S_n b_n - ndS_n = (b_n - nd)S_n,$$

and the maximum traction tension is

$$\sigma = \frac{N}{A} = \frac{T_1}{(b_n - nd)s_n} = F \frac{e^{f\vartheta}}{e^{f\vartheta} - 1} \frac{1}{(b_n - nd)s_n}.$$

To account for the hole effect or fatigue breaking, a maximum theoretical tension of three times the average is calculated before proceeding to the fatigue test.

If the calculation does not take into account the precise operating conditions, the average tensions ($40–60$ N/mm^2) previously calculated for high safety coefficients are used.

As we have said for other types of brakes, to avoid overheating, the product of fpV must be kept within opportune values (Giovannozzi, vol. I, pp. 861, 862, 759). For the sake of calculation p as a function of T, the equilibrium of an infinitesimal length of ribbon must be considered radially using R defined by $M_f = (T_1 - T_0)R$, which coincides with the external radius of the pulley. The equilibrium equation is

$$pb_n Rd\alpha - 2T \sin \frac{d\alpha}{2} - dT \sin \frac{d\alpha}{2} = 0,$$

and linearly:

$$pb_n Rd\alpha - 2T \frac{d\alpha}{2} - dT \frac{d\alpha}{2} = 0.$$

An approximate p can be obtained from the average tension applied to the end of the encircled pulley arc, p being constant:

$$p = (T_0 + T_1) / (2Rb_n).$$

An approximate pressure calculation is justified because of the uncertainties surrounding the friction coefficient especially during operation. Suffice seeing how it varies with temperature and even knowing this variation it would still be difficult to put the design conditions into practice.

The other components needing testing form the leverage which applies the load necessary for braking. Once the sizes of levers and pivots are established, measuring their reactions is easy and consequently the load characteristics.

For example, the left side of Figure 21.26 shows that for a balanced lever reaction, R on the pivot passes through the pivot itself meeting up with T_1 and external force Q. Its modulus is vector resultant of tension T_1 and Q. To obtain this value, the pivot must be sized according to Chapter 11.

The lever must be sized for the maximum bending moment, which equals Ql.

Once the cross-sectional form is chosen, it is easy to evaluate the load. Since ribbon brakes operate in fatigue loading conditions special attention must be paid to any effects on the levers due to the hinge holes.

As regards the particulars of assembling the brake components, please refer to Giovannozzi, vol. I.

22 Case Study
Design of a Differential

22.1 INTRODUCTION

The differential is the most widespread and best known planetary gearing with two degrees of freedom. This mechanical system enables the two drive wheels of the vehicle to become cinematically independent one from the other during a curved trajectory, forcing the internal wheel to have a lower speed than the external outer wheel.

Figure 22.1 shows, in a schematic form, the structure of an ordinary differential (Figure 22.2) used for transmitting power to each axle's pair of wheels. In the box (7) of the differential are housed the two planetary gears (3 and 4), connected with torsional coupling to the two shafts (1 and 2) that transmit the motion to the wheels of the vehicle. The planetary gears are geared with two satellites (5 and 6), that with the central shaft (8) and the box are the carrier of the gearing.

In a vehicle where the engine and transmission are arranged lengthwise, gearing receives the motion from change by the bevel gear (10–9) that transmits the motion directly or indirectly to the arm. In vehicles with engine and transmission transversely arranged, there is an ordinary cylindrical speed reducer.

The peculiarity of the differential, as with all the planetary gearing, is the fact that the axes of the satellites that form the arm are not fixed, but they revolve around the same planetary axis (Figure 22.3). The relationship between the speeds of the gears of an epicyclic gearing is defined by the Willis formula:

$$\tau_0 = \frac{\omega_2 - \omega_c}{\omega_1 - \omega_c},$$

in which ω_1 and ω_2 are the speeds of the planetary, ω_c is the speed of the arm, and τ_0 is the ratio of the equivalent ordinary gearing (with the arm blocked); this depends on the size and arrangement of gears in the gearing. The configuration of the differential is such that τ_0 is always equal to –1.

From the Willis formula, the equation of equilibrium of the torques and the equation of the conservation of power is

$$C_{t1} = C_{t2} = \frac{C_c}{2},$$

$$\omega_c = \frac{\omega_1 + \omega_2}{2}.$$

From this we can define the two basic characteristics of the differential.

The first, that justifies its use, makes the speeds of two planetary gears and therefore the drive wheels of the vehicle totally independent (not linked).

The second rule requires that the two wheels always transmit the same equal torque to the ground.

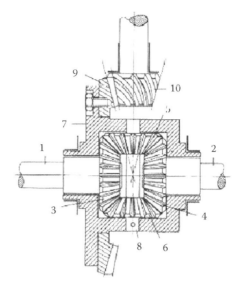

FIGURE 22.1 Schematic structure of an ordinary differential.

This, however, is its more evident limit, because the maximum torque is always equal to twice that of the least of the two wheels. Therefore, the wheel that has the best condition for adherence and/or load will never be able to be fully used. With the introduction of locking differentials it is possible to overcome this problem.

Compared to the structure type of an ordinary differential, the one that has been designed in this project has a further speed reducer between the input bevel gear and the differential itself. This allows increasing the torque transmitted to the wheels, as opposed to reducing their speed when the operating conditions are particularly severe.

22.2 REVIEW FOR THE GEARS PROJECT

The data project specifications do not impose any constraint on the number of gear teeth, and therefore the only restriction is the satisfying of the non-interference during the work conditions. After choosing the pressure angle θ and calculating the transmission ratio τ, to minimize the volume, we can define the minimum number n_{min} of teeth in the smaller gear (pinion) to

FIGURE 22.2 Internal parts of a differential.

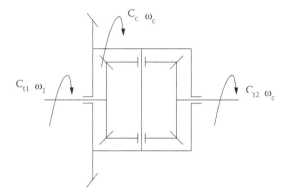

FIGURE 22.3 Scheme for relationship between the speeds.

$$n_{min} = \frac{2k'\tau}{-1+\sqrt{1+\tau(2+\tau)\sin^2\theta}}.$$

Referring to Table S UNI 3522, for the normal proportions of the gears:

$$\text{addendum} = a = m \quad \text{dedendum} = u = 1.25m \quad k' = 1.$$

The calculation proceeds with an initial resistance of teeth followed by a check for wear or pitting, depending on working conditions, and finally a dynamic verification is made to take into account the influence of the rotation speed.

The force acting on each tooth of the gear depends on the dimension, the shape of the teeth, the tip speed, and the conditions of lubrication and cooling. The real load distribution on each pair of teeth comes from the contact ratio ε. For simplicity and safety, we can begin with the classic Lewis formula:

$$m = \sqrt[3]{\frac{2M}{\sigma_{amm}y\lambda n}},$$

where M is the applied moment to the pinion, $\sigma_{amm} = (1/3 - 1/5)\,\sigma_R$ is the allowable stress of the material, $\lambda = b/m$ is the ratio between the width b and the metric module m, y is the coefficient of Lewis that depends on the number of teeth n, on the pressure angle, and the on the type of proportioning.

Generally the decommissioning of a gear is not due to the bending failure of the teeth but to the wear to which the tooth profile is subjected under stress. Taking into account this phenomenon, then if the maximum stress of contact (Hertzian contact) is less than 20%–25% of the smallest surface hardness between the surface hardness of two gear materials, then

$$\sigma_{max} = 0.59\sqrt{\frac{QE(1+\tau)}{bR\sin 2\theta}} < 0.25\,HV$$

where, Q is the force applied to the tooth, $E = 2/(1/E_1 + 1/E_2)$ is the equivalent modulus of elasticity of the gear materials, (E_1 and E_2 are the elastic modules of each gear), R is the pinion pitch radius, and HV is the Brinell hardness (or Vickers if, as often happens, the surface material is heat treated).

Finally, we must take into account the rotational speed of the gearing; if it is high, in addition to the increase of the power of dissipation, friction produces a reduction of the bending resistance and pitting resistance, resulting from manufacturing or installation errors. The dynamic test is to verify that σ_L is less than the allowable stress σ_{amm} reduced by an appropriate α coefficient that is a function of the pitch line velocity v:

$$\sigma_L = \frac{2M}{m^3 \lambda y n} < \alpha \sigma_{amm},$$

in which $\alpha = A/(A + v)$ with the speed factor A, variable from a minimum of 3 and a maximum of 10 m/sec. It also takes into account the accuracy of the working of the teeth and the type of gear (slow or fast).

The above refers to the cylindrical spur gears, but providing some adjustment it can be extended to the bevel straight teeth.

To be precise, as in the previous formulae, it is sufficient to use the virtual number of teeth n^* and the virtual transmission ratio τ^* as indicated by the Tredgold's method:

$$n^* = \frac{n}{\cos \varphi},$$

$$\tau^* = \frac{n_1^*}{n_2^*},$$

in which φ is the cone pitch angle.

The non-interference condition becomes

$$n_{min}^* = \frac{2k'\tau^*}{-1 + \sqrt{1 + \tau^*(2 + \tau^*)\sin^2 \vartheta}}.$$

Furthermore, for the module variation along the profile, the medium module must be used in the formulae. Such approximation is valid when the contact pressures do not vary too much along the tooth profile and the medium module is not so different from the maximum value of the module.

The formulae to calculate bending, wear, and dynamic become:

$$m_{min}^{med} = \sqrt[3]{\frac{2M}{\sigma_{amm} y^* \lambda n^*}},$$

$$\sigma_{max} = 0.59 \sqrt{\frac{QE(1 + \tau^*)}{bR^{med} \sin 2\vartheta}}$$

$$\sigma_L = \frac{2M}{m_{med}^3 \lambda y^* n^*}.$$

The Lewis coefficient must be evaluated starting from the virtual number n^*.

22.3 DIMENSIONING OF THE BEVEL GEARS

22.3.1 INTRODUCTION

The shaft is perpendicular to the axis of the driving wheels, therefore a bevel gear to deflect the motion of 90° is necessary. This bevel gear (Figure 22.4), makes a further reduction in the speed of rotation of the drive wheels in confrontation to the speed engine. It consists of a conductive small

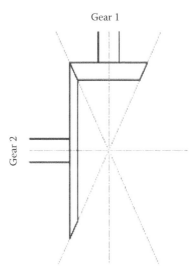

FIGURE 22.4 Scheme of the bevel gears.

diameter gear (1) (pinion) that is keyed on the driveshaft and of a conduct gear (2) (rim) of large diameter, fixed directly or by an additional gearing, to the box of the differential. The box causes the rotating of the drive shafts.

22.3.2 CALCULATION OF THE TEETH NUMBER

The bevel gear transmission ratio is given by

$$\tau_{12} = \frac{\sin \varphi_1}{\sin \varphi_2}.$$

From the geometry of the system and from the project input data $\tau_{12} = 1/4.5$:

$$\tau_{12} = tg\,\varphi_1 \Rightarrow \begin{cases} \varphi_1 = \arctan \tau_{12} = 12.53° \\ \varphi_2 = 90° - \varphi_1 = 77.47° \end{cases}.$$

From the non-interference condition:

$$n^*_{1\min} = \frac{2\tau^*_{12}}{-1 + \sqrt{1 + \tau^*_{12}(2 + \tau^*_{12})\sin^2 \vartheta}} = 16.73,$$

in which $\vartheta = 20°$ and:

$$\tau_{12}^* = \frac{n_1^*}{n_2^*} = \frac{n_1}{n_2} \cdot \frac{\cos\varphi_2}{\cos\varphi_1} = \tau_{12}\frac{\cos\varphi_2}{\cos\varphi_1} = \tau_{12}^2 = 0.0494,$$

therefore:

$$n_{1\min} = n^*_{1\min}\cos\varphi_1 = 16.33.$$

The minimum number of teeth chosen for the pinion is $n_1 = 18$ because, in addition to verifying the non-interference condition, it is a whole number. As the transmission ratio $\tau_{12} = 1/4.5$, the teeth number for the crown is $n_2 = 81$.

22.3.3 BENDING CALCULATIONS

The material required for the differential gears must have high surface hardness and toughness in order to offer adequate resistance to dynamic stress and wear. UNI-treated steel 34 Ni Cr Mo 16 presents these characteristics:

$$\sigma_s = 1275 \text{ MPa} \quad \sigma_R = 1716 \text{ MPa} \quad H_d = 4414 \text{ MPa}.$$

The module must be calculated for the smaller gear because this is the more stressed:

$$m_{1\min}^{med} = \sqrt[3]{\frac{2M_1}{\sigma_{amm}y_1^*\lambda n_1^*}} = 4.748 \text{ mm}.$$

The calculation should be made in the first pair of gears when the torque moment is $M_1 = M_{mot}\tau_I = 2630$ Nm. The virtual number of the teeth is $n_1^* = 18$, and the allowable stress is $\sigma_{a\mu\mu} = \sigma_p/3 = 572$ MPa, the ratio $\lambda = b/m = 15$, and the virtual Lewis coefficient is $y_1^* = 0.31064$.

The value of y_1^* is obtained by a linear interpolation of the values of Table IX in Giovannozzi, Vol. II, p. 135.

For the module, the value 6 mm instead of 5 mm is chosen, closer to the calculated value, to assure the surface durability. Verification, that is generally more difficult compared to the bending verification. Therefore the radius of the two gears is

$$R_1^{med} = \frac{n_1 m_{12}^{med}}{2} = 54 \text{ mm},$$

$$R_2^{med} = \frac{n_2 m_{12}^{med}}{2} = 243 \text{ mm},$$

and the width b is

$$b = \lambda m = 90 \text{ mm}.$$

22.3.4 WEAR CALCULATION

The normal stress σ_{max} is due to the contact of the gear teeth and is equal to

$$\sigma_{max} = 0.59\sqrt{\frac{Q_1 E(1 + \tau_{12}^*)}{bR_1^{med}\sin 2\vartheta}} = 1094 \text{ MPa},$$

where: $Q_1 = 1000 M_1/R_1^{med} = 48{,}712$ N, $E = 220$ GPa, $R_1^{med} = 54$ mm, $b = 90$ mm, $\tau_{12}^* = 0.0494$ and $\vartheta = 20°$, therefore:

$$\sigma_{max} = 1094 \text{ MPa} < \frac{HV}{4} = 1104 \text{ MPa}.$$

22.3.5 Dynamic Verification

The normal bending stress of the teeth is

$$\sigma_{L1} = \frac{2M_1}{m_{12med}^3 \lambda y_1^* n_1^*} = 283 \text{ MPa},$$

Where $M_1 = 2630$ Nm, $m_{12\,med} = 6$ mm, $\lambda = 15$, $y_1^* = 0.31064$ and $n_1^* = 18.44$.

From the design data in the first pair of gears, the angular speed of the pinion is $\omega_1 = \omega\tau_I = 62.72$ rad/s, therefore $v_1 = \omega_1 R_1^{med} = 3.387$ m/s.

For the correct and silent working of the system, the bevel gear must be coupled as accurately as possible. This means the teeth must be made with a high degree of accuracy, and for this reason it must be that

$$\alpha = \frac{A}{A + v_1} = 0.626,$$

and consequently: $\sigma_L = 283$ MPa $< \alpha\sigma_{amm} = 293$ MPa.

22.3.6 Bending Verification of the Crown

If the same material is used for the pinion, no additional verification is needed. As reported at the beginning in the design data, for the construction of the rim and the differential with the exception of the pinion and its shaft, a cemented carbon steel Ni Cr Mo 5 UNI 18 is used, for which:

$$\sigma_R = 1230 \div 1520 \text{ MPa} \quad \sigma_s = 981 \text{ MPa} \quad HV = 720 \div 810.$$

The operating stress for the teeth of the rim is

$$\sigma_{L2} = \frac{2M_2}{\lambda m^3 y_2^* n_2^*} = \frac{2M_1}{\lambda m^3 y_2^* n_1^* 4.5},$$

whereas $4.5\, y_2^* > y_1^*$, therefore:

$$\sigma_{L2} < \sigma_{L1} = 283 \text{ MPa} < 410 \text{ MPa} = \frac{\sigma_R}{3} = \sigma_{amm}.$$

The wear verification is not required because the chosen material for the realization of the crown has a similar contact stress, but the hardness is greater than that of the pinion.

22.3.7 Dynamic Rim Verification

With the same pressure angle and type of proportion, the Lewis coefficient increases with the number of teeth. Given the relationship between the virtual number of teeth of each gear of the bevel gears, the Lewis coefficient is

$$y_2^* > y_1^* \Rightarrow \sigma_{L2} < \frac{\sigma_{L1}}{4.5} = 63 \text{ MPa} < 262 \text{ MPa} = \alpha\sigma_{amm}.$$

With an equal a coefficient used for the pinion verification.

Table 22.1 shows the summary of the characteristics for the bevel gear of the differential.

TABLE 22.1

	n	n_{med} [mm]	m_{12med} [mm]	b [mm]	θ [deg]	Angle cono [deg]
Pinion	18	54	6	90	20	12.53
Rim	81	243	6	90	20	77.47

22.4 PROJECT OF PLANETARY REDUCER

22.4.1 INTRODUCTION

The rim of the bevel gear is not fixed directly to the box of the differential, but transmits the motion through a planetary gearing (Figure 22.5). This system allows a further reduction of the speed of the wheels compared to the engine velocity with a global volume that is less than one ordinary gearing with the same transmission ratio. Also, it has the great advantage of being disengaged (transmission ratio equal to 1) because torque to the wheels is not particularly high. The movement of the gearing transmitted by the rim of the input bevel gear that is internally teethed and at the same time engaged with three cylindrical gears with movable shafts (satellites). The satellites are engaged to a gear (solar) coaxial to the rim. The solar can translate axially by a lever that can be locked to the frame or, (when it is disengaged), it can be engaged at the same time with satellites and with the internal teeth of the closing plate of the arm. In the first case it should have a zero rotation speed, while in the second case it will be free and can rotate with equal speed to the arm.

The arm consists of three satellite pins and the closure plate of gearing that is locked, by a threaded coupling, to the differential box. The motion to the differential arm is transmitted in part by the pins that directly engage with one of the two semi-boxes, and in part by the screws that connect the closure plate to the box.

22.4.2 TEETH NUMBER CALCULATION

When the sun gear (5) is not blocked to the frame, it turns with the rim at the same speed ω_p of the arm, and the transmission ratio is 1.

When $\omega_5 = 0$, instead:

$$\tau_0 = \frac{\omega_5 - \omega_p}{\omega_3 - \omega_p} = \frac{-\omega_p}{\omega_3 - \omega_p} = -\frac{R_3}{R_5},$$

t_0 is the transmission ratio of the equivalent ordinary gearing. Taking $\tau_c = \omega_p / \omega_3 = 1/1.3$ according to the project data the transmission ratio of the epycicloidal reducer is

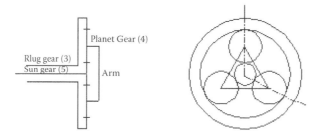

FIGURE 22.5 Planetary gearing reducer.

$$\tau_e = \frac{\tau_0}{\tau_0 - 1} = \frac{R_3}{R_3 + R_5}.$$

For the three gears (3),(4), and (5) it is necessary that

$$R_3 = R_5 + 2R_4.$$

To maintain all the satellites at 120° during the motion, the sum of the numbers of the teeth mover and the solar must be an integer multiple (*k*) of the satellites number, therefore it is

$$\begin{cases} n_3 + n_5 = 1.3n_3 \\ n_5 + 2n_4 = n_3 \quad , \\ n_3 + n_5 = k \end{cases}$$

resolved:

$$\begin{cases} n_3 = \frac{10}{13}k \\ n_4 = \frac{7}{26}k \Rightarrow \tau_{54} = \frac{n_5}{n_4} = 0.857. \\ n_5 = \frac{3}{13}k \end{cases}$$

Thus the gear (5) is the smallest. For this we must start from the sun gear (5), and according to a non-interference condition for cylindrical gear, we have

$$n_{5\min} = \frac{2\tau_{54}}{-1 + \sqrt{1 + \tau_{54}(2 + \tau_{54})\sin^2 \vartheta}} = 12.77.$$

Therefore, $k_{\min} = 55.3$. The real value of *k*, must be a multiple of three and must be chosen so that all the teeth gears are integers.

For $k = 78$, this is

n_3	n_4	n_5
60	22	18

22.4.3   BENDING CALCULATION

To define the module of the gear we must begin from

$$m_{5\min} = \sqrt[3]{\frac{2M_5}{\sigma_{\text{amm}}y\lambda n_5}} = 4.705 \text{ mm}.$$

Where as $M_5 = 1184$ Nm, $\lambda = 10$, $y = 0.308$ and $\sigma_{\text{amm}} = 410$ MPa.

From the equilibrium equation and from the conservation of the power this is

$$\begin{cases} M_p\omega_p = M_3\omega_3 \\ M_3 + M_{sol} + M_p = 0 \end{cases}.$$

As $M_3 = M_2$, $M_5 = M_{sol}/3$ e $\tau_{\varepsilon\pi} = \omega_\pi/\omega_3 = 1/1.3$, the torque moment M_5 is

$$M_5 = 0.1M_3.$$

From the UNI S 3522 table, $m_{ep} = 5$ mm is chosen (the smallest value of the module immediately above the value found). Therefore the dimensions of the gears are

$$R_3 = \frac{n_3 m_{ep}}{2} = 150 \text{ mm},$$

$$R_4 = \frac{n_4 m_{ep}}{2} = 52.5 \text{ mm},$$

$$R_5 = \frac{n_5 m_{ep}}{2} = 45 \text{ mm},$$

$$b = \lambda m_{ep} = 50 \text{ mm}.$$

22.4.4 WEAR CALCULATION

The normal Hertian stress is

$$\sigma_{max} = 0.59\sqrt{\frac{Q_5 E(1 + \tau_{54})}{bR_5 \sin 2\vartheta}} = 1571 \text{ MPa},$$

where as $Q_5 = 1000\, M_5/R_5 = 26{,}311$ N, $E = 220$ GPa, $b = 50$ mm, $\tau_{54} = n_5/n_4 = 0.857$ and $\vartheta = 20°$.

$$\sigma_{max} = 1571 \text{ MPa} < \frac{HV}{4} = 1766 \text{ MPa}.$$

22.4.5 DYNAMIC VERIFICATION

Furthermore, the gears of the epicyclic train must be high precision tooled, therefore it is assumed that $A = 6$ m/s.

For its velocity the relative motion between the sun gear and the satellites must be taken in to account. If it is observed at the motion from a fixed system with the carrier assembly, it will be seen turn the sun gear at ω_{5rel} and therefore:

$$v = \omega_{5rel}R_5 = \omega_p R_5 = \tau_{ep}\omega_3 R_5 = \tau_{ep}\omega_2 R_5 = \tau_{ep}\tau_{p-c}\omega_1 R_5 = 0.482 \text{ m/s}.$$

The velocity factor is

$$\alpha = \frac{A}{A + v} = 0.925,$$

TABLE 22.2

	n	R [mm]	m_{ep} [mm]	b [mm]	θ [deg]
Rim	60	150	5	50	20
Satellite	22	52.5	5	50	20
Sun	18	45	5	50	20

and the stress will be

$$\sigma_L = \frac{2M_5}{m_{54}^3 \lambda y n_5} = 342 \text{ MPa}.$$

The verification is satisfied as being:

$$\sigma_L = 342 \text{ MPa} < \alpha\sigma_{amm} = 379 \text{ MPa}.$$

Table 22.2 reports the characteristics of the epicyclic train gears.

22.5 DIFFERENTIAL GEAR DIMENSIONS

22.5.1 CALCULATION OF THE NUMBER OF TEETH

The calculation must be made considering gearing with only two gears: gear (6) (of the satellite with mobile axle) and gear (7) (of the fixed axle planet) (Figure 22.6).

The gear ratio of the bevel gears is

$$\tau_{67} = \frac{\sin\varphi_6}{\sin\varphi_7} = \tau_{diff}.$$

For $\varphi_7 = 90° - \varphi_6$ and according to the project data $\tau_{diff} = 0.5$:

$$\tau_{diff} = \tan\varphi_6 \Rightarrow \begin{cases} \varphi_6 = \arctan\tau_{diff} = 26.56° \\ \varphi_7 = 90° - \varphi_6 = 63.44° \end{cases}.$$

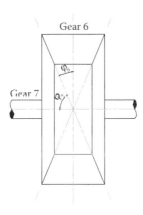

FIGURE 22.6   Scheme of the bevel gears.

According to non-interference under working conditions:

$$n_{6\text{min}}^{*} = \frac{2\tau_{\text{diff}}^{*}}{-1 + \sqrt{1 + \tau_{\text{diff}}^{*}(2 + \tau_{\text{diff}}^{*})\sin^2\vartheta}} \cong 13.44,$$

in which $\vartheta = 20°$ and

$$\tau_{\text{diff}}^{*} = \frac{n_6^{*}}{n_7^{*}} = \frac{n_6}{n_7} \cdot \frac{\cos\varphi_7}{\cos\varphi_6} = \tau_{\text{diff}} \frac{\cos\varphi_7}{\cos\varphi_6} = \tau_{\text{diff}}^{2} = 0.25,$$

then:

$$n_{6\text{min}} = n_{6\text{min}}^{*}\cos\varphi_6 = 13.81.$$

Because the satellite teeth numbers and the planet teeth numbers are integers, and the gear ratio must be respected, the number of teeth will be:

n_6	n_7
14	28

22.5.2 BENDING CALCULATION

The number of planets must be two, but the number of satellites can be either two or four. When one needs to transmit high power, as for trucks, the number four are preferred. The cinematic system of the differential is as that the transmitted torque by the carrier assembly is equally divided between the two planets independently of the wheel velocity. The torque for the bending calculation is

$$M_6 = \frac{M_7 R_6}{n_{\text{sat}} R_7} = \frac{M_7 n_6}{4 n_7} = \frac{M_7}{8} = 962\,\text{Nm},$$

where n_{sat} is the number of satellites of the differential gear.

The module is

$$m_{6\text{min}}^{\text{med}} = \sqrt[3]{\frac{2 M_6}{\sigma_{\text{amm}} y_6^{*} \lambda n_6^{*}}} = 4.4\,\text{mm},$$

in which $n_6^{*} = 15.65$, $\lambda = 12$, $y_6^{*} = 0.2929$ and $\sigma_{\text{amm}} = 410$ MPa.

From the UNI S 3522 table, we choose module $m_{67}^{\text{med}} = 5$ mm, therefore the dimensions of the gears will be

$$R_6^{\text{med}} = \frac{n_6 m_{67}^{\text{med}}}{2} = 35\,\text{mm},$$

$$R_7^{\text{med}} = \frac{n_7 m_{67}^{\text{med}}}{2} = 70\,\text{mm},$$

$$b = \lambda m_{67}^{\text{med}} = 60\,\text{mm}.$$

22.5.3 Wear Calculation

The maximum hertzian stress for the pinion is

$$\sigma_{max} = 0.59 \sqrt{\frac{Q_6 E (1 + \tau_{diff}^*)}{b R_6^{med} \sin 2\vartheta}} = 1364 \text{ MPa},$$

in which $Q_6 = 1000 \, M_6 / R_6^{med} = 27{,}479$ N, $E = 220$ GPa, $b = 60$ mm, $\tau_{\delta t\varphi\varphi}^* = 0.25$ and $q = 20°$.
 The verification is satisfied thus:

$$\sigma_{max} = 1364 \text{ MPa} < \frac{\text{HV}}{5} = 1766 \text{ MPa}.$$

22.5.4 Dynamic Verification

As the past calculation, taken as $A = 6$ m/s.

 Contrary to the bevel gear (ordinary gearing) and the epicyclic speed reducer (this works with the blocked sun gear), in this case it is impossible to define the velocity of any gears beginning only from the input gear. For the dynamic verification we chose the worst situation with one wheel of the truck blocked and the other running at double speed relative to the velocity of the arm (carrier assembly). In this situation the planet wheel velocity v_6 is

$$v_6 = \omega_{6rel} R_{6med} = \omega_{7rel} R_{7med} = \omega_p R_{7med} = 0.75 \text{ m/s}.$$

The velocity factor is therefore:

$$\alpha = \frac{A}{A + v} = 0.888,$$

and the stress is

$$\sigma_L = \frac{2M_6}{m_{67med}^3 \lambda_{y_6}^* n_6^*} = 281 \text{ MPa}.$$

The verification is satisfied as

$$\sigma_L = 281 \text{ MPa} < \alpha \sigma_{amm} = 364 \text{ MPa}.$$

Table 22.3 reports the characteristics of the differential gears

22.6 MEASUREMENT OF TIME GEARING

22.6.1 Introduction

For the design-time of any mechanical part one needs to know *a priori* the cyclic stress or the load history to which it will be subjected during its life. The history represents the trend of loading stress as a function of time or number of cycles. Those assigned by the project data specifications were drawn from experimental measurements of torque delivered by a truck driving a number of routes

TABLE 22.3

	n	$R_{12}{}^{med}$ [mm]	m_{12med} [mm]	b [mm]	θ [deg]	Pitch angle [deg]
Planet wheel	14	35	5	60	20	26.56
Differential bevel	28	70	5	60	20	63.44

considered representative of all normal operating conditions. Each of the ten journeys (App. Chapter 22: pp. 55–59) belong to three different types; two routes are highways, seven are mountainous, and only one is city type. For each of the three categories the project data specifications will allocate a percentage representing the fraction of kilometers (or cycles) of the total expected to be consumed by the truck in any type of journey. The highway journey that is assigned a percentage of 15%, the mountainous journey 70%, and the city journey the remaining 15%. In turn, each journey is given an additional percentage in its category that it is allowed to accumulate for each type of route, using the original ten journeys.

Within each category, for each range of torque, the corresponding number of cycles n_t by mean the weighted average for each journey belonging to that category is obtained:

$$n_t = \sum_1 p_i n_i,$$

where p_i is the weight of the journey I, obtained by multiplying the percentages relative to the journey and the category to which it belongs.

The three cumulatives obtained (App. Chapter 22: pp. 60–62) have been adapted according to the specific project data assigned. It should be remembered that cumulatives represent the shaft load conditions, and this should be taken into account in the normalization phase of torque and number of cycles.

The torques have been "corrected" as though the maximum value of the cumulate were equal to 80% of the maximum torque used for the project to an infinite endurance.

As was done for couples, the scale factors with which it would be possible to "correct" the cycles for each load level inside each journey were determined. The calculation of the total real cycles of the motor shaft follows.

The evaluation of the number of real cycles to the arm is carried out for 900,000 km according to specifications given, and assuming that the wheels have a diameter of 914.4 mm.

To calculate the number real cycles of the shaft, we take into account the gear ratio of the epyciciloil reducer, of the bevel gear, and of the first engaging of gear box. Obviously, it is assumed that the epyciciloil gear reducer only operates on the mountain trails and therefore the gear ratio for the remaining journeys is 1.

After the normalization of the three shafts, cumulates are reported corresponding to each of the gears (pinion of the input bevel gear, solar epyciciloil reducer, and satellite of the differential). Obviously, in the computation of the cycles and the project torques of the epycicloidal reducer, including the differential, account is taken of the number of satellites in the gearing system.

Moreover, the differential assumes that all the cycles occurred in the toughest conditions, in which one wheel of the vehicle is blocked and the other has, therefore, a rotation speed double that of the arm.

Thereafter, for each gearing, a single standard cumulate is obtained that has in ordinate the torque intervals (reordered following passing through the epycicloidal reducer) and in abscissa the sum of the city cycles, mountain cycles, and highway cycles, properly integrated by interpolation of the missing values. For the epycicloidal reducer, the cumulate standard is made to coincide with that of mountain only. Finally, the calculation is made of the contribution of the negative torque in the projecting of the gearing. For the bending calculation, the cumulative effect of the torques of different sign can be seen for every absolute value of torque, summing up the values of the corresponding number of cycles. For the wear calculation, the effect of the negative torque can be ignored because in absolute value it is smaller than the positive, and particularly than those agents on the other side of the teeth. For safety reasons and for easier calculation, the same cumulative effects used for the bending can also be used for wear.

Both UNI and AGMA unification provide indications on the calculation of the gears transmitting constant torque or characterized by small overloads, but do not provide examples of gears subjected to load histories of various types.

For the calculation of fatigue, criteria exist, starting with the load history of the component. These take into account the cumulative damage of the component, by means of lowering the fatigue limit. One of the criteria more used is the Miner criterion:

$$\sum_i \frac{n_i}{N_i} = 1,$$

where n_i is the number of cycles consumed at the stress $\sigma_I > \sigma_0$ (fatigue limit) and N_i is the number of cycles that would be broken for that σ_I value.

The formulae used for gearing calculation follow the UNI 8862 and the Miner criterion.

The UNI 8862 norms are valid only for cylindrical spur gears and helical gears, but experience shows that this can be extended to the dimensioning of the bevel without committing to excessive errors. This can be used only when $\varepsilon_\alpha < 2$; (ε is applied to cylindrical spur gears).

To verify the use of the norm it is necessary to evaluate the contact ratio:

$$\varepsilon = \frac{1}{2\pi} \left[\sqrt{\left(\frac{n' + 2k'}{\cos \vartheta} \right)^2 - n'^2} + \sqrt{\left(\frac{n + 2k}{\cos \vartheta} \right)^2 - n^2} - \left(n + n' \right) tg\vartheta \right].$$

For the bevel gears the teeth number must be the virtual number. For all gearing, the applicability of the norms was verified, resulting in

	E
Piston	1.74
Sun	1.55
Satellite	1.64

The norms propose calculations similar to these of Lewis, with a verification at the bending and at the pitting or wear. UNI 8862 is used to calculate allowable bending stress under working conditions, with the following formulae:

$$\sigma_F = \frac{F_t}{bm} y_{Fa} y_{Sa} y_\varepsilon y_\beta (k_A k_V k_{F\beta} k_{F\alpha}).$$

$$\sigma_{FP} = \frac{\sigma_{F\lim} y_{ST} y_{NT}}{S_{F\min}} y_{\delta relT} y_{RrelT} y_x.$$

Given that:

σ_Φ is the working stress

σ_{FP} is the allowable stress

σ_{Flim} is the possible material stress,

b is the width of the gear

m is the module

F_t is the load transmitted by the tooth.

Because there are gear parameters, working condition parameters, fatigue parameters, and in particular:

y_{Fa} is the size factor of the tooth

y_{Sa} is the stress factor

y_e is the ratio gear factor

y_b is the helical angle factor

k_A is the application factor

k_V is the dynamic factor

k_{Fb} is the load longitudinal distribution factor

k_{Fa} is the load trasversal distribution factor

y_{ST} is the stress correction factor

y_{NT} is the endurance factor

SF_{min} is the assurance bending factor

$y_{\delta rel}T$ is the notch sensitivity factor

$y_{\delta relT}$ is the surface factor

y_χ is the size factor

Because there is no fatigue data for the material, the Bach's criteria is used, therefore:

$$\sigma_{Flim} = \sigma_0 = \frac{\sigma_{SN}}{3},$$

with $y_{ST} = 1.4$, it is possible to correct the fatigue limit to put in account that the load ratio $R = 0$.

y_{NT} is a function of the number of cycles, and of the material hardness. For the hardening steel, it is

$$y_{NT} = \left(\frac{3 \cdot 10^6}{N}\right)^{0.115}, \quad N \le 3 \cdot 10^6.$$

The bending verification is done by comparing the working bending stress, σ_F, and the allowable stress, σ_{FP}:

$$\frac{F_t}{bm} y_{Fa} y_{Sa} y_\varepsilon y_\beta \left(k_A k_V k_{F\beta} k_{F\alpha}\right) = 1.4\sigma_{Flim}\left(\frac{3 \cdot 10^6}{N}\right)^{0.115} y_{\delta relT},$$

in which there are no parameters S_{Fmin} and y_{RrelT} because they are assumed equal to 1.

Writing the previous equation in a different form, we have

$$\left(\frac{F_t}{b}\right)^{8.7} \cdot N = \left(\frac{1.4\sigma_{Flim} y_{\delta relT}}{A}\right)^{8.7} 3 \cdot 10^6 = W,$$

in which

$$A = \frac{y_{Fa} y_{Sa} y_{\varepsilon} y_{\beta} (k_A k_V k_{F\beta} k_{F\alpha})}{m}.$$

The calculus of W is done taking the parameter according to the indications of the norms. In particular, the factor y_{Fa} is calculated in Table 17 of page 24 of UNI8862/2; y_{Sa} is taken from Figure 25 on page 33 of the same norm; and $y_b = 1$ because the gears are cylindrical spur gears, $k_A = 1.5$ according to Table 1 on page 2 of the norm, $k_{F\beta} = 1.05$ because it is assumed that there are no errors in the parallelism of the axes; $k_{Fa} = 1$ because the gears will be hardened, and because the effect of the surface conditions brings G to a value no lower than 7, $y_{\delta relT} = 1$ because it is supposed usual values of the fillet radius. The other factors are calculated according to

$$y_{\varepsilon} = 0.25 + \frac{0.75}{\varepsilon},$$

$$k_V = 1 + BA \text{ where } \begin{cases} A = \dfrac{v z_1}{100}; v = \text{velocity in } m/s \text{ and } z_1 = \text{teeth} \cdot \text{number} \\ \\ B = 0.18(0.7)^{9-G}; G = \text{precision} \cdot \text{rate} \end{cases}.$$

For the application of the UNI norm, we must evaluate a constant load that becomes a variable for the rotation of the gears. It is necessary to determine the equivalent tangential load that is able to take into account the real load history.

Combining the equation of the norm with the Miner hypothesis it is possible to define the equivalent load F_{teq} that is able to produce similar effects to the load cumulate:

$$F_{teq} = \left[\sum_{1} p_i F_{ti}^{8.7} \right]^{1/8.7},$$

in which F_{ti} is the load on the tooth for n_i cycles and $p_i = n_i / N_{TOT}$ is the percentage of cycles relative to the total number N_{TOT}.

The ultimate goal of the time fatigue calculation is to reduce the size of the gears in the gearing. This can be done by reducing the axial width b of the gears or the radial dimension, or by means of reduction of module m.

In this particular case, it was decided to maintain the module m before it was calculated and only change the width b. Therefore λ.

similar to the above calculation for bending. UNI 8862 provides a further verification of wear, following the method already described. In detail:

$$\sigma_H = z_H z_E z_{\varepsilon} z_{\beta} \sqrt{\frac{F_t (u+1)}{d_1 b u}} \sqrt{k_A k_V k_{H\beta} k_{H\alpha}},$$

$$\sigma_{HP} = \frac{\sigma_{H\lim}z_N}{S_{H\min}}z_L z_R z_W z_x z_V,$$

in which σ_H is the maximum local pressure stress, σ_{HP} is the corrected allowable stress of the material, $\sigma_{H\lim}$ is the allowable stress of the material, F_t is the transmitted tangential load, u is the engaging ratio, d_1 is the pitch pinion diameter, and b the width of the tooth. The others are parameters of the gearing, of the working s, and of fatigue conditions. For other indications please refer directly to the UNI norms.

The verification will be confirmed when it is

$$\sigma_H \le \sigma_{HP},$$

that becomes:

$$z_H z_E z_\varepsilon z_\beta \sqrt{\frac{F_t(u+1)}{d_1 bu}} \sqrt{k_A k_V k_{H\beta} k_{H\alpha}} \le \frac{\sigma_{H\lim}z_N}{S_{H\min}}z_L z_R z_W z_x z_V.$$

Applying the same procedure exposed as for the bending calculation, we have

$$\left(\frac{F_t}{b}\right)^{6.6} \cdot N \le W_1,$$

in which

$$W_1 = 5\cdot 10^7 \left(\frac{K\sigma_{H\lim}}{A_1}\right)^{1/0.0756}\left(\frac{d_1 u}{u+1}\right)^{6.6},$$

$$K = \frac{z_L z_R z_W z_x z_V}{S_{H\min}},$$

$$A_1 = z_H z_E z_\varepsilon z_\beta \sqrt{k_A k_V k_{H\beta} k_{H\alpha}}.$$

The calculation of the parameter W_1 is done by means the indications reported in the UNI norms. Those selected in this case are $z_H = 2.495$, $z_E = 190$ (N/mm²)$^{0.5}$, $z_\beta = 1$, $k_A = 1.5$, $k_{Ha} = 1$, $z_L = 1$, $z_R = 1$, $z_W = 1$, $z_c = 1$ and $S_{H\min} = 1$.

For the parameters that depend on the geometric and working conditions, the functions are reported as

$$z_\varepsilon = \sqrt{\frac{4-\varepsilon}{3}},$$

$$k_V = 1 + BA \text{ where } \begin{cases} A = \dfrac{vz_1}{100}; v = \text{velocity in } m/s \text{ and } z_1 = \text{teeth}\cdot\text{number} \\ \\ B = 0.18(0.7)^{9-G}; G = \text{precision}\cdot\text{ratio} \end{cases},$$

$$k_{H\beta} = \left(k_{F\beta}\right)^{\frac{1}{N}} \text{ where } N = \frac{\left(\dfrac{b}{h}\right)^2}{1 + \dfrac{b}{h} + \left(\dfrac{b}{h}\right)^2},$$

$$z_V = C_{zV} + \frac{2(1 - C_{zV})}{\sqrt{0.8 + \dfrac{32}{v}}}; C_{zV} = 0.93,$$

$$z_N = \left(\frac{5 \cdot 10^7}{N}\right)^{0.0756},$$

$$\sigma_{H\lim} = \frac{HV}{4}.$$

For more details on the choice of parameters, please refer directly to the norms.

Clarification is needed regarding calculation of the allowable stress of the material, for which the Brinell hardness (HB) should be used. Due to the high surface hardness of the material used, following r (the heat treatment of hardening), it is physically impossible to measure the hardness by the Brinell test (for the deformability of the steel ball test), and therefore Vickers hardness is used; the numerical results can be considered equivalent.

As for the bending calculation, the Miner rule is applied to evaluate a tangential load equivalent to the cumulative project load:

$$F_{teq1} = \left(\sum_1 p_i F_{ti}^{6.6}\right)^{1/6.6},$$

in which F_{ti} and p_i have the same definitions as those seen for bending.

With reference to the fatigue behavior of metallic materials, particularly steel, it makes sense to consider time resistance of the mechanical components when the life prevision is no fewer than (or equal to) a few million cycles.

The UNI 8862 norm uses time factors to determine the allowable fatigue stress; these are a function of the number of cycles when these are fewer than 3×10^6 for bending (or 10^7 for AGMA norms) and 5×10^7 for wear. The norms indicate a value of 1 for a higher number of cycles (resistance to infinite life).

In our case, from the design data specifications, the resulting number of cycles and the corresponding cumulative standard (App. Chapter 22: 63–66) go on to the limit for which it makes sense to speak of time resistance.

For this reason, in order to have the calculation at finite time, it is assumed to consider a partially initial damage of the material, followed by a non-estimable time of failure.

The combined "cumulate" does not allow us to have the exact time sequence of the loads, therefore, acting for security reasons, the worst condition is chosen, assuming that the loads are applied in descending order from the maximum torque.

We choose to make a "limited time" design, and suggest that, for each gearing, the only damaging torques are those acting in the range 0–3×10^6.

Beginning from these torques, the load F_{teq} is calculated as follows:

$$\beta_i \quad \frac{n_i}{3 \cdot 10^6} \, i$$

The bending verification using the F_{teq} should have a value of the width b of the gear that leads to failure for a number of cycles equal to 3×10^6.

To avoid this, the F_{teq} used in the calculation is increased by 30% to ensure the resistance of the gearing for the time imposed by the project specifications.

For wear, verification F_{teq} is not increased, because in the calculation equivalent load and negaive torque are taken into account, assuming that these act on the same side of the tooth. The verification is performed for a limit of number cycles equal to 5×10^7 instead of 3×10^6.

22.6.2 BENDING CALCULATION

Before the bending calculation at the working conditions, it was performed a bending verification at infinite time. The tangential force F_{tind} is derived according to UNI 8862, with each gearing able to resist, and it is verified that the value of F_{tind} exceeds the maximum value contained in its load spectrum (App. Chapter 22: 64–66).

The following tables, in addition to summarizing the parameters used for the project (see page 582 for the meaning of the symbols), show the strength F_{tind} following the infinite life verification, strength of the load F_{tind} used for the time project, the project width $b_{progetto}$ of the gears, and also the final choice to satisfy the wear verifications.

	y_{Fa}	y_{Sa}	y_ε	k_A	k_V	$k_{F\beta}$	$y_{\delta relT}$	m
Pinion	2.88	1.54	0.681	1.5	1.055	1.05	0.99	6
Sun	2.91	1.53	0.734	1.5	1.0077	1.05	0.99	5
Satellite	3.08	1.51	0.707	1.5	1.012	1.05	0.98	5

	A	σ_{Flim}	$W/3E+6$	b_{ind}	F_{tind}	F_{teq}	$b_{progetto}$	b
Pinion	0.83645	425	5.96E+24	90	63588	31247	44.22567	52
Sun	1.03734	327	9.36E+22	50	22912	18920	43.17346	44
Satellite	1.04818	327	7.82E+22	60	25762	20400	47.51184	48

22.6.3 WEAR VERIFICATION

	z_H	z_E	z_V	z_ε	k_V	$k_{H\beta}$	A_1
Pinion	2.495	190	0.96	0.868	1.055	1.07061	535.589
Sun	2.495	190	0.947	0.904	1.0077	1.06696	544.225
Satellite	2.495	190	0.95	0.886	1.012	1.0627	533.457

	$\sigma_{H lim}$	u^*	d	$W1/5E+7$	F_{teq}	b	$(Ft/b)^{6.6}$
Pinion	1104	20.25	108	1.59E + 17	20348	52	1.29E + 17
Sun	1766	1.16667	90	3.73E + 17	11039	44	6.86E + 15
Satellite	1766	4	70	1.32E + 18	11768	48	5.89E + 15

22.6.4 Bending and Wear Verification of the External Rim

The rim material of the input bevel gear is different to the material of the pinion, and therefore it is necessary to conduct the bending and wear verification. (For the mean of the symbols see page 582).

	y_{Fa}	y_{Sa}	Y_ε	k_A	k_V	$k_{F\beta}$	$y_{\delta relT}$	m
Rim	2.1	1.97	0.681	1.5	1.055	1.05	1.01	6

	A	s_{Flim}	$W/3E+6$	b	F_{tlim}	F_{teq}
Rim	0.78022	327	1.33E+24	52	30915	24036

The verification is satisfied due to the equivalent load F_{teq} applied to the crown that is less than that if it was able to resist $F_{t\,lim}$.

For the wear verification it is not necessary to make any calculations because the local hertzian stress is equal to the pinion stress, for which the verification is satisfied even if it is made of a less hard steel than the steel of the rim.

22.7 DIMENSIONING OF THE SHAFT PINION AND CHOICE OF THE BEARINGS

The pinion shaft is loaded from the torque from the motor and is subjected to significant bending stresses, shear, and normal stress due to contact with the teeth of the crown. The scale of these actions very much depends on the fixed points of the shaft at the frame.

For the bevel gear input at the differential to be smooth and silent, the contact between the teeth of the pinion and the rim must be at the pitch cones. For this reason, the clearance between the teeth is of particular importance in the project, and the position of the bearings must be designed to minimize the deformation induced by the forces exchanged.

This leads to the decision to connect the shaft, as in Figure 22.7.

Upstream of the hinge A, the pinion shaft receives the motion directly from the transmission shaft through a rigid coupling. For the coupling of the differential with the flange, a splined profile is used.

22.7.1 Calculating Splined Profile

Splined profiles are used when the torque to be transmitted through the coupling hubshaft is high, and therefore it is not advisable to use keys, tongues, or shrink-fit. In particular, this solution must

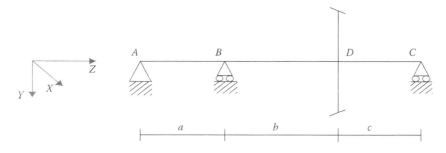

FIGURE 22.7 Forces on the shaft.

be used when the smoothness of the hub axle is required. UNI 229/225 reports the profiles as unified.

Generally, the implementation of the teeth is done by milling the shaft and broaching the hub.

According to the number and depth of the grooves, for equal diameter, the splines can be divided into three categories in terms of the maximum torque to transmit.

To obtain the correct position, coupling can be done with tight tolerances on the circle of diameter d (in this case for the mechanical working a grinding wheel is used), or (only for the profiles with large support) coupling can be done with tight tolerance between the sides of the grooves (presenting greater difficulty for the grinding).

The profile that meets the specific requirements for coupling with the input flange is one with a medium internal support with straight teeth (UNI 222).

The calculation of the diameter of the project shaft is conducted by considering the equivalent torque $M_{1eq} = 1,297,944$ N/mm used for the dimensioning of the input pinion of the bevel gear (App. Chapter 22: 64), therefore:

$$d_{prog} = \sqrt[3]{\frac{16M_{1eq}}{\pi\tau_{amm}}} = 30 \text{ mm},$$

of which

$$\tau_{amm} = \frac{\sigma_{SN}}{3\sqrt{3}} = 245 \text{ MPa},$$

according to the Von Mises hypothesis and the Bach criteria to define the fatigue limit at $R = -1$.

The calculations for determining the equivalent torque M_{1eq} resulted from the application of the Miner rule at the bending behavior of the material and are defined by UNI 8862.

The application of the torque equivalent to the dimensioning of all components of the differential arises from assuming that the behavior of the standard material for the gears is also valid in the design of other mechanical elements.

From the table on page 345 of Giovannozzi, Vol. I, the unified profile with a minimum reference diameter d greater than the design shaft d_{prog} was chosen.

Summarizing

d	D	Ω	M	k
32	38	0.38	2.1	0.96

The m and k factors were obtained from the table on page 343 and 344 of Giovannozzi, Vol. I, considering a fixed coupling, with no, or only one, cemented surface.

As the minimum diameter of the grooved does not coincide with that of the project, and therefore permits a higher torque Mt', it is possible make a correction of the parameter k that produces a reduction of the effective length l of the profile:

$$l = \frac{md\Omega}{k'} = 22 \text{ mm},$$

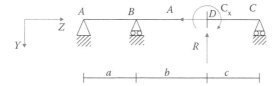

FIGURE 22.8 Scheme for the reactions calculation.

whereas

$$k' = k\frac{M'_t}{M_{1eq}} = 1.166.$$

22.7.2 CALCULATION OF REACTION FORCES

In addition to the torque and for its effect, the pinion shaft is downloaded, by a radial action R, by an axial action associated with a bending transport moment C_x, and the tangential force Q:

$$Q = F_{1eq} = 24,036\,N,$$

$$R = Q\tan\vartheta\cos\phi_1 = 8541\,N,$$

$$A = Q\tan\vartheta\sin\phi_1 = 1898\,N.$$

These actions are orthogonal to each other and thus it is necessary to study the problem in two orthogonal planes (XZ and YZ) that contain all the loads.

Assuming that the pinion gear rotates in a counterclockwise direction when the vehicle is running ahead, and that it engages with the crown in the lower part of the diagram in Figure 22.7, the direction and the versus of the exchanged forces between the teeth are as shown in Figures 22.8 and 22.11.

According to the equilibrium equation in Z direction, the axial reaction is

$$R_{AZ} = A = 1898\,N.$$

For the calculation of the reaction in Y direction, we have a no statically defined problem in statics terms not defined. According to the load method we may consider therefore:

- The system in Figure 22.9 with the point B and the external loads free
- The system in Figure 22.10 loaded by unit force only

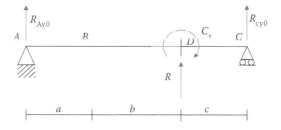

FIGURE 22.9 Scheme for the reactions calculation.

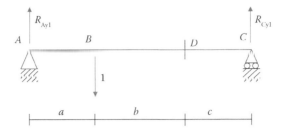

FIGURE 22.10 Scheme for the reactions calculation.

According to the equilibrium equations:

$$\begin{cases} R_{Ay0} = -\dfrac{Rc + C_x}{a+b+c} \\[3mm] R_{Cy0} = \dfrac{C_x - R\cdot(a+b)}{a+b+c} \\[3mm] R_{Ay1} = \dfrac{b+c}{a+b+c} \\[3mm] R_{Cy1} = \dfrac{a}{a+b+c} \end{cases},$$

and according to the virtual work principle:

$$P = -\frac{R_{Ay1}R_{Ay0}\dfrac{a^3}{3} + (R_{Ay1}-1)R_{Ay0}\dfrac{b^3}{3} + (2R_{Ay1}R_{Ay0}-R_{Ay0})\dfrac{ab^2}{2} + R_{Ay1}R_{Ay0}a^2b + R_{Cy1}R_{Cy0}\dfrac{c^3}{3}}{R_{Ay1}^2\dfrac{a^3}{3} + (R_{Ay1}-1)^2\dfrac{b^3}{3} + R_{Ay1}(R_{Ay1}-1)ab^2 + R_{Ay1}^2a^2b + R_{cy1}^2\dfrac{c^3}{3}}.$$

The shear actions are deliberately neglected as irrelevant compared with the flexural components.

Again, applying the effects of the superposition principle, the reaction forces in the YZ plane are evaluated as

$$\begin{cases} R_{Ay} = R_{Ay0} + PR_{Ay1} \\ R_{By} = P \\ R_{Cy} = R_{Cy0} + PR_{Cy1} \end{cases}.$$

For the resolution of the loading configuration (Figure 22.11) in the XZ plane, a similar procedure to that used in the YZ plane is adopted.

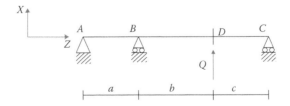

FIGURE 22.11 Forces on the shaft.

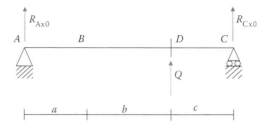

FIGURE 22.12 Scheme for the reactions calculation.

Figures 22.12 and 22.13 report the resolving schemes.
According to the equilibrium equations:

$$
\begin{cases}
R_{Ax0} = -\dfrac{Qc}{a+b+c} \\[4mm]
R_{Cx0} = -\dfrac{Q\cdot(a+b)}{a+b+c} \\[4mm]
R_{Ax1} = \dfrac{b+c}{a+b+c} \\[4mm]
R_{Cx1} = \dfrac{a}{a+b+c}
\end{cases},
$$

therefore:

$$
P_1 = -\frac{R_{Ax1}R_{Ax0}\dfrac{a^3}{3} + (R_{Ax1}-1)R_{Ax0}\dfrac{b^3}{3} + (2R_{Ax1}R_{Ax0}-R_{Ax0})\dfrac{ab^2}{2} + R_{Ax1}R_{Ax0}a^2b + R_{Cx1}R_{Cx0}\dfrac{c^3}{3}}{R_{Ax1}^2\dfrac{a^3}{3} + (R_{Ax1}-1)^2\dfrac{b^3}{3} + R_{Ax1}(R_{Ax1}-1)ab^2 + R_{Ax1}^2a^2b + R_{Cx1}^2\dfrac{c^3}{3}}.
$$

According to effects superposition principle:

$$
\begin{cases}
R_{Ax} = R_{Ax0} + P_1 R_{Ax1} \\
R_{Bx} = P_1 \\
R_{Cx} = R_{Cx0} + P_1 R_{Cx1}
\end{cases}.
$$

The distances a, b, and c between the constraints are defined in order to minimize the reaction forces, making sure that the distance between B and C are not less than the width of the pinion teeth

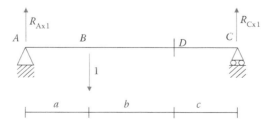

FIGURE 22.13 Scheme for the reactions calculation.

TABLE 22.4

a	b	c	R_{Ay}	R_{By}	R_{Cy}	R_{Ax}	R_{Bx}	R_{Cx}
68	68	51	947	5309	2286	4683	−17659	−11059

(suitably increased to take into account the footprint of the bearings and their shoulders), and at the same time that the total length of the shaft is not excessive.

Table 22.4 shows the distance in millimeters and the reactions on the two planes in *N*.

For the definition of the bearings, the result of reactions at each support is assessed as

$$\begin{cases} R_A = \sqrt{R_{Ax}^2 + R_{Ay}^2} = 4{,}777\,\text{N} \\[2mm] R_B = \sqrt{R_{Bx}^2 + R_{By}^2} = 18{,}440\,\text{N}. \\[2mm] R_C = \sqrt{R_{Cx}^2 + R_{Cy}^2} = 11{,}293\,\text{N} \end{cases}$$

Figures 22.14 and 22.15 show diagrams of the qualitative characteristics of the stresses in the two planes.

22.7.3 CALCULATION OF THE SHAFT

Imagining making a section in *D* of the shaft (Figure 21.16), which according to the stress diagrams is the critical section, it appears that there may be two points where the stress can be maximum.

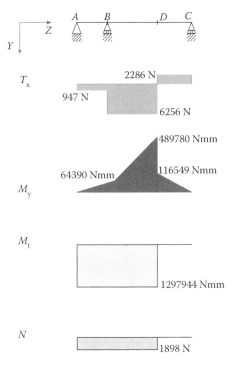

FIGURE 22.14 Diagrams of the qualitative characteristics of the stresses in the Y plane.

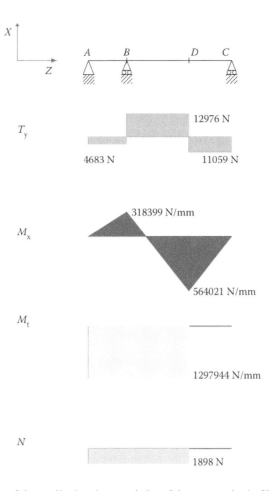

FIGURE 22.15   Diagrams of the qualitative characteristics of the stresses in the X plane.

Assuming for simplicity that the resultants of the shear and the bending are mutually orthogonal, in Q the maximum of the normal stress is $\sigma_{zf\,max} = 32\,M_D/(\pi d^3)$ for bending, and the shear stress is zero. In P the contrary is true. At each point the normal compression stress $\sigma_{zn\,max} = 4\,N_D/(\pi d^2)$ is maximum and the torsional shear stress is $\tau_{zt\,max} = 16\,M_{teq}/(\pi d^3)$.

Generally the normal stress is negligible compared with the action of the bending. Using the hypothesis of Von Mises the equivalent stress is

$$\sigma_{eq} = \sqrt{\sigma_{zf\,max}^2 + 3\tau_{zt\,max}^2} = \sigma_{amm} \Rightarrow \sigma_{eq} = \frac{16}{\pi d^3}\sqrt{4M_D^2 + 3M_{tu}^2} = \sigma_{amm},$$

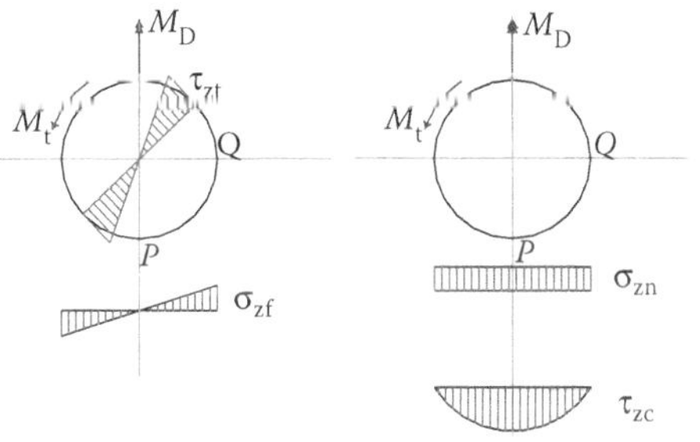

FIGURE 22.16   Qualitative stresses on the shaft.

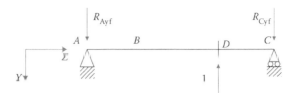

FIGURE 22.17   Scheme for the deflection calculation.

In which $\sigma_{amm} = \sigma_s/3 = 425$ MPa is the allowable material, $M_D = 746{,}997$ N/mm is the total bending moment in D. Therefore, the minimum diameter of the pinion is

$$d_{min} = \sqrt[3]{\frac{16}{\pi\sigma_{amm}}\sqrt{4M_D^2 + 3M_t^2}} = 31.86\,\text{mm}.$$

The compression stress for $N_D = 1898$ N is negligible (2.5 MPa) and therefore the diameter of the shaft chosen is 32 mm.

For the total shear load $T_D = 14{,}405$ N the tangential stress is: $\tau_{zcmax} = 16\,T_D/(3\pi d^2)$. In a similar way in Q it is

$$\sigma_{eq} = \sqrt{\sigma_{znmax}^2 + 3\left(\tau_{ztmax} + \tau_{zcmax}\right)^2} = 420.84\,\text{MPa} \leq \sigma_{amm} = 425\,\text{MPa}.$$

22.7.4   DEFLECTION VERIFICATION

To ensure smooth and quiet working of the bevel gear one needs to verify that at each point of the shaft, particularly at the section where the pinion is located and where it then makes contact with the rim, in conditions of maximum load the deflection is no less than 1% of the module.

Applying the principle of virtual work, the isostatic system equivalent for each load plane (YZ and XZ) is defined. In both, to find the real deflection, the virtual unified force is therefore applied at the section of interest (Figure 22.17 and 22.18).

The equations of equilibrium for the translation and rotation permit us to determine the reactions of the constraints:

$$\begin{cases} R_{Axf} = R_{Ayf} = \dfrac{c}{a+b+c} \\[3mm] R_{Cxf} = R_{Cyf} = \dfrac{a+b}{a+b+c} \end{cases}.$$

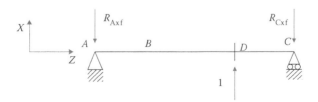

FIGURE 22.18   Scheme for the deflection calculation.

Applying the principle of virtual work:

$$v_{Dy} = -\frac{1}{EI} \left[\frac{R_{Ayf}R_{Ay}a^3}{3} + R_{Ayf}R_{Ay}a^2b + \frac{R_{Ayf}\left(R_{Ay}-R_{By}\right)ab^2}{2} + \frac{R_{Ayf}R_{Ay}ab^2}{2} + \frac{R_{Ayf}\left(R_{Ayf}-R_{Byf}\right)b^3}{3} \right.$$
$$\left. + \frac{R_{Cyf}R_{Cy}c^3}{3} \right],$$

$$v_{Dx} = -\frac{1}{EI} \left[\frac{R_{Axf}R_{Ax}a^3}{3} + R_{Axf}R_{Ax}a^2b + \frac{R_{Axf}\left(R_{Ax}-R_{Bx}\right)ab^2}{2} + \frac{R_{Axf}R_{Ax}ab^2}{2} + \frac{R_{Axf}\left(R_{Axf}-R_{Bxf}\right)b^3}{3} \right.$$
$$\left. + \frac{R_{Cxf}R_{Cx}c^3}{3} \right].$$

The deflection at the section of interest, is given by the resultant of the two components:

$$v_D = \sqrt{v_{Dx}^2 + v_{Dy}^2} = 0.0548\,\text{mm} \le \frac{m}{100} = 0.06\,\text{mm}.$$

To satisfy the verification it is necessary to increase the diameter of the shaft until the value $d = 40$ mm and its stiffness improves.

22.7.5 Choice of Bearings

The choice of the bearings of the pinion shaft is one of the most delicate phases in the design of the entire group differential. In fact, the smooth and silent working of the system depends largely on the clearance of the bearings and that between the teeth of the gears.

As suggested in the order of calculations, the pinion shaft is connected to the frame with two sliding supports and a hinge. The solution to realize the scheme is by means of two roller conical bearings, in correspondence with the constraints in A and in B, and a cylindrical roller bearing for the constraint in C. The correct working is ensured by the fact that one of the two conical bearings reacts with the axial pinion load, while the other balances the load residual that arises as a result of the constructive characteristics of the bearing itself.

The preloading of the system (while making an objective complication) ensures greater accuracy and rigidity and in particular recovers the lost clearance due to wear, therefore always ensuring regularity and quiet operation.

The resistance under certain loading conditions together with certain duration is defined by calculating the coefficient of a dynamic load:

$$C = P \cdot L^{1/\alpha},$$

where P is the load applied, L is the nominal life in millions of cycles, and α is a factor that depends on the type of rolling element; this is 3 for ball bearings and 10/3 for roller bearings.

The dynamic calculation of the bearing does not ensure that in static conditions or small fluctuations excessive deformation of the rings does not occur. This leads to making the static verification determining the safety factor as

$$S_0 = \frac{C_0}{P_0},$$

where C_0 is the coefficient of static load and P_0 is the load, and leads to verify that S_0 is less than the limit imposed by the manufacturer. Obviously, this ratio should be higher for the applications that require great sensitivity at rotation, while it may be even smaller than 1 if it is necessary to minimize overall dimensions and weights. Generally, the verification is satisfied if the equivalent static load is lower than the corresponding load factor.

The axial load distribution between the two tapered roller bearings depends on the assembly configuration (X or O) and the Y value of the coefficient of each bearing. For this reason, initially, the choice was made on the basis of dimension criteria and then confirmed by checking the resistance.

A limit (that should not be less than the diameter of the inner ring) is the external diameter of the entry splined profile equal to 38 mm.

For both supports, bearing (32310) is suggested, having the following characteristics:

$$\begin{cases} d = 50\,\text{mm} \\ C = 172,000\,\text{N} \\ C_0 = 212,000\,\text{N} \\ Y = 1.7 \\ Y_0 = 0.9 \\ e = 0.35 \end{cases}$$

Referring to the configuration as O and considering that the bearings are all equal, it follows that

$$\begin{cases} \dfrac{R_B}{Y_B} > \dfrac{R_A}{Y_A} \\ \dfrac{1}{2Y}(R_B - R_A) > A \end{cases}$$

According to those reported in the general SKF catalog Pag. 522, it follows therefore that

$$\begin{cases} F_{aB} = \dfrac{R_B}{2Y} = 5423\,N \\ F_{aA} = F_{aB} - A = 3525\,N \end{cases}$$

This allows one to determine the equivalent dynamic load for each support:

$$\begin{cases} \dfrac{F_{aB}}{R_B} = 0.294 < e \Rightarrow P_B = R_B \\ \dfrac{F_{aA}}{R_A} = 0.738 > e \Rightarrow P_A = 0.4R_A + YF_{aA} \end{cases}$$

According to the design data specification, the remaining time must be $L = 1706$ million cycles (App. Chapter 22: 64), and then

$$\begin{cases} C_A = P_A L^{3/10} = 73,695\,N \\ C_B = P_B L^{3/10} = 171,931\,N \end{cases} \Rightarrow C_A < C_B < C.$$

Instead, in the working static condition as

$$\begin{cases} P_{0A} = 0.5R_A + Y_0 F_{aA} = 5,561\,N \\ P_{0B} = 0.5R_B + Y_0 F_{aB} = 14,100\,N \end{cases} \Rightarrow P_{0A} < P_{0B} < C_0.$$

Both verifications are verified.

During the assembly a preload on the bearings equal to half of the axial load will be assigned. For the cylindrical roller support C the equivalent dynamic load is

$$P_C = R_C = 11,293\,N,$$

from which it follows that

$$C_C = P_C L^{3/10} = 105,294\,N.$$

Bearing NU2308EC, chosen from the SKF General Catalog, page 344, has the characteristics necessary to meet the requirements of the project:

$$d = 40\,mm \quad D = 90\,mm \quad C = 112,000\,N \quad C_0 = 120,000\,N.$$

Even the static calculation has been verified:

$$P_{0C} = R_C = 11,293\,N < C_0.$$

22.8 CHOICE OF THE DIFFERENTIAL BEARINGS BOX

The box is a rugged steel enclosure designed to house the final parts of the pivots of satellites with which it forms the arm of differential gearing. Fixed to the group, usually by means of bolts, is the rim of the bevel gear input. In this specific case it is made as a unit, always using a threaded coupling to the closure plate of the epicyclic train arm. On one of the two parts of which the box is composed, the supports of the pins of the train are made. When the reducer is inserted the train and the closure plate receive the motion by the pins.

The choice of the bearings of the rim-epicyclic-box, together with the choice of the bearings of the pinion shaft, is one of the most delicate phases in the design of the entire group differential.

For the calculation of reaction forces, the box is represented as a supported beam subject to the system of forces acting on the shaft duplicating that acting on the pinion (Figure 22.19).

For overall reasons, the following dimensions $a = 185$ mm and $b = 187$ mm are chosen. Again the resolution of the structure must be carried out on two orthogonal planes (Figures 22.20 and 22.21).

The external loads are the same as those already calculated for the pinion shaft, except that the radial and axial action loads are reversed:

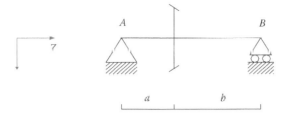

FIGURE 22.19 Scheme of the differential box calculation.

$$\begin{cases} Q = 24{,}036\,\text{N} \\ A = 8{,}541\,\text{N} \\ R = 1{,}898\,\text{N} \\ M = AR_{2\,\text{med}} = 2{,}075{,}236\,\text{N/mm} \end{cases}$$

According to the equations of equilibrium at the rotation, the horizontal and vertical translation are as follows:

$$\begin{cases} R_{Ax} = \dfrac{Qa}{a+b} \\[2mm] R_{Bx} = \dfrac{Qb}{a+b} \end{cases} \quad \begin{cases} R_{Az} = A \\[2mm] R_{Ay} = \dfrac{Rb+M}{a+b} \\[2mm] R_{By} = \dfrac{Ra-M}{a+b} \end{cases}.$$

Finally, the resulting radial and axial reactions of each support are

$$\begin{cases} R_A = \sqrt{(R_{Ax})^2 + (R_{Ay})^2} = 13{,}736\,\text{N} \\[2mm] R_B = \sqrt{(R_{Bx})^2 + (R_{By})^2} = 12{,}821\,\text{N} \\[2mm] R_{Az} = 8541\,\text{N} \end{cases}.$$

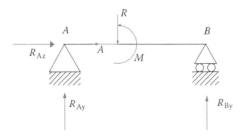

FIGURE 22.20 Loads in ZY plane.

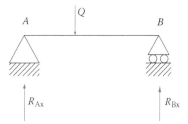

FIGURE 22.21 Loads in ZX plane.

To achieve the configuration of constraints, a ball bearing in A is used (this also resists well with axial loads), and a cylindrical roller bearing in B.

As for tapered roller bearings, so for the ball bearing; first it is necessary to make the choice, according to the dimensional limits, and then calculate the equivalent dynamic load for verification.

From the general catalog SKF, page 196, the characteristics of ball bearing (6228) are

$$d = 140\,\text{mm} \quad D = 250\,\text{mm} \quad C = 165,000\,\text{N} \quad C_0 = 150,000\,\text{N}.$$

Since the value of the axial load is as follows:

$$\frac{A}{C_0} = 0.05694 \Rightarrow 0.24 < e < 0.27,$$

and

$$\frac{A}{R_A} = 0.622 > e \Rightarrow X = 0.56 \quad Y = 1.69,$$

the equivalent dynamic load therefore is

$$P_A = XR_A + YA = 0.56R_A + 1.69A = 22,126\,\text{N}.$$

According to the project data specifications the life of the bearing is $L = 379.1$ million of cycles, thus

$$C_A = 160,137\,\text{N} < C.$$

For the static verification therefore:

$$P_{0A} = 0.6R_A + 0.5A = 12,512\,\text{N} < R_A \Rightarrow P_{0A} = R_A < C_0,$$

consequently a cylindrical roller bearing type NU 227 EC is used for the other support of the box and has the following characteristics:

$$d = 85\,\text{mm} \quad D = 150\,\text{mm} \quad C = 165,000\,\text{N} \quad C_0 = 200,000\,\text{N}.$$

The dynamic equivalent load is

$$P_B = R_B = 12,821 \, \text{N},$$

therefore the load coefficient is

$$C_B = 76,129 \, \text{N} < C.$$

As for the dynamic verification, the static is satisfied as

$$P_{0B} = R_B = 12,821 \, \text{N} < C_0.$$

22.9   DIMENSIONING OF THE PINS

22.9.1   PINS OF THE EPICYCLIC SPEED REDUCER

The pins of the satellites of the epicyclic speed reducer can be summarized as a circular beam supported at the ends and loaded by a concentrated load at the center. To simplify the assembly operation of the system (gear-box-rim) it is assumed that the pin is free to rotate in respect to each of the supports and is coupled with the interference on the hub of its satellite. This involves the installation of an anti-friction bushing on the seat cut out in a semi-box of the differential and another on the closing plate of the arm (carrier assembly).

According to the equilibrium condition of the satellite, a force $F = 2Q$ acts on the pin, whereas Q is the action tangential exchanged between the teeth of the satellites as follows:

$$Q = \frac{C_5}{3R_5}.$$

Making reference to Figure 22.22: a is the width of the closing plate of the rim, $b = 1.25 \, b_{sat}$ is equal to the width of the satellite plus the rate necessary to prevent the sliding between the lateral surfaces of the satellite, the boxes and the plate, and c is the length of the pin inside the box. A small distance satellite-box and satellite-plate are also necessary in order to enable the oil film to arrive early in the bushes. For the symmetry of the structure we can put $a = c$. The dimensioning of the pin provides three types of verification:

- At maximum deflection
- At the specific pressure and power dissipation
- At shear and bending

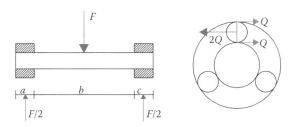

FIGURE 22.22   Scheme for pins of the epicyclic speed reducer.

22.9.2 MAXIMUM DEFLECTION

For the smooth and silent operation of the epicyclic speed reducer it is necessary that the pivots of the satellites are not too deformed under the action of the load. Usually this condition is the most stringent of the three verifications.

Obviously, the calculation (as for the subsequent verification of the specific pressure) must be performed at the more onerous operating conditions, therefore when at the gearing is applied to the maximum project torque.

Having chosen $a = c = 10$ mm, and recalling that the width of the satellite $b_{sat} = 44$ mm, it is possible to define the length of the pin:

$$l = 2a + 1.25b_{sat} = 75 \text{mm}.$$

The deflection at the center is

$$f_{max} = \frac{Fl^3}{48EI},$$

where $I = \pi d^4/64$.

Requiring the maximum deformation value of the deflection $f_{lim} = \min(l/3000,00)$ mm, the minimum diameter of the pin is

$$d_{min} = \sqrt[4]{\frac{64Fl^3}{48\pi E f_{lim}}} = 34.61 \text{mm}.$$

For subsequent verification it is assumed that $d = 36$ mm.

22.9.3 SPECIFIC PRESSURE

To avoid causing interruption or malfunction of support of the lubricant film while working, it is necessary that the specific pressure value be higher than 60–75 MPa:

$$p_s = \frac{F}{(a+c)d} = 58 \text{MPa} < 60 \text{MPa}.$$

This condition is not sufficient to guarantee the regularity of lubrication on the supports. To ensure the efficient disposal of thermal power developed by the sliding of the pin it is necessary that the value W of the dissipated power is less than 40 MPa m/s:

$$W = p_s v_{4\text{rel}} = \frac{d}{2} p_s \omega_{4\text{rel}} = \frac{d}{2} p_s \frac{n_3}{n_4} (\omega_3 - \omega_p) = 10 \text{MPa} \frac{m}{s}.$$

This verification is therefore confirmed.

22.9.4 SHEAR AND BENDING

Contrary to the calculations already performed, verifying the fatigue strength of the pin, it is now necessary to apply an equivalent load (App. Chapter 22 :65) rather than the maximum load.

The loading conditions cause a shear and bending on the pin as diagrams in Figure 22.23. show.

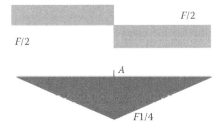

FIGURE 22.23   Diagrams of the qualitative characteristics of the stresses.

Applying the von Mises method at the section that is most stressed (*A*), the equivalent stress is determined as

$$\sigma_{eq} = \sqrt{\sigma_{max}^2 + 3\tau_{max}^2} = 107\,\text{MPa} < \sigma_{amm} = 327\,\text{MPa},$$

whereas:

$$\begin{cases} \sigma_{max} = \dfrac{8Fl}{\pi d^3} = 102\,\text{MPa} \\[2mm] \tau_{max} = \dfrac{8F}{3\pi d^2} = 19\,\text{MPa} \end{cases},$$

and the allowable stress σ_{amm} is defined with the Bach rule.

22.9.5   Pins of the Satellite Differential

The satellites are bound to the differential box and achieve motion through the pins of the satellites, each of which can be designed as a beam fixed at one end and loaded perpendicular to the axis.

In the most severe applications with two satellites, the replacing of the two pins with a single shaft that bears both gears can be considered, but four satellites can achieve a single cross member on the ends of which the four gears are housed.

The configuration of the system suggests that the pin is forced between the two semi-boxes and permits the free rotation of the satellite through the interposition of bushing.

Referring to Figure 22.24, *c* is the thickness of semi-box at the satellite, *a* is the distance where it is supposed to be an applied load, and *b* is the length of the pin coinciding with the width of gear.

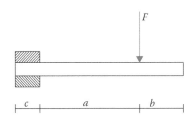

FIGURE 22.24   Scheme for pins of the satellites.

22.9.6 Dimensioning at Deflection

According to the equilibrium of the rotation of the satellite, the load F is equal to

$$F = 2Q = 2C_6 / R_{6med}.$$

The maximum deformation of the pin is evaluated at the section at which it is supposed to apply the load. Assuming that the tangential load Q is applied at the center of the tooth along the direction of the primitive cone, the distance a of the load is equal to the projection along the axis of the gear semi-width:

$$a = \frac{b_{sat} \cos \varphi_6}{2} = 21.47 \, \text{mm}.$$

The project diameter is

$$d = \sqrt[4]{\frac{64Fa^3}{3\pi E f_{amm}}} = \sqrt[4]{\frac{64,000 Fa^2}{9E\pi}} = 21.4 \, \text{mm},$$

where $F = 42,234 \, N$ is calculated from the maximum design load and f_{amm}, and must be less than $0.003 \, a$.

Assuming $d = 26$ mm we can proceed with the following verifications.

22.9.7 Specific Pressure Verification

To ensure that the lubricant film between the pin and bushing will not be broken, specific pressure, corresponding to the support, must be less than 60–75 MPa:

$$p_s = \frac{F}{(a+b) \cdot d} = 37.75 \, \text{MPa}.$$

The thermal power developed following the friction will be effectively dissipated if the index of the power dissipation is more than 40 MPa m/s.

As the motor speed as determined by the data project and therefore chosen for the calculation in the toughest conditions when one wheel of the vehicle is stopped and the other is rotating at a speed twice that of the arm:

$$W = p_s v_{6rel} = \frac{d}{2} p_s \omega_{6rel} = \frac{d}{2} p_s \frac{n_7}{n_6} (\omega_7 - \omega_p) = \frac{d}{2} p_s \frac{n_7}{n_6} \omega_p = 7.2 \, \text{MPa} \frac{m}{s}.$$

Both conditions have been largely verified.

22.9.8 Verification at Shear and Bending

The satellite differential pins are subjected to shear and bending. The corresponding stress diagrams are shown in Figure 22.25.

The ideal stress at the fixed section is

$$\sigma_{ID} = \sqrt{\sigma_{max}^2 + 3\tau_{max}^2} = 414 \, \text{MPa} < 458 \, \text{MPa} = \sigma_{amm},$$

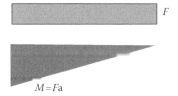

FIGURE 22.25   Stress diagrams.

whereas:

$$\begin{cases} \tau_{max} = \dfrac{4}{3}\dfrac{F}{A} = 79\,\text{MPa} \\[3mm] \sigma_{max} = \dfrac{M}{I}\dfrac{d}{2} = \dfrac{32}{\pi}\dfrac{Fa}{d^3} = 391\,\text{MPa} \end{cases}$$

To determine the allowable stress, the fatigue limit $\sigma_{amm} = \sigma_s/3$, is defined by the criterion of Bach, and added to the duration factor $y_{ST} = 1.4$ of UNI 8862, that takes in to account the value of stress ratio $R = 0$. In fact, during normal operation the pin does not rotate in respect to the load.

The verification performed by applying the load equivalent (App. pag. 619, 620 of this chapter), is therefore verified.

22.10   DIMENSIONING OF THE BOX SCREWS

The box is a rugged steel enclosure that is designed to house the ends of pins with which the satellite forms the arm of differential gearing. It is usually fixed by means of bolts to the rim of the bevel gear input. In this case, due to the presence of the epicyclic reducer, the rim is also fixed to the plate closing of the arm reducer by means of bolts.

The supports of pins of the epicyclic are placed on one of two semi-boxes of which they are composed. The epicyclic, together with the plate of closure, receives the motion by the pins. The two semi-boxes are coupled by tightening screws, usually equal in number and symmetrically arranged, and their axes are perpendicular to those of the pivots of the satellites.

To ensure the transmission, the screws must be proportionate to the semi-box that does not receive the motion of half of the couple at the arm directly. This allows the pins to work symmetrically.

With the aim that the screws are stressed at traction and not at shear, it is necessary to preload the screws to create tangential friction actions on the surface of contact between the two semi-boxes.

The necessary preload to ensure transferability of the friction couple must be

$$F_1 = \frac{C_p}{2fR_m} = 255{,}106\,N,$$

where $C_p = 8800$ Nm is the equivalent torque to the arm, $R_m = 115$ mm is the mean radius of the circle along which the screws are arranged, and $f = 0.15$ is the friction coefficient.

The torque transmitted to the satellites of the differential exists because the pressures develop along the contact surfaces between the pins of the satellites and their supports on the two semi-boxes. The resultant of contact actions on each pin therefore is

$$P = \frac{C_p}{4R_m} = 19{,}132\,N,$$

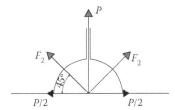

FIGURE 22.26 Scheme for the forces on the differentia box.

as the seats of the pins are placed about equidistant from the screws and the axis of the vehicle wheels.

In reality, the load P is the resultant of the forces F_2 exchanged by each semi-box with the pins, that in consequence give rise to axial actions equal to $P/2$ (assuming that they are arranged symmetrically with each semi-arch of contact (Figure 22.26). These forces tend to keep the semi-boxes away one from the other and therefore are taken into account in the design phase to prevent the normal action of contact between the surfaces. This must be sufficient to transmit the friction torque provided.

Furthermore, it should also be noted that the axial component developed by the planetary during the engaging is equal to the force that tends to drive the semi-boxes apart, generating traction on the screws that is equal to

$$F_3 = \frac{C_{pmax}}{2R_7} tg\vartheta \sin\varphi_7 = 20466\,N.$$

On each of the tightening screws there is a tensile load equal to

$$F = \frac{F_1 + P/2 + F_3}{n},$$

where $n = 12$ is the number of screws.

Assuming, as one might, that the screws are a Saint-Venant solid, it is easy to perform the calculation of the minimum diameter:

$$d_{min} = \sqrt{\frac{4F}{\pi\sigma_{amm}}} = 8.4\,mm,$$

where $\sigma_{amm} = 2\sigma_s/3k_t = 427$ MPa and the allowable tension steel that is heavily stressed 28 MCV 5 (Table XVII page. 38 of Giovannozzi, vol 1).

Since the screw is subject to a pulsing fatigue stress, the allowable stress has been determined according to the Bach criteria and by adding a correction factor equal to $k_t = 1.3$ that the stress concentration factor takes into account.

Finally, the ISO metric thread M10 is chosen because it meets the dimensional requirements.

22.11 DIMENSIONING OF THE SEMI-SHAFT

The entire group differential is supported by a casing that is designed to protect it from mud, water, and dust and permits it to have an oil bath with the various gears. In trucks such a framework will be fixed by bolts to the axle that connects the pairs of wheels. Inside the axle, which together with the differential is called the rear axle, the semi-shafts are coupled at one end to each planet by a splined profile, and at the other end, fixed to the hub of the wheel by means of bolts.

The sliding of the semi-shafts is prevented by a flange that, near the wheel, blocks the external ring of the bearing of the axle at the flange of the axle.

This configuration is designed so that all actions of bending and shear, that the two wheels of the vehicle exchange with each other, act only on the axle, whereas torque acts only on the semi-shafts. Therefore, the minimum diameter of the semi shaft is equal to

$$d_{\text{prog}} = \sqrt[3]{\frac{16M_t}{\pi\tau_{\text{amm}}}} = 49\,\text{mm}.$$

where $M_t = 4400$ Nm and $\tau_{\text{amm}} = 189$ MPa.

The characteristics of the profile that has a diameter closest to the value of the project are:

d	D	Ω	m	k
52	60	0.5	2.1	0.96

The minimum diameter of the previous splined profile shaft is not equal to that of the project and permits a greater torque M_t' therefore it is possible to make a correction of the k parameter at which a reduction of the length l of the splined profile follows as

$$l = \frac{md\Omega}{k'} = 47.6\,\text{mm},$$

whereas:

$$k' = k\,\frac{M_t'}{M_{\text{1eq}}} = 1.147.$$

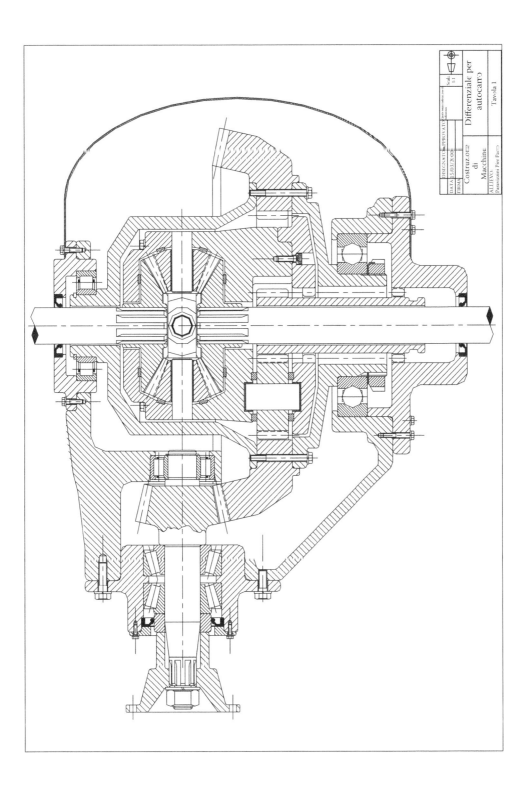

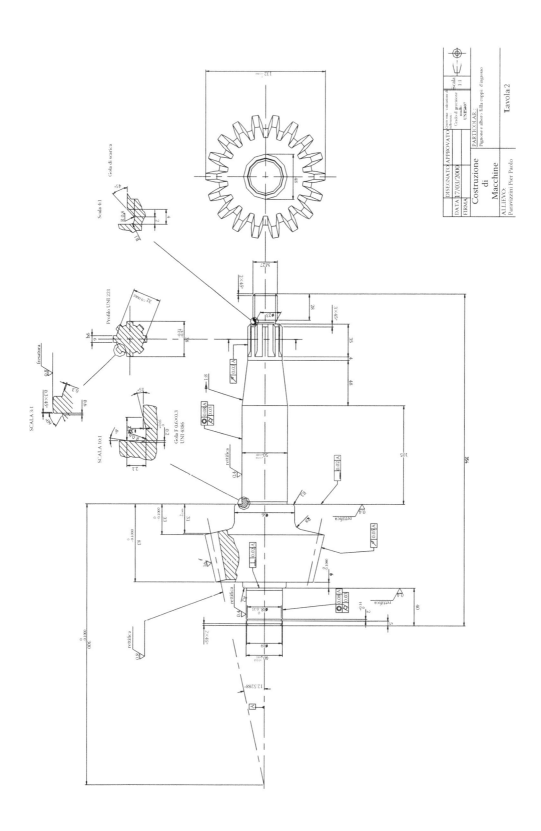

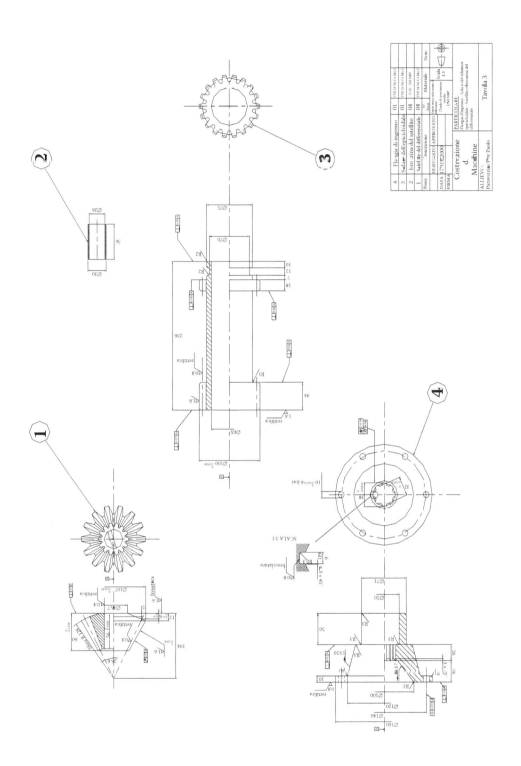

APPENDIX

Project Data Specification

P	ω	C_m	τ_1	$\tau_{p\text{-}c}$	τ_D	$\tau_{\varepsilon\pi}$
165 kW	345 rad/s	478 Nm	1/5,5	1/4,5	0.5	1/1,3

P power
ω angular speed of the motor shaft
C_m torque moment
$\tau_{p\text{-}c}$ bevel gear transmission ratio
τ_D differential transmission ratio
$\tau_{\varepsilon\pi}$ epycicloidal reducer transmission ratio

Distance	Highway	Mountain	City
900.000 km	15%	70%	15%

Itinerary	Name	Type	Percentage
Borgotaro–Pontremoli	Journey 1	Highway	0.25
Torino–Aosta	Journey 2	Highway	0.75
Pontremoli–B.Albiano	Journey 3	Mountain	0.09
Alba–Piana Crixia	Journey 4	Mountain	0.25
Aosta–Pila	Journey 5	Mountain	0.12
Susa–Moncenisio	Journey 6	Mountain	0.08
Pontremoli–P.sso Cisa	Journey 7	Mountain	0.27
Aosta–Lillaz	Journey 8	Mountain	0.16
Dezzo–Presolana	Journey 9	Mountain	0.03
Torino	Journey 10	City	1

Gear1	Gear 2	Gear 3	Gear 4	Gear 5	Gear 6	Gear 7
Pinion	Rim	Int. Rim	Satell. Epic.	Sun Epic.	Satell. Diff.	Sun Diff.

For all mechanical components, apart from the pinion and the bevel gear input, the steel UNI 18NiCrMo5 was used.

E	HV	σ_R	σ_{SN}
220 GPa	7063 MPa	1230 MPa	981 MPa

E modulus of elasticity
HV Vickers hardness
σ_R ultimate traction stress
σ_{SN} allowable traction stress

The UNI 18NiCrMo5 for the construction of the gear must be thermochemically treated (case hardened) at a temperature between 880°C and 930°C to reach the hardness of previous table.

Journey 1: Borgotaro–Pontremoli

Torque (Nm)		Average Torque	Cycles	Cumulate Cycles
–175	–150	–162.5	0	0
–150	–125	–137.5	0	0
–125	–100	–112.5	0	0

Torque (Nm)		Average Torque	Cycles	Cumulate Cycles
−100	−75	−87.5	0	0
−75	−50	−62.5	71.02	71.02
−50	−25	−37.5	332,930.73	333,001.75
−25	0	−12.5	647,867.69	980,869.44
0	25	12.5	27,781.08	1,182,402.60
25	50	37.5	349,648.41	1,154,622.52
50	75	62.5	798,720.87	804,973.11
75	100	87.5	6,227.87	6,252.24
100	125	112.5	34.37	34.37
125	150	137.5	0	0
150	175	162.5	0	0
175	200	187.5	0	0
200	225	222.5	0	0
225	250	237.5	0	0
250	275	262.5	0	0
275	300	287.5	0	0

Journey 2: Torino–Aosta

Torque (Nm)		Average torque	Cycles	Cumulate Cycles
−175	−150	−162.5	0	0
−150	−125	−137.5	0	0
−125	−100	−112.5	0	0
−100	−75	−87.5	0	0
−75	−50	−62.5	34.38	34.38
−50	−25	−37.5	46,170.78	46,205.16
−25	0	−12.5	237,423.84	283,629.00
0	25	12.5	10,20,145.62	1,869,456.25
25	50	37.5	809,977.94	849,310.63
50	75	62.5	38,438.82	39,332.69
75	100	87.5	893.87	893.87
100	125	112.5	0	0
125	150	137.5	0	0
150	175	162.5	0	0
175	200	187.5	0	0
200	225	222.5	0	0
225	250	237.5	0	0
250	275	262.5	0	0
275	300	287.5	0	0

Journey 3: Pontremoli–B. Albiano

Torque (Nm)		Average torque	Cycles	Cumulate Cycles
−175	−150	−162.5	0	0
−150	−125	−137.5	0	0
−125	−100	−112.5	0	0
−100	−75	−87.5	69.09	69.09

(continued)

Torque (Nm)		Average torque	Cycles	Cumulate Cycles
−75	−50	−62.5	584.35	653.44
50	−25	−37.5	83,927.62	84,581.06
−25	0	−12.5	582,280.25	666,861.31
0	25	12.5	397,536.44	1,485,613.14
25	50	37.5	843,513.3	1,088,076.70
50	75	62.5	225,550.28	244,563.40
75	100	87.5	22,847.22	29,013.12
100	125	112.5	4,815.85	6,165.9
125	150	137.5	854.94	1,350.05
150	175	162.5	397.24	495.11
175	200	187.5	51.81	97.87
200	225	222.5	37.42	46.06
225	250	237.5	0	8.64
250	275	262.5	8.64	8.64
275	300	287.5	0	0

Journey 4: Alba–Piana Crixia

Torque (Nm)		Average Torque	Cycles	Cumulate Cycles
−175	−150	−162.5	0	0
−150	−125	−137.5	0	0
−125	−100	−112.5	1,252.79	1,252.79
−100	−75	−87.5	72,628.80	73,881.59
−75	−50	−62.5	285,078.03	358,959.62
−50	−25	−37.5	223,561.23	582,520.85
−25	0	−12.5	329,901.44	912,422.29
0	25	12.5	9,1271.85	1,234,471.29
25	50	37.5	29,0345.81	1,143,199.44
50	75	62.5	312,574.56	852,853.63
75	100	87.5	303,512.03	540,279.07
100	125	112.5	223,540.64	236,767.04
125	150	137.5	1,2236.74	1,3226.4
150	175	162.5	350.02	1,089.66
175	200	187.5	487.81	739.64
200	225	222.5	229.66	251.83
225	250	237.5	7.92	22.17
250	275	262.5	9.5	14.25
275	300	287.5	4.75	4.75

Journey 5: Aosta–Pila

Torque (Nm)		Average Torque	Cycles	Cumulate Cycles
−175	−150	−162.5	18,596.82	18,596.82
−150	−125	−137.5	47,924.39	66,522.22
−125	−100	−112.5	114,342.29	18,0863.5
−100	−75	−87.5	432,038.03	612,901.53
−75	−50	−62.5	130,004.01	742,905.54

Torque (Nm)		Average Torque	Cycles	Cumulate Cycles
−50	−25	−37.5	169,959.02	912,864.56
−25	0	−12.5	118,272.99	1,031,137.55
0	25	12.5	36,592.54	1,126,423.27
25	50	37.5	27,224.29	1,089,830.73
50	75	62.5	95,623.87	1,062,616.44
75	100	87.5	262,567.44	966,992.57
100	125	112.5	242,295.48	704,425.13
125	150	137.5	320,950.16	462,229.65
150	175	162.5	7,2969.04	141,279.49
175	200	187.5	58,232.48	68,310.45
200	225	222.5	9,209.18	10,077.97
225	250	237.5	446.13	868.79
250	275	262.5	244.2	422.66
275	300	287.5	178.46	178.46

Journey 6: Susa–Moncenisio

Torque (Nm)		Average Torque	Cycles	Cumulate Cycles
−175	−150	−162.5	0	0
−150	−125	−137.5	0	0
−125	−100	−112.5	143,703.28	143,703.28
−100	−75	−87.5	669,561.87	813,265.15
−75	−50	−62.5	1,5037.58	828,302.73
−50	−25	−37.5	159,995.02	988,297.75
−25	0	−12.5	67,754.36	105,6052.11
0	25	12.5	18,964.4	1,104,671.71
25	50	37.5	39,487.36	1,085,707.31
50	75	62.5	104002.84	1046229.95
75	100	87.5	109,293.39	942,227.11
100	125	112.5	333,505.56	832,923.72
125	150	137.5	498,638.88	499,418.16
150	175	162.5	706.22	779.28
175	200	187.5	73.06	73.06
200	225	222.5	0	0
225	250	237.5	0	0
250	275	262.5	0	0
275	300	287.5	0	0

Journey 7: Pontremoli–P.sso Cisa

Torque (Nm)		Average Torque	Cycles	Cumulate Cycles
−175	−150	−162.5	0	0
−150	−125	−137.5	48.04	48.04
−125	−100	−112.5	4,228.96	4,267
−100	−75	−87.5	262,900.62	267,167.62
−75	−50	−62.5	103,166.09	370,333.71
−50	−25	−37.5	280,986.6	651,320.31

(continued)

Torque (Nm)		Average Torque	Cycles	Cumulate Cycles
−25	0	−12.5	277,452.28	928,772.59
0	25	12.5	61,788.27	1,208,313.36
25	50	37.5	220,247.86	1,146,613.09
50	75	62.5	360,122.19	936,365.23
75	100	87.5	232952.13	576244.04
100	125	112.5	303,786.91	343,291.91
125	150	137.5	39,189.7	39,505.00
150	175	162.5	262.75	315.3
175	200	187.5	40.54	52.55
200	225	222.5	0	12.01
225	250	237.5	0	12.01
250	275	262.5	7.51	12.01
275	300	287.5	4.5	4.5

Journey 8: Aosta–Lillaz

Torque (Nm)		Average Torque	Cycles	Cumulate Cycles
−175	−150	−162.5	885.73	885.73
−150	−125	−137.5	16,030.44	16,916.17
−125	−100	−112.5	3,854.95	2,0771.12
−100	−75	−87.5	257,188.22	277,959.34
−75	−50	−62.5	152,279.09	430,238.43
−50	−25	−37.5	250,286.86	68,0525.29
−25	0	−12.5	264,807.58	945,332.87
0	25	12.5	69,473.06	1,189,849.03
25	50	37.5	169,477.07	112,0375.97
50	75	62.5	350,787.44	950,898.9
75	100	87.5	256,580.97	600,111.46
100	125	112.5	228,378.39	343,530.49
125	150	137.5	83,262.33	1,15152.1
150	175	162.5	24,978.36	31,889.77
175	200	187.5	6,522.22	6,911.41
200	225	222.5	275.11	389.19
225	250	237.5	83.88	114.08
250	275	262.5	20.13	30.2
275	300	287.5	10.07	10.07

Journey 9: Dezzo–Presolana

Torque (Nm)		Average Torque	Cycles	Cumulate Cycles
−175	−150	−162.50	74,433.15	74,433.15
−150	−125	−137.50	149,950.08	224,383.23
−125	−100	−112.50	39,531.43	263,914.66
−100	−75	−87.50	186,914.13	450,828.79
−75	−50	−62.5	229,549.39	680,378.18
−50	−25	−37.5	180,622.94	861,000.12
−25	0	−12.5	107,788.13	968,788.25
0	25	12.5	34,659.71	1,145,896.97
25	50	37.5	49,106.52	1,111,237.26
50	75	62.5	163,271.83	1,062,230.74

Torque (Nm)		Average Torque	Cycles	Cumulate Cycles
75	100	87.5	263104.47	898858.91
100	125	112.5	220,405.89	635,754.44
125	150	137.5	198,993.47	415,348.55
150	175	162.5	45,644.95	226,355.08
175	200	187.5	109,682.09	170,710.13
200	225	222.5	60,028.44	61,028.04
225	250	237.5	946.99	999.60
250	275	262.5	52.61	52.61
275	300	287.30	0	0

Journey 10: Torino

Torque (Nm)		Average torque	Cycles	Cumulate Cycles
−175	−150	−162.5	0	0
−150	−125	−137.5	0	0
−125	−100	−112.5	0	0
−100	−75	−87.5	132.85	132.85
−75	−50	−62.5	5,782.84	5,915.69
−50	−25	−37.5	96,198.31	10,2214
−25	0	−12.5	610,566.10	712,680.12
0	25	12.5	263,470.80	1,462,878.64
25	50	37.5	420,943.80	1,199,407.83
50	75	62.5	499,035.60	778,464.05
75	100	87.5	153,995.00	279,428.49
100	125	112.5	77,239.96	125,433.49
125	150	137.5	29,234.39	48,193.53
150	175	162.5	12,855.89	18,959.14
175	200	187.5	3,078.99	6,103.25
200	225	222.5	1,195.64	3,024.26
225	250	237.5	1,258.16	1,828.62
250	275	262.5	343.84	570.46
275	300	287.5	226.62	226.62

Cumulate Highway

Torque (Nm)		Average torque	Journey 1		Journey 2		Cumulate Type
da	A		Cumulate cycle	Percentage	Cumulate cycle	Percentage	
−175	−150	−162.5	0	0.25	0	0.75	0
−150	−125	−137.5	0	0.25	0	0.75	0
−125	−100	−112.5	0	0.25	0	0.75	0
−100	−75	−87.5	0	0.25	0	0.75	0
−75	−50	−62.5	71.02	0.25	34.38	0.75	45.54
−50	−25	−37.5	333,001.75	0.25	46,205.16	0.75	117,904.3075
−25	0	−12.5	980,869.44	0.25	283,629	0.75	457,939.11
0	25	12.5	1,182,402.60	0.25	1,869,456.25	0.75	1,697,692.838
25	50	37.5	1,154,622.52	0.25	849,310.63	0.75	925,638.3525
50	75	62.5	804,973.11	0.25	393,32.69	0.75	230,742.795
75	100	87.5	6,252.24	0.25	893.87	0.75	2,233.4625
100	125	112.5	34.37	0.25	0	0.75	8.5925
125	150	137.5	0	0.25	0	0.75	0

(continued)

Torque (Nm)		Average torque	Journey 1		Journey 2		Cumulate Type
da	A		Cumulate cycle	Percentage	Cumulate cycle	Percentage	
150	175	162.5	0	0.25	0	0.75	0
175	200	187.5	0	0.25	0	0.75	0
200	225	222.5	0	0.25	0	0.75	0
225	250	237.5	0	0.25	0	0.75	0
250	275	262.5	0	0.25	0	0.75	0
275	300	287.5	0	0.25	0	0.75	0

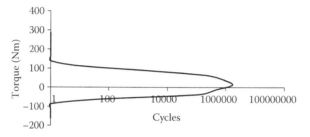

Cumulate Mountain

Torque (Nm)		Average Torque	Cumulate Type
−175	−150	−162.5	4,606.3297
−150	−125	−137.5	17,433.6001
−125	−100	−112.5	45,905.9889
−100	−75	−87.5	287,229.6267
−75	−50	−62.5	434,451.1937
−50	−25	−37.5	652,420.6088
−25	0	−12.5	927,429.2712
0	25	12.5	122,6864.895
25	50	37.5	1,123,545.843
50	75	62.5	883,262.0406
75	100	87.5	607,666.9172
100	125	112.5	377,637.9315
125	150	137.5	140,400.2578
150	175	162.5	29,011.0027
175	200	187.5	14,638.1351
200	225	222.5	3,172.8136
225	250	237.5	162.0584
250	275	262.5	64.7123
275	300	287.5	25.4289

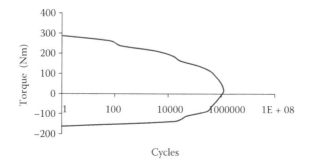

Cumulate City

Torque (Nm)		Average Torque	Journey 10	Cumulate Type	
da	*a*		Cumulate Cycle	Percentage	
−175	−150	−162.5	0	1	0
150	125	−137.5	0	1	0
−125	−100	−112.5	0	1	0
−100	−75	−87.5	132.85	1	132.85
−75	−50	−62.5	5,915.69	1	5,915.69
−50	−25	−37.5	102,214.00	1	102,214.00
−25	0	−12.5	712,680.12	1	712,680.12
0	25	12.5	1,462,878.64	1	1,462,878.64
25	50	37.5	1,199,407.83	1	1,199,407.83
50	75	62.5	778,464.05	1	778,464.05
75	100	87.5	279,428.49	1	279,428.49
100	125	112.5	125,433.49	1	125,433.49
125	150	137.5	4,8193.53	1	48,193.53
150	175	162.5	18,959.14	1	18,959.14
175	200	187.5	6,103.25	1	6,103.25
200	225	222.5	3,024.26	1	3,024.26
225	250	237.5	1,828.62	1	1,828.62
250	275	262.5	570.46	1	570.46
275	300	287.5	226.62	1	226.62

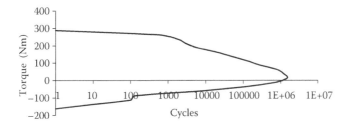

Total Cumulate Correct by Project Data Specifications

Torque	Motorway	Mountain	City'	Cumulate
Media	Cumulate Cycle	Cumulate Cycle	Cumulate Cycle	
−207.025	0	13,233,063.96	0	13,233,063.96
−175.175	0	50,083,246.37	0	50,083,246.37
−143.325	0	131,878,724.90	0	131,878,724.90
−111.475	0	825,124,543.60	32255.98	825,156,799.60
−79.625	13055.469	1248091389	1436329.532	1249540774
−47.775	35353606.6	1874273925	24793279.2	1934420811
−15.925	137313042.1	2664318810	173038733.1	2974670586
15.925	50905397.3	3495809472	355186933.8	4360049603
47.775	277552660	3227722497	291226222.1	3796491378
79.625	69188227.08	2537435190	189011071.3	2795634489
111.475	669703.7306	1745705520	67845237.37	1814220461
143.325	2576.461125	1084878250	30455251.37	1115336077
175.175	0	403341860.6	11701389.08	415043249.7
207.025	0	83342808.56	4603279.192	87946087.75
238.875	0	42052434.52	1481869.1	43534303.62

(continued)

Torque	Motorway	Mountain	City'	Cumulate
Media	Cumulate Cycle	Cumulate Cycle	Cumulate Cycle	
270.725	0	9111868.91	724200.328	9849149.238
302.575	0	465561.3715	443988.936	909550.3075
334.425	0	185905.4954	138507.688	324413.1834
366.275	0	73052.14392	55023.336	128075.4799

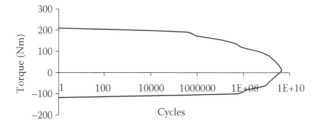

Cumulate of Project at the Pinion of the Bevel Gearing

Torque (Nm)			Cycles	$p_{i\text{-bending}}$	$p_{i\text{-wear}}$
0	175	88	1705902918.24		
175	350.73	263.04	1291467909.44		
350.73	526.09	438.41	887287207.97		
526.09	701.45	613.77	562813130.24		
701.45	876.81	789.13	265603967.62		
876.81	1052.18	964.50	99118111.92		
		1076.13	50000000.00		0.56087036
1052.18	1227.54	1139.86	22956481.99		0.37912964
		1291.28	3000000.00	0.999323504	0.05995941
1227.54	1402.90	1315.22	2029.49	0.000514398	3.08639E-05
1402.90	1578.27	1490.58	486.29	0.000138599	8.31591E-06
1578.27	1753.63	1665.95	70.50	1.55674E-05	9.34044E-07
1753.63	1928.99	1841.31	23.79	4.79271E-06	2.87563E-07
1928.99	2204.36	2016.67	9.42	3.13893E-06	1.88336E-07
		F_{teq} (N)		24035.59	20347.37

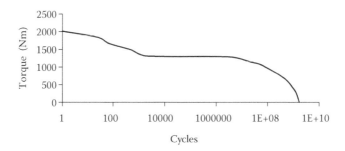

Cumulate of Project at the of the Epicyclic Reducer

Torque (Nm)			Cycles	$p_{i\text{-Flessione}}$	$p_{i\text{-Usura}}$
0.00	236.74	118.370	657923992.8		
236.74	473.48	355.110	544911675.1		
473.48	710.22	591.850	404307922.9		

Torque (Nm)			Cycles	$p_{i\text{-Flessione}}$	$p_{i\text{-Usura}}$
710.22	946.96	828.590	274573962.5		
946.96	1183.70	1065.330	129954050.4		
0.00	0.00	1297.503	50000000		0.031451184
1183.70	1420.44	1302.070	48427440.8		0.968548816
1420.44	1657.18	1538.810	10314634.32		0.116465642
1657.18	1893.92	1775.550	4491352.239		0.029827045
0.00	0.00	1875.913	3000000	0.675499901	0.040529994
1893.92	2230.66	2012.280	973500.2062	0.324500099	0.010470006
2230.66	2367.40	2249.029	49723.14892	0.009955923	0.000597355
2367.40	2604.14	2485.769	19855.3799	0.004017838	0.00024107
2604.14	2840.88	2722.509	7801.865234	0.002600622	0.000156037
		F_{teq} (N)		14553.30595	11038.12004

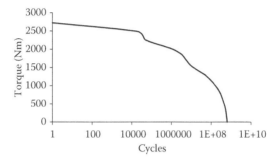

Cumulate of Project at the Satellite of the Differential

Torque (Nm)	Cycles	$p_{i\text{-Flessione}}$	$p_{i\text{-Usura}}$
24.66	1253188552		
32.06	1227313754		
73.98	982266336.2		
96.18	878005452		
123.30	735133305		
160.29	579314667		
172.62	530162247.6		
222.94	383590368.7		
224.41	376603613.2		
271.26	229350998.4		
288.53	176552087		
320.59	120559167.5		
352.64	65486059.09		
369.91	51415750.55		
371.68	50000000	0.719450829	
416.76	14027458.56	0.006238636	
419.23	13715526.77	0.122518116	
468.55	7639620.954	0.03040653	
480.88	6119294.429	0.055751435	
517.87	3331722.658	0.006634453	

(continued)

Torque (Nm)	Cycles	$p_{i\text{-Flessione}}$	$p_{i\text{-Usura}}$
522.37	3000000	0.556	0.033380964
545.00	1330951.784	0.147	0.008794581
567.19	891222.7227	0.275	0.016498504
609.11	66297.53189	0.013	0.000796474
673.23	26473.83987	0.005	0.000322427
737.35	10402.48698	0.003	0.00020805
	F_{tcq} (N)	15716.97823	11767.659436

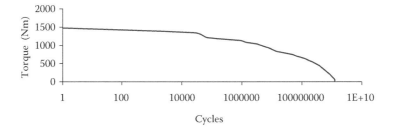

Problems

PART I

CHAPTER 1

Questions

1. What are the principal phases of the product design and development process? (Find the answer in Section 1.2.)
2. What are the principal activities used during the design phase and at which level of the product defining do they come out? (Find the answer in Section 1.3.)
3. What are the differences between the sequential model and the simultaneous/integrated model of the product development process? Which factors push toward the diffusion of the latter? (Find the answer in Section 1.5.)
4. What is concurrent engineering? What are its characteristic points? (Find the answer in Section 1.5.)
5. What is life cycle design? What are its characteristic points? What is the relation to concurrent engineering? (Find the answer in Section 1.6.)
6. What is the DFX approach? (Find the answer in Section 1.7.)
7. What are the better known DFXs? (Find the answer in Section 1.7.1.)
8. What is knowledge-based engineering? (Find the answer in Section 1.9.1.)
9. What is total quality management in product design? (Find the answer in Section 1.9.2.)
10. What is reverse engineering? How could it be integrated into the product development process? (Find the answer in Section 1.9.3.)

CHAPTER 2

Questions

1. What is design for environment? What are its goals? What are the peculiar characteristics in the approach to design? (Find the answer in Section 2.3.)
2. What are the most common rules in the design? (Find the answer in Section 2.5.)
3. What are the principal phases of the life cycle of a product? (Find the answer in Section 2.6.)
4. Which are the principal factors that determine the impact of product life-cycle phases on the environment? (Find the answer in Section 2.7.)
5. What is the life cycle assessment (LCA)? What are its goals? (Find the answer in Section 2.9.)
6. What are the steps for the LCA application procedure? (Find the answer in Section 2.9.1.)
7. What are eco-indicators? (Find the answer in Section 2.9.2.)
8. Which are the principal environmental strategies to be applied in the design for environment process? (Find the answer in Section 2.13.)
9. Which are the main recovery strategies at the end of useful life? How could their potential for environmental benefit be differentiated? (Find the answer in Section 2.13.2.)
10. What is the role of the DFX in the design for environment process? (Find the answer in Section 2.15.)

PART II

CHAPTER 3

Questions

1. What is the percentage of carbon in cast iron and the percentage of carbon in steel?
2. What are the most used cast irons in mechanical construction and why?
3. Which are the chemical elements that give heat resistance to stainless steel?
4. What is the characteristic chemical element in spring steel alloys? What is the effect of this element on the mechanical characteristics of these steels?
5. What are the two chemical base elements of bronze? What is the chemical element capable of increasing their antifriction and workability properties?
6. According to the temperature effects, how can polymeric materials be classified? What are the differences in terms of workability?
7. One parameter used to qualify a material is the ratio between the failure tensile stress and the specific gravity (specific strength). Make a confrontation of this parameter between an AISI 4340 steel, a high-strength carbon fiber and an aluminum alloy AL7178-T6.

CHAPTER 4

4.1   Where M_t is the torque and L is the length, calculate the diameter and evaluate the *material quality index* for a cylindrical twisting shaft if the target is to minimize the mass.

4.2   For a specimen beam of length $L = 175$ mm (see Figure A.4.1), choose the positions of the constraints and the load system in order to have a pure and symmetrical bending on the central zone of the beam. Evaluate the applied load to have a maximum stress σ equal to 200 N/mm².

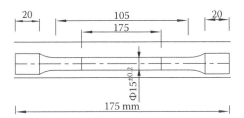

FIGURE A.4.1

4.3   For a cylindrical fixed beam of length $l = 500$ mm, and diameter $d = 10$ mm, as in Figure A.4.2, choose a load system and the constraints necessary to have a pure torsion on the beam. Evaluate the applied load necessary to have a maximum value of the angular deformation θ = 1°. Assume the shear modulus $G = 83.000$ N/mm².

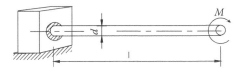

FIGURE A.4.2

4.4   Consider a torsion bar of 0.45 m in length with a circular section (the diameter is 22 mm) as reported in Figure A.4.3; r is 150 mm. Determine the maximum deflection in the P direction that arises as a result of the load P equal to 1.2 kN. Neglect the bending deflection of the connecting rod. The modulus of elasticity $E = 200.000$ N/mm² and the shear modulus $G = 70000$ N/mm².

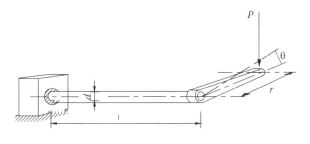

FIGURE A.4.3

4.5   The cylindrical bar constrained in Figure A.4.3 has a length $l = 1.2$ m and a diameter $d = 0.03$ m. For a test, a load $P = 200$ N was applied at the end of the bar as in figure ($r = 0.30$ m). It was found that the ratio between the angular torsion Θ and the angular deflection α was 0.62. Calculate the modulus of Poisson $\nu = 1/m$.

4.6   A test for the Brinell hardness (HB) of 10 specimens of a steel gave the results: 195, 203, 205, 198, 198, 201, 203, 200, 197, and 197, respectively. Calculate the average value and the standard deviation. Estimate the value of the ultimate stress R_m of the steel.

4.7   Consider a metal sheet obeying the von Mises criterion, with the following yield stress: $\sigma_{Eq-0} = 65$ MPa and the following constitutive curve: $\sigma_{Eq}(\varepsilon_{Eq}) = 65 + 200 \cdot \varepsilon_{Eq}^{0.3}$. As in Figure A.4.4, the sheet is gradually loaded by two uniform tensile stresses, in the x direction up to 120 MPa and in the y direction up to 150 MPa;

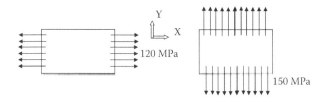

FIGURE A.4.4

Determine the equivalent plastic strain of the plate at the end of the two stress sequences. The elastic strains can be neglected and the plastic strains can be assumed to be infinitesimal.

4.8   As in the previous problem, make the scheme of the yield surfaces during the transformation.

CHAPTER 5

5.1   A steel bar with a diameter $d = 3$ mm is fixed at both ends. The yield tensile stress of the steel is $\sigma_{sn} = 600$ N/mm². What temperature is needed to heat the bar if the stress is equal to the yield tensile stress?

If the temperature variation is 50°C above the value calculated, what could happen to the steel or a material with an equal thermal dilatation coefficient value (α) if perfectly fragile?

What will be the load F on the bar when the temperature variation is 150°C?

Assume that $a = 1.3 \; 10^{-5}/1°C$.

5.2 A steel belt lies on a drum as in Figure 21.26 (page 559). The yield stress of the steel is 1300 N/mm².

The thickness of the belt is 3 mm and the width is 30 mm. The diameter of the drum is 500 mm.

Calculate the stress on the belt and how much the thickness of the belt must change in order to have a stress less of 20%.

5.3 A steel shaft has a constant diameter and is mounted on bearings at both ends (Figure A.5.1). On the gear acts a chain and the resultant force is $F = 1000$ N. Neglecting the shaft weight, calculate the diameter of the shaft. Choose an adequate material for the shaft, considering that the gear is tooled on it.

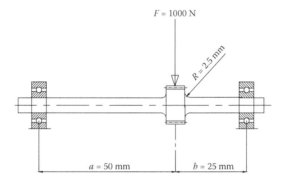

FIGURE A.5.1

5.4 A steel shaft with a diameter $d = 150$ mm and a length $l = 2000$ mm transmits a power P. To reduce the weight, a hollow shaft is adopted with a ratio $\beta = D/d_i = 1.4$ between the external diameter D and the internal diameter d_i. What is the reduction in weight ΔP if the shaft must transmit the same power at the same angular velocity and with equal torsion stress?

5.5 For a normalized AISI 1040 connecting rod of length $l = 160$ mm, the minimum slenderness ratio λ is 45. Evaluate the Eulerian load and the corresponding axial stress, knowing that the average value of the area S of the normal section is 120 mm². Put this value into confrontation with the load F corresponding to the yield stress of the material.

5.6 Evaluate the shear stress for the pin that links the connecting rod of the previous problem to the crankshaft when the load is equal to yield load. The diameter d of the pin is 55 mm, the length l is 35 mm, and the material is a type of steel. Calculate also the axial stress σ when the temperature of the crankpin increases of $\Delta T = 200°C$ under the hypothesis that the ends of the pin are fixed. Assume that $\alpha = 1.3 \; 10^{-5} \; 1/1°C$ and $E = 210.000$ N/mm².

5.7 Calculate the shear stress for the crankpin in the previous problem when the connecting rod is in an orthogonal position with the crank. Assume that the load on the pin is 0.75F, the crank radius r is 60 mm, and the diameter d of the crankpin is 40 mm.

5.8   The equivalent flywheel has a mass moment of inertia $I = 2 \times 10^{-3}$ kg m². The flywheel is assembled to the motor by a shaft (Figure A.5.2) that runs at an angular velocity $\omega = 10$ rad/s. The mass moment of inertia of the motor is $I_m = 2000$ kg mm². The steel shaft length is $l = 700$ mm and the diameter is $d = 15$ mm.

Determine the tangential stress when the motor stops suddenly.

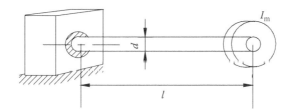

FIGURE A.5.2

CHAPTER 6

6.1   Design the Smith and the Haigh diagram for a steel knowledge where:

- $\sigma_{sn} = 380$ MPa
- $\sigma_0 = 250$ MPa
- $\sigma_{0'} = 350$ MPa
- $\sigma_r = 500$ MPa

Design the diagram for specimens for which the roughness is 0.02 mm.

6.2   Using the same material as Problem 6.1, where:
Constant stress $\sigma_m = 80$ MPa and $\Delta\sigma = 20$ MPa, find the assurance coefficient m.
Assume a roughness of $h = 0,003$ mm and a coefficient of $\beta_i = 1,3$.

6.3   Project a fatigue traction test (load ratio $R = -1$) and evaluate the cost for the dynamic test. Report the Wöhler curve of the material.

Data:
Hourly cost of staff: €70/h
Machine cost: €100,000 with a straight-line depreciation over 20 years
Unit cost of specimen: €30
Business hours: 8 h/d
Profit margin: 0.2
Frequency f of the machine: 50 Hz
The total number of specimens: 15

6.4   For an AISI 4130 normalized (tensile strength 430 MPa, yield strength 317 MPa), project the fatigue test and determine the fatigue limit together with the fatigue curve by means of the Risitano method.

The tests are performed under identical conditions to the previous problem and the area of the specimens is 80 mm².

6.5   Design the railway axle shown in Figure A.6.1 under a load of 25 t/axle for a residual life of 10^5 cycles. Assuming that it has been used for 50×10^3 cycles under the load of 25 t/axle, determine the residual life of the railway axle after applying a load of 20 t/axle ($q = 0.8$).

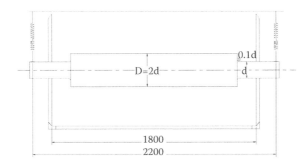

$D=2d$ 0.1d d

1800
2200

FIGURE A.6.1

6.6   The fatigue testing machine in Figure A.6.2 produces an alternate bending on a steel strip having a yield tensile stress of 900 MPa (σ_y). The strip is bent on rollers of diameters 250 and 500 mm, as shown in Figure A.6.2. Using Bach's criterion to estimate the approximate resistance to oscillation (σ_0) as well as the resistance from origin (s01), and assuming for simplicity that the axial tensile load T is only applied in the area of rules (right side of unloaded strip), determine the safety factor using Haigh's diagram.

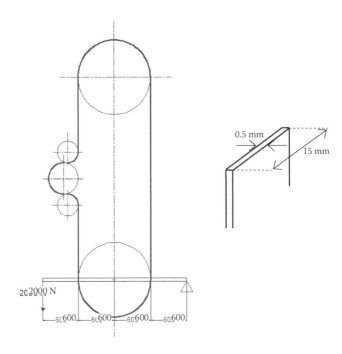

0.5 mm   15 mm

20 2000 N

6 600   6 600   60 600   60 600

FIGURE A.6.2

6.7   The camshaft shown in Figure A.6.3 brings into rotation a cam that drives the opening of a valve by compressing a spring. Determine the safety factor of the shaft, assuming that the minimum and maximum spring compressions are 2 mm and 8 mm, respectively, $K = 1$ kN/mm, $q = 0.8$.

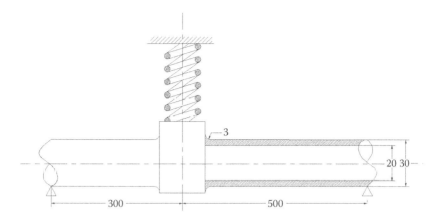

FIGURE A.6.3

CHAPTER 7

7.1 Let us examine the tubes ($p = 50$ MPa) of a shell-and-tube heat exchanger. The inner diameter is 14.85 mm, the thickness is 2.1 mm. Considering only the three materials reported in the table, verify that each one satisfies the primary constraint of structural resistance and use the *normalized and weighed properties method for selection* to find the best solution, taking into account the thermal conductivity, the minimum price, and the minimum weight as objective-properties.

	Thermal Conductivity λ [W/m K]	Density ρ [kg/m³]	Price C_m [€/kg]	Elastic Limit σ_{el} [MPa]
1 – Low alloyed steel	55	7900	0.90	720
2 – Molybdenum	147	10300	24	320
3 – Tungsten	175	19350	15	1400

7.2 With regard to data given in Problem 7.1, define the best material, using the application of the *limits and targets on properties method,* in which the target property is the structural strength and the limits for the other properties are:

$\lambda > 50$ W/m K;
$\rho < 20000$ kg/m³; and
$C_m < 30$ €/kg.

7.3 Referring to Problem 7.1, determine the performance indices of the material when the targets are the minimum weight and the minimum cost of tubes, and when the respect of strength condition is imposed as a constraint. Consider the thickness of the tubes, in addition to the material, as free variable

7.4 Referring to Problem 7.3 and to the results obtained in minimizing the weight of the tubes, use the diagram in Figure 7.5 to represent graphically the performance index. Analyze the best materials required to solve the problem.

7.5 Referring to Problem 7.1, determine the performance index of the material when the target is the maximum tube thermal exchange. The choice has to be made according to the minimum weight and subsequently the minimum cost of the tubes. Impose a strength constraint (security condition for tubes at normal working pressure). Consider the thickness of the tubes, in addition to the material, as free variable.

7.6 Referring to the nomenclature of Section 13.4.2, determine the material performance index for a cylindrical helical spring for which the goal is the minimum weight, when the wire's diameter is a free variable. The spring parameters are: radius r, helical angle α, pitch p, and spire number i. The load acting is F. Impose the fulfilment respect of spring strength as constraint.

7.7 Referring to the data of Problem 7.6, determine the material performance index in the case of maximal axial deformation without contact between the spires. Assume the same characteristics and conditions as in the Problem 7.6.

Questions

1. What are the principal approaches for the best material selection? What are the differences between them? (Find the answer in Section 7.4)
2. In which form should we translate the design problem before operating an integrated materials selection? (Find the answer in Section 7.6.1)
3. How is it possible to make a graphic material selection by using the performance index? (Find the answer in Section 7.6.2)

PART III

CHAPTER 8

8.1 A shaft transmits power $P = 6$ kW at 2000 rpm. The minimum diameter d of the shaft is 20 mm.

Knowing that the axial stresses are negligible, determine the factor of safety m applying the maximum shear stress theory, the maximum normal strain theory, and distortion energy theory (the von Mises theory).

The yield traction stress is $\sigma_{sn} = 200$ MPa, equal to the compression stress.

8.2 In Figure A.8.1 the most stressed part of a shaft is shown on which a torque $M = 200$ Nm is acting. The maximum diameter D of the shaft is 70 mm. The minimum diameter is $d = 40$ mm. The radius r of the notch is 10 mm. If the steel used for the shaft is an AISI 1040 (tensile ultimate stress $S_u = 590$ N/mm^2), estimate the notch coefficient β_i and the maximum shear stress τ_{max}. Consider the other coefficients ($c_1 = c_2 = 1$).

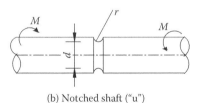

(b) Notched shaft ("u")

FIGURE A.8.1

8.3 The shaft of Figure A.8.2 moves an eccentric with a deformation $f = 5$ mm that loads a spring for which $d = 5$ mm, $R = 10$ mm, $i_u = 6$, $E = 200$ GPa, and $v = 0.3$. The torque applied to the shaft is $M_t = 50$ Nm. Supposing the spring is in the rest condition when the radius of the eccentric is at a minimum, design the circles of Mohr, find the principal stresses, and calculate the ideal stress using the von Mises hypothesis in the fillet zone of the shaft.

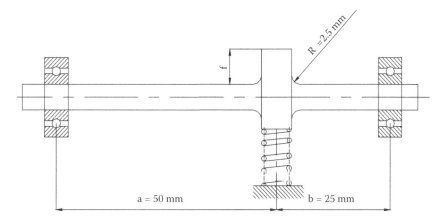

FIGURE A.8.2

8.4 Given a stress state plan with parameters $\sigma_x = 50$ MPa; $\sigma_y = -10$ MPa; $\tau_{xy} = 40$ MPa (as shown in Figure A.8.3), the student will find the value of the ideal σ_i with:

A. The hypothesis of maximum shear stress
B. The hypothesis of deformation energy
C. The hypothesis of energy of distortion

Knowing that the value of m must be assumed to be 10/3.

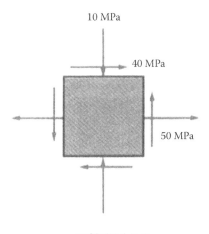

10 MPa

40 MPa

50 MPa

FIGURE A.8.3

8.5 Put in confrontation the Tresca criterion and the von Mises criterion for the stress state defined by $\sigma_r = 450$ MPa and $\tau_{xy} = 200$ MPa.

CHAPTER 9

9.1 A steel roller ($E_1 = 200000$ MPa, $m_1 = 10/3$, $\gamma_1 = 7.81 \cdot 10^{-6}$ kg/mm³) presses on a plastic material strip ($E_2 = 2000$ MPa, $m_2 = 2$, $\sigma_{yield} \cong 80$ MPa) that is subjected to a tensile axial load of 4 kN by means of moving rollers (Figure A.9.1). Assuming that the rollers allow for sliding along the x-axis only, determine the maximum stress on the strip and evaluate the safety factor.

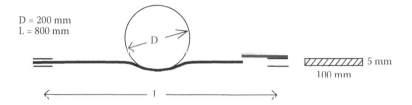

D = 200 mm
L = 800 mm

D

5 mm
100 mm

I

FIGURE A.9.1

9.2 An aluminum spherical indenter ($E_1 = 70$ GPa, $m_1 = 4$) is pressed on a flat plastic surface ($E_2 = 2$ GPa, $m_2 = 2$) leaving a circular mark of radius 0.5 mm (Figure A.9.2). Determine the contact force, the maximum stress, as well as the approach produced.

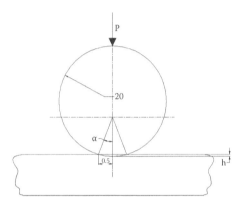

P

20

α

0.5

h

FIGURE A.9.2

9.3 It has a sphere bearing (Figure A.9.3) that must withstand a load P equal to 2000 N. Knowing that the number of spheres z in the buffer is 12, determine the maximum load bearable by the sphere. Assuming, for the bearing, the dimensions of the previous problem, determine the medium stress σ_m and the maximum stress σ_{max}.

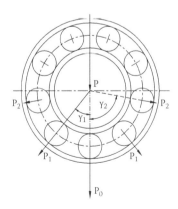

P

P_2

P_2

Y_2

Y_1

P_1

P_1

P_0

FIGURE A.9.3

9.4 For the bearing in Figure A.9.4, determine the maximum load on the ball, the relative curvature ρ, the length of the ellipse of the contact, and the maximum stress.

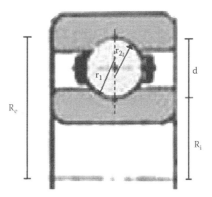

FIGURE A.9.4

Known:

$d =$	14	mm
$r_1 =$	7,7	mm
$R_i =$	26	mm
$R_e =$	40	mm
$P_{trad} =$	9000	N
$z =$	10	
$m_1 = m_2$	3,33	
$E =$	210	GPa

9.5 For the bearing in Figure A.9.5, determine the maximum load on the ball, the relative curvature ρ, and maximum stress of the ball.

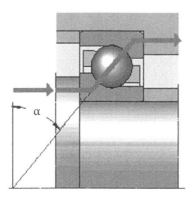

FIGURE A.9.5

Known

$D =$	14	mm	
$R_i = r_e =$	7,7	mm	
$R_m =$	38	mm	
$P_t =$	8000	N	
$Z =$	8		
$\alpha =$	30°	0,52	rad
$m_1 = m_2$	3,3		
$E =$	210000	GPa	

9.6 A steel sphere placed on a plane of the same material is loaded by a force $P = 3000$ N in an orthogonal direction to the plane. The radius R of the ball is 20 mm. Evaluate the maximum Hertzian stress σ_{max} and the value of the approach α. Assume that $E = 210000$ N/mm^2.

CHAPTER 10

10.1 Given the following values, calculate the average diameter d and the lost power:

$P = 1000$ N
$N = 500$ rpm 52,3333 rad/s
T_m of the oil film $= 70°C$
Lubricant SAE-40
$l/b = 1$
$h_m = 1$ μm average roughness
$h_o = 3$ μm minimum film thickness
$\varphi = 0,9$ empty ratio
$i = 8$ number of pads

FIGURE A.10.1

10.2 Using the data below, calculate the minimum film thickness h_0, the maximum value of the pressure P_{max}, the load per unit of projected bearing area P, the flow of oil Q_t, and the power lost due to friction P_a.

l	100	mm
d	200	mm
c	0.26	mm
n	30	g/s
W	10	kN
T_m	55	°C
Lubricant	SAE 30	

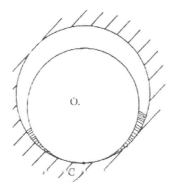

FIGURE A.10.2

10.3 Using the data from the previous problem, calculate the minimum film thickness h_0 according to Bosch indications for $(\varphi_1 - \beta) = 90°$.

10.4 For a journal bearing with:

$r = 60$ mm
$R = 60.25$ mm
$N = 3600$ rpm $= 60$ Hz
$l = 180$ mm
Load per unit of length $P = 7000/180 = 38.9$ N/mm $= 38900$ N/m
Entrance of the oil by the horizontal plane $(\varphi_1 - \beta) = 90°$
Viscosity of the oil $\mu = 0.011$ Ns/m^2

Calculate the coefficient of friction f and the oil flow Q using Bosch indications.

10.5 A journal bearing has the following characteristics:

- Axial length $l = 200$ mm
- Diameter $d = 140$ mm
- Radial clearance $c = 0.28$ mm
- Significant speed $N = 41.66$ rev/s
- Load $P = 6000$ kN
- Load per unit of length $P = W/l = 30000$ N/m^2
- Average temperature of the film $T_{average} = 55°C$
- Oil SAE 30, density $\rho - 8$ kg/m^3, specific heat $c_p - 1700$ J/kgK.
- The inlet of the oil is in a horizontal plane $(\varphi_1 - \beta) = 90°$

According to Bosch indications, find:

- The minimum film thickness
- The maximum film pressure
- The oil flow
- The power loss

10.6 Using the data from Problem 10.2, calculate the minimum coefficient of friction f according to the Schiebel formula $(\varphi_1 - \beta) = 90°$ (from page 234 of the book):

$$f = 2.81 \sqrt{\frac{\mu V}{P} \left[1 + 2 \left(\frac{d}{l} \right)^2 \right]}$$

CHAPTER 11

11.1   A steel shaft in a ship transmits a nominal power $P = 1550$ kW when the angular speed is $N = 800$ rev/min. The axial load of the propeller is $N = 600$ kN. The minimum diameter of the shaft is $d = 130$ mm. Determine the ideal stress using the von Mises theory and the maximum shear stress theory. Knowing that the length l of the shaft is 12 m, determine the angular deformation assuming the hypothesis that the medium diameter is $1.1d$.

   The tangential modulus $G = 83000$ N/mm^2.

11.2   The steel shaft in Figure A.11.1 transmits a nominal power of $P = 8$ kW when the angular speed is $N = 100$ rev/min. Determine the maximum stress and, according to the von Mises hypothesis, the ideal stress.

   Assume that the material is a ductile material, the modulus of elasticity $E = 210000$ N/mm^2 and the tangential modulus $G = 83000$ N/mm^2.

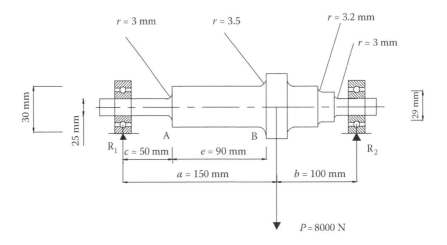

FIGURE A.11.1

11.3   Calculate the maximum stress in the helical spring and the deflection at the end of the cantilever (Figure A.11.2).

   Data:
   $E = 200$ GPa, $G = 70$ GPa, $i_u = 6$, $\lambda' = 1.5$, $\lambda'' = 1$, $d = 5$ mm, $R = 20$ mm, $\tau_{amm} = 200$ MPa, $b = 10$ mm, $h = 6$ mm, $L = 200$ mm.

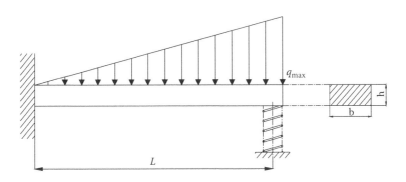

FIGURE A.11.2

11.4 The length *l* of a bearing (material cast bronze) is 40 mm and the radial load is $R = 2400$ N. The diameter *D* of the shaft is 35 mm. The velocity of the shaft is $v = 0.2$ m/s.

The wear ratio *K* of the material declared by the producer is: 2000×10^{-18} [m^3sW].

Knowing that the wear ratio *K* of the material declared by the producer is: 2000×10^{-18} [m^3sW], evaluate:

- The maximum value of the pressure p_0
- The *PV* parameter
- The time *t* to have a reduction of the thickness Δs equal to 0.12 mm.

11.5 Referring to the data of Figure A.11.3, find the diameter φ of the shaft (2). Apply the Tresca hypothesis.

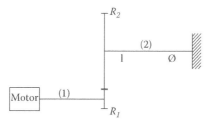

FIGURE A.11.3

Data:

Torque	$M_1 = 15$ Nm
Radius gear 1	$R_1 = 20$ mm
Radius gear 2	$R_2 = 80$ mm
Modulus of elasticity	$E = 210$ GPa
Yield stress	$\sigma_y = 450$ MPa
Factor of safety	$s = 1.29(-)$
Shaft length (2)	$l = 0.4$ m

CHAPTER 13

13.1 A system of two leaf springs (as shown in Figure A.13.1a) has a force-deflection relation as in Figure A.13.1b. Supposing that the two springs are as two bending triangular plates, determine the width b_0 and the thickness *h* at the fixed point of the spring, knowing that the lengths of the leaf springs are $L_1 - 100$ mm and $L_2 = 50$ mm, and that the allowable stress is $\sigma_{amm} = 200$ MPa. The distance between the two leaf springs is 4 mm.

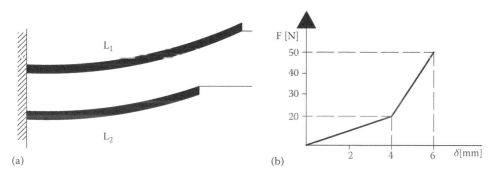

(a)   (b)

FIGURE A.13.1

13.2 Using the data from the table below and the system of two triangular springs with ends distant θ = 5 mm (Figure A.13.2), calculate:

	Right	Left
b (mm)	80	100
h (mm)	10	10
l (mm)	290	310
E (MPa)	78000	210000
v	0,3	0,3

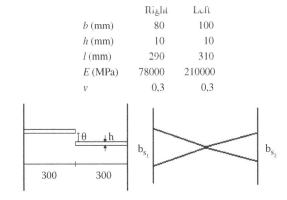

FIGURE A.13.2

The vertical load to apply at the end of the right spring so that the two springs come into contact;
The maximum stress σ of the right spring in this condition;
The stiffness K_1 and K_2 of the two springs and the characteristic diagram of the spring system.

13.3 Using the data from the table for a spiral spring (see Figure A.13.3):

Data:

r	20 mm
R	40 mm
M	400 N*mm
d	4 mm
E	210000 MPa
a	10 mm
$J_C = (\pi d^4)/64$	12,56 mm^4
b	2 mm
h	4 mm
$J_r = (1/12)bh^3$	10,66667 mm^4

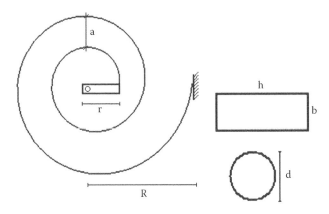

FIGURE A.13.3

Calculate for the two cases (rectangular and circular cross sections) the deformation work L, the relative rotation $\Delta\varphi$ between the two extremities, and the maximum stress σ.

The length l of the spring is:
$$L = (a/2\pi)((\text{In}(\alpha_2 + (1 + \alpha_2{}^2)^{0.5}) + (\alpha_2(1 + \alpha_2{}^2)^{0.5}) - (\text{In}(\alpha_1 + (1 + \alpha_1{}^2)^{0.5}) + (\alpha_1(1 + \alpha_1{}^2)^{0.5})).$$
Where: $\alpha_1 - 2\pi r/a - 12{,}56$; $\alpha_2 = 2\pi R/a = 25{,}12$; and $l = 792{,}78$ ınını.

13.4 Using the data below for a spiral spring (see Figure A.13.4), and supposing an identical length as the spring from the previous problem:

Data:

r	20 mm
R	40 mm
d	4 mm
E	210000 MPa
a	10 mm
$J_C = (\pi d^4)/64$	12,56 mm⁴
b	2 mm
h	4 mm
$J_r = (1/12)bh^3$	10,66667 mm⁴
$P =$	10 N
$V_r = bhl$	6342,235 mm³
$V_c = \pi(d24)l$	9957,309 mm³

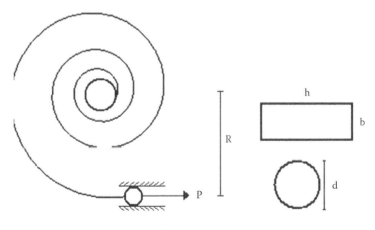

FIGURE A.13.4

Calculate, for the two cases (rectangular and circular cross sections) the deformation work L, the relative rotation $\Delta\varphi$ between the two extremities, and the maximum stress σ.

13.5 Using the data below for a helicoidally spiral spring of a camshaft (see Figure A.13.5), calculate the dynamic loads for the bearing and the dimensions of the springs:

Data:

$\omega =$	2500 rpm
number of the spring turns i	6
$\lambda' =$	1,3
$\lambda'' =$	1
$K =$	10 N/mm
$\sigma_{sn} =$	600 MPa
$\sigma_0' =$	400 MPa
$C =$	1500 N
$f_i =$	6 mm
$E =$	210000 MPa
$v =$	0,3

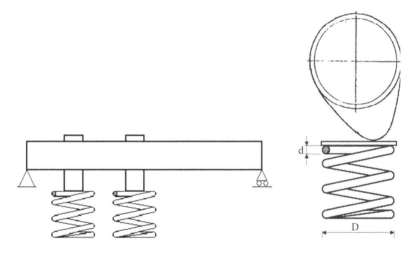

FIGURE A.13.5

CHAPTER 14

14.1   Find the dimensions of a belt for the following data:

Power = 10 kW
Shaft velocity $n = 2000$ rpm $= 2 \pi 2000/60 = 209,33$ rad/s
Diameter of the pulley $d = 100$ mm
Ratio velocity between the two shafts $\tau = 1/3$
Distance from centre to centre $i = 2000$ mm
Temperature of the system is less than 100°C

CHAPTER 15

15.1   For the two gears in Figure A.15.1, calculate from the data given below:

The pitch module
The radii of the gears
The center to center distance
The length
Make a wear verification.

FIGURE A.15.1   Speed reducer.

Data:
$T = 0.33333$
Torque $M = 5.6$ Nm
$\Theta = 20°$ (0.34888 rad)
$\lambda = b/m$ 10
$\omega_1 = 157$ rad/s $\omega_2 = 51.81$ rad/s

Material characteristics:
$\sigma_{sn} = 1127$ N/mm^2
$\sigma_r = 1372$ N/mm^2
Brinell hardness $H_B = 2320.30$ N/mm^2
Module of elasticity $E = 210000$ N/mm^2
Allowable stress $\sigma_{all} = \sigma_{sn}/2 = 563.5$

15.2   For the conical gears, calculate the pitch module and verify the wear stress.

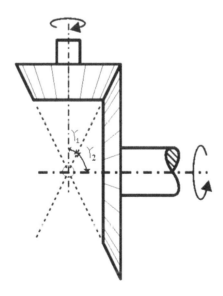

FIGURE A.15.2

Data:
$t = 0.22$
$\theta = 20 = 0.348888889$

$\lambda = 10$

$M = 1465$ Nm

$\omega_i = 61.8$ rad/s

Material 35NCD15, for which:

$H_B = 2750$ N/mm^2

$\sigma_r = 1750 - 2000$ N/mm^2

$\sigma_{sn} = 1300$ N/mm^2

$E = 210000$ N/mm^2

$\sigma_{wall} = H_B/2$

15.3 Using the data below:
  a) Find the dimensions of the cylindrical gear reducer with the minimum teeth number.
  b) Evaluate the ratio ε between the arc and the segment of the contact.

Data:

$\theta = 20 = 0,348888889$	pressure angle
$\tau = 0,2$	transmission ratio
$M = 85$ Nm	torque
$\lambda = 10$	
$\omega_1 = 50$ rad/s	angular velocity
$A = 6$ m/s	
$\sigma_{sn} = 1127$ MPa	yield stress
$\sigma_r = 1372$ MPa	ultimate stress
$H_B = 2256,3$ MPa	Brinell hardness
$E = 210000$ MPa	modulus of elasticity

CHAPTER 16

16.1 For a steel gear pinion ($Ri = 40$ mm, $b = 20$ mm) coupled by hot interference ($i/d = 0.12\%$) to the final part of a full shaft ($d = 40$ mm), calculate:

The maximum torque moment transmissible if the coefficient of friction $f = 0.6$

The shear stress τ for the shaft

The radial stress σ_r and the tangential stress σ_t at the external diameter of the shaft

The thermal variation ΔT necessary to the coupling pinion at the shaft.

Assume a coefficient of friction $f = 0.6$ and a linear thermal coefficient $\alpha_t = 1.3 \times 10^{-5}/1°C$.

16.2 Determine the stresses in a copper disc assembled on a steel ring. The disc diameter $d = 250$ mm and the thickness is 10 mm. The internal ring diameter, d_i, is 249.50 mm and the external diameter, D, is 264 mm. The assembling is done by heating of the ring. Evaluate the stress produced on the ring. Assume the dilatation coefficient of the steel $\alpha_t = 1.3 \times 10^{-5}$.

For the ring: $\alpha = 249.50/264 = 0.945$.

For the disc: $\beta = 0$.

CHAPTER 17

17.1 Determine the stress in a piping used for steam transfer if the pressure p is 1200 N/cm^2, and the steam's temperature is 200°C. The diameter d is 200 mm and the thickness s is 10 mm. Evaluate the variation of the diameter if the piping's material is a steel ($E = 210000$ N/mm^2, $\alpha = 1.2 \ 10^{-5}$). Calculate the diameter variation if the material is glass fibers ($E = 39300$ N/mm^2, $\alpha = 5 \ 10^{-5}$).

Assume that the minimum value of ambient temperature is $-10°C$ and calculate the diameter variation for the temperature effect.

17.2 In normal conditions the pressure in a steam piping is 130 ATA and the temperature of the steam is $550°C$. The external temperature is $500°C$. The dimensions of the pipe are:

$D = 219$ mm,
$s = 20$ mm.
The material is a 14CrMo3 steel ($\sigma_{sn} = 470$ N/mm^2, $\sigma_t = 550$ N/mm^2)
Find the stress for the piping and the assurance coefficient.

17.3 Calculate the critical external pressure, p_{crit} (pressure per unit length), capable of causing the implosion of a tank when the effect of the edge is missing. The dimensions and characteristics are:

Diameter $d = 2r = 7000$ mm
Thickness $s = 12$ mm
Height = 20000 mm
Young's modulus $E = 210000$ N/mm^2
$\sigma_{sn} = 400$ N/mm^2

After you have taken into account the edge effect, confront this value with the value of the minimal internal pressure which is necessary to bring the tank to plastic deformation conditions.

17.4 The edge fund of a circular vessel (Figure A.17.1) must be connected by 36 bolts. The internal diameter D of the vessel is 2000 mm and the internal pressure p is 2 N/mm^2. Find the thickness s of the vessel wall, and the dimensions of the flange connecting the wall of the vessel to the fund.

The yield strength σ_{sn} of the steel is 400 N/mm^2.

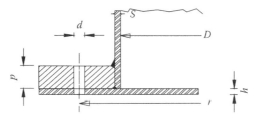

FIGURE A.17.1

17.5 The pressure in a steel vessel is 17 N/mm^2, the internal diameter d of the vessel is 380 mm, and the thickness is 25 mm.

Evaluate the stress σ_t on the wall of the vessel using the hypothesis of narrow pipes and wide pipes (see page 424 of the book).

17.6 A spherical steel tank used to contain fuel gas has diameter $d = 2$ m and thickness $s = 70$ mm. The internal pressure p is 1000 N/cm^2. The steel is an AISI 304 ($\sigma_s = 241$ N/mm^2, $\sigma_r = 568$ N/mm^2, $E = 210000$ N/mm^2). Calculate the maximum stress σ_t and the assurance coefficient m. Estimate the equatorial diameter variation Δd for the pressure effect.

17.7   The tank in the previous problem has an inspection flanged hole $B = 21$ mm (Figure A.17.2). The diameter D_N of the hole is 700 mm, the number n of the bolts is 30, and the minor diameter d of each bolt is 15 mm. Calculate the stress on the bolts.

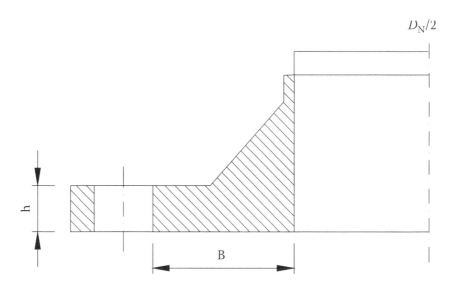

FIGURE A.17.2

17.8   Determine the stress at point A of the gas tank (Figure A.17.3) and evaluate the factor of safety m, when:

Diameter $d_i = 100$ mm
Thickness $s = 30$ mm
Internal pressure $p = 10$ atm
Yield stress of the steel 450 MPa

Neglect the weight of the tank and gas

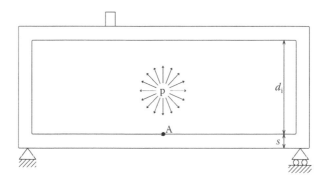

FIGURE A.17.3

CHAPTER 18

18.1   Calculate the stress of the welding joint as in Figure A.18.1 for the static load $P = 500$ N applied at the end.

Data:
The thickness s of the plate is 10 mm
The width of the plate is equal to the fillet length $b = 40$ mm
The length l of the plate is 200 mm
The height h of the fillet is 4 mm

Make a fatigue verification if the joint is a steel component of a railway bridge loaded at the load ratio $R = -0.7$ and determine the factor of safety m. The steel is a C 40 (1050) UNI (SAE-AISI) (tensile strength $\sigma_T = 520$ N/mm², σ_Y yield strength 290 N/mm²).

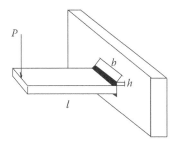

FIGURE A.18.1

18.2   A steel plate is welded by two fillets as in Figure A.18.2. The applied load $P = 100$ kN. Calculate the assurance coefficient m in static conditions.

Data:
$b = 450$ mm; $h = 100$ mm; $L = 200$ mm; $s = p = 20$ mm
$\sigma_R = 500$ MPa; $\sigma_{SN} = 415$ MPa; $\sigma_{amm} = 280$ MPa

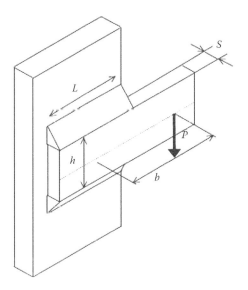

FIGURE A.18.2

18.3   Verify the plate of a full penetration T-joint (Figure A.18.3) used to connect the motor at the test bed when the applied loads are P_1 and P_2 (as in Figure A.18.3).

Data:

Steel: AISI 304; $\sigma_R = 620$ N/mm²; $\sigma_{SN} = 330$ N/mm²; $\sigma_{amm} = 223$ N/mm²

$P_1 = 25$ kN; $P_2 = 40$ kN; $h = 75$ mm; $b = 35$ mm; $s = 12$ mm

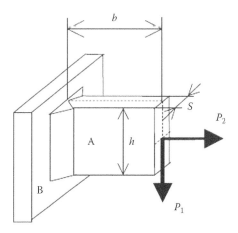

FIGURE A.18.3

18.4 A plate is welded as in Figure A.18.4. The dimensions are: $h = 120$ mm and $s = 10$ mm. The dimensions of the two fillets are: $p = s$ and $L = 95$ mm. The material is: AISI 304; $\sigma_R = 620$ N/mm²; $\sigma_{SN} = 330$ N/mm²; $\sigma_{amm} = 223$ N/mm². Calculate the maximum bending moment M according to the static strength.

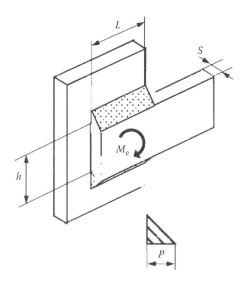

FIGURE A.18.4

18.5 On each rivet ($d = 5$ mm) of a riveted joint, in working condition (Figure A.18.5), acts a shear load of $T = 1000$ N. The rivets are assembled, heating them at a $\Delta T = 70°$C.

Calculate the principal stresses under the hypothesis of infinity stiffness of the two plates ($s = 3$ mm).

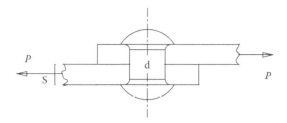

FIGURE A.18.5

PART IV

CHAPTER 19

19.1 Determine the relative rotation between the two parts of a Bibby joint (Figure A.19.1) for a torque $M_t = 800$ Nm under the hypothesis of little deformations, where:

- Curvature radius of the tooth $R_d = 200$ mm
- Rolled strip width = 10 mm
- Rolled strip thickness = 1.5 mm
- Distance l between the medium point of the joint and the beginning of the curved part of the tooth = 25 mm (see Figure A.19.1)
- R_m (distance between point C and the centre of the shaft) = 125 mm
- Teeth number $n = 40$
- Elastic module of the strip $E = 210.000$ N/mm^2

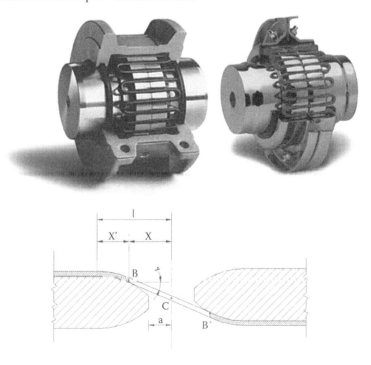

FIGURE A.19.1

19.2 Determine the ratio between the maximum torques transmittable by the disc shaft coupling depicted in Figure A.19.2 in the case of transmission by friction and in the case of transmission by shear force. Consider a maximum allowed von Mises stress of 200 MPa and the following dimensions:

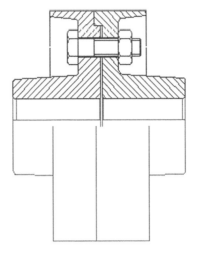

FIGURE A.19.2 Disc shaft coupling

$d = 50$ mm
$D = 190$ mm
Number n of fasteners = 8
Coefficient of friction = 0.3
R_m = mean radius of the contact area = 150
R_b = radius of the circle containing the centers of the fasteners = 150 mm
d_n = minor diameter of fasteners = 6.77 mm
d_b = cross section diameter of fasteners = 8 mm

19.3 Determine the maximum contact pressure p between the two shells and the shafts of the joint as seen in Figure A.19.3. For the transmission, a torque of 800 N · m is necessary, considering a static coefficient of friction of 0,75 (steel-steel, dry conditions) and the following geometrical parameters:

• Diameter of shafts $d = 80$ mm
• Coupling length $L = 150$ mm
• Distance g (gap) between the opposite faces of the shells $g = 2$ mm
• Number of fasteners $n = 3$

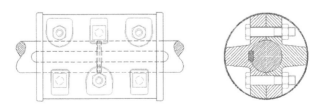

FIGURE A.19.3 Friction shaft coupling

Determine also the maximum tensile stress acting on the ISO M8 treaded fasteners.

CHAPTER 20

20.1   Given the flat multi-disc clutch in Figure A.20.1 and the data of the table, calculate the number of the discs necessary to bring the driven shaft U at velocity $\Delta\omega$ in five s.

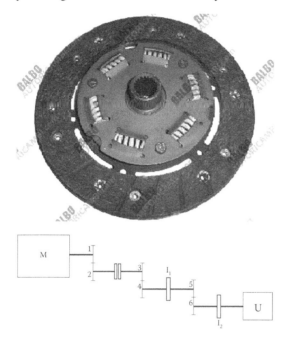

FIGURE A.20.1   Flat multi-disc clutch.

Data

$P_i =$	0,2	MPa
$R_e =$	100	mm
$x = R_e/R_i$	0,7	
$\Delta\omega =$	50	rad/s
t	5	s
$M_r =$	2900	Nm
$f =$	0,2	
$\tau_{1,2}$	0,5	
$\tau_{3,4}$	0,3333333	
$I_1 =$	25	kg*m^2
$I_2 =$	40	kg*m^2

20.2   For a flat multi-disc conic clutch with hydraulic clamping (Figure A.20.2), calculate the conditions of oil pressure when:

$n_d = 3$	number of discs
$f = 0.25$	friction coefficient
$R_i = 70$ mm	internal radius
$R_e = 100$ mm	external radius
$M_{TOT} = 450$ Nm	torque
$d_{cil} = 50.8$ mm	oil cylinder diameter

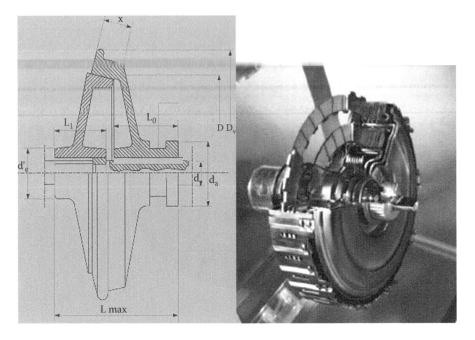

FIGURE A.20.2 Multi-disc conic clutch.

20.3 With the data of the previous problem, calculate the oil pressure in the case of a mono-disc conic clutch.

20.4 The coefficient of friction f (typical applications of clutches or brakes) does not remain constant over time but changes its value depending on the temperature achieved by the system. According to the following variation:

$T\,[°C]$	f
60	0.7
180	0.475
300	0.25

and for the following data values:

M	18 kNm	torque
$x = R_i / R_e$	0.6	radii ratio
p_0	0.8 MPa (uniform)	maximum uniform pressure
n_d	1	disc number
L_s	250 Nm/cm^2	rate of energy dissipation
Δn	2000 r/min	revolutions per minute
$\Delta \omega = \Delta n\, \pi/30$	209.4395 rad/s	angular velocity

find the characteristic dimensions of the clutch.

Also, calculate how much time is needed to the driven shaft in order to accelerate at "Δn," knowing that the maximum work ratio L_s is 250 Nm/cm^2.

20.5 Centrifugal clutches are composed of a bell connected with the driver shaft. The bell houses three hinged mass "m". They follow the driver shaft speed with a centrifugal action.

The contact between the external part of the mass and the surface of the bell happens following the centrifugal forces that win the action "T_0" of the appropriated springs. In this way, thanks to the normal component to the contact surface, a torque "M" on the driven shaft is generated.

For the centrifugal clutch, as in Figure A.20.3 and for the following data, find:

1. The force T_0 of each spring to have the first contact at n_1.
2. The torque M transmitted from each mass m at an angular velocity $n_2 = 1.5\, n_1$.

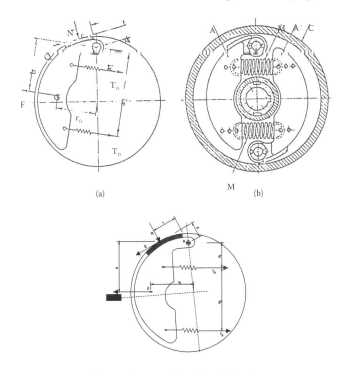

(a)   (b)

FIGURE A.20.3   Centrifugal clutches.

CHAPTER 21

21.1   Considering an initial condition characterized by a longitudinal acceleration equal to $X_{acc} = 1$ ms^{-2}, determine the maximum braking pressure that it is possible to apply on the rear brake pads of the car in Figure A.21.1 before slipping occurs.

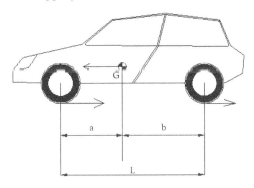

FIGURE A.21.1

Consider the following data:

$L = 2,20$ m
$a = 0.90$ m
$b = 1,30$ m
m (automobile mass) $= 1600$ kg
h_g (height of center of mass) $= 0.7$ m
$m_x =$ maximum coefficient of wheel-road adhesion $= 0.7$
$I_{wheel} = 0.4$ kgm^2
$f =$ coefficient of friction for the pads-disk contact $= 0.5$
$\alpha = 45°$
$R_e =$ external radius of the pad $= 0,18$ m
$R_i =$ internal radius of the pad $= 0,15$ m
$R_{wheel} =$ wheel radius $= 0.20$ m.

21.2 For a Fiat 500 car (shown in Figure A.21.2) with a drum breaking system, determine the minimum stop distance when the car is going at a velocity of 90 km/h on a straight horizontal road and the braking torque on the front and rear wheel.

The data are:

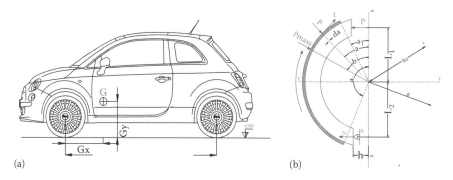

(a)   (b)

FIGURE A.21.2 (a) Fiat 500. (b) Scheme of the brake system.

For the car:
$P = 11$ [KN]
$V_0 = 90$ [Km/h] (car speed)
$f_t = 0.3$ (coefficient of tire-soil adhesion)
$r = 20$ cm (wheel radius)
$G_x = 0.58$ m
$G_y = 0.52$ m
$p = 2.3$ m (distance between the front and rear wheel)

For the brake system:
$f_c = 0.20$ (coefficient of strain drum adhesion)
$L_1 = L_2 = 9$ cm
$h = 4$ cm
$Ra = 12$ cm (Ferodo brake pads radius)
$Rp = 10$ cm (pin radius)
$b = 4$ cm^2
$\beta = 60°$
$\theta_0 = 60°$.

21.3 Using the data from the previous problem, determine the load to apply to the block knowing its geometry is as in Figure A.21.2b.

21.4   Determine the angle of contact between the band and the drum (B in Figure A.21.3) able to stop the mechanical device in 10 seconds. Consider the following data:

- I_1 (rotational inertia of flywheel 1) = 1 kg · m^2
- I_2 (rotational inertia of flywheel 2) = 0.7 kg · m^2
- N_{G1} (number of teeth gear 1) = 30
- N_{G2} (number of teeth gear 2) = 90
- a_1 (bevel angle gear G3) = 30°
- a_2 (bevel angle gear G4) = 60°
- f (coefficient of friction) = 0.35
- u – 20 cm
- l = 60 cm
- Q (force to apply to the brake) = 500 N
- w_4 = 30 rad/s
- R = radius of the drum = 0.3 m

Neglect (for the sake of simplicity) the rotational inertia of all gears (G1–G4).

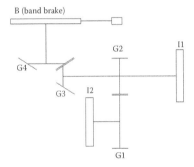

FIGURE A.21.3   Schematic representation of a mechanical device.

21.5   Determine the maximal actuating force F and the corresponding torque M for a disc brake. The shoes are symmetrical, the area S of each pad is 300 mm^2, and the distance of the pressure center of the pads from the rotating axes is 45 mm. The disc radius R is 100 mm. The friction coefficient at ambient temperature is f = 0.4 and the maximal pressure is 2.5 N/mm^2. Assume that the pressure is constant. Determine also the reduction of the torque when the temperature of the disc reaches 300°C and the friction coefficient becomes 50% of the coefficient at the ambient temperature.

21.6   The maximum torque M that must be applied to a wheel of a car is 60 Nm. The wheel radius R_w is 250 mm. The maximal pressure p on the pads is 1.00 N/mm^2. Choose adequate dimensions for the disk and for the pads of this disk brake system.

Assume that the friction coefficient f = 0.15 at 300°C and the pressure p is constant on the surfaces.

References

ADOPTED BOOKS

1. Giovannozzi R., *Costruzione di Macchine,* Vol. 1, Pàtron, 1980.
2. Giovannozzi R., *Costruzione di Macchine,* Vol. 2, Pàtron, 1980.

SOURCES OF CONSULTATION

3. Atzori B., *Appunti delle Lezioni di Costruzione di Macchine,* Facoltà di Ingegneria, Università di Padova, A.A. 1998/99.
4. Belingardi G., Calderale P.M. et al., *Principi e Metodologie della Progettazione Meccanica* (4 vol.), Levrotto & Bella, 1987-1990.
5. Bongiovanni G., Roccati G., *Giunti Articolati [Articulated Joints]* Levrotto & Bella Torino.
6. Dubbel H., *Manuale di Ingegneria Meccanica* (2 vol.), EST/Springer, 15ª ed., 1985.
7. *Il Nuovo Manuale di Meccanica* Zanichelli/ESAC, Edizioni Scientifiche 2ª ed., 1994.
8. Juvinall R.C., Marshek K.M., *Fondamenti della Progettazione dei Componenti Meccanici,* Edizioni ETS, 2001.
9. Lanczos C., *The Variational Principles of Mechanics,* Dover Publications, 4th ed., 1986.
10. Shigley J.E., Mischke C.R., Budynas R.G., *Mechanical Engineering Design,* McGraw-Hill, 7th ed., 2003.

REFERENCES FOR THE SETTING PROJECT METHODOLOGY

11. Giudice F., La Rosa G., Risitano A., *Product Design for the Environment: A Life Cycle Approach,* CRC/ Taylor & Francis, 2006.
12. Ulrich K.T., Eppinger S.D., *Progettazione e Sviluppo di Prodotto,* McGraw-Hill, 2001
13. Ashby M.F., *Materials Selection in Mechanical Design,* 3rd ed., Butterworth Heinemann, Oxford, 2005.

Index

A

ABS resins, 66
ACF 60 aluminum bronzes, 58
Acrylic resins, 66–67
"After-elastic" effect in polycrystalline metals, 127
Allyl resins, 68
Aluminum bronzes, 58
Amino resins, 68
Anti-friction bearings. *See* Rolling contact bearings
Antimonial lead, 60
Aramid organic fibers, 73
Artificial plastics, 63
Assembly models
  bearing assembly
    for bi-directional axial loads, 262–263
    for unidirectional axial loads, 260–261
  convergent conical bearings
    with inner-ring adjustment, 263–264
    with outer-ring adjustment, 264–265
  divergent conical bearing
    with outer-ring adjustment, 265–266
Atesina. *See* Elektron
Axial bearings, 257
Axial thrust bearings, 270. *See also* Friction bearings

B

Bach's coefficient, 249–250
Bailey–Orowan model, 441
Bakelite (PF), 68
Bearing
  adjustment, 259–260
  assembly
    for bi-directional axial loads, 262–263
    rule, 259
    for unidirectional axial loads, 260–261
Bending springs
  coil springs
    applied force, 306
    applied moment, 305
    parallel and qualitative characteristic, 304
  conical disc springs, 311
    qualitative characteristic, 313
    scheme for, 312
    working condition, 312
  cylindrical helical springs
    bending, 309
  diaphragm spring
    deformation calculation, 317
    stress calculation, 316
  leaf springs
    elastic deflection, scheme of, 299
    geometric form, 302
    logarithmic characteristic, 304
    mass spring system, 303
    qualitative characteristic, 297
    scheme for calculation, 298
    with spring shackles, 298–299
    trapezoid spring, 297
    triangular spring, 296
  parabolic bending spring, 294
  rectangular, 291
    qualitative stress in, 292
  trapezoid bending spring, 295
  triangular bending spring, 293
  variable thickness disc springs, 313
    characteristic, 315
    extreme variable thickness, 314
Bevel gears
  bending calculations, 568
  crown bending verification, 569
  dimensioning, 566
    scheme of, 567, 573
  dynamic rim verification, 569–570
  teeth number, calculation of, 567–568
  wear calculation, 568
Bibby joints, 480, 482
  characteristics of, 486–487
Boron fibers, 73
Boundary elements method (BEM), 148
Box screws dimensioning
  pin supports, 600
  resultant of contact actions, 600
  scheme for, 601
  tensile load, 601
  transferability of friction couple, 600
Brakes, 529
  disc
    compressed air braking system, 534
    drum replacing, 530–531
    floating callipers, 531–532
    fluid pressure, 532
    heating, 533
    power and pad area, 534
    solutions for, 532
    twin, 533
    ventilated, 533
  drum, 545
    calculations and sizing, 548–557
    constructional configurations for, 548
    differential of, 560–562
    ribbon brakes, 557–560
    schematic, 547
    servo-brake, 546
  dynamic equilibrium equation, 529–530
  pads, 532
  performance and dimension analyses, 534
    acting forces calculation, 541
    axial force, 535
    characteristics of, 539
    clutch friction plate, 534–535

corrector, 542
disc, 535
friction coefficient and velocity, 536, 543
hypotheses, 535
lineal expansion coefficient and, 538
materials, characteristics of, 539
pneumatic servo-brake, 545
pressure curves, 543
qualitative thermal stresses, 540
system, 535
temperature rise, 538
thermal stresses, qualitative components, 540
traction coefficient, 543
work lost, 536
Buckling, 108
BZn7 alloy, 58
BZn2 and BZn6, low lead copper, 57

C

C Alloy, low lead copper, 57
Carbon fibers, 72
Casigliano's theorem, 252
Cast irons, 48
alloys, 51
classification, 50–52
elements, influence on, 49–50
uses, 49
Cellulose acetate butyrate (CAB), 66
Cellulose nitrate (CN), 66
Cellulose propionate, 66
Cellulose resins, 65–66
Centrifugal clutches
Desch centrex, 526–527
equilibrium equations, 524–526
friction surfaces, 523
hypothesizing, 524
moments, equilibrium of, 522
Piaggio clutch, 523
Ceramic materials, 74–75
Chitonal 24, 61
Chromium steel, 53
Clutches
centrifugal, 521
Desch centrex, 526–527
equilibrium equations, 524–526
friction surfaces, 523
hypothesizing, 524
moments, 522
Piaggio clutch, 523
classification of, 501
cone and double cone, 510
model for, 511
in motion, 511
surfaces, 512
transmittable peripheral force, 511
flat plate
component of, 510
friction material, 506
friction system, 507
mean radius, 508
overheating, 509
Reye's hypothesis, 508–509
transmittable moment, 508

friction, 501
dynamic equilibrium, 503
engine work, 502
mobile friction surfaces and axial shift, 505
motion quantity principle, 505
polar moment of inertia, 503
surfaces of, 505
working conditions, 502
work lost, 503
radial block, 512
bending moment, 515
Dolmen–Leblanc, 513, 515
hypothesis, 516
moment, 514
specific pressure, 513
tension of, 514, 516–517
safety clutches
Hertzian loads, 520
qualitative characteristic of, 521
spring characteristics, 520–521
surplus
cylindrical rollers, 517–518
Nepero's formula, 519
non-cylindrical rollers, 518–520
scheme of, 518–519
Coil springs. *See also* Bending springs
applied force, 306
applied moment, 305
parallel and qualitative characteristic, 304
Combustibility defined, 65
Composites, 69
admissible load and elasticity modulus, 73
aramid organic fibers, 73
boron fibers, 73
carbon fibers, 72
categories, 69
characteristics, 69
components of, 69
epoxy matrices, 71
fibrous stiffening, 71
formation by
brushing and spraying, 72
glass fibers, 71–72
manufacturing, 70
matrices, 70–71
multilayer, 73
parallel continuous and discontinuous fibres, 70
phenolic matrices, 71
polyester matrices, 71
random discontinuous fibres, 70
silicone matrices, 71
Compression, 108
deformation under traction load, 113
qualitative deformation, 114
stability, equation of, 113
steels and, 113
Conceptual design, 5
Concurrent engineering (CE), 28
aim of, 8–9
principles, 8
Cone and double cone
engagement in motion, 511
model for, 511

surfaces, 512
transmittable peripheral force, 511
Conical bearings, 258
Conical disc springs, 311. *See also* Bending springs
qualitative characteristic, 313
scheme for, 312
working condition, 312
Convergent conical bearings. *See also* Assembly models
with inner-ring adjustment, 263–264
with outer-ring adjustment, 264–265
Coordinate measuring machines (CMM), 13
Copper, 56
Copper–lead–tin alloys, 57
Copper–tin binary alloys, 56
Copper–tin–zinc–nickel alloys, 56
Corrofond, 61
Costantan, 62
Couplings
joints
analyzing, 468
elastic and semi-elastic fixed, 467
fixed, 467
mobile, 467
size of, 467–468
Creep tests for hot metal
constant value, 437, 440
crystalline lattice, 441
curves, 437, 439
deformation, 438
graphs, 437
results, 440
stress parameter, 439
temperatures, 440–441
Cylindrical helical springs, 309

D

Decorated DELTA foundry brass, 59
Defects in metals, 48
Deflection stability equation, 116
DELTA and DELTA B foundry brass, 59
Design-centered product development model, 10
Design for environment (DFE), 15
concept, 16
defined, 16
guidelines, implementing design
data gathering, 17
tools for, 18
and LCD, 15
life cycle and
analysis, 18
interactions with ecosphere, 18–19
phases, 18–19
multidisciplinary approach, 16
objectives and characteristics, 16
optimum environmental performance, 17
Design for X (DFX), 10
contextualization of, 12
for environmental strategies
design for disassembly (DFD), 39
design for maintainability, 39
design for recycling, 39
design for remanufacturing, 39
design for serviceability (DFS), 39

in product development
conceptual design, 39
detail design, 40
embodiment design, 40
property objective
environmental quality, 11
for production phase, 11
product retirement/recovery, 11
sustainability, 11
use phase, 11
role in environmental requisites, 36
tools, 11–12
use, 40
Design process of product
conceptual design, 5
detail design, 6
embodiment design, 6
environmental factors, 26–27, 38
methodologies, 7
problem specification, 5
DFE. *See* Design for environment (DFE)
Diaphragm spring
deformation calculation, 317
stress calculation, 316
Differential bearings box, 593
horizontal and vertical translation, 594
load
equivalent dynamic, 595–596
value of axial, 595
scheme of, 594
Differential design
defined, 563
gears project, 564
classic Lewis formula, 565
decommissioning of, 565
pitch line velocity function of, 566
rotational speed of, 565–566
Tredgold's method, 566
internal parts of, 564
ordinary, 564
planetary gears, 563
relationship between speeds, 565
Willis formula, 563
Differential gear dimensions
bending calculation, 574
dynamic verification
planet wheel velocity, 575
number of teeth, calculation of, 573–574
wear calculation
Hertzian stress, 575
Differential ribbon brakes
equilibrium, 560
equation, 562
lever
end travel, 561
rotation, 561
traction
area, 561
tension, 562
Divergent conical bearing with outer-ring adjustment, 265–266. *See also* Assembly models
Drum brakes, 545
bending moment, 551
block resistance calculation, 551

calculations and sizing, 548, 553
    cable car built braking system of, 549
    calculations scheme, 550
    disc brakes comparison, 557
    equilibrium equations, 554
    Giovanozzi's formulae, 552
    constructional configurations for, 548
    differential of, 557
    rotation, direction of, 546
    schematic figure, 547
    servo-brake, 546
Duralumin/Duralite, 61

E

Elastic deflection, deformation limit and
    shift elasticity assessment, 253
        model of shaft, 256–257
        qualitative linear bending moment, 254
        rotation deflection in finite part, 255
Elastic polymers, 64
Elco joints, 480
Elektron, 61
Embodiment design, 6
Environmental impact of product
    depletion effects, 19
    disturbance effects, 19
    pollution effects, 19
Environmental strategies
    for product planning, 30
        end of life, reclamation, 32–35
        recovery planning optimization, 35
        useful life extension strategies, 31–32
Epoxy matrices, 71
Epoxy resins, 68–69
Ergal 55 and Ergal 65, 61
Erichsen drawing test, 95–96
Ethylcellulose, 66
Euler's theorem, 123–124, 199

F

Failure hypothesis, 186
    comparison, 195
    deformation energy (Beltrami's), 191–192
    distortion energy (Von Mises), 192–194
    maximum linear strain hypothesis
        components in, 188
        Saint-Venant's hypothesis, 188
    maximum normal stress, 187
    maximum tangential stress hypothesis, 188
    Mohr's hypothesis, 189–190
    plane dominion, comparative stress, 190
    Rankine's hypothesis, 188
    resistance, 189
    unloaded lateral sides, cube with, 188
Failure theories
    hypothesis
        comparison, ductile steels, 195
        deformation energy (Beltrami's), 191–192
        distortion energy (Von Mises), 192–194
        ideal stress, 187
        pure torsion, 187
        tangential stress, 188–191

main stresses
    three-dimensional case, 184
material strength and design
    stress state limit, 101
Mohr's circles, 185
    bending and shearing, 186
    pure torsion, 186
    tangential stresses, 185
    tensile compression, 186
stresses in three-dimensional object
    elementary surface and forces, 182
    normal and tangential, 181
    prism, 182–183
    solid body, 182
Fatigue limit stress, 139
Fatigue of materials
    concepts
        infinite life, 128
        metal test sample, phases of, 127
        micro-plasticization, 128–129
        oscillation resistant, 128
        polycrystalline metals and, 127
        static stress, 127
        stress-strain curve, 127–128
        tensile stress, 127
    curve determining, 136
        Prot and Locati, limit values of, 138
        rapid Risitano method, 137
        tension–cycle number, 137
    diagrams
        dispersion in, 131
        Goodman-Smith's, 131–132
        Haigh's, 133
        Pohl and Bach's, 132
        Weyrausch–Kommerell curve, 132–133
    factors influencing
        alternating flat bending, 145
        heat treatments and, 143
        micro-plasticization, 144
        reduced limit, 143–144
        shot blasting, 145
        torsion, bending and tensile, 144
    internal damping
        area of ellipse, 156
        fragile and perfectly plastic, 157
        potential elastic energy, 156
        relative yield, 156
        steel, 155–156
    load characteristics
        mono-axial stress, 129
        oligocyclic fatigue, 130–131
        oscillation amplitude, 129
        tension values and, 129
        Wöhler's curve, 130
    mechanical component, designing
        non-resonating rotational bending, 158
        Pax 750, 160
        phases, 158
        resonating operation, 159
        vertical pulsator PV, 159
    mono axial traction test
        limit stress, 139
        load limit, 140
        stress–strain and temperature–strain curves, 139–140

stress–strain law, 138
temperature variations, 139
thermal behavior, 138
thermodynamic law, 138
thermo-elasticity theory and, 138
residual resistance
breakage risk, 140
Gassner's method, 143
load history and cumulate, 142
Manson's curve, 143
Miner's equation, 140
notched steel, 141
safety factor
coefficient, 145
Haigh's curves, 146–147
residual life, 147
Smith's diagram for, 146
stress concentration factor
boundary elements method (BEM), 148
circular notched shaft, 147–148
elasticity theory and, 147
finite elements method (FEM), 148
Neuber diagrams, 148–149
nominal tension, 149
notch sensitivity, 150–152
resistance, 148, 150
Wöhler's curve, 150–151
test machines
applicator and gauge, 157
deformation, 157
fixed drive, 158
fixed loads, 157
thermography
elastic deformation energy, 134
internal energy, 134
plastic deformation energy, 134
quasi-Wöhler curves and, 133
surface temperature, 134
temperature increments and, 136
T–N curve, 135–136
welded machine components
Bobeck's indications, 154–155
fragility, 154
Smith's diagrams, 153–154
Fatigue resistance
welded machine components, 153–155
Fibrous polymers, 64
Finite elements method (FEM), 148
Fixed flange calculation, 432
bending stress, 433
codes, 435
joints, 434
safety tensile traction stress, 434
Fixed semi elastic and elastic joints, 475
angular deformation, 477
behavior, 477
elastic moment, 476
flywheel, 476
mass production of, 477–478
mechanical and hysteresis characteristics, 478
rotation, 476
Rotex, 478
scheme for working, 476
Flange bolts

calculation
diameter, 435
fixed flange, section of, 432
symbols in, 435
Flat plate clutches
component of, 510
friction material, 506
friction system, 507
mean radius, 508
overheating, 509
parts of, 506
Reye's hypothesis, 508–509
transmittable moment, 508
Flexible machine elements
belts, 341
ribbon and pulleys, 340
defined, 341
precautions for
metallic belts, 343
tensile stress, 343
testing durability, 343
sizing belts
pulleys and scheme, 341
timing and pulleys, 340
timing belts, 342
Fluorinated resins, 67
Formaldehyde casein (CS), 66
Forst joint, 488
cross-section variability law, 489
load, 490
neutral radius, 490
spring of, 489
Foundry brass, 59
Free flange
calculation, 434
maximum bending, 435
Friction bearings
anti-friction rings, 268–269
axial thrust bearings, 270
bush design, 267
one-piece bush, 266
requirements, 268
Friction clutches, 501
dynamic equilibrium, 503
engine work, 502
mobile friction surfaces and axial shift, 505
motion quantity principle, 505
polar moment of inertia, 503
surfaces of, 505
working conditions, 502
work lost, 503

G

G Alloy, low lead copper, 57
G-CuAl9Fe3 and G-CuAl11Fe4, aluminum bronzes, 58
G-CuAl11Fe4Ni4 aluminum bronzes, 58
G-CuSn10Pb10 and G-CuSn5Zn5Pb5 alloy, 57
G-CuSn3Zn10Pb7 and G-CuSn7Zn4Pb6 alloy, 58
Gear resistance, verification calculations
tooth, 394, 399
Lewis calculation model, 397
maximum wear's zone, 400
position of load, 396
wear calculation scheme, 400

General assembly rules, 259
Glass fibers, 71–72
Goodman-Smith's diagram, 131–132
Grade A SAE 430, high resistance brass, 59
Gray cast iron, 50–51
Gudgeons, 249

H

Haigh's diagram, 133
Hastelloy, 62
Hertz theory
   auxiliary angle, 201
   average stress, 204
   ball bearings, 213
      distribution in, 214
   coefficient for, 210–211
   contact area, elliptical shape, 198
   curvature
      centre of, 198
      radius of, 201
   cylinder and cylinder contact, 208–209
   cylindrical surface, intersecting principal for, 199
   defined, 201
   elastic behavior, 197
   ellipse equation, 205
   ellipsoid's generic equation, 203
   equivalent ideal stress, 212–213
   Euler's theorem used, 199
   friction forces, 197
   generic angle, 200
   graphic curvature centre, 198
   hypotheses, 197
   infinite sheaf planes, 198–199
   Meusnier's theorems use, 199–201
   principal planes, 198
   qualitative variation, 205, 211
   rolling and ball bearings
      curvatures, 209
   semi-axes, 197
   semi-cylinder volume, 205
   sphere and sphere contact, 208
   sphere/ball, 207
High aluminum zinc alloy, 60
High cycle fatigue, 128
High lead bronzes, 57–58
High resistance brass, 59
Hirth's spur gears, 284
Hot metal creep, 436
   curves, 437–438
   microstructural aspects, 440–441
   physical interpretation, 441–442
      Bailey–Orowan model, 441
      deformation, 442
      recovery, 441
   tests
      constant value, 437
      crystalline lattice, 441
      curves, 437, 439
      deformation, 438
      graphs, 437
      results, 440
      stress parameter, 439
      temperatures, 440
Hyperstatic bearings, 252–253. *See also* Shafts

I

Inafond S 13/Silafond, 61
40-Inconel, 62
90/10 Industrial and Industrial 86/14, low lead copper, 57
Industrial bronze, 59
Integrated design
   tools
      DFXs, 40–41
      finite element analysis (FEA), 42
      quality function deployment (QFD), 41
      and techniques, 41–42
Integrated product development (IPD), 28
Integrated selection tools
   analytical formulation
      constraints, 170
      free variables, 170
      functionality, 170
      indexes, 170
      length section side, 171
      mass, 171–172
      objectives, 170
   application, 175
      objective, 176
      strength–density chart, 176
   graphic selection
      logarithms, 174
      performance indices, 173
      Young's modulus–density, 174
   research development
      normalized properties method and, 177
Integrating environmental aspects
   with product development, 29
      design phases, 30
      post-design planning, 30
Internal material damping
   area of elipse, 156
   fragile and perfectly plastic, 157
   potential elastic energy, 156
   relative yield, 156
   steel for, 155–156
Ionomers, 67
IPD. *See* Integrated product development (IPD)
Iron–carbon alloys, 48
   cast irons
      elements, influence on, 49–50
   steels, 52
      chromium, 53
      heat resistant, 55–56
      manganese, 53
      molybdenum, 54
      nickel, 54
      silicon, 54–55
      stainless corrosion, 55–56
      tungsten, 54
Izod test, 94

K

Knowledge-based engineering (KBE) systems, 12

L

Lattice defect, 47
LCD. *See* Life cycle design (LCD)

Lead alloys, 60
Leaf springs. *See also* Bending springs
  elastic deflection, scheme of, 299
  geometric form, 302
  logarithmic characteristic, 304
  mass spring system, 303
  qualitative characteristic, 297
  scheme for calculation, 298
  with spring shackles, 297–298
  trapezoid spring, 297
  triangular spring, 296
Life cycle
  and environmental impact
    analysis, 18
    phases and interactions with ecosphere, 19
  modeling behavior
    activity modeling, 20
    closed and open loop, 23
    physical–technological approach, 20
  product and material resources, 21–22
  reference activity model, 21
Life cycle assessment (LCA)
  eco-indicators
    Eco-Indicator 99 method, 25–26
    ecosystem quality, 26
    human health damage, 25
    resources, 26
  framework, 24
  goal and scope definition, 24
  impact assessment, 24
  interpretation (ISO)/improvement analysis, 24
  inventory analysis, 24
  mandatory and optional procedures, 25
  property and structure, 23
    environmental effects, analyzing, 23
    environmental interaction analysis, 23
    evaluations and activity comparisons, 24
  SETAC, 24
    improvement analysis, 25
    interpretation, 25
    methodological structure, 23
    property and structure, 23
  valuation, 25
Life cycle design (LCD), 15
  aspects, 9
  concept of, 9
  concurrent engineering (CE), 28
  cost-oriented development process, 27
  environmentally-oriented integrated product
      development, 28
  external integration, 28
  internal integration, 28
  IPD, 28
  phases, 9–10
  role, 29
  tools, 27
Load–elongation curve, 110
Local plasticization, 134
Low lead copper, 57
Lubrication
  Bosch's approximate computation, 236
    peripheral force, 236
  constant pressure and velocity, 218
  contour and hydrodynamic, 217

defined, 219
flow, 219
friction coefficient, 218
infinite and finite length, 220–221
  Funaioli's abacus, 223
  results for, 222
  sliding block scheme, 221
journal-bearing pair, 227
  Bosch's abacus, 230–231
  coefficient of friction, 229
  as function of, 229
  identification, 232
  load, 220
  scheme of, 228
Michell thrust bearings, 223
  scheme of, 224
  surface roughness, 225
Newtonian fluid and incompressible, 218
notes on
  society of automotive engineers (SAE), 246
  Stokes and centiStokes, 247
null inertial forces, 218
Ocvirk's hypothesis
  Reynolds' equation, 237
Petroff's law, 217
Raimondi–Boyd calculation, 238
  bearing characteristic number, 242–243
  film pressure, polar diagram, 241
  pressure distribution, 239
  Sommerfeld number, 244
  viscosity-temperature chart, 240
  Westinghouse Research Labs, 238
Reynolds' equation, 219
  and clearance height, 226–227
Schiebel's formulae, 232
  Euler–Lagrange equation, 233
  flow rate, 235
  Ten Bosch's diagrams, 236
  time, 232
  total drag force, 235
Sommerfeld number, 220, 236, 244
velocity, 218

M

Magnalium, 61
Main stresses
  three-dimensional case, 184
Malleable cast irons, 51–52
Manganese steel, 53
Manganin, 62
Martens degree defined, 65
Materials
  approaches
    coupled project variables, 164
    project requirement, 164
    selection based, 164
  properties
    chemical, 162
    electrical, 162
    mechanical, 162
    physical, 162
    processability, 162
    thermal, 162

Mechanical characterization tests
  bars for, 87
  Brinell hardness and number, 88
    penetrator 89
  compression tests, 86
  curvature tests, 86–88
  diamond pyramid hardness number, 90–91
  hardness tests, 88
  Knoop hardness, 92
  non-ferrous materials, 93
  pendulum charpy, 93
  penetrator for, 88
  resilience tests, 92–93
  rhomboid crater surface, 92
  Rockwell hardness
    HRB, 91–92
    HRC, 91
  shearing tests, 88
  standard bars, 93
  steel, 94
  tensile and compression stress, 87
  test bars, 93–94
  torsion tests, 88
  Vickers method, 90
Mechanical design requirements
  environmental conditions, 162
  failure phenomena, 163
  manufacturing and budget, 162
  mechanical–structural, 162
  physical properties, 162
  planning phase properties
    electrical, 162
    mechanical, 162
    physical, 162
    processability, 162
    thermal, 162
  typologies, 161–162
Melamine resins (MR), 68
Melt index, 65
Metals, 77–105
  mechanical characterization tests (see Mechanical
      characterization tests)
  mechanical components, 106
  metallurgic tests
    colability test, 101
    cylindrical test bar for, 102
    stretch tests, steel for, 101
  plastic constituent link, 83
    cylindrical main stresses domain, 84
    equivalent plastic deformation, 86
    isotropic hardening, 85
    Tresca and Von Mises domains, 84
  static elastoplastic
    spheroidal cast irons, 79
    tension engineering curve, 81
    true and equivalent curve, 83
    types of, 80
  steel wires
    alternate bending, 104
    tensile stress, 104
    torsion, 105
    winding, 105
  technological characterization tests (see Technological
      characterization tests)

  tensile stress tests
    circular section bars, 78
    test bars for, 79
    tube tests (see Tube tests)
Metals and alloys, 45
  Importance of, 46
Solid state and structure
  compact hexagonal cell, 46
  crystal lattices of iron, atomic arrangement, 46
  lattice defects, 47–48
  linear and helical, 48
  solidification, 47
  three-dimensional lattice, 46
  two-dimensional grid, 46
Methodological setup and design tools, 35
  direct reuse-(ES5), 37
  and environmental strategies (ES1), 36–37
  maintenance-(ES2), 37
  parameters and strategies, 37
  repair-(ES3), 37
  reuse of parts-(ES6) recycling, 37
  updating/adaptation (ES4), 37
Meusnier's theorems, 199–201
Micro-plasticization, 128–129
Mixed bearings, 257
Mobile joints, 490
  Cardan (universal), 491
    doubles, 496
    function of, 493, 495
    scheme of, 492
    shaft angular acceleration, 494
    transmission ratio, 492–493
  Oldham joint, 496
    centrifugal force, 497, 499
    intermediate disc centre, 498
    performance, 498–499
    scheme of, 497
Mohr's method, 253
  bending and shearing, 186
  pure torsion, 186
  tangential stresses, 185
  tensile compression, 186
Molybdenum steel, 54
Monel, 62
Monomers, 63
Moplen. See Polypropylene (PP)
"Muntz brass," 59

N

Natural plastics, 63
Neutral plane, 114
Nichromium, 62
Nickeline, 62
Nickel steel, 54
Nimonic, 62
Non-ferrous metals and alloys
  aluminum bronzes, 58
  cables for, 60
    aluminum, 61
    chromium, 62
    cobalt, 62
    magnesium, 61
    molybdenum, 63

nickel, 62
  titanium, 61–62
  tungsten, 62–63
  vanadium, 62
copper, 56
copper–lead–tin, 57
copper–tin binary, 56
copper–tin–zinc, 56
copper–tin–zinc–nickel, 56
foundry brass, 59
high aluminum zinc, 60
high lead bronzes, 57–58
high resistance brass, 59
industrial bronze, 59
lead alloys, 60
low lead copper, 57
pressure die-casting zinc, 60
Non-resonating rotational bending test machine, 158

O

Oil pipe
  diameter, 427
  fuel, 427
  investigation, 428
  non-destructive checks (NDC), 428
  operating conditions, 428
Oligocyclic fatigue, 130–131
One-piece bush, 266. *See also* Friction bearings
Optimal materials selection
  choice approaches
    project variables, 164
    properties, based on, 164
    defined, 164
    failure phenomena, 163
  integrated selection tools
    analytical formulation, 170–173
    application, 175–176
    graphic, 173–175
    research development, 176–177
  manufacturing and budget, 162
  physical properties and, 162
  primary constraint, 164
    Boolean conditions, 164
    quality and quantity limits, 164
  and processability, 162
  properties tools
    limits and objectives, 168–169
    standard and weighted, 165–168
  requirements and correlation, 161
  structural, 162
OTS 1, OTS 2 and OTS 3, high resistance brass, 59

P

Parabolic bending spring, 294
Peg joints
  Elco, 480
  Norton and Rupex, 478–479, 481
  rubber-coated, 479
Performance analysis (PA), 10
Peterson's diagram, 152
Phenolic matrices, 71
Phenolic resins, 68. *See also* Bakelite (PF)

Pins dimensioning
  deflection
    gear semi-width, axis of, 599
    project diameter, 599
  epicyclic speed reducer
    scheme for, 596
    verification, 596
  maximum deflection
    deformation value, 597
  pressure, 597
    verification, 599
  satellites differential
    scheme for, 598
  shear and bending, 597
    equivalent stress, 598
    qualitative characteristics of stresses, 598
  shear and bending verification
    allowable stress, 600
    ideal stress, 599
Planetary reducer
  bending calculation, 571–572
  defined, 571
  dynamic verification, 572–573
  teeth number calculation, 571
    gearing, 570
  wear calculation, 572
Planning process, 5
Plastic polymers, 64
Pohl and Bach's diagram, 132
Polyacetates, 67
Polyacrylonitrile, 67
Polyamide resins, 67
Polyarylether, 67
Polycarbonates (PC), 67
Polychlorotrifluoroethylene (PCTFE), 67
Polyester matrices, 71
Polyester resins, 68
Polyethylene (PE), 67
Polyfluoroethylenepropylene (PFEP), 67
Polymers
  acrylic resins, 66–67
  allyl resins, 68
  amino resins, 68
  cellulose resins, 65–66
  classification
    origin based, 63–64
    physical characteristics and, 64
    and temperature effects, 64
  defined, 63
  epoxy resins, 68–69
  fluorinated resins, 67
  phenolic resins, 68
  polyamide resins, 67
  polycondensation, 63
  polyester resins, 68
  polymerization and polyaddition, 63
  polyurethanes, 69
  properties of
    mechanical, 65
    thermal, 65
  silicones, 69
  styrene resins, 66
  synthetic resins

macromolecule structure, 64
    production, 64–65
Polymethacrylate (PMMA), 66
Polypropylene (PP), 67
Polystyrene (PS), 66
Polytetrafluoroethylene (PTFE), 67
Polyurethanes (PUR), 69
Polyvinylchloride (PVC), 67
Polyvinylfluoride, 67
Pomp drawing test, 96–97
Press and shrink fits
    fixing flywheel components, 407
        acting loads, 409
        cooling process, 409
        displacements, 411
        force, 413
        infinitely rigid, 411
        interference, 408, 410, 413
        loads rim and plate in rest condition, 411
        overloading, 414
        plastic deformation case, forces in, 414
        plate, heating, 407
        rim, gravity of, 409
        section area, 409
        two pairs of plates, 407–408
        working condition, 412
    forced hub-shaft shrinking
        diameter, 414–415
            stress and strain, nomenclature, 416
        interference, 415, 419, 421
        maximum ideal tension, 420
        pressure, 415
        radius, 416
        thick rotating discs, 415
        variation, 420
    interference and thermal variation, 421–422
        expansion coefficient, 421
        thermal variation, 421
    shaft–hub assembly, 408
Pressed pipe
    instability conditions
        critical pressure, 430
        deformation mode in, 429–430
        elastic deformation of, 428
        Eulerian critical load, 428
        inertial moment, 429
        qualitative pipe shapes in, 429
        steel tank vacuum, 430–431
Pressure die-casting zinc, 60
Pressure tubes, 423
    dynamic loads, 435
    fixed flange calculation, 432
        bending stress, 433
        codes, 435
        safety tensile traction stress, 434
    flange bolts, 431, 435
        diameter, 435
        fixed flange, section of, 432
        junction, 431
        symbols in, 435
    free flange calculation, 434
        maximum bending, 435
    hot metal creep, 436
        curves, 437–439

microstructural aspects, 440–441
        physical interpretation, 441–442
        tests, 437–440
    oil pipe
        diameter, 427
        fuel, 427
        investigation, 428
        non-destructive checks (NDC), 428
        operating conditions, 428
    pipe strength/sizing and verifying
        average velocity of fluid, 423
        equilateral hyperbole equation, 425
        forced fittings, case, 424
        longitudinal tension, 426
        Marriotte formula, 424
        narrow pipes, 424
        piston water pump, 423
        pressure, 424–425
        radial movement, 425
        tensile stress, 425
    pressed pipe
        critical pressure, 430
        deformation mode in, 429–430
        elastic deformation of, 428
        Eulerian critical load, 428
        inertial moment, 429
        qualitative pipe shapes in, 429
        steel tank vacuum, 430–431
    pressurised vessels, 426
        design, 426
        end caps, 427
        formula, 426
        maximum axial deformation, 426
        spherical cap, 426–427
        temperature environments, 426
Problem specification, 5
Product
    design and development, 3
        client needs, 4
        planning, 4
        process, 4
        production ramp-up, 4–5
        prototyping and testing, 4
        typologies, 8
Project data specification
    Alba–Piana Crixia, 608
    Aosta–Lillaz, 610
    Aosta–Pila, 608–609
    bevel gearing, 614
    Borgotaro–Pontremoli, 606–607
    cumulate city, 612–613
    cumulate correct, 613–614
    cumulate highway, 611–612
    cumulate mountain, 612
    Dezzo–Presolana, 610–611
    epicyclic reducer, 614–615
    Pontremoli–B. Albiano, 607–608
    Pontremoli–P.sso Cisa, 609–610
    satellite of differential, 615–616
    Susa–Moncenisio, 609
    Torino, 611
    Torino–Aosta, 607
Project definition, 4

Q

QFD. See Quality function deployment (QFD)
QQC 390 aluminum bronzes, 58
Quality function deployment (QFD), 6, 41

R

Radial bearings, 257, 260
Radial block clutches, 512
    Dolmen–Leblanc, 513, 515
    hypothesis, 516
    moment, 514
    specific pressure, 513
    tension of, 514, 516–517
Rapid Risitano method, 137
R Bronze alloy, 58
Red alloy, 58
Resonating operation with electromagnetic stimulation,
    159
Reverse engineering (RE), 12
Reynolds' equation, 219
    and clearance height, 226–227
        fluid film, 227
        friction coefficient, 226
Ribbon brakes
    calculations scheme, 553, 558–559
    groups, 557
Rigid joints
    box coupling/sleeve, 468
    cylindrical
        bolted shell, 469
        bolts and spline with, 469
        bolt tightness, 471
        pressure, 470
        with rings, 469
        Sellers, 471
        semi-shells, 470
    disc, 471–472
        couplings by bolts, 475
        Pomini's formula, 474
    flange couplings, 471–473
Roller-type bearings, 260
Rolling contact bearings
    classification
        by assembly method, 257
        of bearings, 257
        by rolling body shape, 257
        by trust direction, 257
Rotating shafts, 250

S

SAE. See Society of automotive engineers (SAE)
SAE 43, high resistance brass, 59
SAE 430 Grade B, high resistance brass, 59
SAE 660 alloy, 58
Safety clutches
    Hertzian loads, 520
    spring characteristics, 520–521
Saint Venant's theory, 279
SAMO foundry brass, 59
SAN. See Styrene acrylonitrile copolymer (SAN)
Saturated polyesters, 68

Schiebel's formulae, 232
    Euler–Lagrange equation, 232, 233
    flow rate, 235
    Ten Bosch's diagrams, 236
    time, 232
    total drag force, 235
Schnad test, 95
Semi-shaft dimensioning, 601
    diameter of, 602
Shaft pinion and choice of bearings
    calculation of
        qualitative characteristics of stresses, 588–589
        Von Mises equivalent, 589
    choice bearings
        axial load distribution, 592
        characteristics, 592–593
        dynamic load, 591
        preloading of system, 591
        safety factor, 591–592
    deflection verification
        equilibrium equations, 590
        scheme for, 590
        virtual work principle, 591
    diameter of, 590
    forces on, 583
    reaction forces, calculation of
        equilibrium equations, 586
        forces on, 583, 586
        result of, 588
        scheme for, 585–587
        superposition principle, 587
        virtual work principle, 586
    splined profiles, 583
        bevel gear, 584
        coupling, 584
Shafts
    axles and, 249
    elasticity assessment
        deformation, 253
        hypothesis of linearity, 254
        procedural steps, 256
        qualitative linear bending moment, 254
        rotation and deflection, 255–256
    isostatic
        three-stage steam turbine rotor, 251
    measuring, 249
        with more than two bearings (hyperstatic), 252–253
        sizing, 251
        two-bearing shafts (isostatic), 250–252
Sheaf plane, 199–200
Shear, 108
Shearing
    bar, rectangular cross section of, 118–119
    circular sections and, 120
    deformation, 119
    isotropic bodies and, 121
    moment of inertia, 120
    tangenzial stress distribution, 120
    tension, variation in, 119
    unitary tangential stress, 118
Silica brass foundry brass, 59
Silicone matrices, 71
Silicones, 69
Silicon steel, 54–55

Silumin, 61
Simultaneous/integrated (S/I) model, 7
Sinusoidal mono-axial stress, 129
Smeltered press foundry brass, 59
Smith-Goodman's diagram, 249
Sn12Ni1, Industrial 5, low lead copper, 57
Society of automotive engineers (SAE), 246
Socket drawing test, 96–97
Solidification, 47
Sommerfeld number, 220, 236, 244
Special cast irons, 51
Special silicon brass foundry brass, 59
Spheroidal cast irons, 51
Splined couplings
    categories of, 271
    curved-sided shafts, 271
    elastic deformation, 271
    fraction of modulus, 273
    straight-sided, 271
        geometric values for, 272
        tooth procedure, 273
    tangential strength, 271
    transmission of moment, 273
Splines and tongues, 273
    elastic deformability of material, 278
    fatigue calculation for shaft, 279
    key-ways, 279
    lateral surfaces
        forces on, 277
        pressure, 277
    lateral distribution, 278–279
    loads
        on surfaces of, 275
        transmission, 276
        moment, 276
    Saint Venant's
        beams, 284
        theory, 290
    shaft, transmitted from, 278
    side and shaft shearing stress, 277
    strain values used, 278
    surface pressure, 274
    system stress, analysis, 279
    unit compression tension, 277
    used, 274
    working conditions, 274
Spring computation
    linear spring characteristic, 288, 289
    no linear spring characteristic, 290
    scheme of beam, 290
Spring joints
    Bibby, 480, 482
    deflection expression, 484–485
    position of opposing teeth, 482–483
    scheme for calculation, 484
    strip in working conditions, 483
Springs
    bending springs
        characteristics, 303
        coil, 304
        conical disc, 311
        cylindrical helical, 309
        diaphragm, 316

        leaf, 296
        parabolic bending, 294
        qualitative stress, 292
        triangular bending, 291
        with spring shackles, 297
        trapezoid, 295–296
        triangular, 293
        variable thickness disc, 313
    compression springs, 329–330
    computation, 288
        linear spring characteristic, 289
        no linear spring characteristic, 290
        scheme of beam, 290
    defined, 287–288
    resonant spring calculations, 333
        harmonic component, 336
        qualitative displacement, 334, 338
        resonate condition, 337
        and shaft velocity, 337
    torsion springs
        conical helix, 325
        cylindrical helix, 322
        qualitative characteristic curves, 319
        torsion bars, 321
Spur gears
    American norms
        bending test, 402
        fatigue, corrective coefficients for, 403
    axial force, 284
    Brinnel hardness of material, 285
    contact lines
        conjugated involutes, 375
        cutter's profile, 376
        first contact in, 374
        generic contact point, 379
        pinion in, 373
        qualitative shape teeth, 374
        rack cutter and pinion, 375
    contact segment and action arc, 368–369
    creep curves, 378–379
        pitch cut in, 382
    cut and operating conditions, 353
        ordinary sizing, 354
    definition and nomenclature, 347–348
        circles, 352
        involute point, 349–350
        teeth, 351
        test bars for, 349
    design phases for, 405–406
    Euler–Savary theorem
        roulette, curvature centre, 361
    friction, 284
    gear resistance, verification calculations
        tooth, 394–400
    German norms, 391–393
    Hirth's, 284
    interlocking of teeth, 285
    involute equation
        rack cutter, profile, 354
    meshing, action segment, 365
    modified gears, 381
        addendum and dedendum in, 387
        contact arc, determination of, 389

engaging gears, load for, 393
layout, 384
pitch cut circle, position of, 388
non-interference, 371
addendum, maximum value for, 370
minimum teeth number, 373
pinion and rack, 372
transmission ratio, 373
resistance, verification calculation for
teeth wear, 399–402
tooth bending, 394–399
Saint Venant's beam, 284
Schiebel's fillet, 362–365
arc of action, 363
curvature of, 364
sizing, 352–353
stresses calculation in, 284
tooth, 355–356
angular coordinate, 356
base and dedendum circle, 359
contact of, 393
point of tooth, 358
Schiebel's fillet, 360
used, 284
Steel
chromium steel, 53
classification, 52
heat resistant, 55–56
Italian norm, 52
manganese steel, 53
molybdenum steel, 54
nickel steel, 54
properties, 52
silicon steel, 54–55
stainless corrosion, 55–56
tungsten steel, 54
Stress concentration factor
boundary elements method (BEM), 148
circular notched shaft, 147–148
elasticity theory and, 147
finite elements method (FEM), 148
Neuber diagrams, 148–149
nominal tension, 149
notch sensitivity
Peterson diagram, 152
trends, 150–151
resistance, 148, 150
Wöhler's curve, 150
Stress condition
column bars, buckling of
bending moment, 121
compressive and bending, 126
critical load values, 123
Eulerian critical load, 123
Euler's curve, 124–125
force increments and, 121
load vector, non-coincidence of, 121
parameterized curves, 125
scheme of, 122
serial development and, 125
slenderness, 123
Stresses in three-dimensional object. *See also* Failure
theories

elementary surface and forces, 182
normal and tangential, 181
prism, 182–183
solid body, 182
Styrene acrylonitrile copolymer (SAN), 66
Styrene butadiene copolymers (SB), 66
Styrene-elastomer copolymers, 66
Styrene resins, 66
Surplus clutches
cylindrical rollers, 517–518
Nepero's formula, 519
non-cylindrical rollers, 518–520
scheme of, 518–519
Synthetic plastics, 63
Synthetic resins
macromolecule structure, 64
production, 64–65

T

Technological characterization tests
bending test, 97
drawing, 95
Erichsen drawing test, 95–96
forging, 100
heading, 100, 102
Izod test, 94
mandrel bearing bending, 98–100
modified, 96
percentage elongation, 98–101
Pomp drawing test, 96–97
Schnad test, 95
socket drawing test, 96–97
stretch, 100, 101
template bending, 97–98
Tetmajer coefficient and, 98–100
Teflon. See Polytetrafluoroethylene (PTFE)
Ten Bosch's diagrams, 236
Tensile stress, 104
admissible unitary load, 112
elongation percentage, 112
equilibrium condition of, 109
Hooke's law for, 110
ideal distribution, 109
internal response, 108
limits, assess, 110
load–elongation curve, 110
modulus of elasticity, 110
normal and tangential components, 111
qualitative distribution, 109
tangential tension and, 112
uniform distribution and, 111
unitary load, 111
Termofond C12T, 61
Test machines
applicator and gauge, 157
deformation, 157
fixed drive test machine, 158
fixed loads, 157
Tetmajer coefficient
bending test, 98–101
Thermography
elastic deformation energy, 134

Haigh's diagram, 133
  internal energy, 134
  plastic deformation energy, 134
  quasi-Wöhler curves and, 133
  surface temperature, 134
  temperature increments and, 136
  T–N curve, 135–136
Thermoplastic and thermosetting matrices, 69
Three-dimensional object, stresses in
  elementary surface and forces, 182
  normal and tangential, 181
  prism, 182–183
  solid body, 182
Three-stage steam turbine rotor, 251
Time gearing measurement
  allowable stress, 578
  bending calculation, 580, 582
  bending stress, 577
  contact ratio, 577
  cyclic stress, 575
  equivalent load, 579
  external rim and bending, 583
  functions, 580–581
  goal of, 579
  hardening steel, 578
  load, 582
  number of real cycles, evaluation of, 576
  single standard, 576–577
  tangential load equivalent, 581
  UNI and AGMA unification, 577
  verification of, 580
  wear verification, 582–583
TONVAL foundry brass, 59
Tools for selection of properties
  limits and objectives
    merit parameter, 169
    resistance to fracture, 169
    tensile stress of tank, 168
    tension, peak of, 169
    thermal expansion, 169
    Young's modulus, 169
  materials, ranking of, 165
  standard and weighted
    cost containment, 165
    graphic materials comparison, 168
    materials comparison, 167
    pressure vessel, 165
    rigidity, 165
    standardization, 166
    tenacity, 165
    water resistance, 165
    weight reduction, 165
    working temperature, 165
    yield point, 165
Torsion, 108
  distortion, 117
  fiber
    shift, 116
    under pure torque, 116–117
  resistance module, 118
  tangential stress, 117
Torsion springs, 321
  conical helix, 325
  cylindrical helix, 322

qualitative characteristic curves, 319
torsion bars, 321
Total quality management (TQM), 12
Transverse splines
  collar, 282
  conical stem, 283
  force-deformation link, 280
  loads on, 283
  machine components, 280
  maximum and minimum forces, 280
  stem and ring, 281
  tensile stress, resistance, 280
Trapezoid bending spring, 295
Triangular bending spring, 293
Tube tests
  beading, 103–104
  colability, 103
  compression, 104–105
  enlargement, 102, 104
  flanging, 105
  tensile stress, 104, 108
  traction, 106
Tungsten steel, 54
Two-bearing shafts (isostatic), 250–252. See also Shafts

U

Unitary tangential stress, 118
Unsaturated polyesters, 68
Urea resins (UR), 68

V

Valve alloy, 58
Variable thickness disc springs, 313
  characteristic, 315
  extreme variable thickness, 314
  variable thickness, 314
Vertical pulsator (PV), 159
Vicat degree defined, 65
Voith Maurer joint
  deflection, 487
  tension and bending forces, 488
Vulcanized rubber, 66

W

Welded joints
  axial load, 451
  bending moment, 451–452, 465
  compression zones, traction, 466
  confront curve, 458
  corner beads, static calculation, 449–450
  curves, 460
    with stress, 455
  defects
    blowholes, 446
    chemical/structural, 444
    clinker pollution, 445–446
    cold cracks, 446
    geometric discontinuity, 445
    hot cracks, 446
    magnetic repulsion, 446
    penetration and melting, lack of, 445

semi-weld, 446
defined, 443–444
dispersion strip, 459
equivalent method, 456
external surfaces, 445
fatigue calculation
    acceptable thickness, 455–456
    curves with stress, 454–455
    safety coefficient, 453
loads on bead
    dangerous fluids, tubing, 453
    external, balancing, 451
    fluid tubing, 453
    gas pipes, welding, 453
    throat section, 451
Miner's rule, 456
nails
    different rows, 463
    loads, 462
    two rows, 462
parts, 444
pressure distribution, 464
riveted, 459–461
shank, pressure, 465
spot
    automobile industry, 456
    fatigue calculation, 458–459
static calculation, 447–448
    corner beads, 449–450
    head and complete penetration, 448–449

norm CNR-UNI 10011, 448
strength
    hot hammering, 447
    surface finishing, 447
stress
    components, 450
    T joint, 449
tests
    lamina cracks, 446–447
    tubing cracks, 447
transversal load moment, 451–452
types, 445
weld bead, 449
Weyrauch–Kommerell diagram, 453
zone, 446
    fracture, 447
Weyrauch–Kommerell diagram, 132–133
Wide pipes, qualitative tangential stress distribution in, 424
Wöhler's curve
    average tension, 131
    qualitative dispersion in, 131
    test data, 130
    zones in, 130–131

Z

ZAMA 12, ZAMA 13 and ZAMA 15 pressure die-casting zinc, 60
ZA 8, ZA 12 and ZA 27, high aluminum zinc alloy, 60